Lecture Notes in Bioinformatics

Edited by S. Istrail, P. Pevzner, and M. Wate

Subseries of Lecture Notes in Computer Science

Lecture Notes in Bioinformatics

Edited by S. Istrail, P. Pevzner, and M. Waterman

Editorial Board: A. Apostolico S. Brunak M. Gelfand
T. Lengauer S. Miyano G. Myers M.-F. Sagot D. Sankoff
R. Shamir T. Speed M. Vingron W. Wong

Subseries of Lecture Notes in Computer Science

Ben Raphael Jijun Tang (Eds.)

Algorithms in Bioinformatics

12th International Workshop, WABI 2012
Ljubljana, Slovenia, September 10-12, 2012
Proceedings

Volume Editors

Ben Raphael
Brown University
Department of Computer Science
P.O. Box 1910
Providence, RI 02912, USA
E-mail: braphael@cs.brown.edu

Jijun Tang
University of South Carolina
Department of Computer Science and Engineering
301 Main Street
Columbia, SC 29208, USA
E-mail: jtang@cse.sc.edu

ISSN 0302-9743   e-ISSN 1611-3349
ISBN 978-3-642-33121-3   e-ISBN 978-3-642-33122-0
DOI 10.1007/978-3-642-33122-0
Springer Heidelberg Dordrecht London New York

Library of Congress Control Number: 2012945432

CR Subject Classification (1998): F.2, J.3, G.2.2, F.1, I.2.6, H.2.8, G.1.2

LNCS Sublibrary: SL 8 – Bioinformatics

Typesetting: Camera-ready by author, data conversion by Scientific Publishing Services, Chennai, India

Printed on acid-free paper

Springer is part of Springer Science+Business Media (www.springer.com)

Preface

This volume contains papers presented at the 12th Workshop on Algorithms in Bioinformatics (WABI), held September 10–12th, 2012, at University of Ljubljana, Slovenia. WABI 2012 was one of six workshops which, along with the European Symposium on Algorithms (ESA), constituted the ALGO annual meeting. WABI focuses on algorithmic advances in bioinformatics, computational biology, and systems biology with a particular emphasis on discrete algorithms and machine-learning methods that address important problems in molecular biology.

The 35 papers this volume include algorithms for a variety of biological problems including phylogeny, DNA and RNA sequencing and analysis, protein structure, and others. In addition to presentations of these papers, WABI 2012 featured a keynote address "Genome Rearrangement by Double Cut and Join" by Jens Stoye, Universität Bielefeld. The 35 papers presented here were selected from a total of 92 submissions. Each submission was reviewed by at least three members of the Program Committee. Following these initial reviews, submissions were further discussed by the Program Committee to arrive at the final list of selected papers. We sincerely thank the members of the Program Committee and their sub-referees for their hard work in reviewing the submissions. We also thank EasyChair for providing an outstanding online system for managing the review process.

We thank the Organizing Committee, Andrej Brodnik (Co-chair), Uroš Čibej, Gašper Fele-Žorž, Matevž Jekovec, Jurij Mihelič, Borut Robič (Co-chair), and Andrej Tolič, for their tireless efforts in managing all of the local arrangements required for a successful meeting. Finally, we offer special thanks Bernard Moret for his invitation to chair this year's WABI meeting, his advice throughout the process, and his continuing commitment to WABI as a venue for discussing the latest algorithmic advances that impact molecular biology. We were honored to be part of this year's WABI and look forward to the continued success of the meeting in the coming years.

July 2012

Ben Raphael
Jijun Tang

Organization

Program Committee

Tatsuya Akutsu	Kyoto University, Japan
Max Alekseyev	University of South Carolina, USA
Mathieu Blanchette	McGill University, Canada
Guillaume Blin	Université Paris-Est Marne-la-Vallée, France
Paola Bonizzoni	Uuniversità di Milano-Bicocca, Italy
Daniel Brown	University of Waterloo, Canada
Philipp Bucher	Swiss Federal Institute of Technology (EPFL), Switzerland
Sebastian Böcker	Friedrich Schiller University Jena, Germany
Lenore Cowen	Tufts University, USA
Minghua Deng	Peking University, China
Pufeng Du	Tianjin University, China
Nadia El-Mabrouk	University of Montreal, Canada
Guillaume Fertin	LINA, University of Nantes, France
Liliana Florea	Johns Hopkins University, USA
Anna Gambin	Warsaw University, Poland
Olivier Gascuel	LIRMM, CNRS - Université Montpellier 2, France
Katharina Huber	University of East Anglia, UK
Daniel Huson	University of Tübingen, Germany
Carl Kingsford	University of Maryland, College Park, USA
Jim Loobong-Mack	University of Georgia, USA
Xinghua Lu	University of Pittsburgh, USA
Ion Mandoiu	University of Connecticut, USA
Satoru Miyano	University of Tokyo, Japan
Bernard M.E. Moret	Swiss Federal Institute of Technology (EPFL), Switzerland
Burkhard Morgenstern	Universität Göttingen, Germany
Vincent Moulton	University of East Anglia, UK
Luay Nakhleh	Rice University, USA
Macha Nikolski	LaBRI/CNRS, Université Bordeaux, France
Ron Pinter	Technion - Israel Institute of Technology
Nadia Pisanti	Università di Pisa, Italy
Mihai Pop	University of Maryland, USA
Teresa Przytycka	National Institutes of Health (NIH), USA
Tal Pupko	Tel Aviv University, Israel
Sven Rahmann	University of Duisburg-Essen, Germany
Ben Raphael (Co-chair)	Brown University, USA
Marie-France Sagot	INRIA Grenoble Rhône-Alpes and Université de Lyon 1, Villeurbanne, France

David Sankoff	University of Ottawa, Canada
Thomas Schiex	INRA Toulouse, France
Russell Schwartz	Carnegie Mellon University, USA
Charles Semple	University of Canterbury, New Zealand
Roded Sharan	Tel Aviv University, Israel
Jingna Si	Zhejiang University, China
Jörg Stelling	Swiss Federal Institute of Technology Zurich, Switzerland
Leen Stougie	Free University Amsterdam, The Netherlands
Jens Stoye	Bielefeld University, Germany
Krister Swenson	Université de Montréal / McGill University, Canada
Jijun Tang (Co-chair)	University of South Carolina, USA
Glenn Tesler	University of California, San Diego, USA
Jerzy Tiuryn	Warsaw University, Poland
Li-San Wang	University of Pennsylvania, USA
Lusheng Wang	City University of Hong Kong, China
Christopher Workman	Technical University of Denmark, Denmark
Feng Yue	University of California San Diego, USA
Alex Zelikovsky	Georgia State University, USA
Louxin Zhang	National University of Singapore, Singapore
Meng Zhang	Jilin University, China
Xiuwei Zhang	Swiss Federal Institute of Technology (EPFL), Switzerland
Ziding Zhang	China Agricultural University, China

Additional Reviewers

Atias, Nir
Bankevich, Anton
Berry, Vincent
Biecek, Przemyslaw
Brunetti, Sara
Bulteau, Laurent
Burstein, David
Caciula, Adrian
Canzar, Stefan
Catanzaro, Daniele
Chauve, Cedric
Chikhi, Rayan
Christinat, Yann
Cohen, Ofir
Crochemore, Maxime
Csuros, Miklos
D'Addario, Marianna

Daniels, Noah
De Givry, Simon
Della Vedova, Gianluca
Dondi, Riccardo
Dutkowski, Janusz
Dvorkin, Mikhail
Dührkop, Kai
Faller, Beata
Faraut, Thomas
Fassler, Raffael
Gambin, Tomasz
Garde, Christian
Ghiurcuta, Cristina
Glebova, Olga
Gorecki, Pawel
Gottlieb, Assaf
Grossi, Roberto

Halldorsson, Bjarni
Hao, Tong
Hoinka, Jan
Huang, Hailiang
Huang, Yang
Humphries, Peter
Jakobi, Tobias
Jankowski, Aleksander
Jean, Geraldine
Jiang, Minghui
Jiang, Shuai
Kluge, Boguslaw
Kochanczyk, Marek
Kopczynski, Dominik
Koskas, Michel
Kulikov, Alexander
Köster, Johannes
Labarre, Anthony
Lacroix, Vincent
Lavallée-Adam, Mathieu
Lefebvre, Arnaud
Li, Fan
Li, Tingting
Li, Wei
Lin, Yu
Linz, Simone
Liu, Zheng
Mahmoody, Ahmad
Mancuso, Nicholas
Mangul, Serghei
Martin, Marcel
Navlakha, Saket
Nefedov, Alexey
Nurk, Sergey
Oberhardt, Matthew
Oren, Yaara
Pardi, Fabio
Patro, Robert
Peeri, Michael

Peng, Jian
Penn, Osnat
Pirola, Yuri
Popa, Alexandru
Puglisi, Simon
Rizzi, Romeo
Rusu, Irena
Ryvkin, Paul
Sacomoto, Gustavo
Salari, Rahele
Scheubert, Kerstin
Sciortino, Marinella
Scornavacca, Celine
Sczyrba, Alexander
Si, Jingna
Sikora, Florian
Sirotkin, Alexander
Skums, Pavel
Soueidan, Hayssam
Subramanian, Ayshwarya
Sundermann, Linda
Sykulski, Maciej
Thévenin, Annelyse
Tsai, Ming-Chi
van Iersel, Leo
Venturini, Rossano
Wang, Xi
Wang, Ying
Wilczynski, Bartek
Willing, Eyla
Xu, Wei
Yang, Xiao
Zhang, Bo
Zhang, Ting
Zheng, Jim
Zheng, Yu
Zhou, Wanding
Zytnicki, Matthias

Table of Contents

Preserving Inversion Phylogeny Reconstruction

Matthias Bernt[1], Kun-Mao Chao[2], Jyun-Wei Kao[2],
Martin Middendorf[1], and Eric Tannier[3]

[1] Parallel Computing and Complex Systems Group,
Institute of Computer Science, University Leipzig, Germany
{bernt,middendorf}@informatik.uni-leipzig.de
[2] Department of Computer Science and Information Engineering,
National Taiwan University, Taiwan
kmchao@csie.ntu.edu.tw
[3] INRIA Rhône-Alpes; UMR CNRS 5558 "Biométrie et Biologie Évolutive";
Université de Lyon 1; Villeurbanne, France
Eric.Tannier@inria.fr*

Abstract. Tractability results are rare in the comparison of gene orders for more than two genomes. Here we present a linear-time algorithm for the small parsimony problem (inferring ancestral genomes given a phylogeny on an arbitrary number of genomes) in the case gene orders are permutations, that evolve by inversions not breaking common gene intervals, and these intervals are organised in a linear structure. We present two examples where this allows to reconstruct the ancestral gene orders in phylogenies of several γ-Proteobacteria species and Burkholderia strains, respectively. We prove in addition that the large parsimony problem (where the phylogeny is output) remains NP-complete.

1 Introduction

Parsimony problems on trees are challenging computational problems inspired and motivated by molecular evolution. A phylogenetic *character* is a finite set of states equipped with a transition matrix, assigning a weight to every pair of states. Given a rooted binary tree with sets of character states assigned to its leaves, the *small parsimony problem* consists in assigning states at the internal nodes, minimising the sum of the weights of the pair of states at the extremities of every edge. The *large parsimony problem* takes only a set of character states as an input and asks for the phylogenetic tree with these states at the leaves that has the smallest solution to the small parsimony problem.

In phylogeny, *parsimony* describes evolutionary scenarios minimising the number of events modifying a character, which for instance describes nucleotides, amino acids, or the presence/absence of molecular or morphological features. The small parsimony problem can be solved in polynomial-time for a set of

---

* This work was supported by the NSC Taiwan Fellowship of the Alexander von Humboldt Foundation.

B. Raphael and J. Tang (Eds.): WABI 2012, LNBI 7534, pp. 1–13, 2012.

independent characters with a constant number of states [15], whereas large parsimony is NP-complete even if characters can have only two states [10,16].

In recent years *gene order* data, *i.e.* the linear order of genes along the genome, became a popular characteristic of species used for phylogeny reconstruction [17,20]. Gene orders can be formalised as signed permutations, *i.e.* permutations where each element corresponds to a homologous gene and an additional sign at each element indicates the strandedness of the corresponding gene. Gene orders can be seen as characters on a huge state space of size $2^n n!$, where n is the number of genes. Transitions between two states are weighted by distances between gene orders. Parsimony problems may be generalised to this case: if permutations are given at the leaves of a rooted binary tree, find permutations at the ancestral nodes minimising the sum of the distances on every edge. Many distances have been investigated [13]. For all defined distances on permutations (and almost all distances on objects modelling gene order data, with two notable exceptions [12,21]), the small parsimony problem has been proved to be NP-complete, even for a phylogenetic tree with only three leaves.

An *inversion* is a rearrangement which reverses a segment of a permutation and flips the signs of its elements. For a set of permutations, *common intervals* [18] are groups of genes forming a contiguous segment in all permutations. *Preserving inversions* maintain all common intervals. A common interval I is strong if for every other common interval it holds that it is either included in I or includes I. By definition strong common intervals have a nested structure, which can be represented by a tree of minimal inclusion. This tree is called *linear* if every node with at least three children strictly includes a common interval containing more than one of the node's children.

Preserving inversions have been introduced in [14] with the objective to yield biologically relevant results. The idea is that if the genomes at the leaves of a phylogenetic tree have a common structure, *e.g.* as the result of phylogenetic inertia or natural selection, it is desirable that genomes at the ancestral nodes and all intermediate ones share this structure. The preserving inversion sorting problem, which consists in finding a shortest sequence of inversions transforming one gene order into another such that all intermediate gene orders have the common intervals of the input gene orders, was shown to be NP-hard [14], but fixed-parameter tractable [9]. In the special case when the strong interval tree of the common intervals of the two gene orders is linear, there is a linear-time algorithm [2,3], and starting from this, algorithms for the general case have been devised [2,3]. Efficient algorithms for three gene orders have been developed in [5,6], where it is proved that if the strong interval tree is linear, then there exists a linear-time algorithm.

Here we investigate the preserving inversion distance in the special case when the strong interval tree of the gene orders has a linear structure. It is proved that the small parsimony problem is tractable in this case, whereas the large parsimony problem is NP-complete. So we generalise the results of [5,6] by constructing a polynomial-time algorithm solving the small parsimony problem for an arbitrary number of genomes. The algorithm is based on the construction of binary characters on the strong interval tree and the application of Fitch's algorithm [15] on these characters.

The following section contains necessary definitions concerning the small and large parsimony problem for binary characters as well as gene orders, gene orders and preserving inversions, and strong interval trees. In Section 3 we give the polynomial-time algorithm and NP-completeness proof for the small and large parsimony problems, respectively. Preliminary results for γ-Proteobacteria and Burkholderia genomes indicating the utility of the approach are presented in Section 4. All proofs and a detailed description of the data sets are given in the Appendix which is, due to space limitations, not included in the proceedings.

2   Gene Orders and Phylogeny

Phylogenetic Trees. A phylogenetic *character* is a measurable property of species which is in a certain state, called *character state*. The set of all possible character states is called the *state space* of the character. In the following we often use "character" as a synonym for its state space.

For phylogenetic analyses, it is typically assumed that for each character C a transformation relation $F \subset C \times C$ is given that describes the feasible state changes for this character. A *scenario* for two character states is a sequence of feasible transformations transforming one character state into the other. In *maximum parsimony*, shortest scenarios are of interest, because their length can be used as a measure for the evolutionary *distance* between two character states. This is used to define a distance function $d : C^2 \mapsto \mathbb{N}^+$ from the transformation relation where $d(c, c')$ is the minimum number of feasible transformations necessary to transform c into c'.

A character assignment for a set of species X and a character C is a function $l : X \mapsto C$ that assigns a character state to each species in X. A *phylogenetic tree* (also called phylogeny) for species X is a binary tree $P = (V, E)$ with leaf set X. We denote by $L(P)$, $V(P)$, and $E(P)$ the leaf, vertex, and edge set of P, respectively. A *reconstruction* R for a phylogeny P, a character C, and a character assignment $l : L(P) \mapsto C$ is an assignment of character states from C to the nodes of P such that for each leaf $s \in L(P)$ it holds that $R(s) = l(s)$. The *score* of a reconstruction R for a phylogeny P with respect to a distance function d for C is defined as $S(R, P) = \sum_{(e,f) \in E(P)} d(R(e), R(f))$. The *small parsimony problem* is to find $\operatorname{argmin}_R S(R, P)$ for a given phylogeny P, character assignment from $L(P)$ to C, and distance function d over C. In the *large parsimony problem* where the phylogeny for the species X is unknown, $\operatorname{argmin}_{R,P} S(R, P)$ is sought for character assignment from X to C, and distance function d over C.

In this paper, two types of phylogenetic characters are considered: *binary characters* have a state space of size two and other characters where the state space can be described as a set of permutations. Binary characters might be considered as the very basic phylogenetic character and can be used to describe the presence or absence of some property. Observe that, a binary character defines a bipartition of the species, also called *split*. For the reconstruction of a phylogenetic tree with more than four leaves, more than one binary character is necessary. If the characters are independent, the parsimony criterion can be extended by minimising the sum of differences over all characters.

Gene Orders and Rearrangements. A *signed permutation* is a permutation of the set $\{1, \ldots, n\}$ where an additional sign $+1$ or -1 is multiplied to each element. A signed permutation π is denoted by $(\pi(1)\, \pi(2) \ldots \pi(n))$, *i.e.* $\pi(i)$ denotes the i-th element. In the following we consider signed permutations as *unoriented*, *i.e.* π and $-\pi = (-\pi(n) \ldots -\pi(1))$ are considered equal. Differences for the handling of the *oriented* case, *i.e.* $\pi \neq -\pi$ are pointed out in the text. A subset $I \subset \{1, \ldots, n\}$ of the elements of a permutation π is called an *interval* if the elements appear consecutively in the permutation, *i.e.* a pair (s, e) of start and end position with $1 \leq s \leq e \leq n$ exists such that $\{|\pi(i)| : s \leq i \leq e\} = I$. Note that for the definition of intervals, the sign of the elements is not regarded. Two intervals are said to *commute* if they are either disjoint or one is included in the other. An *inversion* ρ applied to a signed permutation π, denoted by $\pi \circ \rho$, is a rearrangement operation formally defined as an interval of π. The effect of ρ on π is that the order of ρ's elements is reversed in π and the sign of the elements is inverted.

A *common interval* [18] of a set of signed permutations Π is a subset of the element set of the permutations that is an interval in all permutations. The set of common intervals of Π is denoted as $\mathcal{C}(\Pi)$. If an inversion ρ applied to a permutation π in a set of permutations Π does not destroy any of its common intervals, *i.e.* $\mathcal{C}(\Pi) = \mathcal{C}(\Pi \cup \{\pi \circ \rho\})$, then it is called *preserving*. A set of permutations Ψ is *consistent* with a set of permutations Π if $C(\Pi) = C(\Pi \cup \Psi)$.

When the set of signed permutations is considered as character, and inversions and preserving inversions, respectively, as the feasible transformations, then a corresponding prefix is used for the terms scenario and distance, *e.g. preserving inversion scenario* and *preserving inversion distance*. Shortest scenarios are also called sorting [17]. In this paper, preserving inversions are the feasible transformations considered for gene orders.

The Linear Structure of Common Intervals. In the following the data structure that is central for this paper, i.e. the strong interval tree [2,4], is introduced. Note that, this data structure is similar to PQ-Trees [8]. A common interval of a set of permutations $\Pi = (\pi_1, \ldots, \pi_k)$ is *strong* if it commutes with any interval in $\mathcal{C}(\Pi)$. Thus strong common intervals form a hierarchy which is captured by the *strong interval tree* (SIT) data structure $\mathcal{T}(\Pi)$ where each node represents one strong common interval and the edges the minimum inclusion relation. SITs are considered as ordered trees where the order of the children of a node is w.l.o.g. given by their order in the first permutation. A node is called *linear* if every union of consecutive child nodes is a common interval and otherwise it is called *prime*. A SIT with linear nodes only is called *linear*. The strength of the SIT that facilitates the development of efficient algorithms is that it can be built in $\mathcal{O}(n)$ time [4] and that it represents all $\mathcal{O}(n^2)$ common intervals using only $\mathcal{O}(n)$ nodes. This is because the children of the linear nodes can be totally ordered in such a way that a set of genes I is a common interval of Π iff it is either a node of $\mathcal{T}(\Pi)$ or the union of consecutive children of a linear node of $\mathcal{T}(\Pi)$ [2,5].

Example 1: Consider the signed permutations $\pi_1 = (1 \ldots 7)$, $\pi_2 = (3\text{-}2\text{-}1\,4\,5\,7\text{-}6)$, $\pi_3 = (\text{-}5\,4\text{-}1\,2\,3\,6\text{-}7)$, $\pi_4 = (\text{-}7\text{-}6\text{-}1\text{-}2\,3\text{-}5\,4)$, and $\pi_5 = (\text{-}6\,7\text{-}1\text{-}2\,3\text{-}5\,4)$ which are used as example throughout the paper. The sets $\{1, 2, 3\}, \{4, 5\}, \{6, 7\}, \{1, \ldots, 5\}$, the singletons, and the set of all elements are the strong common intervals of these permutations. The structure of the SIT is shown in. Most of the nodes have two children and are therefore linear by definition. Since $\{1, 2\}$ and $\{2, 3\}$ are also common intervals, the node $\{1, 2, 3\}$ is linear. Hence the SIT is linear.

In Theorem 1 we prove that if $\mathcal{T}(\Pi)$ is linear then the small parsimony problem for preserving inversions can be solved in time $\mathcal{O}(kn)$ for Π. It is done by defining states of binary characters for the nodes of $\mathcal{T}(\Pi)$.

3 Preserving Inversion Phylogeny on Linear SIT

Here the polynomial-time algorithm for the preserving inversion phylogeny problem is presented. First the SIT and the results for preserving sorting from [2] are extended to more than two permutations. We then define binary characters for permutations with linear SIT – one character for each linear node. In the third part it is shown that preserving inversions for the permutations are in one-to-one correspondence to the state changes of this character. This leads to our main result presented in the last part.

Permutations as Signed Strong Interval Tree. In the following an order on a given set of permutations is required. Therefore, we assume that a sequence Π of signed permutations is given. Let I be an inner node in the strong interval tree $\mathcal{T}(\Pi)$, $I_1, \ldots, I_m$ its m children nodes, and π, λ two permutations consistent with Π. The relative order of $I_1, \ldots, I_m$ in π with respect to a *reference permutation* λ is captured by the *quotient permutation* associated with I for π and λ, denoted by π_I^λ. It is defined as the (unsigned) permutation of the elements $\{1, \ldots, m\}$ such that for $x, y \in [1 : m]$ it holds that x precedes y in π_I^λ iff I_x precedes I_y in π, where $I_1, \ldots I_m$ are numbered by their order in λ. A quotient permutation is called *linear increasing* if it is the identity permutation $(1 \ldots m)$, *linear decreasing* if it is the inverse $(m \ldots 1)$ of the identity permutation, and *prime* otherwise. Note that if a node is linear, then the quotient permutation is linear for every permutation, and conversely [2].

The information about the orientation of linear quotient permutations is represented by a *sign*: increasing as $+1$ and decreasing as -1. A k-tuple of signs, called k-*sign*, is used to represent the information for k (linear) quotient permutations. We denote the i-th element of a k-sign s by $s(i)$. For a reference permutation $\lambda \in \Pi$, the k-*signed strong interval tree* $\mathcal{T}^\lambda(\Pi)$ is defined as the SIT where a k-sign s is added to every node I. Since in the following only the case of a linear SIT is considered we do not specify the assignment to prime nodes here. The assignment to linear nodes is as follows. i) If I is a leaf, $s(i) = +1$ if

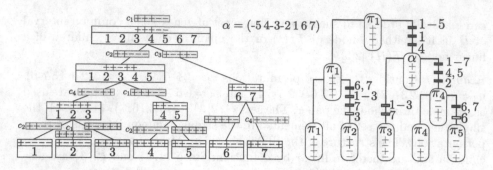

Fig. 1. Left: 5-signed SITfor the signed permutations used in the examples; 5-signs are given in the nodes; the characters are shown on the edges with a unique name; Right: given phylogeny; characters $c_1, \ldots, c_4$ and their reconstructions are shown below the nodes from top to bottom; inferred rearrangements are indicated along the edges; k-signs and characters are shown in a short form omitting "1"

the element has the same sign in π_i and λ and -1 otherwise. ii) If I is linear, $s(i)$ is determined by the orientation of π_{iI}^λ. If Ψ is a sequence of l permutations consistent with Π then $\mathcal{T}(\Pi)$ has the same structure as $\mathcal{T}(\Pi \cup \Psi)$. Hence, the quotient permutations and l-signs can be defined for Ψ by choosing a reference permutation $\lambda \in \Pi$. We denote the resulting l-signed strong interval tree by $\mathcal{T}^\lambda(\Psi, \Pi)$. If λ is clear from the context it may be omitted.

Example 2: Let the reference permutation $\lambda = \pi_1$. Node $\{1, \ldots 5\}$ has two children $I_1 = \{1, 2, 3\}$ and $I_2 = \{4, 5\}$. All but the third quotient permutation are $(1\ 2)$ since I_1 and I_2 are in the same order as in π_1, *i.e.* linear increasing. In π_3 the child intervals are in opposite order, *i.e.* I_2 is left of I_1. Therefore the quotient permutation is $(2\ 1)$, *i.e.* linear decreasing. Consequently the k-sign of this node is $(+1, +1, -1, +1, +1)$.

Because an inversion is preserving iff it commutes with all common intervals [2,14], the set of preserving inversions can be described in terms of the SIT. An inversion ρ is preserving with respect to a set of permutations Π iff it is either a node of $\mathcal{T}(\Pi)$ or the union of the children of a prime node of $\mathcal{T}(\Pi)$ [2,5]. In [2,3], where $k = 2$, a slightly different version of the 2-signed SITwas used to find a sorting inversion scenario from π to σ that is preserving for $\{\pi, \sigma\}$. Instead of 2-signs a single sign is used. This is possible because all 2-signs in $\mathcal{T}^\pi((\pi, \sigma))$ have $+1$ as their first component, *i.e.* only the second component is relevant. Since the preserving rearrangement analysis of more than two permutations cannot be described in terms of changes on one permutation only, in this case, all k components are necessary. Because preserving inversions can only invert the order of the children of a linear node, an algorithm solving the problem of sorting by preserving inversions has to decide if a linear node has to be inverted or not. This was captured in the Parity Lemma of [2] which states that a node is in a preserving inversion scenario if its sign is different from its parent's sign. A reformulation to k-signed SITs is given:

Proposition 1. *Let Π be a set of permutations, $\lambda \in \Pi$, and π and σ two permutations consistent with Π. A node I in $\mathcal{T}^{\lambda}((\pi, \sigma), \Pi)$ is in any sorting inversion scenario from π to σ that is preserving for Π iff the 2-signs of I and its parent differ in exactly one position.*

This gives a linear-time algorithm for sorting by preserving inversions if the SIT is linear [2]. Also the preserving inversion median problem can be solved in linear-time if the SIT is linear [5,6]. For sorting by preserving inversions or preserving inversion median problem instances incorporating prime nodes algorithms that need exponential time in the worst case have been proposed. Since we consider only the case of linear SIT we refer the reader to [2,3,5,6].

Example 3: Comparing the 1st and the 3rd component of the 5-signs of neighbouring nodes in the SITgives five nodes that have exactly one different sign with respect to their parent's sign: $\{1\}, \{1, 2, 3\}, \{4\}, \{1 \ldots 5\}$, and $\{7\}$. Applying the inversion of these sets in any order transforms π_1 into π_3.

Permutations as Binary Characters. A permutation consistent with a SIT defines a sign for each leaf node and a quotient permutation for each of the inner nodes and therefore also a sign for each linear node. Vice versa an assignment of quotient permutations to the nodes and of signs to the leaves of a SIT uniquely determines a consistent permutation. For linear nodes, a sign carries the same information as the quotient permutations. Thus, for a linear SIT, an assignment of signs to the nodes uniquely determines a consistent permutation.

Let $\Pi = (\pi_1, \ldots, \pi_k)$ be a sequence of permutations and U and V a linear node and its linear parent in $\mathcal{T}(\Pi)$ with k-signs s_U and s_V, respectively. The information on the equality or inequality of the two k-signs defines k states of a binary character c_U. Formal convenience suggests to use the character states $\{-1, +1\}$ and to specify the character state assignment to the species, as k-tuple, *i.e.* $c_U \in \{+1, -1\}^k$, where $c_U(i)$, with $i \in [1 : k]$, denotes the character state assigned to species i. Then $c_U(i)$ is defined as $c_U(i) = s_U(i) * s_V(i)$. Thus each linear node with linear parent defines a character, *i.e.* if the SIT is linear each (non root) node defines a character. For oriented permutations a character for the root is defined by comparing its k-sign with $\{+1\}^k$.

A character state assigned to each of the (non root) nodes of a linear SIT determines the signs of all nodes by multiplying the character state and the parent's sign from top to bottom starting at the root node with sign $+1$. Hence, the character state assignment uniquely determines a consistent permutation. Starting with -1 at the root results in inverted signs for all nodes, *i.e.* an inverted permutation which is equivalent for unoriented permutations. Using the root's character as its sign correctly initialises the procedure for oriented permutations.

Example 4: The character states of node $\{1, \ldots, 5\}$ in the SIT are $(+1, +1, -1, -1, -1)$ since the node has 5-sign $(+1, +1, -1, +1, +1)$ which is equal in the first two positions and unequal in the other positions compared to its parent's 5-sign. Assume, that the four nodes on the path from the root to leaf $\{5\}$ have assigned character states $+1, -1, +1$, and $+1$. This implies signs $+1, -1, -1$, and -1 for these nodes. For the root node, which has a $+$ sign, this implies that child

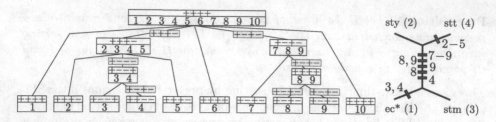

Fig. 2. Left: SIT for the γ-Proteobacteria gene orders; characters are only shown for nodes showing sign differences; Right: the unrooted maximum parsimony tree for the characters and the corresponding rearrangements mapped to the edges; in parentheses the indices within the data set are given

node $\{1,\ldots,5\}$ has to be left of $\{6,7\}$. Each of the nodes $\{1,\ldots,5\}$, $\{4,5\}$, and $\{5\}$ has sign -1. Therefore, nodes $\{1,2,3\}$ and $\{4\}$ have to be right of $\{4,5\}$ and $\{5\}$, respectively, and node $\{5\}$ is negative. Now, assume character states $+1,-1,+1,-1,+1,+1,+1,-1,+1,+1,+1,+1$ for the nodes of the SIT in pre-order. Then, the signs of all nodes in pre-order are $+1,-1,-1,+1,-1,-1,-1,+1,-1,+1,$ $+1,+1$. These signs imply $\alpha = (\text{-}5\,4\text{-}3\text{-}2\,1\,6\,7)$.

Preserving Inversion as a Binary Character State Change. Consider a phylogenetic tree with k leaves that have assigned k permutations $\Pi = (\pi_1,\ldots,\pi_k)$. Each linear node with linear parent in $\mathcal{T}(\Pi)$ defines an assignment of the states of a binary character to the species corresponding to the given permutations. These assignments can be used to assign the character states to the leaves of a given phylogeny. With an algorithm solving the small parsimony problem for binary character states assigned to the leaves, a reconstruction of ancestral binary states for the inner nodes of the phylogeny can be derived. By repeating this for each of the characters derived from $\mathcal{T}(\Pi)$ for each inner node of the phylogeny, an ancestral state of every binary character is known. If $\mathcal{T}(\Pi)$ is linear, then the assignment of binary character states to the nodes of the phylogeny uniquely determines a reconstruction of permutations. But note that there can be multiple optimal solutions to the small phylogeny problem for the binary characters where each may define different ancestral permutations.

Not only there is a one-to-one relation between permutations and binary sequences, but there also is a one-to-one relation between character state changes in the binary alphabet and preserving inversions.

Proposition 2. *Let Π be a sequence of signed permutations and π, σ be two signed permutations consistent with Π. A linear node U in $\mathcal{T}(\Pi)$ is in any sorting inversion scenario for π and σ that is preserving for Π iff the character state defined by U in $\mathcal{T}((\pi), \Pi)$ differs from the character state defined in $\mathcal{T}((\sigma), \Pi)$.*

Example 5: Consider the two cases where the nodes $\{4,5\}$, $\{2\}$, and the root node have assigned character state $+1$ and -1 and all other nodes have equal character states. The signed permutations $\alpha = (\text{-}5\,4\text{-}3\text{-}2\,1\,6\,7)$ and $\pi_4 = (\text{-}7\text{-}6\text{-}1\text{-}2\,3\text{-}5\,4)$ are permutations realising these character states with their corresponding signs in $\mathcal{T}^{\pi_1}((\alpha, \pi_4), \Pi)$. Node $\{4,5\}$ and its parent both have sign -1 for α and -1 and $+1$ for π_4, *i.e.* the 2-signs are $(-1,-1)$ for node $\{4,5\}$ and $(-1,+1)$ for its parent. Nodes $\{2\}$ and its parent $\{1,2,3\}$ have 2-signs $(-1,-1)$ and $(-1,+1)$, respectively, and the root's 2-sign is $(+1,-1)$. Since the characters assigned to the other nodes are equal all other nodes have no sign difference when compared with their parent. Thus exactly $\{4,5\}$, $\{2\}$, and the root node have a sign difference with respect to their parent and therefore these three nodes compose any sorting preserving inversion scenario for α to π_4.

Algorithm 1. Small parsimony problem for preserving inversions.

---

1 Compute k-signed SIT;
2 Determine the k states of a binary character for each of the nodes;
3 Solve the small phylogeny problem for each character;
4 Compute ancestral gene orders from the reconstructed binary states;

Proposition 2 implies a one-to-one correspondence of the changes of the binary characters assigned to the phylogeny and preserving inversions applied along the edges of the phylogenetic tree for the corresponding assignment of preserving permutations. Because the inversions applied to linear nodes do commute by definition the characters are independent. Thus, Algorithm 1 solves the small parsimony problem for preserving inversion and Theorem 1 follows.

Theorem 1. *Let Π be a sequence of k permutations of length n. If $\mathcal{T}(\Pi)$ is linear then the small parsimony problem for preserving inversions can be solved in time $\mathcal{O}(kn)$ for Π.*

Example 6: We now exemplify the parsimonious reconstruction for the phylogeny shown in Figure 1. The sign differences in the SIT define four non-trivial unique binary characters $c_1, \ldots, c_4$. A parsimonious reconstruction for these binary characters is shown in the figure. In the case of ambiguous reconstructions, *i.e.* for c_2 the root and for c_4 the root and its children, we have chosen $+1$ for these character states. The trivial characters defined by nodes $\{3\}$ and $\{6\}$ only have one differing state. Thus the state difference and therefore the corresponding inversion are mapped on the edges leading to leaves π_2 and π_5, respectively. The remaining node $\{5\}$ defines a constant character, which implies no state change or inversion. The reconstructed character state combinations uniquely correspond to ancestral signed permutations, *e.g.* the parent of π_1 and the root node are identical to π_1 and the reconstruction of α for the right child of the root node was presented in Example 4. This reconstruction of the binary characters has a score of 14 corresponding to eight changes in the unique characters. Each of the character changes corresponds to an inversion. The differences between the nodes reconstructed as α and π_4 are due to three inversions corresponding to the change in c_1, see Example 5.

Large Parsimony. Due to the one-to-one correspondence of sorting preserving inversions and binary characters, it follows that there is no efficient algorithm for the large parsimony problem for preserving inversions if $P \neq NP$.

Theorem 2. *The large parsimony problem for preserving inversions is NP-complete even for sequences of permutations with linear SIT.*

Circular Permutations. The gene orders of circular genomes can also be modelled by signed permutations π, but -π and all circular shifts of π and -π are considered equivalent. The described methods yield correct results for circular genomes if the input signed permutations are inverted and shifted such that all start with the same element of the same orientation.

4   Gene Orders of γ-Proteobacteria and Burkholderia

We have determined a maximum subset of the 30 γ-Proteobacteria gene orders from [1] that has a linear SIT (see Figure 2). The four unique gene orders are from *Escherichia coli* and *Salmonella enterica* species (denoted as ec[osec], sty, stm, and stt in [1]). Seven nodes in the SIT have a 4-sign differing from the parent's 4-sign and define a binary character corresponding to an inversion that can easily be mapped to the phylogeny presented in [1]. The unique character corresponding to the most rearrangements defines the only non-trivial split separating ec* and stm from the other two gene orders. The corresponding unrooted tree is parsimonious. Three of these inversions, *i.e.* {8}, {9}, and {8, 9} can be explained alternatively by a transposition.

Burkholderia blocks were generated from gene families of the Hogenom 6 database [19]. Only the genes that are on the main chromosome or the largest plasmid on all species were kept. This yields three different gene sets: (C) containing the genes which are in the chromosome in all species, (P) containing the genes which are in the plasmid in all species, and (CP) containing the genes which are in the chromosome in some species, and in the plasmid in others. Consecutive genes were fused into blocks if they were in the same order in all genomes. Genes not belonging to any block were discarded. This resulted in three block sets: C, P, CP with 63, 12, and 10 markers, respectively. C and P are considered as circular and CP as linear gene order. A phylogeny was computed with BppML from the Bio++ library [11] on a concatenated sequence of 80 genes that have been randomly chosen from those present in one copy in every species.

For each of these gene order data sets a maximum subset that has a linear SIT has been determined. For these subsets the unique gene orders have been determined and co-linear subsets of the blocks are merged. This yielded the final data set with 5 unique gene orders of length 8 for C, 4 unique gene orders of length 7 for P, and 3 unique gene orders of length 3 for the linear fragment CP. None of these subsets covered the complete data set.

Using our method the inversions corresponding to nodes with sign differences to their parent have been mapped to the phylogeny. Only a few inversions could

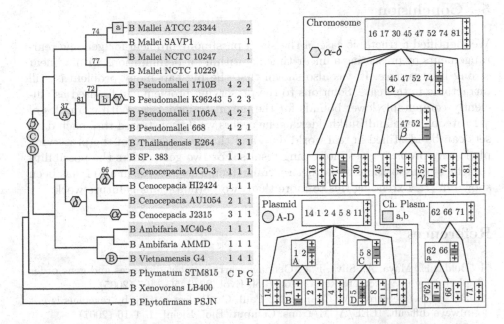

Fig. 3. Results for the Burkholderia data set; Left: maximum likelihood phylogenetic tree; only bootstrap values < 100 are shown; inversions are given on the branches where the shape and a character refers to the actual inversion (see right part); the three columns give the index of the gene order of the corresponding species in the unique gene orders of the maximum subsets with linear SIT for C, CP, and P; Right: SITs for the three block sets; numbers represent the first of co-linear blocks; k-signs are given from top to bottom; sign differences are highlighted; the index of the inversion is given in the node.

not be mapped unequivocally since some gene orders are not included in the subsets. Since the gene orders of the *B. mallei* strains is not included in the maximum subset of P the inversion *A* could be as well on the parent branch. Two of the three genes included in the inversion are at same position in *B. mallei* strains as in *B. thailandensis* which supports the proposed assignment. Also the position of C, D, and γ is unsure because outgroup information is missing. The parsimonious assignment for the chromosomal inversion implies a reversal of inversion δ. In this case this might be caused by a wrong phylogeny which is also indicated by the low bootstrap value for the branch leading to two of the *B. cenocepacia* strains. Interestingly also other inversions are mapped to or near to branches with lower bootstrap support. The presented mapping of inversion *A* gives additional support for the branch that has the lowest bootstrap support in the tree. For the chromosome the SIT shows a deeply nested structure, which implies nested preserving inversions. This is likely related to inversions symmetric around the origin of replication.

5 Conclusion

We identified a tractable case of the small parsimony problem for genome rearrangements by presenting a linear-time algorithm for permutations with a linear strong interval tree. It was also shown that the big parsimony problem is still intractable in this case. Solutions to the preserving problem can be used as efficiently computable lower bounds for the unconstrained case. With examples of γ-Proteobacteria and Burkholderia gene orders we demonstrated that such data sets can be identified in real world data and that the solution of the preserving phylogeny problem gives interesting results. For two gene orders the possibility to detect transpositions and other rearrangements as patterns in a SIT has been shown in [7]. Extending this to more than two permutations is future work.

References

1. Belda, E., Moya, A., Silva, F.J.: Genome rearrangement distances and gene order phylogeny in γ-proteobacteria. Mol. Biol. Evol. 22, 1456–1467 (2005)
2. Bérard, S., Bergeron, A., Chauve, C., Paul, C.: Perfect sorting by reversals is not always difficult. IEEE/ACM Trans. Comput. Biol. Bioinf. 4, 4–16 (2007)
3. Bérard, S., Chauve, C., Paul, C.: A more efficient algorithm for perfect sorting by reversals. Inform. Process. Lett. 106, 90–95 (2008)
4. Bergeron, A., Chauve, C., de Montgolfier, F., Raffinot, M.: Computing common intervals of k permutations, with applications to modular decomposition of graphs. SIAM J. Discrete Math. 22, 1022–1039 (2008)
5. Bernt, M.: Gene Order Rearrangement Methods for the Reconstruction of Phylogeny. PhD thesis, Universität Leipzig (2010)
6. Bernt, M., Merkle, D., Middendorf, M.: Solving the preserving reversal median problem. IEEE/ACM Trans. Comput. Biol. Bioinf. 5, 332–347 (2008)
7. Bernt, M., Merkle, D., Ramsch, K., Fritzsch, G., Perseke, M., Bernhard, D., Schlegel, M., Stadler, P.F., Middendorf, M.: CREx: inferring genomic rearrangements based on common intervals. Bioinformatics 23, 2957–2958 (2007)
8. Booth, K., Lueker, G.: Testing for the consecutive ones property, interval graphs and graph planarity using PQ-tree algorithms. J. Comput. Syst. Sci. 13, 335–339 (1976)
9. Bouvel, M., Chauve, C., Mishna, M., Rossin, D.: Average-Case Analysis of Perfect Sorting by Reversals. In: Kucherov, G., Ukkonen, E. (eds.) CPM 2009. LNCS, vol. 5577, pp. 314–325. Springer, Heidelberg (2009)
10. Day, W.H.E., Sankoff, D.: Computational complexity of inferring phylogenies by compatibility. Syst. Zool. 35, 224–229 (1986)
11. Dutheil, J., Gaillard, S., Bazin, E., Glemin, S., Ranwez, V., Galtier, N., Belkhir, K.: Bio++: a set of C++ libraries for sequence analysis, phylogenetics, molecular evolution and population genetics. BMC Bioinformatics 7(1), 188 (2006)
12. Feijao, P., Meidanis, J.: SCJ: A breakpoint-like distance that simplifies several rearrangement problems. IEEE/ACM Trans. Comput. Biol. Bioinf. 8, 1318–1329 (2011)
13. Fertin, G., Labarre, A., Rusu, I., Tannier, E., Vialette, S.: Combinatorics of Genome Rearrangements. MIT Press (2009)

14. Figeac, M., Varré, J.-S.: Sorting by Reversals with Common Intervals. In: Jonassen, I., Kim, J. (eds.) WABI 2004. LNCS (LNBI), vol. 3240, pp. 26–37. Springer, Heidelberg (2004)
15. Fitch, W.: Toward defining the course of evolution: minimum change for a specified tree topology. Syst. Zool. 20, 406–416 (1971)
16. Foulds, L.R., Graham, R.L.: The Steiner problem in phylogeny is NP-complete. Adv. Appl. Math. 3, 43–49 (1982)
17. Hannenhalli, S., Pevzner, P.A.: Transforming cabbage into turnip: polynomial algorithm for sorting signed permutations by reversals. In: ACM Symposium on Theory of Computing, pp. 178–189 (1995)
18. Heber, S., Stoye, J.: Finding All Common Intervals of k Permutations. In: Amir, A., Landau, G.M. (eds.) CPM 2001. LNCS, vol. 2089, pp. 207–218. Springer, Heidelberg (2001)
19. Penel, S., Arigon, A.M., Dufayard, J.F., Sertier, A.S., Daubin, V., Duret, L., Gouy, M., Perrière, G.: Databases of homologous gene families for comparative genomics. BMC Bioinformatics 10, S3 (2009)
20. Sankoff, D.: Edit Distance for Genome Comparison Based on Non-Local Operations. In: Apostolico, A., Galil, Z., Manber, U., Crochemore, M. (eds.) CPM 1992. LNCS, vol. 644, pp. 121–135. Springer, Heidelberg (1992)
21. Tannier, E., Zheng, C., Sankoff, D.: Multichromosomal median and halving problems under different genomic distances. BMC Bioinformatics 10, 120 (2009)

Fast Phylogenetic Tree Reconstruction Using Locality-Sensitive Hashing

Daniel G. Brown and Jakub Truszkowski

David R. Cheriton School of Computer Science
University of Waterloo
Waterloo ON N2L 3G1 Canada
{browndg,jmtruszk}@uwaterloo.ca

Abstract. We present the first sub-quadratic time algorithm that with high probability correctly reconstructs phylogenetic trees for short sequences generated by a Markov model of evolution. Due to rapid expansion in sequence databases, such very fast algorithms are necessary. Other fast heuristics have been developed for building trees from large alignments [18,1], but they lack theoretical performance guarantees. Our new algorithm runs in $O(n^{1+\gamma(g)} \log^2 n)$ time, where γ is an increasing function of an upper bound on the branch lengths in the phylogeny, the upper bound g must be below $1/2 - \sqrt{1/8} \approx 0.15$, and $\gamma(g) < 1$ for all g. For phylogenies with very short branches, the running time of our algorithm is near-linear. For example, if all branches have mutation probability less than 0.02, the running time of our algorithm is roughly $O(n^{1.2} \log^2 n)$. Our preliminary experiments show that many large phylogenies can be reconstructed more accurately than allowed by current methods, in comparable running times.

1 Introduction

Phylogenetic reconstruction is a core bioinformatics problem. Existing algorithms for the problem often require very long running times, particularly when the number of sequences is large; this problem is especially acute for traditionally slow, yet accurate, methods like maximum likelihood. As biologists start to need trees for hundreds of thousands of taxa [9], even traditionally faster distance methods, such as neighbour joining, become unacceptably slow. Researchers have developed sub-quadratic time heuristics [18,1], but they lack theoretical performance guarantees, so it is unclear whether their use is universally appropriate.

Here, we give an algorithm to correctly reconstruct, with high probability, phylogenetic trees that come from a Markov model of evolution, in sub-quadratic time using sequences of $O(\log n)$ length. King has proved an $\Omega(n^2 \frac{\log \log n}{\log n})$ lower bound on the running time of all distance-based phylogeny reconstruction algorithms using such short sequences; our algorithm avoids the problem by using the sequences directly.

Our algorithm is based on three ideas. First, we use locality-sensitive hashing[13] to find sequences that are near-neighbours in the tree, in sublinear time. This hashing is a first step in choosing which two positions should

B. Raphael and J. Tang (Eds.): WABI 2012, LNBI 7534, pp. 14–29, 2012.

be joined in the tree we incrementally are building. Second, we use ancestral sequence reconstruction to reliably approximate the sequences found at internal tree nodes. Finally, we use reliable estimates of distance, to identify exactly the correct join at each step; this step involves some hoary computation, due to the need to ensure that inferred sequences are independent estimates. Since we start with a forest with each taxon in its own tree, and perform this joining step until only one tree remains in the forest, the overall runtime is sub-quadratic. Specifically, if p is an upper bound on the mutation probability on any edge, and $p < 1/2 - \sqrt{1/8}$, then we show that we can do the locality-sensitive hashing, which is the runtime-determining step, at each step in $O(n^{\gamma(p)} \log^2 n)$ time, where $\gamma(p)$ is always less than 1; the overall runtime is thus $O(n^{1+\gamma(p)} \log^2 n)$.

2   Related Work

Our work spans two recent threads in phylogenetic research: theoretical algorithms with guarantees of performance, and practical algorithms with no guarantees of performance.

2.1   Principled Phylogenetic Algorithms

Erdős et al. [8] gave an $O(n^4 \log n)$ algorithm that reconstructs a phylogeny with high probability, assuming the Cavender-Farris model of evolution, for sufficiently long sequences. For most trees, their algorithm runs in $O(n^2 \text{poly} \log n)$ time and requires $O(\text{poly} \log n)$ sequence length. Csűros [4] provided a $O(n^2)$ algorithm with similar performance guarantees. Recent papers [12,5] give algorithms to identify parts of the tree that can be reconstructed, using quartet queries chosen so that, with high probability, only correct quartets are queried. The only sub-quadratic time algorithm with guarantees on reconstruction accuracy is by King et al. [14]; for most trees, its running time is $O(n^2 \frac{\log \log n}{\log n})$ on sequences of $O(\text{poly} \log n)$ length.

King et al. also showed that any algorithm that reconstructs the true tree with high probability, and uses distance calculations as its only source of information about the phylogeny, will have $\Omega(n^2 \frac{\log \log n}{\log n})$ running time, for sequences of length $O(\text{poly} \log n)$.

Mossel [17] gave a phase transition for phylogenetic reconstruction. Suppose we have a balanced phylogeny with all edges having mutation probability p. If p is less than $1/2 - \sqrt{1/8} \approx 0.15$, then the phylogeny can be reconstructed correctly from sequences of length $O(\log n)$ sequence data. For larger p, sequences must be of polynomial length to allow constant probability of reconstructing the phylogeny. Daskalakis, Mossel and Roch [6] extended the phase transition result to unbalanced trees, and provided an $O(n^5)$ algorithm that reconstructs phylogenies with lengths below the phase transition. Mihaescu, Hill and Rao [16] provided an $O(n^3)$ algorithm for the same problem, which bears some resemblance to our approach.

2.2 Practical Algorithms

In parallel to these theoretical results, many researchers have developed phylogeny reconstruction algorithms that can analyze alignments of tens, or even hundreds, of thousands of taxa. Fast Neighbour Joining [7] runs in $O(n^2)$ time and gives results similar to neighbour joining, which requires $O(n^3)$ time. FastTree [18] reconstructs phylogenies without computing the full distance matrix, which results in $O(n^{1.5} \log n)$ runtime. Our own recent algorithm, QTree [1,2], runs in $O(n \log n)$ time, using an incremental approach to building trees. No theoretical guarantees exist for the quality of solutions obtained from these fast algorithms, however, under realistic assumptions about the evolutionary model, for short sequences.

3 Preliminaries

3.1 Phylogenetic Trees

A phylogeny is an unrooted, weighted tree whose internal vertices all have degree 3 and whose leaves represent extant taxa. The weights represent evolutionary time. Evolution is modelled as a time-reversible Markov process operating on the edges of the tree, where each position of a sequence evolves independently.

A *quartet* is a phylogeny on four taxa. For a set $\{a, b, c, d\}$ of taxa, there are three possible quartets which we denote as $ab|cd, ac|bd$ and $ad|bc$.

We assume the Cavender-Farris model of binary character states over ± 1 evolving according to a continuous Markov process. Each edge e is labelled with length $\ell(e)$, and the probability that the ends of e have different states is $p(e) = \frac{1}{2}[1 - \exp(-2\ell(e))]$. If two sequences differ in a $\hat{p}$ fraction of sites, the maximum likelihood estimator of the distance between them is $\hat{d} = -\frac{1}{2}\log(1 - 2\hat{p})$.

We assume there exist constants f and g such that for each edge e in the phylogeny, we have $f < \ell(e) < g < \ln 2/4$. This gives a minimum length for each edge, and also gives each edge state change probability less than $1/2 - \sqrt{1/8}$, which guarantees a bounded probability of error when reconstructing ancestral sequences [6]. We also assume that all edge lengths are multiples of some constant Δ, consistent with previous work [6]. With these assumptions in place, a surprising fact arises: with sequences of length $O(\log n)$, we can exactly identify the tree distance between close nodes in the phylogeny [6,16].

Theorem 1. *Let* $\Delta \leq f' < g' < \infty$. *Then there exists a constant* $c(f', g', \Delta)$ *such that if* $f' < d(a, b) < g'$ *and* $d(a, b)$ *is a mutliple of* Δ, *we have*

$$\Pr[|d(a, b) - \hat{d}(a, b)| > \frac{\Delta}{2}] \leq \exp(-k/c(f', g', \Delta))$$

where k *is the sequence length and* d *is the true evolutionary distance.*

In particular, if $k = 3c(f', g', \Delta) \ln n$, *we can identify the correct distance with probability at least* $1 - n^{-3}$.

Note also that this theorem applies to any distances in our trees below a constant times g, the upper bound on a single edge length.

3.2 Locality-Sensitive Hashing

Our algorithm requires finding pairs of sequences within a specified small distance from each other, without having to compute all pairwise distances. Indyk and Motwani [13] solved this problem using a collection of randomized hash tables: enough hash tables are chosen so that close sequences likely collide in one of the tables, while keys are long enough that distant sequences do not. This idea, known as *locality-sensitive hashing*, has been applied to many problems in bioinformatics, such as motif finding [3]. Specifically, Indyk and Motwani solve a related problem, the (r_1, r_2)-approximate Point Location in Equal Balls $((r_1, r_2)$-PLEB):

Input: A set of sequences P in $\{0, 1\}^k$, a query sequence q, and radii $r_1 < r_2$

Output: If there exists a sequence $p \in P$ within normalized Hamming distance r_1 from q, output "yes" and a sequence within r_2 of q. If there is no sequence in P within normalized Hamming distance r_2 from q, output "no". Otherwise, output either "yes" or "no".

Indyk and Motwani's solution constructs n^{r_1/r_2} hash tables, each keyed on $O(\log n)$ randomly chosen sequence positions. Given q, a point within distance r_1 of it has a constant probability of colliding with it q each hash table, while points further than r_2 from q have $O(1/n)$ probability of colliding. After inspecting a constant number of collisions with q, we can find, with constant probability, a point whose distance from q is at most r_2; if we boost by running $O(\log n)$ times independently, the success probability is $1 - n^{-\alpha}$, for any choice of α. For more details, see [13]. Overall, finding an (r_1, r_2)-approximate near neighbour for a query point q with high probability takes $O(n^{r_1/r_2} \log n)$ hash table lookups, each on a key of length $O(\log n)$ bits.

Indyk and Motwani solve the approximate PLEB problem so that their hash table solution is not overwhelmed by points within the region greater than r_1 but less than r_2 away from q. In our domain, we can avoid this problem, and find all sequences within distance exactly r from a given query: we choose r_2 to be small enough that at most $O(\log n)$ points are found within the r_2 distance, so we can examine all of them and still have fast runtimes. We choose r_2 to be $1/2 - 1/2(\exp(-2cf \log \log n)$, the relative Hamming distance corresponding to all sequences within evolutionary distance $cf \log \log n$, for a constant c that incorporates errors arising from reconstructing internal sequences of the tree (see Section 3.3 and the Appendix for details). In $\log \log n$ edges, we can reach $O(\log n)$ nodes. The distance r_2 converges to $1/2$ as n grows (though it is quite a bit smaller for smaller values of n), so in the limit, the number of hash tables grows to n^{2r_1}.

Finding all neighbours within r_1 normalized Hamming distance thus takes $O(n^{2r_1+\epsilon} \log^2 n)$ time with high probability, where $\epsilon \to 0$ as n increases: we use $O(n^{2r_1+\epsilon} \log n)$ hash tables, each of which requires $O(\log n)$ time to examine, and we take $O(\log^2 n)$ time examining the hash table hits.

We use the hashing algorithm to find sequences within evolutionary distance d, implicitly relying on the simple correspondence between Hamming and evolutionary distances: our procedure *FindAllClose*(q, d) finds all sequences within evolutionary distance d of q with probability $1 - o(1/n^3)$, and so with high probability makes no errors during the course of running our entire algorithm.

3.3 Four-Point Method

To identify the correct place to join two trees, we will use the four-point method, which reconstructs quartets from the six pairwise distance estimates. The method computes $\hat{d}(a,b)+\hat{d}(c,d)$, $\hat{d}(a,c)+\hat{d}(b,d)$, and $\hat{d}(a,d)+\hat{d}(b,d)$, and outputs $ab|cd$, $ac|bd$ or $ad|bc$, respectively, depending on which is the minimum. If all pairwise distances were estimated exactly, the two sums corresponding to incorrect topologies would both be $2\ell(e)$ greater than the sum corresponding to the correct topology, where e is the middle edge of the true quartet.

Because in our setting, we can estimate distances exactly with high probability, we can assume that all quartets are properly computed, provided that the sequences used to compute them satisfy the *error-independence* property, defined in the next subsection.

Theorem 2. *Let f and g be the upper and lower bounds on the edge length in a quartet tree. Then there exists a constant $c_2(f,g,\Delta)$ such that we can reconstruct the lengths of the edges of the quartet exactly using sequences of length $c_2(f,g,\Delta)\log n$ with probability at least $1 - \frac{6}{n^3}$.*

Proof. The claim follows from Corollary 1, by setting $f' = 2f$ and $g' = 3g$.

3.4 Ancestral States

When all edge lengths are below the phase transition threshold, we can correctly infer the ancestral state of a character at any internal node of the tree with probability greater than $\frac{1}{2} + \beta$ for some constant β [10,17,6]. We identify nodes which should be near-neighbours in the tree, join them together, and infer new ancestral node sequences, until we have only a single tree. The following theorem, which is an adaptation of a result by Daskalakis *et al.* [6], describes the probability of correctly reconstructing ancestral states.

Theorem 3. *Let T be a binary tree with root ρ and edges e all satisfying $p(e) < 1/2 - \sqrt{1/8}$. Let δT denote the leaves of T. Let σ be a Cavender-Farris character on T. The maximum likelihood algorithm A for ancestral state reconstruction infers the correct ancestral state with probability at least $1/2 + \beta$, for some constant β independent of T.*

The maximum likelihood algorithm [11] computes the posterior distribution of a state at an internal node based on previously computed posterior distributions at its children. For constant-sized alphabets, it takes constant time. We pick the state of largest posterior probability. For known edge lengths, this algorithm has optimal probability of correctly reconstructing the ancestral state, among all possible algorithms.

The runtime of our algorithm depends on an upper bound to β, which is hard to obtain from the proof of [6]. We can use the following result by Evans *et al.* [10].

Theorem 4. *Let* T *be a phylogenetic tree with edge mutation probabilities bounded by* $p_q < 1/2 - \sqrt{1/8}$. *Assign to each edge* e *a resistance* $(1 - 2p)^{-2|e|}$, *where* $|e|$ *is the number of edges on the path from root to* e, *including* e *itself. The probability* p_{err} *of incorrectly reconstructing the root state is bounded by* $p_{err} < 1/2 - 1/(1 + \mathcal{R}_{eff})$, *where* $\mathcal{R}_{eff}$ *is the effective resistance between the root and the leaves of* T.

This bound is quite loose. For this reason, we will use a better bound, originally developed for the simpler Fitch parsimony algorithm. The following bound is sharper for $p_q < 0.118$.

Theorem 5. *Let* T *be a phylogenetic tree where all mutation probabilities across edges are equal to* $p_q < 1/8$. *The probability* p_{err} *of incorrectly reconstructing the root state using Fitch parsimony is bounded by*

$$p_{err} < \frac{1}{2} - \frac{\sqrt{(1 - 4p_g)(1 - 8p_g)}}{2(1 - 2p_g)^2} < 1 - 4p_g$$

The original result was stated to hold for variable edge lengths. This was corrected by Zhang *et al.* [19].

This bound applies to the maximum likelihood algorithm, even for variable edge lengths, as long as they are bounded by p_q. For constant length edges, the result holds by the optimality of maximum likelihood. If we shrink an edge in tree T, creating a tree T', the mutual information between the leaves and the root of T' is greater than the mutual information between the leaves and the root of T, and the probability of reconstructing the root of T' must be higher. Applying this argument to all edges except the longest, we obtain that the bound holds for trees of variable edge lengths when maximum likelihood is used.

The bound on p_{err} is important in our algorithm because we will use locality-sensitive hashing on these sequences: it determines the number of hash tables required. Better bounds on p_{err} give faster algorithms with the same performance. Let g_{err} be the distance corresponding to a mutation probability of p_{err}.

Suppose that we reconstruct ancestral sequences for subtrees T_1, T_2 of T, rooted at ρ_1, ρ_2, respectively. Moreover, suppose that the path connecting T_1 and T_2 has ends ρ_1, ρ_2. By the Markov property, reconstructing σ_{ρ_1} correctly is independent of reconstructing σ_{ρ_2} correctly. We call such two sequences *error-independent*. The distance estimate $\hat{d}(\sigma_{\rho_1}, \sigma_{\rho_2})$ will converge to $g + g_1 + g_2$, where g_1 and g_2 are edge lengths corresponding to the probabilities of incorrectly reconstructing states in the two sequences. When comparing independently reconstructed sequences, we can effectively treat these errors in the reconstructed sequences as "extra edges" [6], whose length can be bounded using Theorems 4 and 5. This observation has been used extensively in many theoretical algorithms [6,17,16].

Theorem 3 combined with Theorem 2 and the Markov property enable us to estimate internal branch lengths from reconstructed ancestral sequences, and also correct quartets.

Theorem 6. *Let* i, j, k, l *be ancestral sequences reconstructed by maximum likelihood from four disjoint subtrees, and such that no path between any two of them in the true phylogeny shares an edge with any of the subtrees. Let* f' *and* g' *be the upper and lower bounds on the edge lengths in the quartet* $ijkl$. *Then there exists a constant* $c(f', g', \Delta)$ *such that given reconstructed ancestral sequences of length* $c \log n$, *we can estimate the length of the middle edge of* $ij|kl$ *exactly, with probability* $1 - 6n^{-3}$.

4   The Algorithm

The algorithm starts with a forest F of n trees, each with one taxon. It progressively merges subtrees into larger subtrees of the true tree, finding two nodes that are quite close using locality-sensitive hashing, identifying where they should be joined, and inferring ancestral sequences. The core idea is complicated by the requirement in Theorem 6 that the sequences be reconstructed from disjoint subtrees of the true phylogeny.

Algorithm 1 presents a simplification of the algorithm. Our algorithm maintains four invariants:

1. Every tree in F is a subtree of the true tree T
2. No two trees in F overlap as subtrees of T
3. For each tree T' in F, all edges except at most one have length at most g. The remaining edge has length at most $2g$. We call it the *long edge* of T'.
4. The length of every path in every subtree in F is reconstructed correctly

The first three are the same as in the work of Mihaescu *et al.* [16]; the fourth, we maintain using Theorem 6. The invariants, together with routine *CheckErrorIndependence*, ensure its preconditions are satisfied.

Algorithm 1. SimplifiedReconstruct($\{\sigma\}, f, g$)
___
Start with a forest with each node in its own tree.
Use locality-sensitive hashing to find all sequences whose pairwise distance is less than $3g + 2g_{err}$. Put them in a priority queue, *DistQueue*.
while the forest has more than one tree **do**
  Find two sufficiently close nodes x and y that are not currently in the same tree of F.
  Identify the nearby edge (i, j) to x and (k, l) to y that should be joined to connect the trees containing x and y in the forest
  Create two new nodes, a and b in the middle of (i, j) and (k, l), and join a and b together with a new edge.
  Estimate the lengths of all five edges in the quartet $ijkl$.
  Reconstruct the ancestral sequences at a and at b.
  Find all sequences within $3g + 2g_{err}$ of the newly inferred sequences, and add these distances to *DistQueue*.
end while

In what follows, we will expand the details of this algorithm, focusing on ensuring invariants 3 and 4. We will assume the existence of three procedures: $FindAllClose(q, d)$ uses locality-sensitive hashing to identify all sequences within relative Hamming distance d of q. $Quartet(a, b, c, d)$ uses the four-point method

to identify the correct topology of the quartet *abcd*. And *MiddleEdge(ab|cd)*
computes the length of the middle edge in the quartet *ab|cd*. Assuming the pre-
conditions to Theorem 3, these procedures work with high probability. Most of
the subroutines are presented for the case where all their arguments are internal
nodes. The cases where some nodes are leaves are analogous; we omit them for
brevity. We will often treat subtrees with long edges as rooted, with the root
located somewhere on the long edge.

4.1   Independent Inferences

If reconstructed sequences in quartet queries are not independent, the quartet
middle edge length estimates and inferred topology might be incorrect. This
could lead to a wrong choice of which edges to join.

  The order in which ancestral sequences are reconstructed defines a partial
order of the nodes in F. We call it the *reconstruction order*. For two nodes a and
x in F, at least one of the children of b with respect to reconstruction order is not
on the path from a to x in T. *CheckErrorIndependence* uses this observation
to detect cases when lack of error-independence impacts the middle edge length
estimate for the quartet.

Algorithm 2. CheckErrorIndependence(x, y, a, b)

Require: a and b are error-independent, x and y are error-independent, $T(a)$ and $T(x)$ do
  not overlap
  for all $z \in \{a, b, x, y\}$ **do**
    Let z_1, z_2 be the children of z in reconstruction order(if they exist)
  end for
  $d \leftarrow MiddleEdge(xy|ab)$
  for $i, j \in \{1, 2\}$ **do**
    $d_{a_i b_j} \leftarrow MiddleEdge(xy|a_i b_j)$
    $d_{x_i y_j} \leftarrow MiddleEdge(x_i y_j|ab)$
  end for
  if any of the $d_{a_i b_j}$ or $d_{x_i y_j}$ differs from d by more than $\Delta/2$ **then**
    return false
  end if
  return true

Lemma 1. *Let $ME(xy|ab)$ denote the true length of the middle edge in the
quartet $xy|ab$ in T. If $|MiddleEdge(xy|ab) - ME(xy|ab)| > \Delta/2$, then CheckEr-
rorIndependence returns false. Otherwise, it returns true with high probability.*

Proof. Let $T(a)$ and $T(x)$ be the subtrees that contain a and x, respectively.
Sequences a and b are independent and lie on the opposite sides of some edge e.
If x and y join the tree at e, we have $ME(xy|ab) = ME(xy|a_i b_j)$ for any choice
of i and j. If x, y, a, b are independent, all middle edge estimates are within $\Delta/2$
of each other and correct within $\Delta/2$. Suppose some two of these sequences are
not independent (say a and x). Without loss of generality, $T(x)$ joins $T(a)$ at
some edge in subtree of $T(a)$ consisting of all nodes from which we reconstruct
the sequence at a. One of the sequences a_1, a_2 is then independent of x, so its
corresponding call to $MiddleEdge$ will return a value that is correct within $\Delta/2$.

4.2   Detecting Overlapping Subtrees

To maintain Invariant 2, we need to prevent merging two trees if the new edge created between them overlaps some other subtree in F. The procedure *CheckOverlaps* detects overlapping subtrees. We use $R(T(x))$ to indicate the set of sequences in $T(x)$ and $d_{T'}$ for the path metric associated with T'.

Algorithm 3. CheckOverlaps(x,y)

$S = (FindAllClose(x, 2g + 2g_{err}) \cup FindAllClose(y, 2g + 2g_{err})) - R(T(x))$
for each sequence a in S **do**
   for each sequence b in $T(a)$ that is independent of a and such that $d_{T(a)}(a, b) < 5g$ **do**
      if $Quartet(a, b, x, y) \neq ab|xy$ and $CheckIndependence(Quartet(a, b, x, y)) = true$
      then
         return true
      end if
   end for
end for
return false

Lemma 2. *If the edge (x, y) overlaps some other edge in F and the sequences at x and y are independent, CheckOverlaps will return true. Otherwise, it will return false.*

Proof. Suppose $T(x)$ overlaps with some other subtree T'. One can easily show by case analysis that there exists a reconstructed sequence in T' that is within distance $2g + 2g_{err}$ from x or y and such that the two sequences are independent. Since $T(x)$ overlaps with T', there has to exist a node b' in T' such that the induced topology on x, y, a, b' is $ax|b'y$ (w.l.o.g. we assume that a forms a clade with x). The reconstructed sequence at b' need not be independent from a, x and y, but then there must be a node b within distance $2g$ from b' whose reconstructed sequence is independent of the other sequences, which means $ax|by$ will pass the independence test. On the other hand, if $T(x)$ does not overlap with any other subtree, all quartet queries that pass the independence test will return $ab|xy$. Searching for b can be done by breadth-first search on $T(a)$.

4.3   Three-Way Ancestral Sequence Reconstruction

In order to connect edges from different subtrees, we need to have independent sequence reconstructions at both ends of the new edge. To ensure this can be achieved, we will maintain, where possible, three separate sequence reconstructions at each internal node of the subtree, each based on two subtrees of $T(x)$ created by removing x, but independent of the third subtree.

Theorem 3 applies when all subtree edges have length less than g. We treat subtrees with a long edge as rooted at a node on that edge with a single sequence reconstruction. When that node is joined with another tree without creating a new long edge, we then create new three-way reconstructions; such a tree can only be joined with another tree via the long edge, in order to maintain Invariant 3. The routine *ThreeWayReconstruction* takes a tree with sequences reconstructed by maximum likelihood, adding to each vertex the two remaining reconstructions of its sequence. It must be started from a node that has no successors in reconstruction order.

Algorithm 4. ThreeWayReconstruction(r)

Let x, y, z be the neighbours of r.
Reconstruct sequences $\sigma_{xy}(r), \sigma_{yz}(r), \sigma_{xz}(r)$ conditioned on $\{x, y\}, \{y, z\}$ and $\{x, z\}$, respectively.
Let S be the set of vertices in $T(r)$ with only one sequence reconstruction.
Visit vertices in $T(r)$ in Breadth-First Search order, reconstructing each sequence conditioned on all choices of 2 neighbours. Stop branching if a node visited during an earlier call of *ThreeWayReconstruction* is encountered.

Algorithm 5. CandidateEdges(x)

---

if x is the root of a tree T in F **then**
    Return the set containing the edge e that contains x. (This may be a long edge.)
else
    Return the set containing all edges in T within distance $3g + 2g_{err}$. (This can be determined by a breadth-first search.)
end if

4.4 Long Edges Must Be Joined

To maintain Invariant 3, when either of the two closest sequences is in a subtree with a long edge, we must break that edge. We will be finding the shortest pairwise distance between trees in F, so we will certainly find one of the endpoints of the long edge, but we must not consider its other neighbouring edges.

Procedure *CandidateEdges(x)* identifies valid edges that can be broken, given a node x in the tree.

4.5 The Full Algorithm

With these minor issues resolved, we can present the full algorithm.

Lemma 3. *During the execution of the Reconstruct, there always exists a pair of subtrees that can be merged.*

Proof. There are two cases when two subtrees with internal nodes within distance $2g$ from each other cannot be merged. One is when one of the hits is not a root and *ThreeWayReconstruction* has not been called on its subtree. The other is when merging two trees would give rise to a tree with two long edges.

Let us consider the second case first (the other case is analogous). For each two subtrees T_1, T_2 with that property, remove them from T together with the path connecting them. This gives rise to a forest F_1 where at least two components border exactly one of the removed subtrees. Call one such component T' and let T'' be its unique adjacent removed subtree. Let e' be the edge between T' and T''.

If T' contains a reconstructed subtree that is incident to e' and can be joined with the subtree at the other end of e', we are done. Otherwise, there exist two subtrees in T' that can be merged. Consider the forest $F^c = T' - F$. We will refer to components of F^c as *antitrees* to avoid confusion with reconstructed trees from F. Since F cannot contain leaves that are internal nodes in T, each antitree in F^c is either a single edge or contains a cherry. Therefore, each antitree in F^c contains two leaves at distance less than $2g$.

Table 1. The approximate runtime for different values of p_g

p_g	0.01	0.02	0.05	0.075	0.10
runtime	$n^{1.10} \log^2 n$	$n^{1.19} \log^2 n$	$n^{1.47} \log^2 n$	$n^{1.67} \log^2 n$	$n^{1.85} \log^2 n$

Take any such two leaves x and y from antitree T_1^c. If they cannot be joined, then at least one of the trees (say $T(x)$) has a long edge that doesn't include x. The root r at this long edge belongs to an antitree T_2^c that must be different from T_1^c (otherwise we would have a cycle in T). T_2^c has at least one pair of leaves within distance $2g$. If they cannot be merged, then one of its leaves is incident to another tree in F with a long edge. Its root is incident to another tree T_3^c in F^c. The claim holds by induction on the size of F^c.

Algorithm 6. Reconstruct($\{\sigma\}$,f,g)

Start with a forest F where each sequence is in its own tree.
Initialize hash tables for nearest neighbour search
Use *FindAllClose* to identify all sequence pairs at distance less than $3g + 2g_{err}$; put these in a queue DistQueue.
WaitList $\leftarrow \emptyset$
while F has more than one tree **do**
  $(x, y) \leftarrow DistQueue.pop()$
  if $T(x) = T(y)$ **then**
    continue
  end if
  $X \leftarrow CandidateEdges(x)$
  $Y \leftarrow CandidateEdges(y)$
  $Joins \leftarrow X \times Y$.
  if $X = \emptyset$ or $Y = \emptyset$ **then**
    $WaitList.add((x, y))$
  end if
  Filter *Joins* to only include pairs $((i, j), (k, l))$ where $Independent(i, j | k, l)$
  Use *Quartet* and *MiddleEdge* to find d_{min}, the smallest middle edge length among *Joins*.
  Let it be the result of joining (i^*, j^*) with (k^*, l^*).
  create an edge between the two edges (i^*, j^*) and (k^*, l^*) that give rise to d_{min}
  use *MiddleEdge* to calculate the lengths $d_1, \ldots, d_5$ of the five new edges
  if $\max_i d_i \geq 2g$ or $d_i > d_j > g$ for some $i \neq j$ or $CheckOverlaps(x, y)$ **then**
    undo this loop iteration
    $WaitList.add((x, y))$
  end if
  if the new tree has a long edge **then**
    create a root on the new long edge
  else
    reconstruct the sequence at the new internal nodes r_1, r_2
  end if
  if the new tree has no long edge **then**
    $ThreeWayReconstruction(r_1)$
  end if
  Use *FindAllClose* to add all newly-created sequence pairs whose distances are below $3g + 2g_{err}$ to *DistQueue*.
  For all hits (x, y) in *WaitList* at distances less than $3g + 2g_{err}$ from any of the newly created sequences, move (x, y) from *WaitList* to *DistQueue*
end while

Theorem 7. *Each iteration of the* **while** *loop maintains invariants 1-4.*

Proof (Sketch). Invariants 1 and 4 are maintained by Theorem 6 and the quartet query independence checks. Invariant 2 is maintained by *CheckOverlaps*. Invariant 3 is maintained by the conditions on d_{min}.

4.6 Runtime Analysis

At each iteration of the **while** loop, all operations except nearest neighbour search (which is invoked a constant number of times per iteration) take constant time, since the ratio g/f is constant (and thus the number of nodes within distance g of any node is a constant). The complexity is therefore dominated by the use of the hash tables. The evolutionary distance $3g + 2g_{err}$ corresponds to a Hamming distance of at most $h = 1/2(1 - (1 - 2p_g)^3(1 - 2p_{err})^2)$, where $p_g = (1 - e^{-2g})/2$, and the bound on p_{err} is given by Theorem 5. *FindAllClose* is used a constant number of times per loop. Each loop iteration requires $O(n^{2h+\epsilon} \log^2 n)$ time. The loop is run $O(n)$ times, since the number of sequences within distance $3g + 2g_{err}$ of any sequence is constant. Overall, the runtime is bounded by

$$Cn^{2-(1-2p_g)^3(1-2p_{err})^2+\epsilon} \log^2 n < Cn^{2-(1-2p_g)^3 \frac{(1-4g)(1-8g)}{(1-2g)^4}+}$$

$$\epsilon \log^2 n < Cn^{2-(1-2p_g)^3(8g-1)^2+\epsilon} \log^2 n,$$

which is always $o(n^2)$. Table 1 shows the runtime for selected values of p_g.

5 Experiments

5.1 A Practical Algorithm

Many assumptions made by our theoretical algorithm are impractical. The hash tables required for *FindAllClose* to work with high probability may require a prohibitive amount of memory. Using maximum likelihood for ancestral sequence reconstruction requires much memory to store conditional probabilities. We have developed a simpler and more memory-efficient practical algorithm.

Our implementation uses a number of hash tables required to find near neighbours with constant probability, not high probability. For reasons of memory efficiency, we also do not perform three-way reconstruction; instead, we join non-root nodes of different subtrees without requiring the sequences in quartet queries to be error-independent. Note that we still require error-independent sequences for estimating branch lengths in an existing subtree. After two subtrees are joined, the sequences in the smaller subtree are re-estimated according to an ordering compatible with that of the larger tree.

For simplicity, the practical algorithm does not use routines *CheckOverlaps* and *CheckErrorIndependence*. Instead, we perform a small number of Nearest Neighbour Interchange (NNI) moves after each join, to ameliorate the problems originally addressed by these two functions. After an edge has been added, we re-estimate the length of all edges whose length might have been affected by the merge. If any quartet gives a topology that is not consistent with the edge, we perform an NNI move to fix the topology and re-estimate the lengths of adjacent edges. We note that this procedure is very different from traditional local search algorithms using NNI's, as only the edges in the close vicinity of the new edge are affected and the associated computational cost is much lower.

The algorithm tries to merge pairs of trees, starting from collisions with the lowest estimated evolutionary distance. If no collisions are found within distance $(r_1 + r_2)/2$, a new hash table is added, until the maximum number of $2n^{r_1/r_2}$ hash tables is reached. If no full tree is produced after examining all the hits in the hash tables, we use a simple heuristic that picks representative ancestral sequences from each tree and finds closest pairs among them, so the algorithm terminates even if the tree contains a small number of long edges.

The current prototype does not attempt to find optimal LSH parameters r_1 and r_2. In our experiments, we set them to 0.2 and 0.6, respectively. This choice leads to memory inefficiency for trees with very short edges, as we will see later. We leave the automatic adjustment of these values for future work.

5.2 Data Sets

We used a data set from our previous paper on QTree [2], to compare our new algorithm with two other fast phylogenetic algorithms, QTree and Fast-Tree. We simulated 10 trees on 20000 taxa from the pure-birth process. Following Liu *et al.* [15], we then multiplied each branch length by a factor chosen uniformly at random from interval [0.5, 2] to deviate the trees from ultrametricity. We then scaled the branch lengths of the entire tree by several choices of constant factors, resulting in trees with mean branch lengths equal to 0.0312, 0.0625, 0.1250, and 0.25, respectively. For each choice of tree and scaling factor, we generated alignments whose length varied between 250 and 4000 positions from the Jukes-Cantor model with variable rates across sites distributed exponentially. No indels were introduced. The data sets can be downloaded at http://cs.uwaterloo.ca/~jmtruszk/datasets.tar.gz.

5.3 Results

Figure 1 shows the performance of our algorithm, compared with QTree and FastTree. We use the Robinson-Foulds metric, defined as the fraction of splits in the true tree that are found in the reconstructed tree. We only ran the Neighbour Joining phase of FastTree, without its concluding local search phase, since similar local search procedures can be applied to the output of any algorithm.

Our algorithm achieves higher accuracy than both FastTree and QTree in most settings. The main exception is trees with long branches, where its accuracy is substantially lower than both QTree and FastTree. The poor performance of our algorithm for trees with long branches is not surprising given that it relies so heavily on reconstructed ancestral sequences, whose accuracy diminishes as branch lengths approach the phase transition. For very short sequences, FastTree appears somewhat more accurate, possibly because of aggregating information from a greater number of distance estimates.

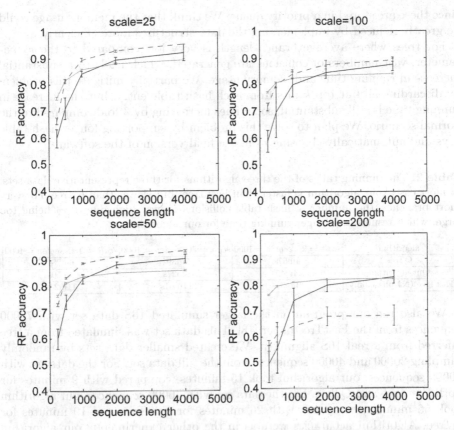

Fig. 1. The performance of the LSH algorithm(red, dashed) compared to QTree(dark blue), and FastTree(light green), as a function of the length of the sequences. The four graphs represent the performance on 10 trees with branch lengths scaled by constant factors 25, 50, 100, and 200, corresponding to data sets with mean branch lengths equal to 0.0312, 0.0625, 0.1250, and 0.25, respectively. The accuracy of the LSH algorithm is superior to both QTree and FastTree in most settings, except for phylogenies with very long branches(scale=200), where it performs worse than the other two, presumably due to poor ancestral sequence reconstruction.

5.4   Running Times and Scalability

On most instances, our program runs in times competitive with FastTree and somewhat longer than QTree (see Table 2). We believe the running times could be improved by a more careful choice of parameters, and note that our work is a preliminary prototype.

For 32-bit machines with up to 4GB RAM, our program does not scale to alignments larger than $2 \cdot 10^7$ letters. This is mostly due to the amount of memory required to store probability vectors for maximum likelihood, but also due to the hash tables. Memory usage may also increase if the number of collisions is high, since these are stored in a priority queue. We think that the memory usage could be greatly reduced by implementing the data structures more efficiently.

For trees where average branch length is very low compared to the r_2 parameter, vast numbers of collisions are generated, which leads to a substantial increase in running time and memory usage. We partially mitigate this problem by discarding all but top k hits from each hash table entry, but the increase in running time is still substantial, sometimes increasing by 3-fold compared to the normal scenario. We plan to solve this problem by supporting longer hash table keys and automatically choosing r_2 in the final version of the software.

Table 2. The running times of the three algorithms for three representative data sets. In most cases, our algorithm is faster than FastTree, but slower than QTree. For very short branches, the number of hash table collisions is very high due to r_2 being too large, which results in a longer running time for our algorithm.

algorithm	$scale = 25, seqlen = 1000$	$scale = 50, seqlen = 1000$	$scale = 100, seqlen = 1000$
QTree	4m57s	5m39s	6m28s
Our algorithm	24m49s	8m27s	6m50s
FastTree (NJ phase only)	10m31s	11m01s	11m23s

We also ran our program on the larger simulated 16S data set with 78000 sequences from the FastTree paper [18]. This data set was simulated from a tree inferred from a real 16S alignment. We created smaller data sets by randomly sampling 20000 and 40000 sequences from the full data set. For the data set with 20000 sequences, our algorithm took 15 minutes, compared with 9 minutes for both FastTree and QTree. For the data set with 40000 sequences, our algorithm took 56 minutes, compared with 26 minutes for FastTree and 19 minutes for QTree. Algorithm accuracies were as in the other experiments: our algorithm was more accurate than QTree and FastTree. The runtime of our algorithm was likely affected by the wrong choice of r_2.

References

1. Brown, D.G., Truszkowski, J.: Towards a Practical $O(n \log n)$ Phylogeny Algorithm. In: Przytycka, T.M., Sagot, M.-F. (eds.) WABI 2011. LNCS (LNBI), vol. 6833, pp. 14–25. Springer, Heidelberg (2011)
2. Brown, D.G., Truszkowski, J.: Towards a practical O(n logn) phylogeny algorithm. Algorithms for Molecular Biology (special issue on selected papers from WABI 2011 (submitted, 2012)
3. Buhler, J., Tompa, M.: Finding motifs using random projections. J. Comp. Biol. 9(2), 225–242 (2002)
4. Csűrös, M.: Fast recovery of evolutionary trees with thousands of nodes. J. Comp. Biol. 9(2), 277–297 (2002)
5. Daskalakis, C., Mossel, E., Roch, S.: Phylogenies without Branch Bounds: Contracting the Short, Pruning the Deep. In: Batzoglou, S. (ed.) RECOMB 2009. LNCS, vol. 5541, pp. 451–465. Springer, Heidelberg (2009)
6. Daskalakis, C., Mossel, E., Roch, S.: Evolutionary trees and the Ising model on the Bethe lattice: a proof of Steel's conjecture (July 27, 2005), http://arxiv.org/abs/math/0509575

7. Elias, I., Lagergren, J.: Fast Neighbor Joining. In: Caires, L., Italiano, G.F., Monteiro, L., Palamidessi, C., Yung, M. (eds.) ICALP 2005. LNCS, vol. 3580, pp. 1263–1274. Springer, Heidelberg (2005)
8. Erdös, P.L., Steel, M.A., Székely, L.A., Warnow, T.: A few logs suffice to build (almost) all trees: Part II. Theor. Comput. Sci 221(1-2), 77–118 (1999)
9. Erdös, P.L., Steel, M.A., Székely, L.A., Warnow, T.: Greengenes, a chimera-checked 16s rrna gene database and workbench compatible with arb. Appl. Environ. Microbiol. 72, 5069–5072 (2006)
10. Evans, W., Kenyon, C., Peres, Y., Schulman, L.J.: Broadcasting on trees and the Ising model. The Annals of Applied Probability 10(2), 410–433 (2000)
11. Felsenstein, J.: Inferring Phylogenies. Sinauer (2001)
12. Gronau, I., Moran, S., Snir, S.: Fast and reliable reconstruction of phylogenetic trees with very short edges. In: Proceedings of SODA 2008, pp. 379–388 (2008)
13. Indyk, P., Motwani, R.: Approximate nearest neighbors: Towards removing the curse of dimensionality. In: Proceedings of STOC 1998, New York, pp. 604–613 (1998)
14. King, V., Zhang, L., Zhou, Y.: On the complexity of distance-based evolutionary tree reconstruction. In: Proceedings of SODA 2003, pp. 444–453 (2003)
15. Liu, K., Raghavan, S., Nelesen, S., Linder, C.R., Warnow, T.: Rapid and accurate large-scale coestimation of sequence alignments and phylogenetic trees. Science 324(5934), 1561–1564 (2009)
16. Mihaescu, R., Hill, C., Rao, S.: Fast phylogeny reconstruction through learning of ancestral sequences (December 08, 2008), http://arxiv.org/abs/0812.1587
17. Mossel, E.: Phase transitions in phylogeny. Trans. Amer. Math. Soc. 356, 2379–2404 (2004)
18. Price, M.N., Dehal, P.S., Arkin, A.P.: FastTree: Computing large minimum evolution trees with profiles instead of a distance matrix. Mol. Biol. E 26(7), 1641–1650 (2009)
19. Zhang, L., Shen, J., Yang, J., Li, G.: Analyzing the fitch method for reconstructing ancestral states on ultrametric phylogenetic trees. Bulletin of Mathematical Biology 72, 1760–1782 (2010)

Efficient Computation of Popular Phylogenetic Tree Measures

Constantinos Tsirogiannis[1], Brody Sandel[1], and Dimitris Cheliotis[3]

[1] MADALGO* and Department of Bioscience
Aarhus University, Denmark
[2] Department of Mathematics
University of Athens, Greece

Abstract. Given a phylogenetic tree $\mathcal{T}$ of n nodes, and a sample R of its tips (leaf nodes) a very common problem in ecological and evolutionary research is to evaluate a distance measure for the elements in R. Two of the most common measures of this kind are the Mean Pairwise Distance (MPD) and the Phylogenetic Diversity (PD). In many applications, it is often necessary to compute the expectation and standard deviation of one of these measures over all subsets of tips of $\mathcal{T}$ that have a certain size. Unfortunately, existing methods to calculate the expectation and deviation of these measures are inexact and inefficient.

We present analytical expressions that lead to efficient algorithms for computing the expectation and the standard deviation of the MPD and the PD. More specifically, our main contributions are:

- We present efficient algorithms for computing the expectation and the standard deviation of the MPD exactly, in $\Theta(n)$ time.
- We provide a $\Theta(n)$ time algorithm for computing approximately the expectation of the PD and a $O(n^2)$ time algorithm for computing approximately the standard deviation of the PD. We also describe the major computational obstacles that hinder the exact calculation of these concepts.

We also describe $O(n)$ time algorithms for evaluating the MPD and PD given a single sample of tips. Having implemented all the presented algorithms, we assess their efficiency experimentally using as a point of reference a standard software package for processing phylogenetic trees.

1 Introduction

Background and Motivation. Ecologists are increasingly using information on phylogenetic relationships among species to gain new insights into both fundamental and applied questions. This is motivated in part by the observation that closely related species often share similar phenotypic and ecological characteristics. Species with high phylogenetic distinctiveness also represent particularly

---

* Center for Massive Data Algorithmics, a Center of the Danish National Research Foundation.

B. Raphael and J. Tang (Eds.): WABI 2012, LNBI 7534, pp. 30–43, 2012.
© Springer-Verlag Berlin Heidelberg 2012

unique evolutionary histories, and are therefore important conservation targets. Phylogenetic relationships among species can be used to understand biogeographic patterns [9], to infer processes underlying local community assembly [2] and to inform conservation decision making [3].

Given a particular phylogenetic tree $\mathcal{T}$, many measures have been proposed to describe the phylogenetic composition of a set of tips, that is a set of leaf nodes of the tree that represent the finest taxonomic unit in the analysis (for example animal species, languages etcetera). Here, we focus on two widely used concepts. The first is *Phylogenetic Diversity* (PD), which measures the total edge weight of the spanning subtree that connects all tips of the sample. Different variants of this metric have been considered in the related literature [4,7,11], mostly as building blocks for the analysis of fundamental combinatorial problems that arise in evolutionary research.

The second is *Mean Pairwise Distance* (MPD), which is equal to the mean cost among all simple paths between pairs of sample tips.

Both PD and MPD depend on the number of tips of the studied sample. Hence, analyses that do not account for this relationship risk conflating patterns of species richness with patterns of phylogenetic community composition. A common solution to this problem is to calculate an index that standardizes PD or MPD based on the expectation and standard deviation over all possible sets of tips of a certain size [12].

In the case of MPD, this index is called the Net Relatedness Index (NRI) [12]. For PD, we call this the phylogenetic diversity index (PDI). Both the expectation and the standard deviation depend on the topology of the tree in potentially complex ways, hence it has been standard to estimate these values using methods that are based on random sampling of tips in $\mathcal{T}$ [13,8]; for a given sample size r, a large number of samples is extracted (often a thousand, but sometimes fewer), each sample consisting of exactly r distinct tips. In most cases, the tips of each sample are selected using a uniform random distribution, that is each tip has the same probability to be selected in a sample as every other tip. Then, for each of these samples, the value of the considered measure (PD or MPD) is computed and the expectation and standard deviation are calculated among the computed values. This expectation and deviation are then used as an estimate of the actual expectation and standard deviation of the measure.

However, this approach produces inexact estimates; even for a tree $\mathcal{T}$ of a few thousand tips, and for a tip sample size r of a few hundreds, the total number of the subsets of tips in $\mathcal{T}$ that have size exactly r is astronomically large. Therefore, it is debatable if the above random method can provide a reasonable approximation by selecting only a limited, yet computationally feasible, number of tip subsets.

More than that, this approach can be quite slow; even for trees that consist of only a few thousand tips, existing software packages that use this method can take several minutes, or hours, to produce an output.

Thus, it is interesting to ask the question if there can be algorithms that compute the expectation and the standard deviation of either the MPD and PD precisely, and at the same time efficiently, without having to tediously check each possible subset of tips of a certain size.

Our Results. In this paper we prove computationally feasible analytical expressions for the expectation and standard deviation of the MPD and PD. Based on these expressions, we describe efficient algorithms that speed up the computation of the NRI and PDI. We also provide efficient algorithms that calculate the basic MPD and PD measures given an individual sample of tips.

In Section 2, we present algorithms for computing analytically the expectation and standard deviation of MPD and PD for a phylogenetic tree $\mathcal{T}$ and for a given tip-sample size, assuming a uniform probability distribution on the selection of the tips of the tree. We show that computing analytically the expectation and the standard deviation of the MPD can be done in $\Theta(n)$ time where n is the combinatorial size of the input tree. We also indicate the fundamental computational problems that arise in the analytical computation of these concepts for the PD measure. We also provide efficient algorithms that compute the MPD and PD measures given a specific tip sample in $O(n)$ time. The proofs of the theorems that are presented in that section can be found in the full version of the paper.

In Section 3 we test our implementation of all of our algorithms against those that appear in the package `picante` [8] of the software library R [10] (one of the standard tools for phylogenetic analyses in ecology). Our tests are designed to assess, first, the improvement in computation time we have achieved, and second, the size of the error induced when estimating the expectation and standard deviation using random sampling.

Significance. Our approach drastically reduces computation time for NRI and PDI on large trees. This is significant because the time required to perform an estimation by random sampling can be a limiting factor in large ecological studies. For example, it may be desirable to not only calculate a global NRI for some set of species samples, but to recalculate it under phylogenies of different sizes, different assumptions about divergence times within unresolved clades (often genera [9]), or different subsets of tips. As the number of different comparisons increases, ecological interpretations can become increasingly refined and sophisticated. However, the size of the induced computational problem also grows radically. To date, incremental improvements in the maximum size of the problem that can be considered have come primarily from increases in computing power. We anticipate that the results presented here will allow a breakthrough to considering a much larger parameter spaces, greatly improving resulting ecological insights.

2   Analytical Expressions and Algorithms

Preliminaries. For a phylogenetic tree $\mathcal{T}$ we denote the set of its edges by E. For an edge $e \in E$, we indicate the (always positive) weight of this edge as w_e.

We denote the set of leaf nodes of $\mathcal{T}$ by S. From here on we will refer to these nodes also as the *tips* of the tree. We indicate the number of these nodes by s, that is $s = |S|$, and we indicate the number of all the nodes of the tree by n.

A phylogenetic tree is a rooted tree, a specific node in the tree is defined as the root. Hence, for any edge $e \in E$ we can distinguish the two nodes adjacent to e into a *parent* node and a *child* node; the child node of e is the adjacent node for which we have to cross e in order to reach the root of the tree. For a node $u \in \mathcal{T}$, we denote the set of the edges for which u is the parent node by $desc(u)$. We use $\mathcal{T}(e)$ to indicate the subtree of $\mathcal{T}$ whose root is the child node of edge e. We denote the set of tips that appear in $\mathcal{T}(e)$ as $S(e)$, and we use also $s(e)$ to denote the number of these tips.

For an edge $e \in E$, we denote the set of all tree edges that appear in the subtree of e by $\mathrm{Off}(e)$. We denote the set of all edges $e' \in E$ for which e appears in the subtree of e' by $\mathrm{Anc}(e)$. We consider that $e \in \mathrm{Anc}(e)$. We also indicate the set $E - (\mathrm{Off}(e) \cup \mathrm{Anc}(e))$ by $\mathrm{Ind}(e)$.

Given any two nodes u, v of $\mathcal{T}$, we call a *simple path* between these nodes a sequence of edges in E that we have to traverse at most once so as to reach u from v. We indicate this path by $p(u,v)$. We call the *cost* of this path the sum of the weights of all the edges that constitute the path. We denote this cost by $cost(u,v)$. As $\mathcal{T}$ is a tree, there exists a unique simple path for any pair of nodes in $\mathcal{T}$. Let $R \subseteq S$ be any subset of the tips of a phylogenetic tree $\mathcal{T}$. We denote the set of all pairs of elements in R, that is the set of all combinations that consist of two distinct tips in R, by $\Delta(R)$. We indicate the set of all paths that connect two elements in R by $\mathrm{Paths}(R)$, that is:

$$\mathrm{Paths}(R) = \{p(u,v) : \{u,v\} \in \Delta(R)\}$$

We denote the set whose elements are all the subsets of S that have cardinality exactly r by $\mathrm{Sub}(S,r)$. For an edge $e \in E$ and a subset R of the tips of $\mathcal{T}$, we denote the elements of $S(e)$ that are also elements of R by $S_R(e)$, that is $S_R(e) = S(e) \cap R$. We indicate the the number of these tips as $s_R(e)$. Consider the subset of $\mathcal{T}$ that is the union of the edges of all paths in $\mathrm{Paths}(R)$. We call this subset the *subtree of $\mathcal{T}$ induced by* R. We denote this subset by $\mathcal{T}(R)$ and we indicate the number of edges in $\mathcal{T}(R)$ by $t(R)$.

2.1 The Mean Pairwise Distance Method

Let R be a subset of the tips of a phylogenetic tree $\mathcal{T}$ and let $r = |R|$. The *Mean Pairwise Distance* (MPD) of the tips in R is defined as:

$$\mathrm{MPD}(\mathcal{T}, R) = \frac{2}{r(r-1)} \sum_{\{u,v\} \in \Delta(R)} cost(u,v) \tag{1}$$

More specifically, the mean pairwise distance of a set R of tips is the sum of the costs of all simple paths between two distinct tips in R, divided by the total number of these paths; since R contains r nodes and since there is a unique simple path between any pair of distinct nodes, then the number of all different paths is equal to the number of all different pairs of elements in R, that is $r(r-1)/2$.

Speeding Up the Computation of the MPD. Given a tree $\mathcal{T}$ and a subset of its tips R, we can derive a simple algorithm for computing MPD$(\mathcal{T}, R)$ directly from the expression in (1); for each $\{u, v\} \in \Delta(R)$ compute $cost(u, v)$ by summing the weights of the edges that form the path between u and v, and add this value to the total sum. However, this approach would be quite inefficient in terms of the number of computational steps involved. Recall that there are $r(r-1)/2$ distinct pairs of tips, and therefore as many paths whose costs we need to compute explicitly. In this manner, we need $\Theta(r^2)$ time only to enumerate those paths. Yet, we can compute this sum more efficiently, using the following lemma.

Lemma 1. *Consider a phylogenetic tree $\mathcal{T}$ and let R be a sample of $|R| = r$ tips of $\mathcal{T}$. For any edge e of this tree, the number of paths in* Paths(R) *that contain e is equal to:*

$$|\{p(u, v) : p(u, v) \in \text{Paths}(R) \text{ and } e \in p(u, v)\}| = s_R(e) \cdot (r - s_R(e))$$

Therefore, the MPD *of R is equal to:*

$$\text{MPD}(\mathcal{T}, R) = \frac{2}{r(r-1)} \sum_{e \in E} w_e \cdot s_R(e) \cdot (r - s_R(e)) \tag{2}$$

Proof. Let e be an edge of $\mathcal{T}$ and let u, v be two distinct tips in R. Edge e appears in the path between u and v if one of these nodes appears in the subtree of e and the other does not. Thus, for each tip in $S_R(e)$ there are as many as $r - s_R(e)$ paths that contain edge e. That means that exactly $s_R(e)(r - s_R(e))$ paths in Paths(R) contain e, and therefore we prove the first part of this theorem.

Instead of computing the cost of each possible path in R explicitly, we can express the MPD in terms of the weight of the edges of $\mathcal{T}$; the weight of an edge in $\mathcal{T}$ is counted in MPD$(\mathcal{T}, R)$ as many times as the number of paths in Paths(R) which contain this edge. Therefore, MPD$(\mathcal{T}, R)$ can be expressed as:

$$\text{MPD}(\mathcal{T}, R) = \frac{2}{r(r-1)} \sum_{e \in E} w_e \cdot occur_R(e)$$

where $occur_R(e)$ is the number of paths in Paths(R) that contain e, and the second part of the theorem follows. □

The expression in (2) can be computed in $\Theta(t(R))$ time in the following manner; first we extract the set of edges that appear in at least one path in Paths(R), that is the edges of $\mathcal{T}(R)$. This can be easily done in $\Theta(t(R))$ time by tracing $\mathcal{T}(R)$ bottom-up starting from the tips in R. Then, we apply a simple recursive algorithm to compute the value $s_R(e)$ for each edge of $\mathcal{T}(R)$.

The Net Relatedness Index. For many applications on phylogenetic trees, given a tree $\mathcal{T}$ and a subset $R \subseteq S$ of $|R| = r$ tips it is important to measure how much the MPD of this set differs from the MPD of any other subset R' of exactly r tips in $\mathcal{T}$. To express this difference, the following quantity is usually calculated:

$$\text{NRI} = \frac{\text{MPD}(\mathcal{T}, R) - \text{E}_{\text{MPD}}(\mathcal{T}, r)}{sd_{\text{MPD}}(\mathcal{T}, r)},$$

$\text{E}_{\text{MPD}}(\mathcal{T}, r), sd_{\text{MPD}}(\mathcal{T}, r)$ are the expected value and the standard deviation respectively of the random variable $\text{MPD}(\mathcal{T}, R)$, where the random set R is picked uniformly out of all subsets of S with r elements.

In what follows, we will compute analytically this expected value and standard deviation. The result for the expected value is stated in the next theorem.

Theorem 1. *Let $\mathcal{T}$ be a phylogenetic tree that contains s tips, and let r be a natural number with $r \leq s$. The expected value of the MPD for a sample of exactly r tips of $\mathcal{T}$ is equal to:*

$$\text{E}_{\text{MPD}}(\mathcal{T}, r) = \frac{2}{s(s-1)} \sum_{e \in E} w_e \cdot s(e) \cdot (s - s(e)), \tag{3}$$

and can be computed $\Theta(n)$ time, where n is the total number of nodes of the tree.

Remark 1. From (3) we see that the expected value of the MPD is independent of the size of R, which is an interesting, yet not surprising, result by itself. However, as we show later on in this paper, this is not the case for the standard deviation of the MPD.

The Standard Deviation of the MPD. Before describing how we can compute analytically the standard deviation of the MPD on a given tree $\mathcal{T}$ and for a given sample size, we introduce a few quantities that relate to groups of paths on $\mathcal{T}$. Our goal is to simplify the computation of the standard deviation of the MPD by expressing the standard deviation in terms of these quantities. We show that we can compute these quantities efficiently by just scanning the tree a constant number of times and, thus, derive a $\Theta(n)$ algorithm for computing the standard deviation.

For a given phylogenetic tree $\mathcal{T}$ we define the *total path cost of $\mathcal{T}$* as the sum of the costs of all distinct simple paths that connect tips of $\mathcal{T}$. We denote this quantity by $TC(\mathcal{T})$, thus:

$$TC(\mathcal{T}) = \sum_{\{u,v\} \in \Delta(S)} cost(u, v).$$

According to Lemma 1, we get that:

$$TC(\mathcal{T}) = \sum_{e \in E} w_e \cdot s(e) \cdot (s - s(e)). \tag{4}$$

Let e be an edge of $\mathcal{T}$. We define the *total path cost of e* as the sum of the costs of all those distinct simple paths between tips of $\mathcal{T}$ that contain e. We denote this quantity by $TC(e)$, thus:

$$TC(e) = \sum_{\substack{\{u,v\} \in \Delta(S) \\ e \in p(u,v)}} cost(u,v).$$

It is easy to show that the latter quantity can be expressed as follows:

$$TC(e) = (s - s(e)) \sum_{l \in \mathrm{Off}(e)} w_l \cdot s(l) + s(e) \sum_{l \in \mathrm{Anc}(e)} w_l \cdot (s - s(l)) + s(e) \sum_{l \in \mathrm{Ind}(e)} w_l \cdot s(l)$$

$$= (s - s(e)) \sum_{l \in \mathrm{Off}(e)} w_l \cdot s(l) + s(e) \sum_{l \in \mathrm{Anc}(e)} w_l \cdot (s - s(l)) \tag{5}$$

$$+ s(e) \left(\sum_{e \in E} w_l \cdot s(l) - \sum_{\mathrm{Off}(e) \cup \mathrm{Anc}(e)} w_l \cdot s(l) \right)$$

For a node u that is a tip of $\mathcal{T}$, we define the *total path cost of u* as the sum of the costs of all simple paths between u and any other tip of $\mathcal{T}$. We indicate this quantity by $TC(u)$, and it is obvious that $TC(u) = TC(e)$, where e is the unique tree edge that is adjacent to u.

From (4) we can derive directly an algorithm for computing $TC(\mathcal{T})$ in $\Theta(n)$ time; scanning the tree in a recursive manner, we compute for each edge e the value $s(e)$ from the respective values of the edges adjacent to its child node, and from that we calculate the number of occurences of e in a path, multiplied by the edge weight.

Based on (5), we use the combination of the following two simple algorithms for computing $TC(e)$ for all the edges of $\mathcal{T}$:

Algorithm. *AllEdgesPathCosts($\mathcal{T}$)*
Input: A phylogenetic tree $\mathcal{T}$.
Output: An array $tc[1 \ldots |E|]$ such that $tc[e] = TC(e), \forall e \in E$.
1.   Initialise array $tc[1 \ldots |E|]$ with all values set to zero.
2.   Set global variable AllWeights $\leftarrow \sum_{l \in E} w_l \cdot s(l)$
3.   **for** every $e \in desc(root(\mathcal{T}))$
4.   **do** *SingleEdgeCosts*$(e, w_e(s - s(e)), w_e \cdot s(e), tc)$.
5.   **return** $tc[\cdot]$

Algorithm. *SingleEdgeCosts(e, SumAnc$_1$, SumAnc$_2$, tc)*
Input: A tree edge e, real numbers SumAnc$_1$ and SumAnc$_2$, and (a reference to) the array tc that stores the computed $TC(\cdot)$ values of the tree edges
Output: A real number which is equal to $w_e \cdot s(e) + \sum_{l \in \mathrm{Off}(e)} w_l \cdot s(l)$.
1.   ▷ Precondition 1: SumAnc$_1$ = $\sum_{l \in \mathrm{Anc}(e)} w_l(s - s(l))$
2.   ▷ Precondition 2: SumAnc$_2$ = $\sum_{l \in \mathrm{Anc}(e)} w_l \cdot s(l)$

3.   SumOff $\leftarrow 0$

4. $u \leftarrow$ child node of e

5.   **for** every $l \in \text{desc}(u)$

6.   **do** SumOff $\leftarrow$ SumOff $+$
           $SingleEdgeCost(l, \text{SumAnc}_1 + w_l(s - s(l)), \text{SumAnc}_2 + w_l \cdot s(l), tc)$

7. $SO \leftarrow (s - s(e)) \cdot \text{SumOff}$

8. $SA \leftarrow s(e) \cdot \text{SumAnc}_1$

9. $SI \leftarrow s(e) \cdot (\text{AllWeights} - \text{SumAnc}_2 - \text{SumOff})$

10. $tc[e] \leftarrow SO + SA + SI$

11. $\triangleright$ Postcondition: SumOff $= \sum_{l \in \text{Off}(e)} w_l \cdot s(l)$

12. **return** SumOff $+ w_e \cdot s(e)$

In the above algorithm, we consider that values $s(e)$ for every $e \in E$ have been already computed; this can be easily done in $\Theta(n)$ time in total. The recursive routine $SingleEdgeCosts$ computes the value $TC(e)$ for the tree edge e for which we call this routine. $SingleEdgeCosts$ is called once for each edge $e \in E$, and the time spent for each call is proportional to the number of edges that are adjacent to the child node of e. Given that the preconditions and the postcondition in $SingleEdgeCosts$ are maintained with each recursive call of this routine, the correctness of the algorithm follows.

Theorem 2. *Let $\mathcal{T}$ be a phylogenetic tree that consists of n nodes of which s are tips, and let r be a natural number with $r \leq s$. The standard deviation of the MPD for a sample of exactly r tips of $\mathcal{T}$ is equal to:*

$$sd_{\text{MPD}}(\mathcal{T}, r) = \sqrt{c_1 \cdot TC^2(\mathcal{T}) + (c_2 - c_1) \sum_{u \in S} TC^2(u) + (c_1 - 2c_2 + c_3) \sum_{e \in E} w_e \cdot TC(e) - E^2_{\text{MPD}}(\mathcal{T}, r)},$$

where $c_1 = \dfrac{4(r-2)(r-3)}{r(r-1)s(s-1)(s-2)(s-3)}$, $c_2 = \dfrac{4(r-2)}{r(r-1)s(s-1)(s-2)}$, and $c_3 = \dfrac{4}{r(r-1)s(s-1)}$.

We can compute $sd_{\text{MPD}}(\mathcal{T}, r)$ in $\Theta(n)$ time, or in $\Theta(1)$ time for each different value of r after running a preprocessing algorithm that runs in $\Theta(n)$ time.

2.2 The Phylogenetic Diversity

Let R be a subset of the tips of a phylogenetic tree $\mathcal{T}$. The *Phylogenetic Diversity* (PD) of the tips in R is defined as:

$$\text{PD}(\mathcal{T}, R) = \sum_{e \in \mathcal{T}(R)} w_e$$

That is, the phylogenetic diversity of a sample R of tips is the sum of the weights of the edges of $\mathcal{T}$ that appear in at least one path in $\text{Paths}(R)$; the weight of each distinct edge $e \in \mathcal{T}(R)$ is counted only once in this sum, even if e appears in more than one path in $\text{Paths}(R)$. In the related literature, the above definition of the PD is known as the *unrooted* PD. The *rooted* version of the PD, instead of just the edge-weights of $\mathcal{T}(R)$, considers the weights of all the edges that have at least one element of R in their subtree. For the analytical expressions of the expectation and the standard deviation of the rooted PD, the reader may refer to the work of Steel [11].

PD$(\mathcal{T}, R)$ can be computed in $\Theta(t(R))$ time with the following simple algorihtm; starting from the tips of R, we trace bottom-up $\mathcal{T}(R)$ marking all the edges that appear in $\mathcal{T}(R)$ and adding their weights to the total sum.

The Phylogenetic Diversity Index. Given a tree $\mathcal{T}$ and a subset $R \subseteq S$ of $|R| = r$ tips we can express how much the PD$(\mathcal{T}, R)$ of this set differs from PD$(\mathcal{T}, R')$ of any other subset $R' \subseteq S$ of exactly r tips by calculating the following index:

$$\mathrm{PDI} = \frac{\mathrm{PD}(\mathcal{T}, R) - \mathrm{E}_{\mathrm{PD}}(\mathcal{T}, r)}{sd_{\mathrm{PD}}(\mathcal{T}, r)},$$

where the $\mathrm{E}_{\mathrm{PD}}(\mathcal{T}, r)$ is the expected value of the PD for all possible subsets that consist of exactly r tips of $\mathcal{T}$, and $sd_{\mathrm{PD}}(\mathcal{T}, r)$ is the corresponding standard deviation for this group of subsets of tips. Assuming again that each tip can be included in a sample of r tips with equal probability, we present next how we can compute the above expected value and standard deviation analytically. In the following theorem, and for the rest of this paper, we consider that $\binom{n}{k} = 0$ if $k > n$.

Theorem 3. *Let $\mathcal{T}$ be a phylogenetic tree that contains s tips, and let r be a natural number with $r \leq s$. The expected value of the PD for a sample of exactly r tips of $\mathcal{T}$ is equal to:*

$$\mathrm{E}_{\mathrm{PD}}(\mathcal{T}, r) = \sum_{e \in E} w_e \left(1 - \frac{\binom{s(e)}{r} + \binom{s-s(e)}{r}}{\binom{s}{r}} \right)$$

From Theorem 3 we get an expression for $\mathrm{E}_{\mathrm{PD}}(\mathcal{T}, r)$ that involves the hypergeometric probability function, which is a ratio of binomial coefficients. The explicit numerical computation of binomial coefficients is something that must be avoided due to the number of bits that are needed to represent their values. In order to achieve a fast computation, this probability function can be implemented using methods that lead to an approximate evaluation of the function. Although such methods can guarantee a fixed error bound for a single coefficient, this does not imply directly a fixed error bound for computing $\mathrm{E}_{\mathrm{PD}}(\mathcal{T}, r)$; the result of summing n numbers, such that each number contains a bounded error value, is not guaranteed to be of bounded error as well. Therefore, although it is possible to devise an algorithm that computes $\mathrm{E}_{\mathrm{PD}}(\mathcal{T}, r)$ in $\Theta(n)$ time, assuming a constant time approximate evaluation of the hypergeometric probability function, the output of this algorithm is not guaranteed to be a good approximation of the actual expected value of the PD. Hence, it is a challenging open problem to devise a numerical method with which we can compute $\mathrm{E}_{\mathrm{PD}}(\mathcal{T}, r)$ both efficiently and with guaranteed precision. Next we provide an analytical expression for the standard deviation of the PD for all tip samples of a certain size.

Theorem 4. *Let $\mathcal{T}$ be a phylogenetic tree that contains s tips, and let r be a natural number with $r \leq s$. The standard deviation of the PD for a sample of exactly r tips of $\mathcal{T}$ is equal to:*

$$sd_{\mathrm{PD}}(\mathcal{T},r) = \sqrt{\sum_{e \in E}\sum_{l \in E} w_e \cdot w_l \cdot (1 - \mathcal{F}_{\mathrm{PD}}(S,e,l,r)) - \mathrm{E}_{\mathrm{PD}}^2(\mathcal{T},r)}, \qquad (6)$$

where:

$$\mathcal{F}_{\mathrm{PD}}(S,e,l,r) = \begin{cases} \dfrac{\binom{s(e)}{r} + \binom{s-s(l)}{r} - \binom{s(e)-s(l)}{r}}{\binom{s}{r}} & \text{if } l \in \mathcal{T}(e). \\[2ex] \dfrac{\binom{s(l)}{r} + \binom{s-s(e)}{r} - \binom{s(l)-s(e)}{r}}{\binom{s}{r}} & \text{if } e \in \mathcal{T}(l). \\[2ex] \dfrac{\binom{s-s(e)}{r} + \binom{s-s(l)}{r} - \binom{s-s(e)-s(l)}{r}}{\binom{s}{r}} & \text{otherwise.} \end{cases}$$

Remark 2. The computation of the analytical expression of the standard deviation of the PD in Theorem 4, suffers from the same problems as the computation of the expected value of the PD; we still need to develop an efficient method that can approximate the hypergeometric quantities that appear in the formula of $sd_{\mathrm{PD}}(\mathcal{T},r)$ and can also guarantee the precision of the final result. However, even if such a method was available, it is still not clear if it is possible to compute the double sum in (6) in subquadratic time with respect to the size of $\mathcal{T}$. This poses one more strong computational constraint on the calculation of the PDI.

3   Experimental Results

We have implemented all the algorithms that we present in Section 2 and we conducted experiments in order to assess their efficiency. The implementation was done in C++, using template programming that allows us to use number types of different precision. All the experiments that appear in the current version of the paper were executed using the **double** built-in C++ type. The experiments were executed on an Intel i5 four-core CPU where each core is a 2.67 GHz processor. The main memory of this computer is 4 Gigabytes.

The trees that we used in the experiments are subtrees of varying size that we extracted from a phylogenetic tree data set that contains 4510 tips and which represents the phylogenetic relations between all mammals [1]. We refer to the complete data set as the **mammals** data set. The input trees are provided in Newick tree format [5] in a **txt** file. As a result, each separate execution of one of our algorithms with a specific input tree took linear time with respect to the input size, in order to read and parse the tree. As a point of reference for the efficiency of our implementation, we did the same experiments using the **picante** software package, which is an extension of the R software environment [10].

In the experiments that we conducted, we examined the sensitivity of running time to variation in tree size and sample size by generating random prunings of the full `mammals` tree, and random communities assembled from those subtrees. For each tree and tip sample, we computed MPD, PD, NRI and PDI and recorded the time required to make this computation using our algorithms and `picante`.

In the first set of experiments we measure the running time of our algorithms, versus the `picante` implementation, given trees of different sizes while using a query sample of a fixed number of tips. More precisely, from the complete `mammals` data set we constructed nine tree instances, with the size of each instance being equal to $n = 200 + 500k$ with k ranging from 0 to 8. For each instance we computed the MPD, PD, NRI and PDI given a sample of exactly 100 tips–see Figure 1, graphs a) and b).

In the second set of experiments, we studied the performance of the algorithms for various sample sizes; we fixed the input tree to be the full `mammals` tree (4510 tips) and sampled $10 + 50k$ tips from it, where k ranges from 0 to 10. When calculating NRI and PDI in `picante`, we used 100 random draws to estimate the expectation and deviation –see Figure 1, graphs c) and d). For individual MPD and PD queries, our algorithms are faster than those provided by `picante`. This is particularly true for MPD, for which the `picante` computation time scales superlinearly with respect to tree size. This behaviour appears because `picante` asks for the calculation of a pairwise distance matrix among *all* possible pairs of tips, requiring quadratic time with respect to the total tips size (not sample size) of the tree. Similarly, the analytical solutions for the expectation and deviation of MPD values allowed substantial speed increases, particularly for large trees. In contrast, our algorithm for computing PDI displays better scaling than `picante` with respect to increasing sample size but worse scaling with respect to increasing tree size. This was expected since the algorithm that handles the analytical computation of the standard deviation of the PD runs in quadratic time with respect to the tree size. As mentioned in Remark 2, it seems quite difficult to design, or disprove the existence of, an algorithm with a subquadratic running time for this problem. However, even for large tree sizes and small sample sizes, our algorithm provides better running time, if one uses the standard 1000 random draws rather than 100 as illustrated.

Much of the running time for our algorithms for individual MPD and PD queries is spent reading the input tree. By sending 100 queries at once, using for all the full `mammals` tree, we estimate that the fixed time to perform MPD and PD queries is approximately 0.3 seconds, while the per-query time is between 0.001 and 0.002 seconds and depends weakly on sample size. In contrast, the fixed cost of a MPD query in `picante` is approximately 10.2 seconds, and a PD query is 0.4 seconds. The per-query time for MPD and PD in `picante` increases with tip sample size, from 0.05 to 0.3 seconds for PD and from 0.005 to 0.02 seconds for MPD (for 10 to 510 species).

In the next set of experiments, we simulated a real ecological data set using the full `mammals` tree with 1000 randomly assembled sets of species (e.g. communities), each with a number of species between 2 and 201, and calculated NRI and PDI for this data set. Our algorithms allowed us to calculate NRI for all samples in 2.3 seconds, while the `picante` functions (using 1000 random draws)

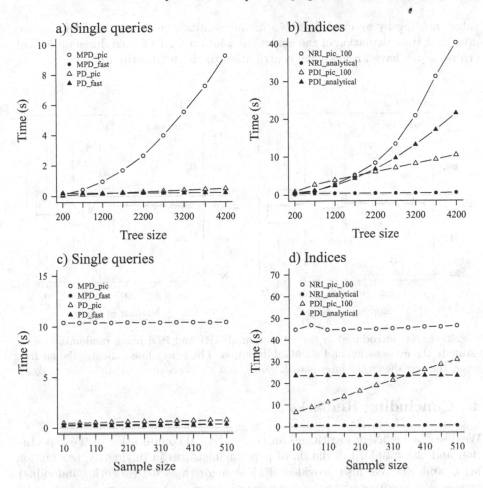

Fig. 1. The running times of the examined algorithms for trees of various size of tips, and a fixed sample of one hundred tips. In the two graphs on the left, MPD_fast, PD_fast are our implementations of the individual MPD and PD queries, while MPD_pic, PD_pic are the respective `picante` routines. In the two graphs on the right, NRI_analytical, and PDI_analytical refer to our implementations for computing the NRI and PDI indices, while NRI_*pic*_100 and PDI_*pic*_100 are the `picante` processes.

required 2024.8 seconds. Our algorithms took 4581.4 seconds to calculate PDI, while `picante` functions took 10524 seconds. In this last case, we used just 200 random draws, 1000 would take roughly five times as long.

The analytical approach provides advantages beyond running time improvements. This is particularly true for NRI, as we are able to calculate exactly the expectation and standard deviation of MPD for a tree. We used this to assess the error introduced by randomized estimation of these values, using the full mammal tree and sample sizes of 20, 100 and 500 tips–see Fig. 2. The random method (using 100 draws) produced accurate estimates of the expectation of MPD, but estimates of the standard deviation varied widely around the true value, ranging by as much as 20%. Similar results were found for PDI, but we note that the calculation of the analytical solution for PDI introduces numerical errors, which have an unknown contribution to the final estimate.

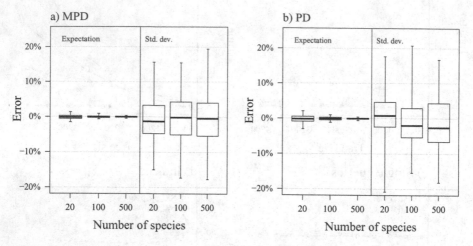

Fig. 2. Errors introduced in the estimation of NRI and PDI using randomization to estimate the expectation and standard deviation. The heavy line indicates the median error, the box shows the interquartile range, and whiskers show the full error range.

4   Concluding Remarks

We described efficient algorithms for the analytical computation of the expectation and the standard deviation of popular measures in phylogeny, namely for MPD and PD. We also provided efficient algorithms for executing individual queries of these meausures on phylogenetic trees. It seems very interesting to extend these results to other kinds of measures that are used in phylogenetic studies; for example computing the expectation and the standard deviation of the mean nearest phylogenetic distance [12]. Also, an important open question that follows from our results is to show if it is possible to derive a method for computing the PDI both efficiently and with arbitrary precision.

References

1. Bininda-Emonds, O.R.P., Cardillo, M., Jones, K.E., MacPhee, R.D.E., Beck, R.M.D., Grenyer, R., Price, S.A., Vos, R.A., Gittleman, J.L., Purvis, A.: The delayed rise of present-day mammals. Nature 446, 507–512 (2007)
2. Cavendar-Bares, J., Ackerly, D.D., Baum, D., Bazzaz, F.A.: Phylogenetic overdispersion in the assembly of Floridian oak communities. American Naturalist 163, 823–843 (2004)
3. Faith, D.P.: Conservation evaluation and phylogenetic diversity. Biological Conservation 61, 1–10 (1992)

4. Faller, B., Pardi, F., Steel, M.: Distribution of phylogenetic diversity under random extinction. Journal of Theoretical Biology 251, 286–296 (2008)
5. Felsenstein, J.: PHYLIP: Phylogeny inference package, version 3.57c. Distributed by the author, Department of Genetics. Univ. of Washington (1995)
6. Graham, C.H., Fine, P.V.A.: Phylogenetic beta diversity: linking ecological and evolutionary processes across space and time. Ecology Letters 11, 1265–1277 (2008)
7. Hartmann, K., Steel, M.: Phylogenetic diversity: From combinatorics to ecology. In: Gascuel, O., Steel, M. (eds.) Reconstructing Evolution: New Mathematical and Computational Approaches. Oxford University Press (2007)
8. Kembel, S.W., Ackerly, D.D., Blomberg, S.P., Cornwell, W.K., Cowan, P.D., Helmus, M.R., Morlon, H., Webb, C.O.: Documentation for picante R package (2011)
9. Kissling, W.D., Eiserhardt, W.L., Baker, W.J., Borchsenius, F., Couvreur, T.L.P., Balslev, H., Svenning, J.-C.: Cenozoic imprints on the phylogenetic structure of palm species assemblages worldwide. Proc. National Academy of Sciences 109, 7379–7384 (2012)
10. R Development Core Team. R: A language and environment for statistical computing. R Foundation for Statistical Computing, Vienna (2010)
11. Steel, M.: Tools to construct and study big trees: A mathematical perspective. In: Hodkinson, T., Parnell, J., Waldren, S. (eds.) Reconstructing the Tree of Life: Taxonomy and Systematics of Species Rich Taxa, pp. 97–112. CRC Press (2007)
12. Webb, C.O., Ackerly, D.D., McPeek, M.A., Donoghue, M.J.: Phylogenies and community ecology. Annual Review of Ecology and Systematics 33, 475–505 (2002)
13. Webb, C., Ackerly, D., Kembel, S.: Phylocom Users Manual, version 4.2 (2012)

SibJoin: A Fast Heuristic for Half-Sibling Reconstruction

Daniel G. Brown and Daniel Dexter

Cheriton School of Computer Science, University of Waterloo
Waterloo, ON N2L 3G1, Canada
{browndg,ddexter}@uwaterloo.ca

Abstract. Kinship inference is the task of identifying genealogically related individuals. Questions of kinship are important for determining mating structures, particularly in endangered populations. Although many solutions exist for reconstructing full-sibling relationships, few exist for half-siblings. We present SibJoin, a heuristic-based clustering approach based on Mendelian genetics, which is reasonably accurate and thousands of times faster than existing algorithms. We also identify issues with partition distance, the traditional method for assessing the quality of estimated sibship partitionings. We prefer an information theoretic alternative called variation of information, which takes into account the degree to which misplaced individuals harm sibship structures.

1 Introduction

SibJoin is a fast heuristic which uses constructive and destructive aspects of Mendelian genetics to reconstruct maternal and paternal half-sibships. Sibship reconstruction has many important applications in conservation biology and molecular ecology, and is particularly important to biologists who analyze mating and reproductive patterns in wild populations [14]. Half-sibling reconstruction also provides specific insight about pollination patterns of plant populations. In plant populations, mothers are pollinated from potentially distant fathers. The diversity of pollinating fathers can be used to measure the degree of isolation, due to deforestation, which threatens many forests [9].

Previously, kinship algorithms have focused on reconstructing full-sibling families from microsatellite data. Current methods for reconstructing full-sibships are very powerful [2, 5, 17], but few algorithms exist for reconstructing half-sibships [10,18], in part because confirming that individuals are not half-siblings is much more difficult than the parallel problem for full-siblings. Furthermore, existing half-sibling prediction algorithms search through large solution spaces and have exponential runtimes, making them unusable for large populations.

For diploid species, children inherit one maternal and one paternal chromosome at each locus from their parents. Mendelian genetic properties can identify which individuals are not related, but they also allow us to make assumptions about individuals who are related. Our algorithm acts on the assumption that if the genotypes of two individuals are very similar, we can be more confident that

B. Raphael and J. Tang (Eds.): WABI 2012, LNBI 7534, pp. 44–56, 2012.
© Springer-Verlag Berlin Heidelberg 2012

they are related than we can of two individuals with much different genotypes. The accuracy and speed of our algorithm allows us to infer half-sibling relationships for population sizes which have been unsolvable by current methods. We demonstrate SibJoin's speed by reconstructing half-sibship partitionings for a real population of 672 kelp rockfish which previous half-sibling reconstruction algorithms failed to solve [15].

Finally, although widely used, the traditional partition distance metric for measuring the quality of kinship algorithms' solutions is very coarse. We will provide examples of instances where partition distance fails to distinguish between two solutions where one solution is clearly better than the other. In place of partition distance, we suggest an information theoretic metric, previously used in clustering [13], called variation of information. The proposed metric is more accurate because it accounts for the structure of the incorrectly assigned individuals.

2   Related Work

Many groups have produced algorithms for constructing full and half-sibling partitionings. Current sibling reconstruction techniques fall into three categories: likelihood estimation, combinatorial objective optimization, and heuristics. While full-sibling reconstruction is a well studied problem with many very accurate algorithms, half-sibling algorithms are relatively few.

2.1   Likelihood Estimation

Likelihood methods estimate the probability of the data under different partitionings of a population. An optimal solution maximizes this probability. Most likelihood methods use Markov Chain Monte Carlo, simulated annealing, or other search strategies to find their solutions. These strategies are often very slow, making them ill-suited for sibling reconstruction on large data sets. On the other hand, because this class of algorithm establishes a probabilistic model, it is often possible to directly incorporate error handling and prior assumptions about the population to increase accuracy. For a more detailed discussion of likelihood methods, see Jones and Wang [11].

COLONY [17] and COLONY 2.0 [18] are likelihood methods which construct half-sibling families. COLONY reconstructs full-sibling families with high accuracy, but allows for polygamy in only one sex. COLONY 2.0 performs half sibship reconstruction when both sexes are polygamous. Both of these programs use a likelihood function and simulated annealing to find an optimal sibling structure for a population. However, results by Sheikh *et al.* [15], as well as our own results, show that COLONY and COLONY 2.0 become prohibitively slow for even medium-sized populations. Additionally, as demonstrated in Almudevar and Anderson [1], COLONY 2.0 often splits true sibgroups into smaller groups, leading to an incomplete reconstruction.

2.2 Combinatorial Optimization

Combinatorial optimization solutions seek to provide a sibship partitioning which minimizes or maximizes some objective function, such as number of families, matings, or parents. As with likelihood methods, finding global optima for large populations can be computationally demanding. However, many optimization techniques are easily parallelizable.

KINALYZER [2] seeks a minimum set cover, by using an integer programming (IP) formulation where each set is subject to restrictions of Mendelian compatibility for full-siblings. KINALYZER yields decent results [6]; however, like the COLONY programs, does not scale well with population size. The minimum set cover objective used by KINALYZER is NP-hard [6]. Recent work has proposed half-sibling IP strategies, similar to the full-sibling strategies in KINALYZER, though they are unsuccessful for large populations [15]. The most viable of these is the half-sibling minimum set cover (HS-MSC) IP. Both COLONY and the HS-MSC failed to estimate half-sibling groups for large populations due to slow runtimes. Additionally, there is no evidence that minimizing the number of sibgroups is the right thing to do in all instances [1].

2.3 Fast Heuristics

Heuristics have been applied to the sibgroup reconstruction problem so that researchers may obtain sibgroups for large populations. By making use of simplifying observations, heuristics can produce reasonably accurate results thousands of times faster than pure likelihood or combinatorial methods. The trade off is that the algorithm does not assign quality to different feasible solutions.

Brown and Berger-Wolf propose a clustering algorithm which joins two individuals based on the number of genetically compatible third partners [5]. The assumption is, if two individuals form a large number of compatible full-sibling triplets, then they are likely to be full-siblings. For a population of n individuals with m loci, this algorithm has an $O(n^3m)$ runtime and gives accurate results for modest numbers of alleles and loci. Another heuristic, called PRT [1] enumerates a list of maximal sibgroups: sibgroups for which no additional population may be added. PRT makes the assumption that it is unlikely to find unrelated individuals in a large sibgroup of this form. A set cover of the maximal sibgroups is selected using a likelihood function. Although it is claimed that PRT supports half-siblings, half-sibling groups are never directly computed. Instead, full-sibling groups are presented with a list of which pairs of groups can form valid half-siblings. This is problematic in instances where both sexes are highly polygamous because there will be many pairs of half-sibling compatible full-sibling families. However, PRT does not indicate which compatibilities are true half-sibling groups nor which are maternal and which are paternal. This makes PRT an ineffective half-sibling tool for all but the most simple instances.

3   Notation

Information about individuals' genotypes are collected and expressed through the measurement of *microsatellites*, sequences of repeating DNA base pairs, such as ATATATAT, on a chromosome. The number of repeats gives an integer value denoting the *allele* for an individual. Microsatellites are collected from two chromosomes, though it is impossible to distinguish the two chromosomes with inexpensive technology. Each measurement site is called a microsatellite *locus*. In practice, scientists identify and report alleles at multiple loci in a population.

SibJoin requires that each individual be diploid, meaning that population members possess two of each type of chromosome. Exactly one chromosome is inherited from each of the individual's parents; therefore, each locus will have a maternal and paternal allele. Let m be the number of measured loci for a population. Each locus will have a variable number of alleles, k, which we represent as $A_l = \{a_0, a_1, \ldots, a_{k-1}\}$.

When maternal and paternal alleles are combined, they give an individual's genotype, which is unordered: (a_i, a_j) is equivalent to (a_j, a_i). Unfortunately, it is not always possible to reconstruct an individual's alleles for a given locus. *Allelic dropout* is a term that refers to a common error in genotyping where information about a locus cannot be confidently determined and is omitted. We express sites with allelic dropouts as $(*, *)$.

The half-sibling problem is, given a population of n offspring, to reconstruct a maternal and paternal partitioning $\mathcal{M}$ and $\mathcal{P}$ respectively that obey the Mendelian laws for half-siblings. For each, $M \in \mathcal{M}$ and $P \in \mathcal{P}$, the individuals in $F := M \cap P$ must respect the laws of Mendelian genetics for full-siblings, as they correspond to the offspring of a single mother and father. In order to find such a clustering, SibJoin relies on measurements of similarity between individuals. We denote the similarity between individuals x and y as s_{xy} and the similarity between clusters C_i and C_j as $sim(C_i, C_j)$. The terms partitioning and clustering may be used interchangeably.

4   Clustering Half-Sibling Groups

Sibship reconstruction finds a population clustering which obeys Mendelian genetics. In the full-sibling clustering $\mathcal{F}$, each individual appears only once. For half-siblings, an algorithm must construct $\mathcal{M}$ and $\mathcal{P}$ when both sexes are polygamous or only one of the two partitions when one sex is monogamous.

4.1   Mendelian Compatibility

Berger-Wolf *et al.* [3] give two Mendelian properties of diploid full-siblings. Refer to their article for the concrete mathematical expression; we give a short exegesis. In any full-sibgroup, at all alleles, at most four alleles appear since there are two parents each with at most two alleles. This is the 4-allele property. The 2-allele property enforces the rule that for each full-sibling group, there is a partitioning of the alleles at each locus into a maternal and paternal group, such that each individual obtains exactly one allele from the maternal set and one from the paternal set, at each locus. Sheikh *et al.* [15] extend these rules to half-siblings. The *half-sibship property* states that for each locus in a half-sibling family, there exists two alleles $\{a_i, a_j\}$, which are the alleles of the shared parent, such that each individual possesses either a_i or a_j at each respective locus.

4.2 Measuring Similarity

For a given half-sibling input, SibJoin relies on an $n \times n$ similarity matrix which expresses the similarity between each pair of individuals in the offspring population set. Brown and Berger-Wolf [5] use a similarity measure which, for each pair of individuals, is the count of third individuals in the population that form a compatible full-sibling triplet with the pair. They prove that any incompatible candidate full-sibling group must contain an incompatible triplet and give a probabilistic argument that pairs of individuals with large numbers of compatible triplets are likely siblings. Unfortunately, the half-sibling property is much weaker because it only operates on one parent. Ruling out a potential half-sibling group can take as many as six individuals, compared to the three that is required for full-siblings.

Theorem 1. *There exist incompatible half-sibling groups for which their smallest incompatible subgroup has six members*

The proof of the upper bound is omitted due to length. However, a simple example will prove that examining sextets of individuals is necessary to rule out half-sibling incompatibility. Consider the sextet of individuals with one locus and four alleles $\{[(1, 2)], [(1, 3)], [(1, 4)], [(2, 3)], [(2, 4)], [(3, 4)]\}$. Any five of the individuals form a valid half-sibship under the half-sibling property, but the incompatibility appears when all six individuals are examined together: the common half-sibling parent would need to have three alleles.

The lower bound suggests that examining triplets for half-siblings could yield a falsely high count when individuals are not actually related. Additionally, the probability that three random individuals form a valid half-sibship is much higher than that of three individuals forming a valid full-sibship. By enumeration of all possible triplets, we see that with five alleles, 96.62% of all triplets were compatible under the half-sibling property. Only 56.61% of all triplets were compatible under the full-sibling properties. If the number of alleles is set to ten, then 75.46% of half-sibling triplets remain compatible, while the number of full-sibling compatible triplets shrinks to 20.94%.

In the place of a triplets-based similarity function, SibJoin uses a pairwise measure. Given two individuals, each with m loci, the similarity function computes the count of shared alleles at each locus independently, among the two individuals. For example, the pair of individuals $x = [(1, 2), (2, 2), (1, 3)]$ and $y = [(1, 1), (2, 2), (2, 3)]$ has a similarity of $s_{xy} = 4$.

To see why this approach is useful, let X be the random variable that represents the number of shared alleles between two individuals at a single locus. We can calculate the expected number of alleles for each relationship type assuming an even allele distribution.

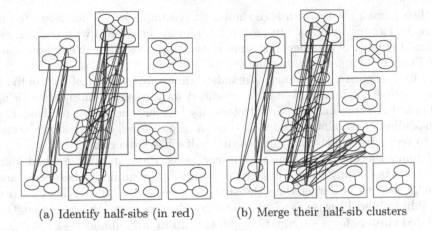

(a) Identify half-sibs (in red)   (b) Merge their half-sib clusters

Fig. 1. Demonstration of a successful iteration of SibJoin. Nodes represent individuals, edges represent a half or full-sibling relationship constructed by the algorithm, and nodes which share a box represent true full-siblings.

$$\mathbf{E}[X|\text{full siblings}] = \frac{8k^2 + 4k + 1}{8k^2} \tag{1}$$

$$\mathbf{E}[X|\text{half siblings}] = \frac{2k^3 + 5k^2 + k + 1}{4k^3} \tag{2}$$

$$\mathbf{E}[X|\text{unrelated}] = \frac{3k^2 - k - 1}{k^3} \tag{3}$$

Eliminating allele double-counting, we end up with Eq. 1, 2, and 3. For full-siblings, the expected number of shared alleles approaches 1 as the number of alleles grows, for half-siblings, the expectation approaches $\frac{1}{2}$, and the expectation approaches 0 for unrelated individuals. For a population with m loci, the expected number of shared alleles for any two individuals is $m \cdot \mathbf{E}[X]$ and strongly concentrated around that value. Additionally, computing allele similarity takes $O(n^2 m)$ time, compared to the $O(n^3 m)$ time to enumerate and compare triplets.

4.3   Joining Individuals

To perform half-sibling clustering, we use the fact that individuals with high allele similarity are very likely to be half or full-siblings. SibJoin begins with $2n$ clusters, each of which contains a single individual. Every individual appears in exactly two clusters, representing the maternal and paternal half-sib groups. We then use a variant of single linkage clustering to join clusters. Single linkage clustering is a form of agglomerative clustering which determines the similarity of two clusters C_i and C_j by computing $sim(C_i, C_j) = \max_{x \in C_i, y \in C_j} s_{xy}$, and then joining groups with high similarity. A sample join is demonstrated in Figure 1. Ties in similarity are broken by joining the groups with the highest combined number of members first since large compatible half-sibling groups are more likely to be related than small groups.

Traditional clustering techniques mandate that only one copy of each individual is allowed. SibJoin implements a modified form of single linkage clustering which places restrictions on which clusters may be joined according to Mendelian compatibility constraints and handles the multiple copies of individuals necessary to reconstruct maternal and paternal half-sibling structures.

SibJoin's success comes from two observations. First, in order for bad joins to occur between any pair of individuals i and j, the similarity between i and j would need to be larger than the similarity between i and each of i's real half-siblings, and likewise for j. Secondly, as clusters grow, the odds that two unrelated clusters form a compatible half-sibship rapidly diminishes.

Joining must only occur if two clusters form a valid half-sibship. At the initialization of the algorithm, each individual is assigned a feasible parent set with size at most $O(k)$ per locus. Each join results in a parent set which is the intersection of the parent set from the two joined clusters. If the intersection produces the null set, then there is no parent which can explain the new cluster and the join is rejected. Therefore, testing whether or not a join is valid takes $O(km)$ time. When a site experiences allelic dropout, SibJoin makes no assumptions about its parental restrictions; however, sites with $(*, *)$ are never counted toward allele similarity between individuals.

Unlike crisp clustering methods which mandate that each individual appear in exactly one cluster, a half-sibling problem contains both a maternal and paternal group for each individual. We enforce the restriction that any set of individuals sharing both a maternal and paternal cluster must be compatible full-siblings under the 4-allele and 2-allele properties by maintaining a clustering of full-siblings. Because incompatible full-sibling groups are less likely than incompatible half-sibling groups of the same size, at each similarity step SibJoin joins clusters which form valid full-sibships first.

Microsatellites give no information about which alleles are maternal and which are paternal. Since SibJoin constructs families in an iterative manner, part of a maternal family could be reconstructed on the maternal side, while the other part is constructed on the paternal side. If we are strict about which sets we call maternal and paternal, then the two halves will never be joined and the half on the paternal side will likely force incorrect future joins. The best solution is to implement a bipartite graph $G = (V, E)$ where each cluster is a vertex and edges exist between clusters which share an individual. Let a join between clusters C_i and C_j be an event which combines C_j into C_i and let $E(v)$ be the set of edges that touch v. In our graph, $join(C_i, C_j)$ results in $E(v_i) := E(v_i) \bigcup E(v_j)$ followed by the removal of v_j and all edges in $E(v_j)$. Enforcing bipartiteness as a postcondition of the join operation allows flexibility while ensuring that the solution results in each individual having one parent of each sex.

4.4 Allowing Candidate Parents

Identifying candidate parents can drastically increase the correctness of sibship reconstructions. SibJoin allows for the inclusion of candidate parents for either or both sexes. If candidate parents are given, a first round of clustering will attempt to join individuals using only parent sets which contain only candidate parents. Once no more joins can be made with the restricted parent set, SibJoin will then continue to join clusters as described in the general case.

5 Experimental Results

SibJoin's performance is evaluated with simulated and real population data. The experimental results from real populations are contrasted with the the HS-MSC and COLONY half-sibling approaches.

5.1 Accuracy Measure

Partition distance is a metric which measures the distance between two partitions as the minimum number of individuals that must be removed from a population until the two clusterings are identical. This metric is widely used in sibship reconstruction literature and in bioinformatics in general [8, 12]. When true partitionings are known, partition distance may be used to compute the true accuracy of an algorithm; however, it may also be used to assess changes between candidate sibships for which ground truth is not known [7].

Despite its prominence in sibship literature, partition distance offers only a coarse estimate of correctness because it disregards how the excluded individuals are constructed within the partitioning. For example, consider a partitioning which contains a split, leaving two halves of a true sibling group unjoined. Intuitively, this split is much more benign than if one of the halves was incorrectly joined with another sibling group: at least the split does not introduce false positives. However, in both instances, the partition distance is identical, which violates our intuition about the correctness of sibling reconstructions. A concrete example is given in Meilă [13].

Instead, we use an information theoretic metric called variation of information (VI) [13]. The VI measures how much the information given by two clusters differ and is preferable because it quantifies the amount of uncertainty introduced by misplaced individuals.

VI requires knowing the *entropy* of each cluster. Define the real partitioning of a population into families as $\mathcal{P}$ and an algorithm's partitioning as $\mathcal{P}'$ with $p := |\mathcal{P}|$ and $p' := |\mathcal{P}'|$. Compute the probability of an individual selected uniformly at randomly from our population of size n as $P(i) = |P_i| \backslash n$ for each $P_i \in \mathcal{P}$ and similarly for $\mathcal{P}'$. Using the two random variables, we can compute the entropy of $\mathcal{P}$ and $\mathcal{P}'$, denoted $H(\mathcal{P})$, and $H(\mathcal{P}')$.

$$H(\mathcal{P}) = -\sum_{i=1}^{p} P(i) \log P(i) \tag{4}$$

Mutual information, denoted $I(\mathcal{P}, \mathcal{P}')$, measures how much information is shared between the two partitionings of the population. It is dependent on the joint distribution of the random variables for $\mathcal{P}$ and $\mathcal{P}'$, given by $P(i, i') = (|P_i \bigcap P_i'|) \backslash n$.

$$I(\mathcal{P}, \mathcal{P}') = \sum_{i=1}^{p} \sum_{i'=1}^{p'} P(i, i') \log \frac{P(i, i')}{P(i)P'(i')} \qquad (5)$$

The variation of information between partitionings $\mathcal{P}$ and $\mathcal{P}'$, which we will call $VI(\mathcal{P}, \mathcal{P}')$, may now be computed in $O(n + p \cdot p')$ time.

$$VI(\mathcal{P}, \mathcal{P}') = (H(\mathcal{P}) - I(\mathcal{P}, \mathcal{P}')) + (H(\mathcal{P}') - I(\mathcal{P}, \mathcal{P}')) \qquad (6)$$

The VI between two partitionings is 0 if and only if the two partitionings are identical and smaller VI corresponds to more similar partitionings. Like entropy, the VI is always non-negative. It has a tight upper bound of $\log n$; therefore, we will normalize VI to a value in $[0, 1]$ before reporting the score for each of our trials.

For half-siblings, calculating the VI is less straight forward due to construction of both maternal and paternal clusterings $\mathcal{M}$ and $\mathcal{P}$ respectively. Since there are two clusterings, we compute the average variation of information between the two maternal clusters, $\mathcal{M}$ and $\mathcal{M}'$, and the two paternal clusters $\mathcal{P}$ and $\mathcal{P}'$, where $\mathcal{M}$ and $\mathcal{P}$ are the true partitionings. To account for the fact that the sex of each partitioning could be mismatched, we calculate two different VI values and choose the smallest.

$$HS_{VI} = \frac{\min \left(\frac{VI(\mathcal{M},\mathcal{M}')+VI(\mathcal{P},\mathcal{P}')}{2}, \frac{VI(\mathcal{M},\mathcal{P}')+VI(\mathcal{P},\mathcal{M}')}{2} \right)}{\log n} \qquad (7)$$

5.2   Simulated Data Set Results

Simulation sets were constructed to test various parameters. Our model generates individuals from an equal number of mothers and fathers. Parents are chosen randomly, and children are generated from mother-father pairs according to an even allele distribution. Simulated data had default parameter values of 6 alleles, 6 loci, half-sibling family sizes of 5 individuals, and a population size of 40 individuals. The results are an average of ten trials per parameter value. Trials which failed to complete in 1 day are reported as '-'. The population size was increased to 80 individuals for family size trials so that the partitionings did not become trivial. The loci count was increased to 10 and family size to 20 when testing population sizes above 200 individuals. A summary of our parameter tests and their results may be found in Table 1. Testing occurred on a 2.66 GHz machine, containing 8 GB of RAM, and running Python 2.7.

In most cases, the reported VI score approximates the ratio of partition distance to population size. Overall, COLONY 2.0 was more accurate, but took thousands of times longer, often with only small gains in accuracy. SibJoin does much worse than COLONY 2.0 on the 10 allele test set, but the discrepancy is due to a single trial for which SibJoin receives a VI of 0.084 while COLONY 2.0

Table 1. Simulated test results for SibJoin and COLONY 2.0 averaged over 10 trials. Trials which did not complete in 24 hours are marked '-'.

Fixed parameter	Parameter settings	SibJoin		COLONY 2.0	
		Runtime	VI (normalized)	Runtime	VI (normalized)
	2	2.8 ms	0.396	48.9 min	0.553
	5	13.2 ms	0.222	19.7 min	0.110
k: number of alleles	10	6.7 ms	0.014	12.8 min	0.004
	15	5.1 ms	0.014	10.2 min	0.006
	20	5.7 ms	0.003	10.0 min	0.000
	2	8.7 ms	0.469	10.7 min	0.524
	5	10.1 ms	0.156	17.2 min	0.130
m: number of loci	10	11.1 ms	0.035	14.2 min	0.001
	15	12.7 ms	0.002	20.4 min	0.000
	20	12.1 ms	0.000	21.3 min	0.000
	10	0.4 ms	0.042	2.29 min	0.343
	50	16.8 ms	0.104	17.1 min	0.078
n: population size	100	82.5 ms	0.201	73.5 min	0.086
	200	3.31 sec	0.230	-	-
	500	34.68 sec	0.013	-	-
	1000	2.84 min	0.015	-	-
	2000	12.43 min	0.018	-	-
	1	51.9 ms	0.546	-	-
	5	51.1 ms	0.183	29.6 min	0.051
f. family size	10	46.2 ms	0.040	19.6 min	0.017
	20	58.4 ms	0.009	21.7 min	0.042

produces a perfect reconstruction. For the 10 loci test set, SibJoin's VI is again higher, but in practice the false positive difference between it and COLONY 2.0 is about one individual per trial.

SibJoin does worst when the population size is large and the family size is small. For instance, when tested with a 100 individual population and families of 5 individuals, SibJoin rendered a VI of 0.201 compared to COLONY 2.0's VI of 0.086. When family sizes are small and population sizes are large, it is much more likely for two unrelated individuals to be mistakenly labeled as half-siblings due to the explanations given in section 4.3. However, SibJoin's accuracy rapidly improves with modest increases in family size. In fact, SibJoin is more accurate than COLONY 2.0 in trials with families containing 20 individuals. Unsurprisingly, both methods poorly reconstruct populations where only two alleles are present. With only two alleles, all individuals can be full or half-siblings.

We may also use SibJoin to explore populations with extreme numbers of individuals. SibJoin was able to reconstruct populations of 500, 1000, and 2000 individuals in under 10 minutes, yet problems of this magnitude are intractable for the HS-MSC and both of the COLONY programs.

Table 2. Tests for biological data. A '-' indicates that an algorithm did not complete after 24 hours. SibJoin was the only algorithm able to construct a solution for a 672 individual population of rockfish. The variation of information is not computed for the HS-MSC since it allows instances of the same individual, which causes ill-defined VI scores.

Data Set	Algorithm	Runtime	VI (normalized)	False Positives
	COLONY 2.0	35.7 min	0.000	0
112 crickets	HS-MSC	-	n/a (see caption)	2
	SibJoin	19.3 ms	0.014	1
	COLONY 2.0	624.5 min	0.000	0
288 kelp rockfish	HS-MSC	-	n/a (see caption)	0
	SibJoin	87.5 ms	0.000	0
	COLONY 2.0	-	-	-
672 kelp rockfish	HS-MSC	-	-	-
	SibJoin	5.02 sec	0.108	78

5.3 Biological Data Set Results

SibJoin was tested on two biological data sets. The first data set is a population of 112 field crickets with 7 mothers and 6 sampled loci [4]. The second data set is a population of 672 kelp rockfish with 7 mothers and 7 sampled loci [16]. Neither COLONY 2.0 nor the HS-MSC produced a solution for the 672 rockfish population, so samples from three of the parents were taken to reduce the population size to 288 individuals. In both populations, only maternal parentage was available. For all trials, SibJoin was run in a configuration that only attempts to reconstruct the maternal sex.

Our results are compared to the HS-MSC results in [15] and to our own benchmarks on COLONY 2.0. Because the HS-MSC is not yet publicly available, we could not assess runtime information for the program. However, the authors do note that the HS-MSC IP finished in under one day. The difference between the two runtimes is not explained merely by CPU speed increases across a small number of years. Additionally, neither COLONY 2.0 nor the HS-MSC's half-sibling minimum set cover approach constructed a feasible answer for the 672 rockfish data set: COLONY 2.0 was stopped after running for three days. SibJoin constructs an accurate solution in under 10 seconds.

The HS-MSC ILP does not enforce that individuals must have one parent of each sex and both partition distance and variation of information are ill-defined when the result is not a true partitioning. In the population of 112 crickets, the HS-MSC had two false positives and was otherwise correct. In the test set containing 288 rockfish, HS-MSC had 4 false positives and was otherwise correct. COLONY 2.0 was correct in all instances. SibJoin correctly reconstructed the half-sibship for the 288 rockfish and only misplaced one individual in the cricket test. SibJoin was the only algorithm to complete for the population of 672 rockfish. Overall, SibJoin is as accurate as the HS-MSC and nearly as accurate as COLONY 2.0, but is much faster than either: SibJoin solves the small rockfish instance over 42,000 times faster than COLONY 2.0.

6 Future Work

We have provided a half-sibship algorithm which is nearly as accurate as full-likelihood algorithms, but is thousands of times faster. SibJoin is the only half-sibling reconstruction algorithm which can partition large populations. However, because of the greedy nature of the heuristic, incorrect joins can have a big impact, especially if they are made in early iterations. In practice, a common error was for a group of individuals to be joined with a single individual who was not actually a half-sibling, causing splits. These anomalies should be detectable, since the incorrect individual ought to look much different than real siblings. We have demonstrated that runtime is not an issue. Because our algorithm is already so much faster than any other software, future work will focus on identifying and correcting errors to obtain more accuracy.

Acknowledgements. This work is supported by the Natural Sciences and Engineering Research Council of Canada and the Government of Ontario. We would like to thank Tanya Berger-Wolf and Saad Sheikh for providing real population data, and also thank Tanya Berger-Wolf, Marina Meilă, and Rita Ackerman for helpful discussions.

References

1. Almudevar, A., Anderson, E.: A new version of PRT software for sibling groups reconstruction with comments regarding several issues in the sibling reconstruction problem. Mol. Ecol. Resour. 12(1), 164–178 (2011)
2. Ashley, M., Caballero, I., Chaovalitwongse, W., Dasgupta, B., Govindan, P., Sheikh, S.I., Berger-Wolf, T.Y.: KINALYZER, a computer program for reconstructing sibling groups. Mol. Ecol. Resour. 9(4), 1127–1131 (2009)
3. Berger-Wolf, T.Y., DasGupta, B.: Combinatorial reconstruction of sibling relationships. In: Proceedings of CGBI 2005, pp. 3–6 (2005)
4. Bretman, A., Tregenza, T.: Measuring polyandry in wild populations: a case study using promiscuous crickets. Mol. Ecol. 14(7), 2169–2179 (2005)
5. Brown, D.G., Berger-Wolf, T.: Discovering Kinship through Small Subsets. In: Moulton, V., Singh, M. (eds.) WABI 2010. LNCS (LNBI), vol. 6293, pp. 111–123. Springer, Heidelberg (2010),
   http://www.springerlink.com/index/L719361L34H0K73M.pdf
6. Chaovalitwongse, W.A., Chou, C.A., Berger-Wolf, T.Y., DasGupta, B., Sheikh, S., Ashley, M.V., Caballero, I.C.: New Optimization Model and Algorithm for Sibling Reconstruction from Genetic Markers. INFORMS J. Comp. 22(2), 180–194 (2009)
7. Coombs, J.A., Letcher, B.H., Nislow, K.H.: PedAgree: software to quantify error and assess accuracy and congruence for genetically reconstructed pedigree relationships. Conserv. Genet. Resour. 2(1), 147–150 (2010)
8. Gusfield, D.: Partition-distance: A problem and class of perfect graphs arising in clustering. Info. Proc. Lett. 82(3), 159–164 (2002)
9. Isagi, Y., Kanazashi, T., Suzuki, W., Tanaka, H., Abe, T.: Highly variable pollination patterns in Magnolia obovata revealed by microsatellite paternity analysis. Int. J. Plant Sci. 165(6), 1047–1053 (2004)

10. Jones, B., Grossman, G.D., Walsh, D.C.I., Porter, B.A., Avise, J.C., Fiumera, A.C.: Estimating Differential Reproductive Success From Nests of Related Individuals, With Application to a Study of the Mottled Sculpin, Cottus bairdi. Genetics 176(4), 2427–2439 (2007)

11. Jones, O.R., Wang, J.: Molecular marker-based pedigrees for animal conservation biologists. Animal Conservation 13(1), 26–34 (2010)

12. Lin, Y., Rajan, V., Moret, B.: A Metric for Phylogenetic Trees Based on Matching. IEEE/ACM Trans. Comp. Bio and Bioinf. 9(4), 1014–1022 (2011)

13. Meilă, M.: Comparing Clusterings by the Variation of Information. In: Schölkopf, B., Warmuth, M.K. (eds.) COLT/Kernel 2003. LNCS (LNAI), vol. 2777, pp. 173–187. Springer, Heidelberg (2003)

14. Painter, I.: Sibship reconstruction without parental information. J. Agric. Biol. Envir. S. 2(2), 212–229 (1997)

15. Sheikh, S.I., Berger-Wolf, T.Y., Khokhar, A.A., Caballero, I.C., Ashley, M.V., Chaovalitwongse, W., Chou, C., Dasgupta, B.: Combinatorial reconstruction of half-sibling groups from microsatellite data. J. of Bioinf. Comput. Biol. 8(2), 337–356 (2010)

16. Sogard, S.M., Gilbert-Horvath, E., Anderson, E.C., Fisher, R., Berkeley, S.A., Garza, J.C.: Multiple paternity in viviparous kelp rockfish. Sebastes Atrovirens. Environ. Biol. Fishes 81(1), 7–13 (2008)

17. Wang, J.: Sibship reconstruction from genetic data with typing errors. Genetics 166(4), 1963–1979 (2004)

18. Wang, J., Santure, A.W.: Parentage and sibship inference from multilocus genotype data under polygamy. Genetics 181(4), 1579–1594 (2009)

Reconstructing the Evolution of Molecular Interaction Networks under the DMC and Link Dynamics Models

Yun Zhu and Luay Nakhleh

Department of Computer Science, Rice University, Houston, TX, USA
{yun.zhu,nakhleh}@rice.edu

Abstract. Molecular interaction networks have emerged as a powerful data source for answering a plethora of biological questions ranging from how cells make decisions to how species evolve. The availability of such data from multiple organisms allows for their analysis from an evolutionary perspective. Indeed, work has emerged recently on network alignment, ancestral network reconstruction, and phylogenetic inference based on networks.

In this paper, we address two central issues in the area of evolutionary analysis of molecular interaction networks, namely (1) correcting genetic distances derived from observed differences between networks, and (2) reconstructing ancestral networks from extant ones. We address both issues computationally under the *link dynamics* and *duplication-mutation with complementarity* (DMC) evolutionary models. We demonstrate the utility and accuracy of our methods on biological and simulated data.

1 Introduction

Molecular interaction networks capture the interactions among molecules (genes, mRNA, proteins, etc.) within and across cells, and govern how cells make decisions about differentiation, proliferation, etc. Advanced biotechnologies are amassing data on such networks at a fast rate and large scale, in particular from different organisms. Consequently, one powerful method to understand these networks is to subject them to evolutionary analyses [15,1]. In addition, these networks provide a new source of data that can be used to reconstruct evolution. Indeed, just as genetic distances between organisms can be measured based on genes and genomes, they can, in theory, be measured based on molecular interaction networks. In this paper, we address two problems related to the evolution of molecular interaction networks, namely the *distance correction* problem and the *ancestral network reconstruction* problem.

Estimating the evolutionary distance between two species is an important problem in phylogenetics. In the *distance correction* problem, we are given the networks of two organisms and would like to compute the genetic distance between the two organisms while accounting for unobserved changes. Due to convergent evolution at the network level, the observed changes between two networks provide a lower bound on their genetic distance (if we assume the

B. Raphael and J. Tang (Eds.): WABI 2012, LNBI 7534, pp. 57–68, 2012.

networks to be complete). Similar to distance correction among sequences (e.g., [5]), it is important to correct the genetic distances between a pair of networks. To the best of our knowledge, no methods for such distance corrections exist. In the *ancestral network reconstruction* problem, we are given the networks of two organisms and would like to compute the ancestral network of both networks in a probabilistic manner. Parsimony methods for solving this problem, e.g., [9,11], infer ancestral networks that minimize the number of evolutionary events allowed under an assumed model of evolution. Probabilistic methods, e.g., [16,17,13,7], take into consideration all possible trajectories during the course of evolution under a given model. Other methods make use of the gene genealogy information for ancestral reconstruction [12,4].

The contributions of our paper are two-fold. First, we provide methods for correcting distances between two molecular interaction networks under the *link dynamics* and *duplication-mutation with complementarity* (DMC) [8] models. We show that the correction provides better estimates of the divergence between two networks than the (uncorrected) parsimony-based distance. Further, we conduct phylogeny reconstruction under the conditions of the "long branch attraction" problem; we show that the corrected distance results in consistent estimates of the phylogeny, whereas the uncorrected distances result in wrong estimates. Finally, we apply the correction to estimating a phylogeny of six species to investigate the divergence pattern of *D. melanogaster* and *C. elegans*, which remains unsolved using genomic and morphological data. The second contribution is probabilistic algorithms for inferring the ancestral network of two given networks under the link dynamics and DMC models. The algorithms further assume knowledge of the gene families. We demonstrate the accuracy of our algorithm on synthetic and biological data.

2   Methods

Assume that we have two networks N_1, N_2 and their ancestral network N_0, as well as a network growth model $\mathcal{M}$ that describes the operations allowed on the networks (node/edge addition or deletion) along with their rates. Evolutionary times t_1 and t_2 are the evolutionary time from N_0 to N_1 and N_2, respectively. The probability of observing N_1 and N_2, given the ancestral network, the network growth model, and the two evolutionary times is denoted by

$$\mathbf{P}(N_1, N_2 \mid \mathcal{M}, N_0, t_1, t_2). \tag{1}$$

Then, assuming knowledge of the network growth model and given two extant homologous networks, the distance correction problem can be formulated as:

$$(t_1^*, t_2^*) = \mathrm{argmax}_{t_1, t_2} \mathbf{P}(N_1, N_2 \mid \mathcal{M}, t_1, t_2). \tag{2}$$

Further, a version of the ancestral network reconstruction, when the evolutionary times and growth model are assumed to be known and two extant homologous networks are given, can be formulated as:

$$N_0^* = \mathrm{argmax}_{N_0} \mathbf{P}(N_1, N_2 \mid \mathcal{M}, N_0, t_1, t_2) \tag{3}$$

2.1 Distance Correction

Under the DMC model of network evolution, the divergence time can be effectively obtained from the increase in the size (number of nodes) of the networks, since the model assumes a constant duplication rate, which can be easily used to infer the time (time would simply be the number of observed duplication events divided by the duplication rate). For the remainder of this section, we assume the *link dynamics* model $\mathcal{M}$, where the sets of nodes in the two extant networks are identical and the evolutionary changes can be estimated from the changes in connectivity (we show below that this can be used to analyze real data, even when the sets of nodes are not identical). If networks N_1 and N_2 have the same set of nodes, and we have a one-to-one correspondence between all the nodes, we can use a bit vector to represent the connections between the nodes, where this bit vector is a "flattening" of the networks' adjacency matrices. We denote by $V = \{v_1, \cdots, v_n\}$ the set of nodes in each of the networks. Then, the bit vector of the connections can be written as $B_{i(i \in \{1,2\})} = \{b_{11}, b_{12}, \cdots, b_{1n}, b_{21}, \cdots, b_{2n}, \cdots, b_{n1}, \cdots, b_{nn}\}$, where $b_{ij} = 1$ if there is an edge between v_i and v_j, and 0 otherwise.

The uncorrected distance between the two networks is the Hamming distance between B_1 and B_2. Next, we show how to obtain a more accurate estimate of the true distance between B_1 and B_2 from this uncorrected distance. Suppose that each existing edge has probability d of being deleted and that for each pair of unconnected nodes, they can be connected with probability a. If nodes v_i and v_j originally have no edge connecting them, the probability that the pair are not connected one generation later is $P_{b_{ij}(t=1)=0} = 1 - a$. The probability that the pair is not connected two generations later is $P_{b_{ij}(t=2)=0} = (1-a)^2 + a \cdot d$. More generally, we have $P_{b_{ij}(t+1)=0} = (1-a)P_{b_{ij}(t)=0} + d(1 - P_{b_{ij}(t)=0})$. From this we can obtain $\Delta P_{b_{ij}(t)=0} = (1-a)P_{b_{ij}(t)=0} + d(1 - P_{b_{ij}(t)=0}) - P_{b_{ij}(t)=0}$ which simplifies to $\Delta P_{b_{ij}(t)=0} = d - (a+d)P_{b_{ij}(t)=0}$. If we treat time as continuous, we get $dP_{b_{ij}(t)=0}/dt = d - (a+d)P_{b_{ij}(t)=0}$. The solution to this differential equation is

$$P_{b_{ij}(t)=0} = \frac{d}{a+d} + \left(P_{b_{ij}(t=0)=0} - \frac{d}{a+d}\right)E$$

where $E = e^{-(a+d)t}$. Similarly,

$$P_{b_{ij}(t)=1} = \frac{a}{a+d} + \left(P_{b_{ij}(t=0)=1} - \frac{a}{a+d}\right)E.$$

If initially $b_{ij} = 0$, then

$$P_{b_{ij}(t)=0} = \frac{d}{a+d} + \frac{a}{a+d} \cdot E \quad \text{and} \quad P_{b_{ij}(t)=1} = \frac{a}{a+d} - \frac{a}{a+d} \cdot E.$$

If initially $b_{ij} = 1$, then

$$P_{b_{ij}(t)=0} = \frac{d}{a+d} - \frac{d}{a+d} \cdot E \quad \text{and} \quad P_{b_{ij}(t)=1} = \frac{a}{a+d} + \frac{d}{a+d} \cdot E.$$

Assume two networks such that $B_1(t = 0) = B_2(t = 0)$, and the proportion of pairs of nodes that are connected is $P_{b_{ij}(t=0)=1} = p$. At some later time t, the probability that a given pair of nodes will have the same connectivity in both networks (they are either disconnected in both networks or connected in both networks) is

$$P_{I(t)} = \frac{1-p}{(a+d)^2}[(a^2+d^2)+(2ad-2a^2)E+2a^2E^2] + \frac{p}{(a+d)^2}[(a^2+d^2)+(2ad-2d^2)E+2d^2E^2].$$

The probability that a given pair of nodes will have different connectivity in the two networks (they are connected in one but not the other) is

$$P_{D(t)} = \frac{2(1-p)}{(a+d)^2}[ad+(a^2-ad)E-a^2E^2] + \frac{2p}{(a+d)^2}[ad+(d^2-ad)E-d^2E^2].$$

Combined with $P_{b_{ij}(t)=1} = \frac{a}{a+d} + (p - \frac{a}{a+d})e^{-(a+d)t}$ we obtain

$$P_{D(t)} = \frac{2}{(a+d)^2}[-adE^2 + ((a^2-d^2)P - a(a-d))E + (a^2 - (a^2-d^2)P)], \quad (4)$$

where $E = e^{-(a+d)t}$, $P = P_{b_{ij}(t)=1}$. When $a = b$, it can be simplified as $P_{D(t)} = \frac{1}{2}(1 - e^{-4at})$ and $t = -\frac{1}{4a}ln(1 - 2P_{D(t)})$.

2.2   Ancestral Reconstruction under Link Dynamics

Using the same notation as above, let $b_{ij}(t = 0)$ represent the connectivity of node i and j at the ancestral species ($t = 0$), and $(b_{ij}^1, b_{ij}^2)(t)$ be a random variable that represents the connectivity of node i and j in the extant networks N_1 and N_2 after diverging from the same ancestral networks t generations in the past. The values that this random variable can take, at any given time t, are $(0,0)$, $(0,1)$, $(1,0)$ and $(1,1)$, and their probabilities, given either of the two initial states, are given in Table 1. Using these probabilities, the problem of ancestral

Table 1. The values of $\mathbf{P}((b_{ij}^1, b_{ij}^2)(t) = \mathbf{x})$. The parameters a and d are the probabilities of edge addition and deletion, respectively, and $E = e^{-(a+d)t}$.

	$\mathbf{x} = (0,0)$	$\mathbf{x} = (0,1)$ or $(1,0)$	$\mathbf{x} = (1,1)$
$b_{ij}(t = 0) = 0$	$\frac{1}{(a+d)^2}(d+aE)^2$	$\frac{1}{(a+d)^2}(d+aE)(a-aE)$	$\frac{1}{(a+d)^2}(a-aE)^2$
$b_{ij}(t = 0) = 1$	$\frac{1}{(a+d)^2}(d-dE)^2$	$\frac{1}{(a+d)^2}(d-dE)(a+dE)$	$\frac{1}{(a+d)^2}(a+dE)^2$

reconstruction under link dynamics becomes straightforward to solve: For every pair of nodes i and j in the ancestral network, given their connectivity in two extant networks, find the connectivity value $b_{ij}(t = 0)$ that would maximize the probability in the appropriate column in Table 1.

2.3   Ancestral Reconstruction under DMC

We now consider ancestral network reconstruction under the DMC model as in [8], except that, we assume each node has a label and a duplicated node will get the same label from its anchor (we use the same terminology as that of [8], where they used the term anchor to denote the gene that gets duplicated, but not the duplicate copy). This corresponds to the concept of gene family and how duplication increases the number of copies in a family. Before we describe the algorithm, we begin with some notations and definitions. We use (a, b) to denote an undirected edge between a and b.

Definition 1 (Node-labeled Undirected Network) *Let Σ be an alphabet of node labels. A node-labeled undirected network over Σ is a tuple $N_\Sigma = (V, E, \lambda)$, where V is the finite set of nodes, $E \subseteq \{(x, y) \mid x, y \in V, x \neq y\}$ is the set of edges, and $\lambda : V \to \Sigma$ is the node labeling function. We write N instead of N_Σ when it is clear from the context. The components of N are denoted as V_N, E_N, and λ_N, respectively. The set of node-labeled undirected network on alphabet Σ is denoted as $\mathcal{N}_\Sigma$.*

For $\lambda : V \to \Sigma$, $\lambda' = ext(\lambda, x \to l)$ is an extension of the map λ which also maps x to l for $x \notin V$; and $\lambda'' = red(\lambda, x)$ is a reduction of the map λ for $x \in V$ such that $\lambda''(y) : V - \{x\} \to \Sigma$ satisfies $\lambda''(y) = \lambda(y)$ for $u \in V - \{x\}$.

PPI networks are node-labeled undirected networks, where the Σ of these networks are the set of protein class names. (i.e., nodes that correspond to homologous proteins are labeled the same).

Definition 2 (Basic Transformation Operations) *The allowed transformations on a given node-labeled undirected network $N_\Sigma = (V, E, \lambda)$ are:*

- node duplication $dup_N : V \to \mathcal{N}_\Sigma$. For $x \subset V$, $dup_N(x) = N' = (V', E', \lambda')$, where $V' = V \cup \{x'\}$ for any $x' \notin V$. If $(v, v) \in E$, then $E' = E \cup \{(y, x') \mid (y, x) \in E\} \cup \{(x', x')\}$, else $E' = E \cup \{(y, x') \mid (y, x) \in E\}$. $\lambda' = ext(\lambda, x' \to \lambda(x))$.
- edge addition $ea_N : (V \times V - E) \to \mathcal{N}_\Sigma$. For $x, y \in V, e = (x, y) \notin E$, $ea_N(e) = N' = (V, E', \lambda)$, where $E' = E \cup \{e\}$.
- edge deletion $ed_N : E \to \mathcal{N}_\Sigma$. For $e \in E$, $ed_N(e) = N' = (V, E', \lambda)$, where $E' = E - \{e\}$.

Definition 3 (DMC model) *In each step of the DMC model, growth proceeds as follows:*

1. *Duplicate a node u and get the new node u'.*
2. *For each neighbor v of u, decide to remove one edge from (u, v) and (u', v) with probability q_{mod}. Both edges have equal chance to be removed.*
3. *Add edge (u, u') with probability q_{con}*

Thus, the probability of obtaining network N after duplication of u is:

$$L_{dup}(N, u, u') = (((u, u') \in E)q_{con} + ((u, u') \notin E)(1 - q_{con}))$$
$$\prod_{\{v \mid (u,v) \in E \wedge (u',v) \in E\}} (1 - q_{mod}) \prod_{\{v \mid (u,v) \in E \vee (u',v) \in E = 1\}} q_{mod} \quad (5)$$

Here, we assume Boolean evaluation of the terms $((u, u') \in E)$ and $((u, u') \notin E)$ gives 0 or 1.

Definition 4 (Reverse DMC model: node merging) *In each step, if we decide to merge nodes u and u' with the same label, do:*

1. *For each neighbor v of u', if v is not a neighbor of u, add edge (u, v).*
2. *Remove node u'.*

The algorithm [8] (we refer to it as **NK**) attempts to find in each step backward the most probable duplicated node using

$$u^* = \operatorname{argmax}_u L_{dup}(N, u). \tag{6}$$

In each step, to construct G_{t-1} from G_t **NK** merges nodes u^* and u'^* that satisfy

$$(u^*, u'^*) = \operatorname{argmax}_{\{(u, u') | \lambda(u) = \lambda(u')\}} L_{dup}(N_t, u, u') \tag{7}$$

and have the same label.

While algorithm **NK** was not devised for ancestral network reconstruction, we now devise such an algorithm (Algorithm 1) that makes use of **NK**. The

---

Algorithm 1. DMCAncRec.

Input: Two homologous molecular interactions networks $N_1 = (V_1, E_1)$ and $N_2 = (V_2, E_2)$, where each node in both networks is labeled by a gene family $1 \leq i \leq m$.
Output: Molecular interaction network $N_0 = (V_0, E_0)$ to optimize Equation (3).

1 Let $\mathcal{U} = \{U_1, U_2, \ldots, U_m\}$ be the partition of V_1 based on gene family membership;
2 Let $\mathcal{W} = \{W_1, W_2, \ldots, W_m\}$ be the partition of V_2 based on gene family membership;
3 $\mathcal{P} \leftarrow \emptyset$;
4 **for** $i = 1$ *to* m **do**
5 $\quad \lfloor \; \mathcal{P} \leftarrow \mathcal{P} \cup \{\textbf{AncestralPair}(U_i, W_i)\}$;

6 Let B be a list of the elements of $\mathcal{P}$ ordered in increasing order by their rightmost entry t;
7 $(Y_1, Y_2, x, t) \leftarrow B[1]$;
8 $V_0 \leftarrow Y_x$;
9 $E_0 \leftarrow \{\{u, v\} : \{u, v\} \in E_x, \; u, v \in Y_x\}$;
10 **for** $i = 2$ *to* m **do**
11 $\quad (Y_1, Y_2, x, t) \leftarrow B[i]$;
12 $\quad V_0 \leftarrow V_0 \cup Y_x$;
13 $\quad \lfloor \; E_0 \leftarrow E_0 \cup \{\{u, v\} : \{u, v\} \in E_x, \; u \in Y_x, \; v \in V_0\}$;
14 **return** $N_0 = (V_0, E_0)$;

---

algorithm has the following general outline. It first finds partitions the nodes of the two networks into equivalence classes by their memberships in different gene families (Lines 1—2). It then proceeds to reconstruct ancestral networks as follows. For each pair of equivalence classes from the two partitions that belong to the same gene family, the algorithm computes the set of ancestral nodes. If the two sets of nodes in the two classes are of equal size, the ancestral set has the same set of nodes; otherwise, the ancestral set has a size equal to that of the smaller of the two sets. Algorithm **NK** is invoked here to compute ancestors of single networks under the DMC model. This is all achieved by the auxiliary algorithm **AncestralPair**. In order to guarantee that the ancestral reconstructions converge onto a single ancestral network, the analysis of pairs of equivalence classes is done in increasing order of times (the order in which genes were created by duplication) computed by **NK**, and the connectivity of the networks is established in an interleaved mode with the ancestral reconstruction (Lines 10—13). We use $(V', t) = \mathbf{NK}(V, E, k)$ to denote the application of the algorithm of [8] on the graph whose nodes and edges are V and E, respectively, until it reaches an ancestral graph with k nodes; this ancestral graph has set V' of nodes, and t is the number of generations taken to reach it. Notice that in the reverse DMC model, the combined node will always have the union of neighbors from the nodes being merged.

Algorithm 2. AncestralPair.

Input: Two networks $N_1 = (V_1, E_1)$ and $N_2 = (V_2, E_2)$, and two sets of nodes $U_1 \subseteq V_1$ and
$U_2 \subseteq V_2$ that correspond to all the members of a single gene family in these two
networks.
Output: Two sets of nodes U'_1 and U'_2, $x \in \{1, 2\}$, and time t.

1 $U'_1 \leftarrow U_1$;
2 $U'_2 \leftarrow U_2$;
3 $x \leftarrow 1$;
4 $t \leftarrow 0$;
5 **if** $|U_1| < |U_2|$ **then**
6 $\quad\lfloor\ (U'_2, t) \leftarrow \mathbf{NK}(U_2, |U_1|);$

7 **if** $|U_2| < |U_1|$ **then**
8 $\quad\lfloor\ (U'_1, t) \leftarrow \mathbf{NK}(U_1, |U_2|);$
9 $\quad\lfloor\ x \leftarrow 2;$

  return (U'_1, U'_2, x, t)

Some practical issues might arise when applying the algorithm to real data due to the fact that the DMC model might not be suitable to bring two networks eventually to the same ancestral network. For example, the set of edges might have diverged too much between the two extant networks, or too many new nodes or edges have been added. For each of these cases, we employ *ad hoc* heuristics. For example, if a node in N_1 has label that does not appear in N_2, this means there is a new node added in the network N_1. We ignore the node and its links when applying the algorithm.

2.4 Ancestral Reconstruction under DMC with Link Dynamics

The **NK** algorithm is not the only algorithm for estimating the node duplication order. For example, [6] also presents a method for estimate the gene duplication history. Further, gene trees can also be used to infer the gene duplication order. We now discuss how to modify Algorithm **DMCAncRec** to solve ancestral reconstruction assuming the network evolved under a combined DMC / link dynamics model, and assuming that the gene duplication order is known.

The algorithm under this combined model is similar to **DMCAncRec** and differs only in establishing the connectivity of the ancestral network in the interleaved mode. That is, the only difference is: E_x (Line 13 in Algorithm 1) is now a set of edges estimated under link dynamics, as discussed in Section 2.2. Again, we first partition the nodes of two networks into equivalence classes (gene families). Then, for each pair of equivalence classes from two networks, whenever both networks reach the minimum number of members in the equivalence class, we compute the ancestral connectivities using the algorithm of Section 2.2 for all one-to-one node correspondences in the equivalence class. The node correspondence and connectivities that result in the largest probability are used as the ancestral network connectivity.

3   Results and Discussion

We show here the performance of our methods on biological and synthetic data.

Distance Estimation on Simulated Data. We simulated the evolution of an ancestral network N_0 (with 100 nodes) along two paths for t steps each to obtain N_1 and N_2. We then applied equation (4) to estimate t, and computed uncorrected distances as well. We computed the difference between the true distance and the estimated corrected distance as well as between the true distance and the uncorrected estimated distance. We conducted 50 runs and averaged results over these runs; results and parameter settings are shown Table 2. As the results

Table 2. Average error in the corrected (t_{corr}) vs. uncorrected (t_{pas}) distances

a	d	t	$E(|t_{pas} - t|/t)$	$E(|t_{corr} - t|/t)$
1×10^{-5}	1×10^{-5}	100	12.3232%	12.354%
1×10^{-5}	1×10^{-5}	1000	3.67677%	3.55597%
1×10^{-5}	1×10^{-5}	10000	3.2139%	0.4268%
1.2×10^{-5}	0.8×10^{-5}	100	12.0404%	11.5320%
1.2×10^{-5}	0.8×10^{-5}	1000	5.21212%	3.44943%
1.2×10^{-5}	0.8×10^{-5}	10000	1.6545%	0.4400%
0.8×10^{-5}	1.2×10^{-5}	100	12.2828%	12.5227%
0.8×10^{-5}	1.2×10^{-5}	1000	7.33333%	3.8738%
0.8×10^{-5}	1.2×10^{-5}	10000	4.9303%	0.4706%

show, the correction achieves much more accurate estimates (notice that these values are normalized by t which can get as large as 10,000). To test our method on data that deviate slightly from the link dynamic model, we also conducted simulations where we allowed nodes to duplicate or get deleted. We then applies the method to the intersection of the two resulting extant networks. We observed trends that are similar to those in Table 2.

The Long Branch Attraction Problem. Using the true 4-taxon phylogenetic tree on the left in Fig. 1(a), we simulated network evolution starting from network N_0 of sizes $10, 25, 50, 100, 125, 150$, with edge addition rate and deletion rate both be 0.00001. We evolved $N0$ for 100 steps to get network $N1$ and $N2$ separately, and evolved $N0$ for 10000 steps to get network $N3$. We also evolved $N1$ for 100 steps to get network $N4$ and evolved $N1$ for 10000 steps to get network $N5$.

That is, we simulated the conditions of the long branch attraction problem. We then computed the pairwise distances among the four networks at the leaves, using both uncorrected distances as well distances corrected by our method above. We fed these distances to the neighbor joining (NJ) method to construct a phylogeny. We each network size we repeated this setup 100 times. For each network size, we computed the number of times (out of 100) that NJ obtained the true tree using both distance estimates. The results in Fig. 1(b) show that the distance correction provides a statically consistent estimate of the distance under our simulation settings, as measured by the accuracy of the inferred phylogeny, especially as the size of the network grows.

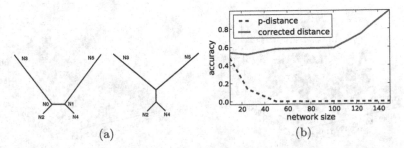

(a)   (b)

Fig. 1. (a) Left: true species tree that we used in the simulation study and the one that NJ converges on when using corrected distances. Right: the tree that NJ converges towards when using uncorrected distances. (b) Accuracy of reconstructed phylogeny using uncorrected distances and corrected distances. Accuracy is measured as the fraction of 100 runs in which NJ inferred the true tree.

Bilateral Phylogeny Inference Using Network Data. The phylogeny of bilateria continues to raise debate about the divergence patterns of species; in particular, different genomic regions and morphological characters support different groupings of *Drosophila melanogaster* and *Caenorhabditis elegans* [2,3,10,14]. Here, we used the network distances to reconstruct the phylogenetic tree of four species from the bilateria group, *Drosophila melanogaster*, *Caenorhabditis elegans*, *Mus musculus* and *Homo sapiens*, as well as two other species, *Escherichia coli* (which would serve as an outgroup) and *Saccharomyces cerevisiae*. Network data were retrieved from the STRING database. Orthologs information were obtained from the eggNOG database. Both of them use the same alias for proteins. The eggNOG database has multiple downloadable files for orthologous groups, most of which are for a specific category of species (such as bacteria or fish). But there are three general ones, which are "COG.members", "KOG.members", and "NOG.members". The orthologous groups of the six species are from the union set of all three files. We computed all pairwise corrected distances on the six species, and built a phylogenetic tree, using the Neighbor Joining method. The phylogenetic tree constructed, rooted at the branch incident with *E. coli*, is: (*E. coli*,(*S. cerevisiae*,(*C. elegans*,(*D. melanogaster*,(*M. musculus*,*H. sapiens*)))))). The network-based phylogeny, which incorporates information from across the genomes, supports the placement of *D. melanogaster* closer to mammals than to nematodes.

Performance of Ancestral Reconstruction. Here, we assume an ancestral network N_0, evolve it forward in time under the DMC model on two different paths for t_1 steps to get N_1 and t_2 steps to get N_2, apply **DMCAncRec** to N_1 and N_2, and compare the reconstructed ancestral network N_0' with N_0. We count the edge differences between the true ancestral network N_0 and the ancestral networks N_{01} and N_{02} constructed from N_1 and N_2 individually using the **NK** algorithm, as well as ancestral network N_{012} constructed from from N_1 and N_2 together using our algorithm. We use a neighborhood of size 100 for N_0, t_1 and t_2 take values from 10 to 150 with internal 10. The results are shown in Fig. 2 (left). We also repeated the above experiment, but when get N_1, N_2

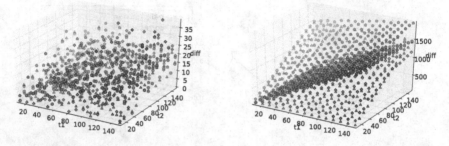

Fig. 2. Ancestral reconstruction on simulated data. Left: Data was simulated under the DMC model; Right: Data was simulated under the DMC model with noise. Markers in blue, cyan, green, and red correspond to edge set differences in the comparisons N_0 v. N_{01}, N_0 v. N_{02}, N_{01} v. N_{02}, and N_0 v. N_{012}, respectively.

from N_0, we randomly changed the connectivity of ten pairs of nodes after each duplication (to simulate adding noise); results are shown in Fig. 2 (right). The results clearly show that using the ancestral reconstruction algorithm we obtain much more accurate ancestral networks than if we had reconstructed ancestral networks from each extant network individually. This is not to cast the **NK** algorithm in negative light, since it is not intended for this type of ancestral network reconstruction. Nonetheless, we used it as a baseline for comparison.

We also applied the algorithm to three networks of *S. enterica*, *E. coli*, and *M. capsulatus*, and a rooted species tree $((S.\ enterica, E.\ coli), M.\ capsulatus)$. From the EggNog database, we found all the orthology groups that contain proteins from each of the species, and used these proteins and the corresponding subnetworks to run our algorithm. The size of the networks of the three species are 1011, 1015, 974 respectively. We constructed their ancestral networks in three ways: (1) constructed ancestral network of *S.enterica* and *E.coli*, then constructed the ancestral network of the three from the constructed network and network of *M.capsulatus*; (2) constructed the ancestral network of *S.enterica* and *M.capsulatus*; and, (3) constructed the ancestral network of *E.coli* and *M.capsulatus*. The ancestral networks from each of the three ways are all of size 885, The total number of node pairs with size 885 is $(885 * 884/2 = 391170)$. we compared the edge differences between ancestral networks constructed in these different ways, and found the differences are 18394(4.7%), 20555(5.2%) and 12729(3.2%). The results on simulated data indicate good accuracy of the method, and the results on biological data indicate consistency of the method (accuracy cannot be assessed in the case of biological data).

4 Conclusions and Future Work

In this paper, we introduced new computational methods for correcting genetic distances and reconstructing ancestral networks of molecular interaction networks under the link dynamics and DMC models of network evolution. Our algorithms produce good results on biological and synthetic data.

While our results are obtained under restrictive models of network evolution, there are three points worth addressing. First, the methods are applicable even when the data does not evolve under these restrictive models, and slight deviations from the models do not seem to affect the results significantly. Second, these problems are computationally very hard; solutions under restrictive models might give insight into how to develop algorithms for more general models. Last but not least, we believe using network data can help resolve some deep phylogenetic relationships where molecular sequences alone might not contain sufficiently informative phylogenetic signal.

Emerging multi-locus approaches for inferring species trees take the approach of reconciling gene trees with species trees to obtain accurate inferences of evolutionary histories. However, such approaches face extreme challenges when the sizes of gene families and coverage of species in these gene families vary significantly. On the other hand, approaches for inferring species phylogenies from molecular interaction data make use of additional powerful source of data—the interactions among genes. This is analogous to the motivation of using gene-order data to infer species, even though each presents a set unique set of challenges.

For future work, we plan to address both problems under other models of network growth. Further, we will curate networks from organisms across the Tree of Life, estimate phylogenies based on the corrected distances we compute, and compare them to phylogenies inferred from molecular sequences and morphological characters.

Acknowledgement. This work was supported in part by NSF grant CCF-0622037 and an Alfred P. Sloan Research Fellowship. The contents are solely the responsibility of the authors and do not necessarily represent the official views of the NSF or the Alfred P. Sloan Foundation.

References

1. Atias, N., Sharan, R.: Comparative analysis of protein networks: hard problems, practical solutions. Commun. ACM 55(5), 88–97 (2012)
2. Bourlat, S., Nielsen, C., Economou, A.D., Telford, M.J.: Testing the new animal phylogeny: a phylum level molecular analysis of the animal kingdom. Molecular Phylogenetics and Evolution 49, 23–31 (2008)

3. Dunn, C.W., Hejnol, A., Matus, D.Q., Pang, K., Browne, W.E., Smith, S.A., Seaver, E., Rouse, G.W., Obst, M., Edgecombe, G.D., Sorensen, M.V., Haddock, S.H., Schmidt-Rhaesa, A., Okusu, A., Kristensen, R.M., Wheeler, W.C., Martindale, M.Q., Giribet, G.: Broad phylogenomic sampling improves resolution of the animal tree of life. Nature 452, 745–749 (2008)
4. Gibson, T.A., Goldberg, D.S.: Reverse engineering the evolution of protein interaction networks. In: Pacific Symposium on Biocomputing, pp. 190–202 (September 2008)
5. Jukes, T.H., Cantor, C.R.: Evolution of protein molecules. In: Mammalian Protein Metabolism, pp. 21–132 (1969)
6. Li, S., Choi, P., Wu, T., Zhang, L.: Reconstruction of network evolutionary history from extant network topology and duplication history. Quantitative Biology, arXiv:1203.2430 (March 2012)
7. Mithani, A., Preston, G.M., Hein, J.: A Bayesian approach to the evolution of metabolic networks on a phylogeny. PLoS Computational Biology 6(8), e1000868 (2010)
8. Navlakha, S., Kingsford, C.: Network archaeology: uncovering ancient networks from present-day interactions. PLoS Computational Biology 7(4), e1001119 (2011)
9. Pál, C., Papp, B., Lercher, M.: Adaptive evolution of bacterial metabolic networks by horizontal gene transfer. Nature Genetics 37, 1372–1375 (2005)
10. Paps, J., Baguna, J., Riutort, M.: Bilaterian phylogeny: a broad sampling of 13 nuclear genes provides a new lophotrochozoa phylogeny and supports a paraphyletic basal a coelomorpha. Molecular Biology and Evolution 26, 2397–2406 (2009)
11. Patro, R., Sefer, E., Malin, J., Marçais, G., Navlakha, S., Kingsford, C.: Parsimonious Reconstruction of Network Evolution. In: Przytycka, T.M., Sagot, M.-F. (eds.) WABI 2011. LNCS (LNBI), vol. 6833, pp. 237–249. Springer, Heidelberg (2011)
12. Pinney, J.W., Amoutzias, G.D., Rattray, M., Robertson, D.L.: Reconstruction of ancestral protein interaction networks for the bZIP transcription factors. PNAS 104(51), 20449–20453 (2007)
13. Ratmann, O., Jørgensen, O., Hinkley, T., Stumpf, M.P.H., Richardson, S., Wiuf, C.: Using likelihood-free inference to compare evolutionary dynamics of the protein networks of H. pylori and P. falciparum. PLoS Computational Biology 3(11), e230 (2007)
14. Rogozin, I.B., Thomson, K., Carmel, L., Koonin, E.V.: Homoplasy in genome-wide analsis of rare amino acid replacemeents: the molecular-evolutionary basis for Vavilov's law of homologous series. Biology Direct 3, 7 (2008)
15. Sharan, R., Ideker, T.: Modeling cellular machinery through biological network comparison. Nature Biotechnology 24(4), 427–433 (2006)
16. Stumpf, M.P.H., Ingram, P.J., Nouvel, I., Wiuf, C.: Statistical Model Selection Methods Applied to Biological Networks. In: Priami, C., Merelli, E., Gonzalez, P., Omicini, A. (eds.) Transactions on Computational Systems Biology III. LNCS (LNBI), vol. 3737, pp. 65–77. Springer, Heidelberg (2005)
17. Wiuf, C., Brameier, M., Hagberg, O., Stumpf, M.P.H.: A likelihood approach to analysis of network data. PNAS 103(20), 7566–7570 (2006)

Estimating Population Size via Line Graph Reconstruction

Bjarni V. Halldórsson[1], Dima Blokh[2], and Roded Sharan[2]

[1] School of Science and Engineering, Reykjavík University, Menntavegur 1,
101 Reykjavik, Iceland
`bjarnivh@ru.is`
[2] Blavatnik School of Computer Science, Tel Aviv University, Tel Aviv 69978, Israel
`roded@post.tau.ac.il`

Abstract. We propose a novel graph theoretic method to estimate haplotype population size from genotype data. The method considers only the potential sharing of haplotypes between individuals and is based on transforming the graph of potential haplotype sharing into a line graph using a minimum number of edge and vertex deletions. We show that the problems are NP complete and provide exact integer programming solutions for them. We test our approach using extensive simulations of multiple population evolution and genotypes sampling scenarios. Our computational experiments show that when most of the sharings are true sharings the problem can be solved very fast and the estimated size is very close to the true size; when many of the potential sharings do not stem from true haplotype sharing, our method gives reasonable lower bounds on the underlying number of haplotypes. In comparison, a naive approach of phasing the input genotypes provides trivial upper bounds of twice the number of genotypes.

1 Introduction

A fundamental problem in population studies is the estimation of the size of the underlying haplotype pool. In these studies, such as genomewide association studies, a number of individuals are genotyped at some single nucleotide polymorphism (SNP) locations. Since we cannot observe the haplotypes of an individual, a common approach to the size estimation problem is to phase the genotype data, i.e., computationally predict the underlying haplotypes. However, haplotype phasing is a notoriously difficult problem [8] and can be optimally solved for small instances only [4].

Here we propose a novel approach that does not require the phasing of the haplotypes. It is based on starting from the easy-to-compute information on potential haplotype sharing between individuals. Specifically, we can test if two individuals have the potential to share a haplotype by checking if their genotypes are consistent with such sharing (i.e., whenever one of the genotypes is homozygous at a certain site, the other is homozygous to the same allele or heterozygous). This information can be encoded in a graph, known as the Clark

B. Raphael and J. Tang (Eds.): WABI 2012, LNBI 7534, pp. 69–80, 2012.

consistency (CC) graph [7], where each individual (genotype) is represented by a node and an edge is added between two individuals if they can share a haplotype.

The line graph of a graph, is a new graph that has nodes for each edge in the root (original) graph and an edge between two nodes of the line graph if the corresponding edges are adjacent in the root graph. If data were perfect, i.e., the only observed potential sharings were true sharings, then the resulting CC graph would be a line graph whose root graph has haplotypes as nodes and edges connect pairs of haplotypes corresponding to the observed genotypes. The reconstruction of this root graph would then enable us to compute the haplotype population size. In practice, the graph is only close to being a line graph and contains "extra" edges that do not represent true sharing. These extra edges are due to the fact that the genotypes are consistent with each other while there is no common shared haplotypes. For reasonable population structure (as reflected in the simulations presented below), we expect such sharing of genotypes to be rare. Thus, our goal is to find a smallest number of edge deletions that will make the CC graph a line graph. Once we arrive at a line graph we can reconstruct its root graph, thereby getting an estimate of the number of underlying haplotypes, as well as predictions of which individuals share a haplotype (those pairs that remain connected by an edge). We also consider a variant of the line graph deletion problem where node deletions are allowed; those correspond to cases where there are significant genotyping errors or missing data in one of the individuals or genotype pools.

CC graphs may also be determined from data generated using other technologies or experimental protocols, including pooled sequencing [12]. In pooled sequencing, pools of DNA are constructed, each containining the DNA of multiple individuals with each individual typically belonging to multiple pools. Each pool is then genotyped and a conflated genotype for all the individuals in the pool is constructed. To create a CC graph we add an edge between two individuals if they can share a haplotype, that is, if for every pair of pools containing these individuals, one in each pool, the two pools are consistent with sharing a haplotype.

We study the complexity of the resulting line graph modification problems. We observe that both problems (edge- and vertex-deletion) are NP-complete and provide polynomial integer programming formulations to solve them. We test our approach using extensive simulations of multiple population evolution and genotypes sampling scenarios. Our computational experiments show that when most of the sharings are true sharings the problem can be solved very fast and the estimated size is very close to the true size. In cases where many of the potential sharings do not stem from true haplotype sharing, due to spurious edge or node insertions, our method gives reasonable lower bounds on the underlying number of haplotypes. In comparison, a naive approach of phasing the input genotypes provides trivial upper bounds of twice the number of genotypes.

2 Preliminaries

Let $G = (V, E)$ be a graph with a set V of n vertices and a set E of m edges. The line graph of G, denoted by $L(G)$, is constructed by having a node in $L(G)$

for each edge in G and an edge between two nodes of $L(G)$ if the corresponding edges are adjacent in G.

2.1   Line Graph and CC Graphs

If $L(G)$ is the line graph of G, we refer to G as the *root graph* of $L(G)$. Line graphs have been studied by a number of authors. Whitney [16] showed that, apart from a single exception, the root graph of a line graph is unique. Lehot [10] and Roussopoulos [13] have independently given linear-time algorithms for detecting whether a graph is a line graph and outputting its root graph. Van Rooij and Wilf [15] gave a characteriziation of line graphs in terms of nine forbidden subgraphs. The characterization can be stated as follows; A triangle in a graph is *even* if every other node is adjacent to 0 or 2 nodes in the triangle; it is *odd* otherwise. A graph is a line graph iff it contains no two odd triangles that share an edge is claw-free (a claw is a node connected to three nodes not connected to each other). This characterization can also be stated as a list of nine distinct subgraphs that are forbidden from the line graph. Each one of the nine different subgraphs has at most six nodes and ten edges.

A key component of our approach is the graph of potential sharing of haplotypes between individuals. This graph, called the Clark consistency (CC) graph, was first suggested in the context of a method for haplotype phasing by Andrew Clark [5]. Given the genotypes of a set of individuals, the CC graph has one node for each individual and an edge between two individuals if their genotypes are consistent with sharing a haplotype, i.e., if for every site where one of the individuals is homozygous, the other individual is either homozygous for the same allele or heterozygous. As the haplotypes of individuals that are homozygous for the whole region being considered are easily determined, we assume that every given individual has two different haplotypes (i.e., it is heterozygous for at least one of the genotyped markers) and no two individuals have the same pair of haplotypes.

As we show below, CC graphs and line graphs are closely related. Let H be the CC graph of a given set of genotypes. We say that H is *allelable* if there is an assignment of pairs of colors to its nodes so that every two adjacent nodes share exactly one color and every two non-adjacent nodes have distinct color sets. Under this assignment every color represents a haplotype or an allele in the population. The number of colors is the estimated number of genotypes. The following observation stands at the basis of our computational approach.

Lemma 1. *H is allelable iff it is a line graph.*

Proof. If H is allelable then construct a graph G in which every allele of H represents a node and edges connect alleles that are paired together in the nodes of H. It is easy to see that H is the line graph of G.

Conversely, if $H = L(G)$ is a line graph then one can assign distinct labels to the nodes of the corresponding root graph G and use these labels to label the edges of G and, hence, the nodes of H. This assignment clearly shows that H is allelable.   □

Lemma 1 implies that a set of haplotypes providing a valid phasing of genotypes will always form a line graph. The converse, however, need not be true as the SNP data of the original genotypes are not encoded in our formulation.

2.2   Problem Definition and Its Complexity

As the input data may have more edges than the underlying line graph, we do not expect real data to yield line graphs. Rather it is desirable to transform a given CC graph into a line graph using a minimum number of node and edge deletions. In the following we formulate three versions of the problem.

Problem 1. Given a graph $G = (V, E)$, find a line graph $G' = (V, E')$, such that $E' \subseteq E$ and $| E - E' |$ is minimized.

Problem 2. Given a graph $G = (V, E)$, find a line graph $G' = (V', E)$, such that $V' \subseteq V$ and $| V - V' |$ is minimized.

We also consider a combined problem where both nodes and edges may be removed from a graph:

Problem 3. Given a graph $G = (V, E)$ and a constant α, find a line graph $G' = (V', E')$, such that $E' \subseteq E$, $V' \subseteq V$ and $| E - E' | + \alpha | V - V' |$ is minimized.

Yannakakis [17] has shown that the problems of deleting a minimum number of edges or nodes of a graph in order to make it into a line graph are both NP-hard.

As the problems can be formulated as a problem of avoiding particular subgraphs which have at most 6 nodes and 10 edges, by [2] both problems are fixed parameter tractable and can be solved in time $\max\{m, n\}O(10^i)$ and $\max\{m, n\}O(6^j)$, respectively, where i (resp., j) is the minimum number of edges (resp., vertices) that need to be deleted. Using the techniques of Niedermeier and Rossmanith [11], the node deletion variant can be solved in $\max\{m, n\}O(5.19^j)$ time.

Problem 2 can be formulated as a hitting set problem, where all the sets that need to be hit are of size $4, 5$ or 6. Trevisan [14] showed that the hitting set problem, when the size of the input sets is bounded by $B \geq 2$ ($B = 6$ in our case), is hard to approximate to within $B^{-\frac{1}{19}}$. The fact that all sets are of size at most six also implies a 6-approximation algorithm for the line graph node deletion problem [6].

3   Integer Programming Formulation

Here we provide integer programming formulations for Problems 1, 2 and 3. The formulations rely on the characterization given in Lemma 1. We let $[i, j]$ represent $\{i, i + 1, ..., j\}$.

3.1   Edge Deletions

Let $G = (V, E)$ be the input CC graph. Our basic program has the following sets of binary variables: (i) A variable d_e for every edge $e \in E$, where $d_e = 1$ iff e is deleted. (ii) A variable $x_{v,j}$ for every node $v \in V$ and every possible allele $j \in [1, 2n]$, where $x_{v,j} = 1$ iff individual v has allele j. (iii) A variable $s_{e,j}$ for every edge $e \in E$ and every possible allele $j \in [1, 2n]$, where $s_{e,j} = 1$ iff the two endpoints of e share allele j. It is formulated as follows:

$$
\begin{aligned}
\min \quad & \textstyle\sum_e d_e \\
s.t. \quad & \textstyle\sum_{j=1}^{2n} x_{v,j} = 2 && \forall v \in V \\
& x_{v,j} + x_{u,j} \leq 1 && \forall (u,v) \notin E, j \in [1, 2n] \\
& \textstyle\sum_{j=1}^{2n} s_{e,j} = 1 - d_e && \forall e \in E \\
& x_{u,j} + x_{v,j} - 1 \leq s_{e,j} && \forall e = (v,u) \in E, j \in [1, 2n] \\
& s_{e,j} \leq x_{u,j} && \forall j \in [1, 2n], u \in e \\
& d_e, x_{v,j}, s_{e,j} \in \{0, 1\} && \forall e \in E, v \in V, j \in [1, 2n]
\end{aligned}
$$

In this formulation, the first constraint ensures that each node is assigned two distinct colors; the next two constraints ensure that non-adjacent nodes (before or after edge deletion) do not share a color; and fourth and fifth constraints ensure that the edge sharing variables are consistent with the node coloring variables. A potential problem with the formulation, however, is that different permutations of the colors yield equivalent solutions. To tackle this problem, we use the symmetry breaking techniques of [3]. Specifically, we start by ordering the nodes from 1 through n. Each node "owns" two colors: node i owns colors $(2i - 1)$ and $2i$. The fact that $x_{v,j} = 1$, where $j = 2v - 1$ or $j = 2v$ means that v is the first node to use color j. v is required to use color $2v - 1$ before it uses color $2v$.

Overall, node v has access to colors $1, 2, \ldots, 2v$ and can use some of the previously used colors or be the first ordered node to use new colors. These restrictions can be represented by the following additional constraints:

$$
\begin{aligned}
x_{v,2v} &\leq x_{v,2v-1} && \forall v \in [1, n] \\
x_{j,2i-1} &\leq x_{i,2i-1} && \forall j \in [i+1, n] \\
x_{j,2i} &\leq x_{i,2i} && \forall j \in [i+1, n]
\end{aligned}
$$

We observe that a number of the variables in the above formulation can be automatically set to 0 and removed from the formulation. In particular, if $(i, j) \notin E, i < j$ then $x_{j,2i} = x_{j,2i-1} = 0$, as j and i cannot share a color when they do not share an edge. If $e = (u, v)$, we also note that $s_{e,j} = 0$ unless both u, v can be colored with color j, that is if $(v, \lfloor \frac{j+1}{2} \rfloor), (u, \lfloor \frac{j+1}{2} \rfloor) \in E$.

3.2   Node Deletions

If a graph does not contain one of the forbidden substructures then the graph is alleleable. Otherwise, a vertex has to be removed from each of the forbidden subgraphs. Our algorithm relies on listing all occurrences of the forbidden substructures and then solving the hitting set problem on the set of forbidden substructures.

We observe that the deletion of a node will not create a new forbidden substructure. The node deletion problem can thus be formulated as an integer program by listing all forbidden substructures and removing one node from each one of them. We let t_n be a binary variable representing whether node i is removed and denote by $\mathcal{F}$ the set of all forbidden induced subgraphs of a line graph. Then the program is formulated as follows:

$$
\min \quad \sum_n t_n
$$
$$
\sum_{n \in F} t_n \geq 1 \; \forall F \in \mathcal{F}
$$
$$
t_n \in \{0,1\} \quad \forall n \in V
$$

3.3   Edge and Node Deletions

In real data, both genotype errors and spurious sharing relations may happen at the same time, leading to a combined node- and edge-deletion problem. We define d,x, s and t as before. We define the variable r_e as a boolean variable representing that an edge has been removed, which occurs when either one of its adjacent nodes or the edge itself is deleted. We let α be the relative cost of node deletion with respect to edge deletion. The combined program is as follows:

$$
\begin{aligned}
\min \quad & \sum_e d_e + \alpha \sum_n t_n \\
s.t. \quad & \sum_{j=1}^{2n} x_{v,j} = 2 - 2t_v & \forall v \in V \\
& x_{v,j} + x_{u,j} \leq 1 & \forall (u,v) \notin E, j \in [1, 2n] \\
& \sum_{j=1}^{2n} s_{e,j} = 1 - r_e & \forall e \in E \\
& x_{u,j} + x_{v,j} - 1 \leq s_{e,j} & \forall e = (v, u) \in E, j \in [1, 2n] \\
& s_{e,j} \leq x_{u,j} & \forall j \in [1, 2n], u \in e \\
& r_e \leq t_v + t_u + d_e & \forall e = (u, v) \in E \\
& d_e \leq r_e & \forall e \in E \\
& t_u \leq r_e & \forall u \in e, e \in E \\
& d_e, t_v, x_{v,j}, s_{e,j}, r_e \in \{0,1\} \; \forall e \in E, v \in V, j \in [1, 2n]
\end{aligned}
$$

We further augment this integer program with the symmetry breaking techniques described in Section 3.1.

4   Experimental Results

We comprehensively test our algorithm for deleting edges and vertices under two simulated scenarios, corresponding to a bottleneck population isolate and a population that has continuously undergone recombination and mutation. For the population isolate we show that we can almost fully recover the haplotype structure from the sharing information alone. For the population that has undergone continuous recombination and mutation we get a bound on the number of haplotypes that is close to the true number of haplotypes in the population. Our experiments further reveal that the occurrence of genotypes showing a large degree of sharing does not materially affect our ability to estimate the number of haplotypes in the remaining population. The computational experiments were done using CPLEX 12, making use of a single CPU processor core with 4GB of memory.

4.1   Bottleneck Population

Haplotypes that are shared across a long region are with high probability identical by descent, i.e. they are the result of the genetic material of a single forefather being passed to a number of his descendents. Some of the haplotypes, however, will be identical by state only, i.e., the haplotypes are the same but cannot be traced to a single common forefather. The probability of identical by state sharing decreases rapidly with the length of the haplotype being considered. The probability of identity by state sharing depends on the size of the ancestral population. We simulate graphs that might occur from the detection of identity by state sharing.

We use Hudson's ms simulator [9] to simulate realistic genotype populations. We assume that there is an initial small bottleneck population that then rapidly expanded. We simulate genotypes for the initial population and then generate the current population as a random sample with replacement from this initial population. The parameters for the simulation are chosen such as to sample approximately a 5 cM locus, or 3027 SNPs, a size for which it is reasonable to expect that no recombinations would take place in the region between two individuals being studied. We consider two sets of experiments for this scenario. In the first experiment we vary the size of the population but keep the number of times that each haplotype is sampled on expectation fixed. In the second experiment we vary the number of times that each haplotype is sampled on expectation and keep the size of the population fixed.

In Table 1 we fix the number of times that a haplotype is sampled on average as 1 and we vary the size of the population, which we present as the number of haplotypes in the ancestral population. It can be seen that the number of edges grows roughly linearly with the size of the population being sample. Our simulations show that the number of estimated haplotypes is in close agreement with the true number of haplotypes in the sample. As some of the haplotypes are sampled multiple times the number of haplotypes in the sample is typically

Table 1. Performance evaluation. Genotypes are generated by sampling with replacement two haplotypes at a time from a population of size twice the number of genotypes. The size of this population is varied. The estimated and true numbers of haplotypes both refer to the set of haplotypes underlying the sampled genotypes.

Genotypes sampled	# edges	# edges removed	Estimated # haplotypes	True # haplotypes	Compute time (s)
25	21	0	33	33	0
50	46	1	64	63	0.04
75	74	0	95	95	0.01
100	105	0	125	126	0.02
125	129	2	154	153	0.05
250	294	1	300	300	0.07
500	634	3	589	594	0.29

smaller than the actual population size. Notably, all instances are solved very fast (less than a second).

Table 2. Performance evaluation with respect to a sample drawn from an initial population of 100 haplotypes. The number of individuals drawn from the sample is varied.

# Genotypes	Coverage	# edges	# edges removed	Estimated # haplotypes	True # haplotypes	Compute time (s)
25	0.5	17	0	37	37	0.00
50	1	46	1	64	63	0.04
75	1.5	98	0	82	82	0.04
100	2	186	0	87	88	0.02
125	2.5	301	0	92	92	0.09

In Table 2 we look at an initial bottleneck population of 100 haplotypes while varying the number of genotypes sampled from this population from 25 to 125. This implies that each haplotype in the ancestral population is sampled between 0.5 and 2.5 times (on average). As the sample size increases, our estimate of the number of haplotypes tightly follows the true number of haplotypes sampled from the population, with both approaching the actual population size of 100. We observe that the number of edges in the graph grows roughly as the square of the sample size. The CPU time used for all these instances is minimal.

4.2   Recombinant Population

The second scenario that we simulate is a population that is undergoing mutation and recombination. We fix mutation rate at 10^{-9} per generation and recombination rate between two adjacent basepairs at 10^{-8} per generation. In all of our experiments we simulate a 1MB region, varying the number of individuals in the ancestral population (N_0) and the number of individuals sampled in the current population.

Table 3. Performance evaluation for a recombinant population while varying population size

N_0	# SNPs	# genotypes	# edges	# edges removed	Estimated # haplotypes	True # haplotypes	Compute time (s)
500	116	100	484	245	83	144	1216
1000	231	100	235	105	114	172	120
2500	584	100	23	0	178	189	0.0
5000	1215	100	15	0	185	195	0.0
10000	3027	500	135	1	880	894	5

First, we vary the size of the initial population, while leaving the number of individuals considered mostly constant (see Table 3). We observe that when the size of the initial population grows the probability that two individuals can share a haplotype decreases rapidly. Many of the edges observed in these graphs are due to sharing between genotypes that are not due to the sharing of haplotypes. The graphs being considered are therefore far from being line graphs and many edges need to be removed in order to make them into ones. Nevertheless, we are able to give estimates of the number of haplotypes in a population even in this setting.

The quality of our estimate is, not surprisingly, dependent on the number of edges that need to be removed to create a line graph. Our estimated number of haplotypes is consistently smaller than the true number of haplotypes, but the number of estimated haplotypes is never below 57% of the true number of haplotypes. Similar conclusions can be drawn from the second set of experiments in which we fix the size of the initial population and vary the size of the sample that we draw from that population (see Table 4).

Table 4. Performance evaluation on a recombinant population while varying sample size

N_0	# SNPs	# genotypes	# edges	# edges removed	Estimated # haplotypes	True # haplotypes	Compute time (s)
1000	193	50	38	5	73	87	5
1000	231	100	235	105	114	172	120
1000	289	200	738	374	170	287	300

4.3 Combined Node and Edge Deletions

In our final set of experiments we simulate a scenario in which there are genotypes showing a high degree of excess sharing of haplotypes. To this end, we focus on the data set of 50 genotypes presented in Table 4, and designate increasing subsets of the genotypes as being partially observed, i.e., only a subset of their markers is observed supposedly due to failure or otherwise missing data.

Table 5. Performance evaluation in a population where some of the genotypes are only partially observed

# Partial genotypes	# Extra edges	α	# genotypes removed	# partial genotypes removed	# edges removed	Estimated # haplotypes	True # haplotypes
5	40	1.5	5	4	1	68	78
5	40	4	2	2	8	68	78
10	70	1.5	10	9	1	63	71
10	70	4	4	3	18	63	71
15	102	1.5	13	9	0	55	62
15	102	4	7	6	14	59	62

In more detail, in Table 5 we modify a graph consisting of 50 genotypes, consisting of 38 edges where 5 edges have to be removed to transform the graph into a line graph. We let $5, 10$ or 15 of the genotypes be partially observed, i.e., genotyped at only at 30% of their markers. We show the number of edges added to the graph due to these partial observations. Next, we apply our combined node and edge deletion algorithm and use two settings for the alpha parameter controlling the relative penalty of node vs. edge deletion. The first value that we use, $\alpha = 1.5$, prefers a node deletion whenever more than one of the node's adjacent edges has to be deleted. The second value of 4 allows up to 4 edge deletions before the node deletion is preferred. In each setting we provide the the number of edges and genotypes removed and the number of removed genotypes that were partially observed. We evaluate the size estimate given by the algorithm against the true number of haplotypes in the data set without counting the haplotypes in the partially observed genotypes.

We observe that the occurrence of the partial genotypes does not appear to materially affect our estimate of the number of haplotypes in the remaining population. We also observe that the largest number of nodes removed in each experiment are from the subset corresponding to the partially observed genotypes. When $\alpha = 1.5$, $60 - 90\%$ of the partially observed genotypes are removed and less then 12% of the other genotypes. When $\alpha = 4$, $30 - 40\%$ of the partial genotypes are removed and less then 3% of the other genotypes. All computations presented in Table 5 finished in under one second.

4.4   Comparison to a Phasing-Based Estimation

Apart from the preprocessing, the complexity of our method does not depend on the number of genotyped markers. In contrast, many haplotype phasing methods are not able to handle the large number of markers dealt with in our approach [4]. We thus opted to compare our method to phasing-based estimates derived from the application of the BEAGLE phaser [1] to our data. Surprisingly, in all the simulated settings, the number of haplotypes estimated by BEAGLE was twice the number of genotypes, which is a trivial upper bound for the number of haplotypes in a population.

5 Conclusions

We show that the problem of assigning alleles to individuals when only information about the sharing of alleles between individuals is known is equivalent to the problem of determining whether a graph is a line graph. When sharing information is not perfect we give polynomial size integer programming algorithms for determining allele sharing.

Researchers are frequently interested in knowing not only the number of haplotypes for a particular instance but also the haplotypes themselves. While the method presented here does not address the problem of determining haplotypes, we hope it may be used in conjunction with or as a part of such a method. In particular, one could modify our approach to remove edges in the graph such that the resulting graph is a line graph with a minimum number of haplotypes. The line graph minimizing the number of haplotypes will always provide a lower bound on the true number of haplotypes in the sample; thus, the approach presented here could provide the basis for a constraint generation technique for minimizing the number of haplotypes.

The model considered here does not deal with noise due to genotype miscalls. Genotyping errors may lead to haplotypes not sharing an edge when they in fact should. There are two ways in which this problem can be handled by our framework: Edges can be added when the genotypes are compatible in all but a fixed number of error locations, or we may allow for edge insertions, possibly with weights related to the number of errors that will need to be corrected for such edges to occur. Our integer programming formulation can be easily extended to allow for such edge insertions, solving the more general problem of editing a graph (by edge insertions and deletions) to become a line graph.

Acknowledgments. RS was supported by a research grant from the Israel Science Foundation (grant no. 241/11).

References

1. Browning, B.L., Browning, S.R.: A unified approach to genotype imputation and haplotype-phase inference for large data sets of trios and unrelated individuals. American Journal of Human Genetics 84(2), 210–223 (2009)
2. Cai, L.: Fixed-parameter tractability of graph modification problems for hereditary properties. Information Processing Letters 58, 171–176 (1996)
3. Campelo, M., Campos, V., Correa, R.: On the asymmetric representatives formulation for the vertex coloring problem. Discrete Applied Mathematics 156(7), 1097–1111 (2008)
4. Catanzaro, D., Godi, A., Labbé, M.: A class representative model for pure parsimony haplotyping. Informs Journal of Computing 22(2), 195–209 (2009)
5. Clark, A.: Inference of haplotypes from PCR-amplified samples of diploid populations. Molecular Biology and Evolution 7, 111–122 (1990)
6. Even, S., Bar-Yehuda, R.: A linear-time approximation algorithm for the weighted vertex cover problem. Journal of Algorithms 2(2), 198–203 (1981)

7. Halldórsson, B.V., Aguiar, D., Tarpine, R., Istrail, S.: The Clark Phaseable Sample Size Problem: Long-Range Phasing and Loss of Heterozygosity in GWAS. Journal of Computational Biology 18(3), 323–333 (2011)

8. Halldórsson, B.V., Bafna, V., Edwards, N., Lippert, R., Yooseph, S., Istrail, S.: A Survey of Computational Methods for Determining Haplotypes. In: Istrail, S., Waterman, M.S., Clark, A. (eds.) DIMACS/RECOMB Satellite Workshop 2002. LNCS (LNBI), vol. 2983, pp. 26–47. Springer, Heidelberg (2004)

9. Hudson, R.R.: Generating samples under a Wright-Fisher neutral model of genetic variation. Bioinformatics 18(2), 337–338 (2002)

10. Lehot, P.G.H.: An optimal algorithm to detect a line graph and output its root graph. J. ACM 21, 569–575 (1974)

11. Niedermeier, R., Rossmanith, P.: An efficient fixed-parameter algorithm for 3-hitting set. Journal of Discrete Algorithms 1(1), 89–102 (2003)

12. Prabhu, S., Pe'er, I.: Overlapping pools for high-throughput targeted resequencing. Genome Research 19, 1254–1261 (2009)

13. Roussopoulos, N.: A max(m, n) algorithm for determining the graph H from its line graph G. Information Processing Letters 2, 108–112 (1974)

14. Trevisan, L.: Non-approximability results for optimization problems on bounded degree instances. In: Proceedings of the Thirty-Third Annual ACM Symposium on Theory of Computing, pp. 453–461. ACM (2001)

15. Van Rooij, A., Wilf, H.: The interchange graphs of a finite graph. Acta Math. Acad. Sci. Hungar. 16, 263–269 (1965)

16. Whitney, H.: Congruent graphs and the connectivity of graphs. American Journal of Mathematics 54, 150–162 (1932)

17. Yannakakis, M.: Node-and edge-deletion NP-complete problems. In: Proceedings of the Tenth Annual ACM Symposium on Theory of Computing, STOC 1978, pp. 253–264. ACM, New York (1978)

Extracting Conflict-Free Information
from Multi-labeled Trees*

Akshay Deepak[1], David Fernández-Baca[1], and Michelle M. McMahon[2]

[1] Department of Computer Science, Iowa State University, Ames, IA 50011, USA
[2] School of Plant Sciences, University of Arizona, Tucson, AZ 85721, USA

Abstract. A multi-labeled tree, or MUL-tree, is a phylogenetic tree where two
or more leaves share a label, e.g., a species name. A MUL-tree can imply multiple conflicting phylogenetic relationships for the same set of taxa, but can also
contain conflict-free information that is of interest and yet is not obvious. We
define the information content of a MUL-tree T as the set of all conflict-free
quartet topologies implied by T, and define the maximal reduced form of T as
the smallest tree that can be obtained from T by pruning leaves and contracting edges while retaining the same information content. We show that any two
MUL-trees with the same information content exhibit the same reduced form.
This introduces an equivalence relation among MUL-trees with potential applications to comparing MUL-trees. We present an efficient algorithm to reduce a
MUL-tree to its maximally reduced form and evaluate its performance on empirical datasets in terms of both quality of the reduced tree and the degree of data
reduction achieved.

1 Introduction

Multi-labeled trees, also known as MUL-trees, are phylogenetic trees that can have
more than one leaf with the same label [5,7,9,14,19] (Fig. 1). MUL-trees arise naturally
and frequently in data sets containing multiple genes or gene sequences for the same
species [18], but they can also arise in bio-geographical studies or co-speciation studies
where leaves represent individual taxa yet are labeled with their areas [6] or hosts [11].

MUL-trees, unlike singly-labeled trees, can contain conflicting species-level phylogenetic information due, e.g., to whole genome duplications [12], incomplete lineage
sorting [17], inferential error, or, frequently, an unknown combination of several factors. However, they can also contain substantial amounts of conflict-free information.
Here we provide a way to extract this information; specifically, we have the following
results.

- We introduce a new quartet-based measure of the information content of a
  MUL-tree, defined as the set of conflict-free quartets the tree displays (Section 2).
- We introduce the concept of the maximally-reduced form (MRF) of a MUL-tree,
  the smallest MUL-tree with the same information content (Section 3), and show
  that any two MUL-trees with the same information content have the same MRF
  (Theorem 3).

* This work was supported in part by the National Science Foundation under grant DEB-
0829674.

B. Raphael and J. Tang (Eds.): WABI 2012, LNBI 7534, pp. 81–92, 2012.
© Springer-Verlag Berlin Heidelberg 2012

- We present a simple algorithm to construct the MRF of a MUL-tree (Section 4); its running time is quadratic in the number of leaves and does not depend on the multiplicity of the leaf labels or the degrees of the internal nodes.

- We present computational experience with an implementation of our MRF algorithm (Section 5). In our test data, the MRF is often significantly smaller than the original tree, while retaining most of the taxa.

We now give the intuition behind our notion of information content, deferring the formal definitions of this and other concepts to the next section. Quartets (i.e., sets of four species) are a natural starting point, since they are the smallest subsets from which we can draw meaningful topological information. A singly-labeled tree implies exactly one topology on any quartet. More precisely, each edge e in a singly-labeled tree implies a bipartition (A, B) of the leaf set, where each part is the set of leaves on one the two sides of e. From (A, B), we derive a collection of bipartitions $ab|cd$ of quartets, such that $\{a, b\} \subseteq A$ and $\{c, d\} \subseteq B$. Clearly, if one edge in a singly-labeled tree implies some bipartition $q = ab|cd$ of $\{a, b, c, d\}$, then there can be no other edge that implies a bipartition, such as $ac|bd$, that is in conflict with q. Indeed, the quartet topologies implied by a singly-labeled tree uniquely identify it [21].

The situation for MUL-trees is more complicated, as illustrated in Fig. 1. Here, the presence of two copies of labels b and c — $b(1)$ and $b(2)$, and, $c(1)$ and $c(2)$ — leads to two conflicting topologies on the quartet $\{b, c, d, e\}$. Edge (u, v) implies the bipartition $bc|de$, corresponding to the labels $\{b(1), c(1), d, e\}$, while edge (v, w) implies $bd|ce$ corresponding to the leaves $\{b(2), c(2), d, e\}$. On the other hand, the quartet topology $af|bc$, implied by edge (t, u), has no conflict with any other topology that the tree exhibits on $\{a, b, c, f\}$. We show that the set of all such conflict-free quartet topologies is compatible (Theorem 1). That is, for every MUL-tree T there exists at

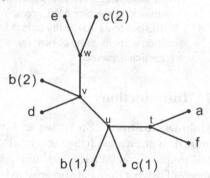

Fig. 1. A MUL-tree. Numbers in parenthesis next to labels indicate the multiplicity of the respective labels and are not part of the labels themselves

least one singly-labeled tree that displays all the conflict-free quartets of T — and possibly some other quartets as well. Motivated by this, we only view conflict-free quartet topologies as informative, and define the information content of a MUL-tree as the set of all conflict-free quartet topologies it implies.

We should note that conflicting quartets may well provide valuable information, whether about paralogy, deep coalescence, or mistaken annotations. In some cases, species-level phylogenetic information can be recovered from conflicted quartets through application of, e.g., gene-tree species-tree reconciliation, an NP-hard problem. However, this is not feasible when the underlying cause of multiplicity is unknown or when conducting large-scale analyses. Our definition of information content is deliberately designed to make no assumptions about the cause of conflict. It is also conservative with respect to species relationships, i.e., it does not introduce quartets not originally supported by the data. Further, knowing the information content of a MUL-tree allows us to easily identify its conflicting quartets as well.

A MUL-tree may have leaves that can be pruned and edges that can be contracted without altering the tree's information content, i.e., without adding or removing conflict-free quartets. For example, in Fig. 1, every quartet topology that edge (v, w) implies is either in conflict with some other topology (e.g., for set $\{b, c, d, e\}$) or is already implied by some other edge (e.g., $af|ce$ is also implied by (t, u)). Thus, (v, w) can be contracted without altering the information

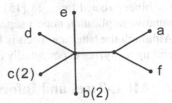

Fig. 2. The MRF for the Mul-tree in Fig. 1

content. In fact, the information content remains unchanged if we also contract (u, v) and remove the leaves labeled $b(1)$ and $c(1)$. We define the MRF of a MUL-tree T as the tree that results from applying information-preserving edge contraction and leaf pruning operations repeatedly to T, until it is no longer possible to do so. The MRF of the tree in Fig. 1 is shown in Fig. 2. In this case, the MRF is singly-labeled; however, this is not true in general (see the example in Section 4.4). If the MRF is itself a MUL-tree, it is not possible to reduce the original to a singly-labeled tree without either adding at least one quartet that did not exist conflict-free in T or by losing one or more conflict-free quartets.

Since any two MUL-trees with the same information content have the same MRF, rather than comparing MUL-trees directly, we can instead compare their MRFs. This is appealing mathematically, because it focuses on conflict-free information content, and also computationally, since an MRF can be much smaller than the original MUL-tree. Indeed, on our test data, the MRF was frequently singly-labeled. This reduction in input size is especially significant if the MUL tree is an input to an algorithm whose running time is exponential in the label multiplicity, such as Ganapathy et al.'s algorithm to compute the contract-and-refine distance between two area cladograms [6] or Huber et al.'s algorithm to determine if a collection of "multi-splits" can be displayed by a MUL-tree [8].

For our experiments, we also implemented a post-processing step, which converts the MRF to a singly-labeled tree, rendering it available for analyses that require singly-labeled trees, including supermatrix [3,23] and supertree methods [2,16,1,22]. On the trees in our data set, the combined taxon loss between the MRF computation and the postprocessing was much lower than it would have been had we simply removed all duplicate taxa from the original trees.

Previous work on MUL-trees has concentrated on finding ways to reduce MUL-trees to singly-labeled trees (typically in order to provide inputs to supertree methods) [19], and to develop metrics and algorithms to compare MUL-trees [6,15,13,10]. In contrast to our approach — which is purely topology-based and is agnostic with respect to the cause of label multiplicity —, the assumption underlying much of the literature on MUL-trees is that taxon multiplicity results from gene duplication. Thus, methods to obtain singly-labeled trees from MUL-trees usually work by pruning subtrees at putative duplication nodes. Although the proposed algorithms are polynomial, they are unsatisfactory in various ways. For example, in [19] if the subtrees are neither identical nor compatible, then the subtree with smaller information content is pruned, which seems to discard too much information. Further, the algorithm is only efficient for binary rooted trees. In [15] subtrees are pruned arbitrarily, while in [13] at each putative duplication node a separate analysis is done for each possible pruned subtree. Although the latter approach is better than pruning arbitrarily, in the worst case it can end up analyzing exponentially many subtrees.

2   MUL-Trees and Information Content

A *MUL-tree* is a triple (T, M, ψ), where (i) T is an unrooted tree[1] with leaf set $\mathcal{L}(T)$ all of whose internal nodes have degree at least three, (ii) M is a set of labels, and (iii) $\psi_T : \mathcal{L}(T) \to M$ is a surjective map that assigns each leaf of T a label from M. (Note that if ψ is a bijection, T is singly labeled; that is, singly-labeled trees are a special case of MUL-trees.) For brevity we often refer to a MUL-tree by its underlying tree T. In what follows, unless stated otherwise, by a tree we mean a MUL-tree.

An edge (u, v) in T is *internal* if neither u nor v belong to $\mathcal{L}(T)$, and is *pendant* otherwise. A *pendant node* is an internal node that has a leaf as its neighbor.

Let (u, v) be an edge in T and T' be the result of deleting (u, v) from T. Then T_u^{uv} (T_v^{uv}) denotes the subtree of T' that contains u (v). M_u^{uv} (M_v^{uv}) denotes the set of labels in T_u^{uv} (T_v^{uv}) but not in T_v^{uv} (T_u^{uv}). C^{uv} is the set of labels common to both T_u^{uv} and T_v^{uv}. Observe that M_u^{uv}, M_v^{uv} and C^{uv} partition M. For example, in Fig. 1, $M_u^{uv} = \{a, f\}$, $M_v^{uv} = \{e, d\}$, $C^{uv} = \{b, c\}$.

A (resolved) *quartet* in a MUL-tree T is a bipartition $ab|cd$ of a set of labels $\{a, b, c, d\}$ such that there is an edge (u, v) in T with $\{a, b\} \in M_u^{uv}$ and $\{c, d\} \in M_v^{uv}$. We say that (u, v) *resolves* $ab|cd$. For example, in Fig. 1, edge (t, u) resolves $af|bc$.

The *information content of an edge* (u, v) of a MUL-tree T, denoted $\Delta(u, v)$, is the set of quartets resolved by (u, v). An edge (u, v) in tree T is *informative* if $|\Delta(u, v)| > 0$; (u, v) is *maximally informative* if there is no other edge (u', v') in T with $\Delta(u, v) \subset \Delta(u', v')$. The *information content* of T, denoted $\mathcal{I}(T)$, is the combined information content of all edges in the tree; that is $\mathcal{I}(T) = \bigcup_{(u,v) \in E} \Delta(u, v)$, where E denotes the set of edges in T.

The next result shows that the quartets in $\mathcal{I}(T)$ are conflict-free.

Theorem 1. *For every MUL-tree T, there is a singly labeled tree T' such that $\mathcal{I}(T) \subseteq \mathcal{I}(T')$.*[2]

Note that there are examples where the containment indicated by the above result is proper.

To conclude this section, we give some results that are useful for the reduction algorithm of Section 4. In the next lemmas, (u, v) and (w, x) denote two edges in tree T that lie on the path $P_{u,x} = (u, v, \dots, w, x)$ as shown in Fig. 3.

Lemma 1. *If $|M_u^{uv}| = |M_w^{wx}|$ then $M_u^{uv} = M_w^{wx}$. Otherwise, $M_u^{uv} \subset M_w^{wx}$.*

---

[1] The results presented here can be extended to rooted trees, using triplets instead of quartets, exploiting the well-known bijection between rooted and unrooted trees [20, p. 20]. We do not discuss this further here for lack of space.

[2] All proofs are available in [4].

Together with Lemma 1, the next result allows us
to check whether the information content of an
edge is a subset of that of another based solely
on the cardinalities of the M_u^{uv}s.

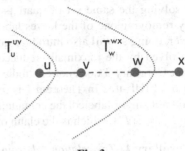

Lemma 2. $\Delta(u, v) \subseteq \Delta(w, x)$ *if and only if*
$M_v^{uv} = M_x^{wx}$.

Lemma 3. *Suppose* $\Delta(u, v) \subseteq \Delta(w, x)$. *Then,*
for any edge (y, z) *on* $P_{u,x}$ *such that* v *is closer to*
y *than to* z, $\Delta(u, v) \subseteq \Delta(y, z) \subseteq \Delta(w, x)$.

Fig. 3.

3 Maximally Reduced MUL-Trees

Our goal is to provide a way to reduce a MUL-tree T as much as possible, while preserving its information content. Our reduction algorithm uses the following operations.

Prune(v)**:** Delete leaf v from T. If, as a result, v's neighbor u becomes a degree-two
  node, connect the former two neighbors of u by an edge and delete u.
Contract(e)**:** Delete an internal edge e and identify its endpoints.

A leaf v in T is *prunable* if the tree that results from pruning v has the same information content as T. An internal edge e in T is *contractible* if the tree that results from
contracting e has the same information content as T. T is *maximally reduced* if it has
no prunable leaf and no contractible internal edge.

Theorem 2. *Every internal edge in a maximally reduced tree T resolves a quartet that
is resolved by no other edge.*

Note that a quartet $ab|cd$ that is resolved by edge (u, v), but by no other edge must have
the form illustrated in Fig. 4.

   Next, we show that the set of quartets resolved by a
maximally reduced tree uniquely identifies the tree.

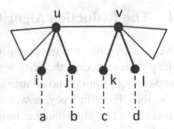

Theorem 3. *Let T and T' be two maximally reduced
trees such that $\mathcal{I}(T) = \mathcal{I}(T')$. Then, T and T' are
isomorphic.*

The *maximally reduced form* (MRF) of a MUL-tree T
is the tree that results from repeatedly pruning prunable leaves and contracting contractible edges from T
until this is no longer possible. Theorem 3 shows that
we can indeed talk about "the" MRF of T.

Fig. 4. Quartet $ab|cd$ is resolved
only by edge (u, v). Here, $a \in$
M_i^{ui}, $b \in M_j^{uj}$, $c \in M_k^{vk}$ and
$d \in M_l^{vl}$.

Corollary 1. *Every MUL-tree has a unique MRF.*

Corollary 2. *Any two MUL-trees with the same information content have the same MRF.*

Corollary 3. *If a maximally reduced MUL-tree T is not singly-labeled, there does not
exist a singly-labeled tree having the same information content as T.*

Fig 5 illustrates the last result. Any singly-labeled tree resolving the same set of quartets must be obtained by removing one of the leaves labeled with f. However, doing so will also introduce quartets that are not resolved by the maximally reduced MUL-tree. Note that Corollary 3 does not contradict with Theorem 1. If the MUL-tree in Theorem 1 is maximally reduced and not singly-labeled, the containment is proper; i.e., $\mathcal{I}(T) \neq \mathcal{I}(T')$, which is the claim of Corollary 3.

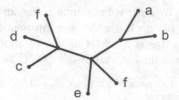

Fig. 5. A maximally reduced MUL-tree

Corollary 4. *The relation "sharing a common MRF" is an equivalence relation on the set of MUL-trees.*

The last result implies that MUL-trees can be partitioned into equivalence classes, where each class consists of the set of all trees with the same information content. Thus, instead of comparing MUL-trees directly, we can compare their maximally reduced forms.

   To end this section, we give some results that help to identify contractible edges and prunable leaves. The setting is the same as for Lemmas 2 and 3: (u, v) and (w, x) are two edges in tree T that lie on the path $P_{u,x} = (u, v, \ldots, w, x)$ (see Fig. 3). We say that subtree T_z^{yz} *branches out* from the path $P_{u,x}$ if $y \in P_{u,x} - \{u, x\}$, and $z \notin P_{u,x}$.

Lemma 4. *Suppose $\Delta(u, v) \subseteq \Delta(w, x)$ then*

   (i) *every internal edge on a subtree branching out from $P_{u,x}$ is contractible, and*
   (ii) *if $\Delta(u, v) = \Delta(w, x)$, every leaf on a subtree branching out from $P_{u,x}$ is prunable. Thus, the entire subtree can be deleted without changing the information content of the tree.*

4   The Reduction Algorithm

We now describe a $O(n^2)$ algorithm to compute the MRF of an n-leaf MUL-tree T. In the previous section, the MRF was defined as the tree obtained by applying information-preserving pruning and contraction operations to T, in any order, until it is no longer possible. For efficiency, however, the sequence in which these steps are performed is important. Our algorithm has three distinct phases: a preprocessing step, redundant edge contraction, and pruning of redundant leaves. We describe these next and then give an example.

4.1   Preprocessing

For every edge (u, v) in T, we compute $|M_u^{uv}|$ and $|M_v^{uv}|$. This can be done in $O(n^2)$ time as follows. First, traverse subtrees T_u^{uv} and T_v^{uv} to count number of distinct labels n_u^{uv} and n_v^{uv} in each subtree. Then, $|M_u^{uv}| = |M| - n_v^{uv}$ and $|M_v^{uv}| = |M| - n_u^{uv}$. We then contract non-informative edges; i.e., edges (u, v) where $|M_u^{uv}|$ or $|M_v^{uv}|$ is at most one.

4.2 Edge Contraction and Subtree Pruning

Next, we repeatedly find pairs of adjacent edges (u, v) and (v, w) such that $\Delta(u, v) \subseteq \Delta(v, w)$ or vice-versa, and contract the less informative of the two. By Lemmas 1 and 2, we can compare $\Delta(u, v)$ and $\Delta(v, w)$ in constant time using the precomputed values of $|M_u^{uv}|$ and $|M_v^{uv}|$. Lemma 4(i) implies that we should also contract all internal edges incident on v or in the subtrees branching out of v. Further, by Lemma 4(ii), if $\Delta(u, v) = \Delta(v, w)$, we can in fact delete these subtrees entirely, since their leaves are prunable. Lemma 3 implies that all such edges must lie on a path, and hence can be identified in linear time. The total time for all these operations is linear, since at worst we traverse every edge twice.

4.3 Pruning Redundant Leaves

The tree that is left at this point has no contractible edges; however, it can still have prunable leaves. We first prune any leaf with a label ℓ that does not participate in any resolved quartet. Such an ℓ has the property that for every edge (u, v), $\ell \notin M_u^{uv}$ and $\ell \notin M_v^{uv}$. All such leaves can be found in $O(n^2)$ time and $O(n)$ space.

Next, we consider sets of leaves with the same label ℓ that share a common neighboring pendant node. Such leaves can be found in linear time. For each such set, we delete all but one element.

After such leaves are removed, let T be the resulting tree. Now, the only kind of prunable leaf with a given label ℓ that might remain are leaves attached to different pendant nodes. The next result identifies such redundant leaves in T.

Lemma 5. *Let ℓ be a multiply-occurring label in T and let T' be the minimal subtree that spans all the leaves labelled by ℓ. Then, any leaf in T labeled ℓ attached to a pendant node in T' of degree at least three is prunable.*

Thus, to identify and prune redundant leaf nodes of the latter type:

1. For each label ℓ, consider the subgraph on the leaves labeled by it.
2. In this subgraph, delete any leaf not attached to a degree 2 pendant node as it is a redundant leaf.

This takes $O(n)$ time per label and $O(n^2)$ time total. The space used is $O(n)$. Hence, the overall time and space complexities are $O(n^2)$ time and $O(n)$, respectively.

The resulting tree has no contractible edges nor prunable leaves. Therefore, it is the MRF of the orginal MUL-tree.

4.4 An Example

We illustrate the reduction of the unrooted MUL-tree shown in Fig. 6 to its MRF.

1. In the preprocessing step, we find that $M_t^{tu} = \emptyset$, $M_s^{su} = \emptyset$ and $M_x^{wx} = \emptyset$, so edges (t, u), (s, u) and (w, x) are uninformative. They are therefore contracted, resulting in the tree shown in Fig. 7.

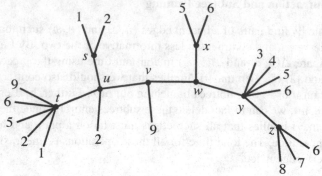

Fig. 6.

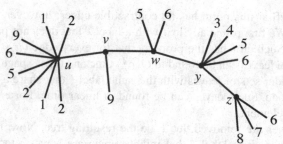

Fig. 7.

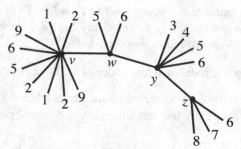

Fig. 8.

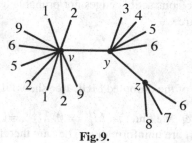

Fig. 9.

2. Since $\Delta(u, v) \subset \Delta(v, w)$, contract (u, v). The result is shown in Fig. 8.
3. Since $\Delta(v, w) = \Delta(w, y)$, delete the subtree branching out at w from the path from v to y and contract (v, w). The result is shown in Fig. 9.
4. Prune taxon 6, which does not participate in any quartet, and all duplicate taxa at the pendant nodes. The result, shown in Fig. 10, is the MRF of the original tree.

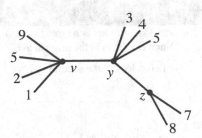

Fig. 10.

5   Evaluation

We implemented our MUL tree reduction algorithm, as well as a second step that restricts the MRF to the set of labels that appear only once, which yields a singly-labeled tree. We tested our two-step program on a set of 110,842 MUL-trees obtained from the PhyLoTA database [10] (http://phylota.net/; GenBank eukaryotic nucleotide sequences, release 184, June 2011), which included a broad range of label-set sizes, from 4 to 1500 taxa.

There were 8,741 trees (7.8%) with essentially no information content; these lost all resolution either when reduced to their MRFs, or in the second step. The remaining trees fell into two categories. Trees in set A had a singly-labeled MRF; 65,709 trees (59.3%) were of this kind. Trees in set B were reduced to singly-labeled trees in the second step; 36,392 trees (32.8%) were of this kind. Reducing a tree to its MRF (step 1), led to an average taxon loss of 0.83% of the taxa in the input MUL-tree. The total taxon loss after the second step (reducing the MRFs in set B to singly-labeled trees), averaged 12.81%. This taxon loss is not trivial, but it is far less than the 41.27% average loss from the alternative, naïve, approach in which all mul-taxa (taxa that label more than one leaf) are removed at the outset. Note that, by the definition of MRFs, taxa removed in the first step do not contribute to the information content, since all non-conflicting quartets are preserved. On the other hand, taxa removed in the second step do alter the information content, since each such taxon participated in some non-conflicting quartet.

Taxon loss is sensitive to the number of total taxa and, especially, mul-taxa, as demonstrated in Figure 11a. The grey function shows the percentage of mul-taxa in the original input trees, which is the taxon loss if we had restricted the input MUL-trees to the set of singly-labeled leaves. The black function shows the percentage of mul-taxa lost after steps 1 and 2 of our reduction procedure.

In addition to the issue of taxon loss, we investigated the effect of our reduction on edge loss, i.e., the level of resolution within the resulting singly-labeled tree. Input

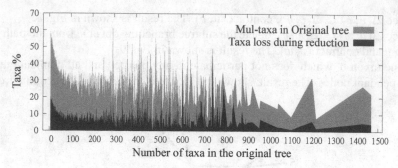

(a) Taxon loss in the second step

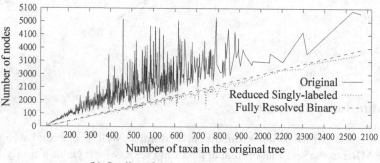

(b) Quality of reduced singly-labeled trees

Fig. 11. Experimental results

MUL-trees were binary and therefore had more nodes than twice the number of taxa (Fig 11b, solid line), whereas a binary tree on singly labeled taxa would have approximately as many nodes as twice the number of taxa (Fig. 11b, dashed line). We found that, although there was some edge loss, the number of nodes in the reduced singly-labeled trees (Fig. 11b, dotted line) corresponded well to the total possible, indicating low levels of edge loss. Note that each point on the dotted or solid lines represents an average over all trees with the same number of taxa.

We have integrated our reduction algorithm into SearchTree (available at `http://searchtree.org/`), a phylogenetic tree search engine that takes a user-provided list of species names and finds matches with a precomputed collection of phylogenetic trees, more than half of which are MUL-trees, assembled from GenBank sequence data. The trees returned are ranked by a tree quality criterion that takes into account overlap with the query set, support values for the branches, and degree of resolution. We have added functionality to provide reduced singly-labeled trees as well as the MUL-trees based on the full leaf set.

6   Conclusion

We introduced an efficient algorithm to reduce a multi-labeled MUL-tree to a maximally reduced form with the same information content, defined as the set of non-conflicting quartets it resolves. We showed that the information content of a MUL-tree uniquely identifies the MUL-tree's maximally reduced form. This has potential application in comparing MUL-trees by significantly reducing the number of comparisons as well as in extracting species-level information efficiently and conservatively from large sets of trees, irrespective of the underlying cause of multiple labels. Our algorithm can easily be adapted to work for rooted trees.

Further work investigating the relationship of the MRF to the original tree under various biological circumstances is also underway. We might expect, for example, that well-sampled nuclear gene families reduce to very small MRF trees, and that annotation errors in chloroplast gene sequences (in which we expect little gene duplication), result in relatively large MRF trees. Comparing the MRF to the original MUL-tree may well provide a method for efficiently assessing and segregating data sets with respect to the causes of multiple labels.

It would be interesting to compare our results with some of the other approaches for reducing MUL-trees to singly-labeled trees (e.g., [19]) or, indeed, to evaluate if our method can benefit from being used in conjunction with such approaches.

Acknowledgements. We thank Mike Sanderson for helping to motivate this work, for many discussions about the problem formulation, and for our ongoing collaboration in the SearchTree project. Sylvain Guillemot listened to numerous early versions of our proofs and offered many insightful comments. We also thank the anonymous reviewers for their comments, which helped to improve this paper.

References

1. Bansal, M., Burleigh, J.G., Eulenstein, O., Fernández-Baca, D.: Robinson-Foulds supertrees. Algorithms for Molecular Biology 5(1), 18 (2010)
2. Baum, B.R.: Combining trees as a way of combining data sets for phylogenetic inference, and the desirability of combining gene trees. Taxon 41(1), 3–10 (1992)
3. de Queiroz, A., Gatesy, J.: The supermatrix approach to systematics. Trends in Ecology & Evolution 22(1), 34–41 (2007)
4. Deepak, A., Fernández-Baca, D., McMahon, M.: Extracting conflict-free information from multi-labeled trees (2012), http://arxiv.org/abs/1205.6359
5. Fellows, M., Hallett, M., Stege, U.: Analogs & duals of the mast problem for sequences & trees. Journal of Algorithms 49(1), 192–216 (2003); 1998 European Symposium on Algorithms
6. Ganapathy, G., Goodson, B., Jansen, R., Le, H., Ramachandran, V., Warnow, T.: Pattern identification in biogeography. IEEE/ACM Trans. Comput. Biol. Bioinformatics 3, 334–346 (2006)
7. Grundt, H., Popp, M., Brochmann, C., Oxelman, B.: Polyploid origins in a circumpolar complex in draba (brassicaceae) inferred from cloned nuclear dna sequences and fingerprints. Molecular Phylogenetics and Evolution 32(3), 695–710 (2004)
8. Huber, K., Lott, M., Moulton, V., Spillner, A.: The complexity of deriving multi-labeled trees from bipartitions. Journal of Computational Biology 15(6), 639–651 (2008)
9. Huber, K., Moulton, V.: Phylogenetic networks from multi-labelled trees. Journal of Mathematical Biology 52, 613–632 (2006)

10. Huber, K., Spillner, A., Suchecki, R., Moulton, V.: Metrics on multilabeled trees: Interrelationships and diameter bounds. IEEE/ACM Transactions on Computational Biology and Bioinformatics 8(4), 1029–1040 (2011)
11. Johnson, K.P., Adams, R.J., Page, R.D.M., Clayton, D.H.: When do parasites fail to speciate in response to host speciation? Syst. Biol. 52, 37–47 (2003)
12. Lott, M., Spillner, A., Huber, K., Petri, A., Oxelman, B., Moulton, V.: Inferring polyploid phylogenies from multiply-labeled gene trees. BMC Evolutionary Biology 9(1), 216 (2009)
13. Marcet-Houben, M., Gabaldón, T.: Treeko: a duplication-aware algorithm for the comparison of phylogenetic trees. Nucleic Acids Research 39, e66 (2011)
14. Popp, M., Oxelman, B.: Inferring the history of the polyploid silene aegaea (caryophyllaceae) using plastid and homoeologous nuclear dna sequences. Molecular Phylogenetics and Evolution 20(3), 474–481 (2001)
15. Puigbò, P., Garcia-Vallvé, S., McInerney, J.: Topd/fmts: a new software to compare phylogenetic trees. Bioinformatics 23(12), 1556 (2007)
16. Ragan, M.: Phylogenetic inference based on matrix representation of trees. Molecular Phylogenetics and Evolution 1(1), 53–58 (1992)
17. Rasmussen, M.D., Kellis, M.: Unified modeling of gene duplication, loss, and coalescence using a locus tree. Genome Research 22, 755–765 (2012)
18. Sanderson, M., Boss, D., Chen, D., Cranston, K., Wehe, A.: The PhyLoTA browser: processing GenBank for molecular phylogenetics research. Systematic Biology 57(3), 335 (2008)
19. Scornavacca, C., Berry, V., Ranwez, V.: Building species trees from larger parts of phylogenomic databases. Information and Computation 209(3), 590–605 (2011); Special Issue: Dediu, A.H., Ionescu, A.M., Martín-Vide, C. (eds.): LATA 2009. LNCS, vol. 5457. Springer, Heidelberg (2009)
20. Semple, C., Steel, M.: Phylogenetics. Oxford University Press, Oxford (2003)
21. Steel, M.: The complexity of reconstructing trees from qualitative characters and subtrees. Journal of Classification 9(1), 91–116 (1992)
22. Swenson, M., Suri, R., Linder, C., Warnow, T.: Superfine: fast and accurate supertree estimation. Systematic Biology 61(2), 214–227 (2012)
23. Wiens, J.J., Reeder, T.W.: Combining data sets with different numbers of taxa for phylogenetic analysis. Systematic Biology 44(4), 548–558 (1995)

Reducing Problems in Unrooted Tree Compatibility to Restricted Triangulations of Intersection Graphs

Rob Gysel, Kristian Stevens, and Dan Gusfield

Department of Computer Science, University of California, Davis, 1 Shields Avenue,
Davis CA 95616, USA
{rsgysel,kastevens,gusfield}@ucdavis.edu

Abstract. The compatibility problem is the problem of determining if
a set of unrooted trees are compatible, i.e. if there is a supertree that
represents all of the trees in the set. This fundamental problem in phylogenetics is NP-complete but fixed-parameter tractable in the number
of trees. Recently, Vakati and Fernández-Baca showed how to efficiently
reduce the compatibility problem to determining if a specific type of
constrained triangulation exists for a non-chordal graph derived from
the input trees, mirroring a classic result by Buneman for the closely related Perfect-Phylogeny problem. In this paper, we show a different way
of efficiently reducing the compatibility problem to that of determining
if another type of constrained triangulation exists for a new non-chordal
intersection graph. In addition to its conceptual contribution, such reductions are desirable because of the extensive and continuing literature
on graph triangulations, which has been exploited to create algorithms
that are efficient in practice for a variety of Perfect-Phylogeny problems.
Our reduction allows us to frame the compatibility problem as a minimal
triangulation problem (in particular, as a chordal graph sandwich problem) and to frame a maximization variant of the compatibility problem
as a minimal triangulation problem.

1   Introduction

A *phylogenetic tree* T is an unrooted tree without nodes of degree two and whose
leaves are bijectively labeled by a set of *species* $\mathcal{L}(T)$. Two phylogenetic trees T
and T' with leaves labeled by $\mathcal{L}$ are *isomorphic* if there is a graph isomorphism
from T to T' that preserves the labeling of $\mathcal{L}$. Let T be a phylogenetic tree and
$X \subseteq \mathcal{L}(T)$. The *restriction* $T|X$ of T to X is the phylogenetic tree obtained from
the minimal subtree of T containing the leaves labeled by X and suppressing
vertices of degree two. Let T and T' be phylogenetic trees such that $\mathcal{L}(T') \subseteq
\mathcal{L}(T)$. A fundamental consideration in phylogenetics is to determine when T is
consistent with the phylogenetic information of T'. Formally, if it is possible
to contract edges of $T|\mathcal{L}(T')$ to obtain T', then T *displays* T' [3,18]. A *profile*
is a collection of phylogenetic trees $\mathcal{P} = \{t_1, t_2, \ldots, t_k\}$ with *species* $\mathcal{L}(\mathcal{P}) =
\bigcup_{i=1}^{k} \mathcal{L}(t_i)$. Two trees in a profile need not have the same species. T is a *supertree*

B. Raphael and J. Tang (Eds.): WABI 2012, LNBI 7534, pp. 93–105, 2012.
© Springer-Verlag Berlin Heidelberg 2012

of $\mathcal{P}$ if $\mathcal{L}(T) = \mathcal{L}(\mathcal{P})$ and T displays every $t_i \in \mathcal{P}$. If $\mathcal{P}$ has a supertree the set $\mathcal{P}$ is *compatible*. Given a profile $\mathcal{P}$, the *unrooted compatibility problem* asks if there exists a supertree T of $\mathcal{P}$. It is NP-Hard to determine if such a tree exists [19].

Bryant and Lagergren showed that this problem is fixed-parameter tractable in k by using a supertree to construct a tree decomposition of width k of a structure they call the *display graph* [3]. Tree decompositions are related to *triangulations* of a graph (e.g. see [2]). An important first step in providing a practical triangulation approach to a problem in phylogenetics requires a graph definition and identification of the triangulations that characterize solutions. Vakati and Fernández-Baca [21] showed that specific triangulations of the display graph characterize the existence of a supertree. Their result is in the same spirit as an earlier result on triangulations and perfect phylogeny [4,19] but those results will not be needed in this paper. Recently, [10,11,20,12] practical triangulation approaches to perfect phylogeny problems have been studied. Those approaches were possible due to the extensive research that has been done on *minimal triangulations* [13].

In this paper, we introduce an *intersection graph* derived from a profile and show that a restricted triangulation of it gives a solution to the compatibility problem. Our restricted triangulation criteria is simpler and we believe our approach more clearly exposes the mathematical relationship between supertrees and compatible profiles than the approach in [21]. It is also simpler for our method to construct supertrees from triangulations and allows the solution to a related compatibility question. In Section 4, we show that solutions to the *maximum compatibility problem* for incompatible profiles are found by weakening our triangulation restriction, analogous to the approach in [11] used to remove characters to find a perfect phylogeny. Finally, our approach naturally relates the compatibility problem to the *chordal sandwich problem* which has an existing and recent literature [6,14].

The *edge-label intersection graph* $L^*(\mathcal{P}) = \text{int}(E(\mathcal{P}) \cup \mathcal{L}(\mathcal{P}))$ (and its restricted triangulations) studied in this paper is distinct from the display graph (and its restricted triangulations) studied in [3,21]. For example, the display graph has *treewidth* k when $\mathcal{P}$ is compatible [3], but $L^*(\mathcal{P})$ does not. Treewidth is well studied (it is the reason unrooted compatibility was found to be fixed parameter tractable [3]) and that literature may provide clues for practical triangulation approaches using the display graph. Alternatively, the triangulation condition in [21] does not seem to lend itself to a restricted triangulation / chordal sandwich problem, as is the case here. In Section 3 we will show a relationship between $L^*(\mathcal{P})$ and the display graph.

All graphs $G = (V, E)$ in this paper are undirected and simple. We will often discuss a graph G and its relationship to a tree T. For clarity, we call $V(G)$ the *vertices* of G and $V(T)$ the *nodes* of T. A pair of vertices $u, v \in V(G)$ are *adjacent in* G if $(u, v) \in E(G)$, and the edge (u, v) is *incident* to u and v. $N_G(v)$, the *neighborhood* of v in G, consists of vertices adjacent to v in G. When $X \subseteq V(G)$ the *induced subgraph* $G[X]$ has vertex set X and $u, v \in X$ are adjacent iff $(u, v) \in E(G)$. When G and H have the same vertex set but

$E(G) \subseteq E(H)$ we write $G \subseteq H$ to express this relationship. X is a *clique* if every pair of nodes in X are adjacent in G, and it is a *maximal clique* if no superset of X is also a clique.

2   A Phylogenetic Tree as a Tree Representation

The purpose of this section is to provide background for the intersection graphs in Section 3. A graph $G = (V, E)$ is *chordal* or *triangulated* if every cycle of length four or more in G has a *chord* (i.e. two non-consecutive vertices in the graph are adjacent). Chordal graphs are characterized by having the following tree representations. A *clique tree* for a (chordal) graph G is a tree $\mathcal{T}$ with the following properties.

1. The nodes of $\mathcal{T}$ are in one-to-one correspondence with the maximal cliques of G. It is convenient to think of the vertices of a maximal clique as being contained in its corresponding node of $\mathcal{T}$, and we will do so frequently.
2. For each vertex v of G, the maximal cliques containing v define a set of nodes that induce a subtree $\mathcal{T}_v$ of $\mathcal{T}$.

Let S be a set and $S_1, S_2, \ldots, S_n$ a collection of subsets of S. The *intersection graph* $\mathrm{int}(S_1, S_2, \ldots, S_n)$ has vertices $S_1, S_2, \ldots, S_n$ and edges (S_i, S_j) iff $S_i \cap S_j \neq \emptyset$. G is a *subtree intersection graph* if it is isomorphic to $\mathrm{int}(T_1, T_2, \ldots, T_n)$ where each T_i is a subtree of a tree T (subtrees intersect if they share at least one node). T and $\{T_1, \ldots, T_n\}$ form a *tree representation* of G.

Theorem 1. *[4,7,22] The following statements are equivalent.*

1. G *is a chordal graph.*
2. G *has a clique tree.*
3. G *is a subtree intersection graph.*

A collection of sets satisfies the *Helly property* if, whenever every pair of sets in the collection have a nonempty intersection, there is an element that is contained in every set of the collection. It is well known that subtrees of a tree satisfy the Helly property [7,9], which may be extended as follows.

Lemma 1. *[5] Let T be a tree and $T_1, \ldots, T_n$ be subtrees of T. Then if every pair of subtrees T_i, T_j share a node, then $\bigcap_{i=1}^{n} T_i$ is a non-empty subtree of T.*

When G is a non-chordal graph, we may add *fill edges* F to obtain a supergraph $G_F = (V(G), E(G) \cup F)$ that is chordal. G_F is called a *triangulation* of G. A *restricted triangulation* of G is a chordal graph in which the fill edges are *restricted fill edges*, i.e. they belong to a restricted subset of pairs of nonadjacent vertices. If T is a phylogenetic tree, the *edge label intersection graph* $L^*(T) = \mathrm{int}(E(T) \cup \mathcal{L}(T))$ is obtained as an intersection graph (here, $a \in \mathcal{L}(T)$ is treated as a set $\{a\}$). For clarity, the vertices of $L^*(T)$ are denoted $\{u, v\}$ when $(u, v) \in E(T)$ and $\{a\}$ when $a \in \mathcal{L}(T)$. $L^*(T)$ is chordal by Theorem 1.

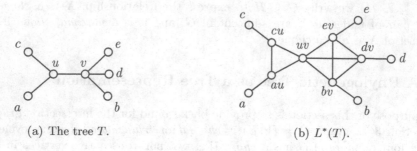

(a) The tree T.   (b) $L^*(T)$.

Fig. 1. A phylogenetic tree and its edge label intersection graph. In the depiction of $L^*(T)$, we use a to mean $\{a\}$ (which is simultaneously a subset of $V(T)$ and a vertex of $L^*(T)$) and au to mean $\{a, u\}$, etc.

Theorem 2. *Let T be a phylogenetic tree. Then T with subtrees $E(T)$ and $\mathcal{L}(T)$ form a tree representation of $L^*(T)$. Further, $L^*(T)$ is a chordal graph.*

We will show that $L^*(T)$ has a unique clique tree that is almost exactly T (clique trees and phylogenetic trees are distinct objects). While some of the results in this section can be derived from results on *line graphs* [15] (the intersection graph defined by the edge set of a graph), we avoid this in order to be as self contained as possible. To discuss $L^*(T)$'s clique tree, we first investigate the maximal cliques of $L^*(T)$. For an internal node v of T, define $K_T(v) = \{\{v, u\} \mid u \in N(v)\}$, and for $a \in \mathcal{L}(T)$ define $K_T(a) = \{\{a\}, \{a, u\}\}$ where u is the unique neighbor of a in T. By intersection, both $K_T(v)$ and $K_T(a)$ are cliques in $L^*(T)$.

Lemma 2. *For a phylogenetic tree T, the maximal cliques of $L^*(T)$ are*

$$\{K_T(v) \mid v \text{ is an internal node of } T\} \cup \{K_T(a) \mid a \in \mathcal{L}(T)\} \ .$$

Proof. Let K be a maximal clique of $L^*(T)$. $L^*(T)$ is connected so K has two or more vertices. Each vertex of K is a subtree of T so $X = \bigcap_{x \in K} x \subseteq V(T)$ is nonempty by Lemma 1. Every vertex of K is an edge or leaf of T, hence it is a set of size at most two and $|X| < 2$. Hence X is a single node, say, v. The clique $K_T(v)$ consists of every vertex of $L^*(T)$ that, when treated as a set, contains v. Therefore $K \subseteq K_T(v)$, and because K is maximal, $K = K_T(v)$. Conversely, suppose $u \in V(T)$ and $K_T(u) \subseteq K'$ is a clique. Then $Y = \bigcap_{y \in K'} y$ is a single vertex by Lemma 1. This vertex must be u because the intersection of any two vertices of $K_T(u)$ is exactly u. Therefore $K = K_T(u)$, so $K_T(u)$ is a maximal clique of $L^*(T)$. □

Lemma 3. *$L^*(T)$ has a unique clique tree $\mathcal{T}$ and $\mathcal{T}$ is graph isomorphic to T.*

Proof. Let G be the intersection graph obtained from the maximal-cliques of $L^*(T)$. We proceed by showing that G is isomorphic to T. There is a one-to-one correspondence between maximal cliques of $L^*(T)$ and nodes of T (Lemma 2). Explicitly, we map $v \in V(T)$ to $K_T(v)$. For each pair $u, v \in V(T)$ the maximal cliques $K_T(u)$ and $K_T(v)$ intersect iff $(u, v) \in E(T)$. Hence $(K_T(u), K_T(v)) \in E(G)$ iff $(u, v) \in E(T)$. Therefore mapping v to $K_T(v)$ induces a graph isomorphism between T and G. The clique trees of $L^*(T)$ are in one-to-one correspondence with the spanning trees of G [1,8]. But G is a tree, so $\mathcal{T} = G$ is the unique clique tree for $L^*(T)$ and $\mathcal{T}$ is isomorphic to T.   □

An important consideration for the following theorem is how to construct a phylogenetic tree from a clique tree (a similar construction will be useful in Section 3). Let $\mathcal{T}$ be the unique clique tree of $L^*(T)$. The $\mathcal{L}$−tree of $\mathcal{T}$ is obtained by labeling $K_T(a)$ with a for each $a \in \mathcal{L}(T)$. The following theorem is an immediate consequence of the proof of Lemma 3.

Theorem 3. *The $\mathcal{L}$−tree of $L^*(T)$ and T are isomorphic as phylogenetic trees.*

3   Compatibility and Restricted Triangulations

In this section, we describe a new formulation for the joint problem of determining if a set of input trees in a phylogenetic profile are compatible and, if they are, constructing a supertree. Our strategy involves creating an intersection graph defined on subtrees of the input phylogenetic profile and then determining if it has a restricted triangulation. Our first intersection graph generalizes the previously introduced *edge-label intersection graph* to a phylogenetic profile $\mathcal{P}$. This formulation has the desirable property that any supertree for a phylogenetic profile will correspond to a tree representation of a restricted triangulation of the intersection graph. Subsequently, we show it is sufficient to consider the problem on a simpler intersection graph and briefly discuss its relationship to the display graph of [3,21].

We begin by extending our definition of an edge-label intersection graph to a phylogenetic profile. The edge-label intersection graph of a profile, denoted $L^*(\mathcal{P})$, is analogous to $L^*(T)$, and is obtained as the intersection graph where the vertices are $E(\mathcal{P}) = \bigcup_{i=1}^{k} E(t_i)$ and $\mathcal{L}(\mathcal{P})$ treated as sets. That is, $L^*(\mathcal{P}) = \text{int}(E(\mathcal{P}) \cup \mathcal{L}(\mathcal{P}))$. In particular, if $b \in \mathcal{L}(t_i), \mathcal{L}(t_j)$ for two trees in the profile and $(b, u) \in E(t_i), (b, u') \in E(t_i)$ are the external edges incident to b then $\{b, u\}$ and $\{b, u'\}$ are adjacent in $L^*(\mathcal{P})$ because their intersection is $\{b\}$ (Fig. 2). Also observe that for $t_i \in \mathcal{P}$, $L^*(t_i)$ is an induced subgraph of $L^*(\mathcal{P})$. Figure 2 shows a profile and its edge-label intersection graph.

A fill edge is *valid* if it is not incident to a vertex $\{a\}$ for any $a \in \mathcal{L}(\mathcal{P})$ and it is not incident to two vertices of $L^*(t_i)$ for a single tree $t_i \in \mathcal{P}$. A *restricted triangulation* of $L^*(\mathcal{P})$ is a chordal graph $L^*(\mathcal{P})_F$ in which the fill edges are all valid. The existence of a restricted triangulation may be posed as a *chordal sandwich problem*. That is, we are seeking a graph $L^*(\mathcal{P})_F$ such that $L^*(\mathcal{P}) \subseteq L^*(\mathcal{P})_F \subseteq L^*(\mathcal{P})_{V(F)}$ where $V(F)$ is the set of valid fill edges. The remainder of this section proves and builds on the following theorem.

Theorem 4. *A profile $\mathcal{P}$ is compatible iff $L^*(\mathcal{P})$ has a restricted triangulation.*

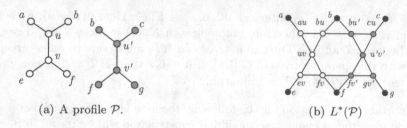

(a) A profile $\mathcal{P}$.   (b) $L^*(\mathcal{P})$

Fig. 2. For illustration purposes, the vertices of $L^*(\mathcal{P})$ have been colored. Valid fill edges of $L^*(\mathcal{P})$ must be incident to vertices of different colors and are never incident to a black vertex. $L(\mathcal{P})$ (see Theorem 7) is obtained by removing the black vertices.

Proof. Immediate from Theorems 5 and 6, stated and proven below.

The proof of Theorem 4 uses a modification of the functions defined in [21] that describe the relationship between a tree in a compatible profile and a supertree that displays it. We will use a definition that is closely related but distinct, so to avoid confusion we refer to the functions that appear in [21] as *VF-embedding functions*. In this paper, an *embedding function* ψ_i embeds t_i in T if it maps the vertices of t_i to subtrees of T with the following properties.

1. **(label preservation)** If v is labeled, then $\psi_i(v)$ is a single node with the same label. Otherwise, no node of $\psi_i(v)$ is labeled.
2. **(disjoint subtrees)** Distinct nodes u, v of t_i have disjoint subtrees $\psi_i(u)$, $\psi_i(v)$.
3. **(adjacency)** If $(v_1, v_2) \in E(t_i)$ there is a unique $(u_1, u_2) \in E(T)$ where u_j is a node of the subtree $\psi_i(v_j)$ for $j = 1, 2$.

Lemma 4. *Profile $\mathcal{P}$ is compatible if and only if there exists a supertree T for $\mathcal{P}$ and embedding functions $\psi_1, \psi_2, \ldots, \psi_k$ where ψ_i embeds t_i in T for $i = 1, \ldots, k$.*

Proof. In [21] it was shown that T is a supertree for $\mathcal{P}$ if and only if there is a VF-embedding function mapping a subgraph of T to nodes of t_i. Due to space considerations we omit the definition of a VF-embedding function and outline a proof here. If ψ_i is an embedding function, then ψ_i^{-1} is a VF-embedding function. The inverse ψ_i' of a VF-embedding function is almost an embedding function. The only issue is that there may be a leaf a where $\psi_i(a)$ is a subtree containing multiple nodes. Moving the nodes of $\psi_i(a)$ not labeled by a to $\psi_i(v)$ where v is the neighbor of a in t_i fixes this problem.   □

Theorem 5. *Let T be a supertree for a profile $\mathcal{P}$. Then there is a restricted triangulation $L^*(\mathcal{P})_F$ of $L^*(\mathcal{P})$ and subtrees of T that form a tree representation for $L^*(\mathcal{P})_F$.*

Proof. Let $\mathcal{P} = \{t_1, \ldots, t_k\}$ be compatible with supertree T. By Lemma 4 we also have a corresponding set of embedding functions $\{\psi_1, \ldots, \psi_k\}$. We will use the embedding functions to define subtrees of T that form a tree representation of $L^*(\mathcal{P})_F$. The subtrees of T for each vertex of $L^*(\mathcal{P})$ are defined as follows.

- For $a \in \mathcal{L}(\mathcal{P})$, define the subtree $T_{\{a\}}$ to be the node of T labeled a. Note that if $a \in \mathcal{L}(t_i)$, then $T_{\{a\}} = \psi_i(a)$ by the label preservation property of embedding functions.
- For $(u, v) \in E(t_i)$ for some $t_i \in \mathcal{P}$, define the subtree $T_{\{u,v\}} = \psi_i(u) \cup \psi_i(v)$.

The adjacency property of embedding functions guarantees that $T_{\{u,v\}}$ will be connected. As an intersection graph of subtrees of a tree, the graph

$$G = \mathrm{int}(\{T_{\{a\}} \mid a \in \mathcal{L}(\mathcal{P})\}, \{T_{\{u,v\}} \mid (u, v) \in E(t_i) \text{ for some } t_i \in \mathcal{P}\})$$

is chordal (Property 3 of Theorem 1). Now we show that G is isomorphic to a supergraph of $L^*(\mathcal{P})$. By construction there is a one-to-one correspondence between vertices of G and vertices of $L^*(\mathcal{P})$. That is, each $x \in V(L^*(\mathcal{P}))$ corresponds uniquely to a subtree $T_x \in V(G)$ of T, and each of the subtrees defines a vertex of G. Let $(x, y) \in E(L^*(\mathcal{P}))$. Each of x and y are either a singleton node or an edge of some $t_i \in \mathcal{P}$ and their intersection is a node v of t_i. Hence $T_x \cap T_y = \psi_i(v)$ so $(T_x, T_y) \in E(G)$. Therefore G is isomorphic to a supergraph of $L^*(\mathcal{P})$. Define the fill F as $(x, y) \notin E(L^*(\mathcal{P}))$ such that $(T_x, T_y) \in E(G)$. Then $L^*(\mathcal{P})_F$ is isomorphic to G and since G is chordal it is a triangulation of $L^*(\mathcal{P})$.

Next we show $L^*(\mathcal{P})_F$ is a restricted triangulation. If $(x, y) \in E(L^*(\mathcal{P})_F)$ then T_x and T_y are intersecting subtrees of T. We will argue by cases that either (x, y) is an edge of $L^*(\mathcal{P})$ or a valid fill edge. First suppose there is a $t_i \in \mathcal{P}$ such that $x, y \in V(L^*(t_i))$. By the disjoint subtrees property of embedding functions there is a node v of t_i such that $T_x \cap T_y = \psi_i(v)$. This only happens if, as sets, x and y contain v, so $(x, y) \in E(L^*(\mathcal{P}))$. So for any $x, y \in V(L^*(t_i))$, $(x, y) \in E(L^*(\mathcal{P})_F)$ if and only if $(x, y) \in E(L^*(\mathcal{P}))$.

The second and only other case is when there is no tree t_i of $\mathcal{P}$ where both $x, y \in V(L^*(t_i))$. For $a \in \mathcal{L}(\mathcal{P})$, the label preserving property of embedding functions ensures that $t_{\{a\}}$ will only intersect trees of the form $t_{\{a,v\}}$ where (a, v) is incident to a in some tree of $\mathcal{P}$. In that case, $(\{a\}, \{a, v\})$ is an edge of $L^*(\mathcal{P})$ and hence not in F. We have shown that if $(x, y) \in F$ then $x, y \neq \{a\}$ for any $a \in \mathcal{L}(\mathcal{P})$ and x, y are edges from different trees of $\mathcal{P}$. Hence (x, y) is a valid fill edge, so $L^*(\mathcal{P})_F$ is a restricted triangulation of $L^*(\mathcal{P})$. □

Now we would like to prove the converse of Theorem 5. Before continuing, we must discuss a construction similar to the $\mathcal{L}$–tree of Section 2. The following lemma will assist our definition of an $\mathcal{L}(\mathcal{P})$-tree. Recall that $L^*(t_i)$ is an induced subgraph of $L^*(\mathcal{P})$, and $K_{t_i}(v)$ is the unique maximal clique of $L^*(t_i)$ whose vertices contain v. For a node v of $t_i \in \mathcal{P}$, the clique $\hat{K}(v)$ of $L^*(\mathcal{P})$ is defined as follows.

- If v is an internal node of t_i, define $\hat{K}(v) = K_{t_i}(v)$.
- If v is a leaf labeled a, define $\hat{K}(a) = \bigcup_{j \mid a \in \mathcal{L}(t_j)} K_{t_j}(a)$.

Note that $\hat{K}(v)$ are precisely the vertices of $L^*(\mathcal{P})$ that contain the node v.

Lemma 5. *Let $L^*(\mathcal{P})_F$ be a restricted triangulation of $L^*(\mathcal{P})$ and $a \in \mathcal{L}(\mathcal{P})$. Then $\hat{K}(a)$ is a maximal clique of $L^*(\mathcal{P})_F$ and $\hat{K}(a)$ is the only maximal clique of $L^*(\mathcal{P})_F$ containing $\{a\}$.*

Proof. By definition $\hat{K}(a)$ has every vertex of $L^*(\mathcal{P})$ that, when viewed as a set, contains a. Hence $\hat{K}(a)$ is a clique through intersection and is a maximal clique of $L^*(\mathcal{P})$. No fill edge of $L^*(\mathcal{P})_F$ is incident to $\{a\}$, so $\hat{K}(a)$ is a maximal clique of $L^*(\mathcal{P})_F$. If K is a maximal clique of $L^*(\mathcal{P})_F$ and it contains $\{a\}$, then every vertex of K is either $\{a\}$ or intersects $\{a\}$ and is therefore a vertex of $\hat{K}(a)$. By maximality, K must be $\hat{K}(a)$, completing the proof. □

Let $L^*(\mathcal{P})_F$ be a restricted triangulation of $L^*(\mathcal{P})$ and $\mathcal{T}$ a clique tree of $L^*(\mathcal{P})_F$. The $\mathcal{L}(\mathcal{P})$-tree of $\mathcal{T}$ is a phylogenetic tree constructed as follows. For each $a \in \mathcal{L}(\mathcal{P})$, $\hat{K}(a)$ is a node of $\mathcal{T}$. If $\hat{K}(a)$ is a leaf, label this node a. Otherwise add a leaf labeled a to this node. After labeling the tree for every species in $\mathcal{L}(\mathcal{P})$, suppress any internal nodes of degree two (i.e. remove the degree two node, its incident edges, and add an edge between its two adjacent nodes). This process produces a unique phylogenetic tree. The following technical lemma will be of use when we prove the converse of Theorem 5.

Lemma 6. *Suppose $L^*(\mathcal{P})_F$ is a restricted triangulation of $L^*(\mathcal{P})$ and u, v are distinct nodes from the same tree $t_i \in \mathcal{P}$. Then no clique of $L^*(\mathcal{P})_F$ contains two or more vertices from both $\hat{K}(u)$ and $\hat{K}(v)$.*

Proof. For the sake of contradiction, assume K is a clique of $L^*(\mathcal{P})_F$ and both $\hat{K}(u)$ and $\hat{K}(v)$ are subsets of K. Every vertex of $L^*(\mathcal{P})$ is a set of size at most two, so $\hat{K}(u) \cap \hat{K}(v) = \{\{u, v\}\}$ or it is empty. K is a clique, so every pair $x \in \hat{K}(u)$ and $y \in \hat{K}(v)$ nonadjacent in $L^*(\mathcal{P})$ defines a fill edge $(x, y) \in F$. However, $\hat{K}(u)$ and $\hat{K}(v)$ have at least two vertices from $L^*(t_i)$ and so there must be an $x' \in \hat{K}(u) - \{\{u, v\}\}$ and $y' \in \hat{K}(v) - \{\{u, v\}\}$ both of which are vertices of $L^*(t_i)$ that are nonadjacent in $L^*(\mathcal{P})$. (x', y') is an invalid fill edge, contradicting the validity of $L^*(\mathcal{P})_F$'s fill edges, so K cannot exist. □

Theorem 6. *If $\mathcal{T}$ is a clique tree of a restricted triangulation $L^*(\mathcal{P})_F$ of $L^*(\mathcal{P})$, then its $\mathcal{L}(\mathcal{P})$-tree is a supertree for $\mathcal{P}$.*

Proof. Let $L^*(\mathcal{P})_F$ be a restricted triangulation with clique tree $\mathcal{T}$, and suppose T is the labeled tree obtained by labeling $\mathcal{T}$ and possibly adding labeled leaves to $\mathcal{T}$ while constructing the $\mathcal{L}(\mathcal{P})$-tree of $\mathcal{T}$. $\mathcal{T}$ is a subgraph of T, a fact that we will use freely. First we define an embedding function that embeds t_i in T and then we will suppress nodes of degree two. We define the embedding function in two phases (ψ_i^1 and ψ_i^2). Define ψ_i^1 as follows.

Leaves: For a leaf a of t_i, $\psi_i^1(a)$ is the leaf of T labeled a.
Internal Nodes: For an internal node v of t_i, $\psi_i^1(v)$ are the internal nodes K of $\mathcal{T}$ such that K and $\hat{K}(v)$ share two or more vertices.

Initially, define $\psi_i^2 = \psi_i^1$. If $(u, v) \in E(t_i)$ and $\psi_i^1(u)$ and $\psi_i^1(v)$ fail the adjacency property, we add to ψ_i^2 as follows. Let $P(u, v) = \{w_1, w_2, \ldots, w_r\}$ be the unique shortest path with $w_1 \in \psi_i^1(u)$ and $w_r \in \psi_i^1(v)$. Assume without loss of generality that v is an internal node. We will continually refer to the internal path $\overline{P}(u, v) =$

$\{w_2, \ldots, w_{r-1}\}$ of $P(u,v)$. Observe that $\overline{P}(u,v)$ is a path in $\mathcal{T}$. Add $\overline{P}(u,v)$ to $\psi_i^2(v)$ (because v is internal). Note that if v is an internal node, it must be that $\psi_i^2(v) \subseteq \mathcal{T}$ because $\psi_i^1(v) \subseteq \mathcal{T}$ by definition and $\overline{P}(u,v) \subseteq \mathcal{T}$ for any neighbor u of v. Next we show that ψ_i^2 embeds t_i in T.

(**Label Preservation**). If $a \in \mathcal{L}(t_i)$ then $\psi_i^2(a) = \psi_i^1(a)$ is the single node of T labeled a. Now let v be an internal node of t_i. Recall that $\psi_i^2(v)$ contains only nodes from $\mathcal{T}$; if K is a node of $\psi_i^2(v)$ that is labeled b, then $\hat{K}(b) = K$ (by definition of the $\mathcal{L}(\mathcal{P})$-tree) and K and $\hat{K}(v)$ share two or more vertices (by definition of $\psi_i^1(v)$). This violates Lemma 6, so every node of $\psi_i^2(v)$ is an internal node of T and therefore not labeled.

(**Mapping to Subtrees**). Now we show that $\psi_i^2(v)$ is a subtree (i.e. is connected). As we have seen, if $a \in \mathcal{L}(t_i)$ then $\psi_i^2(a)$ is a single node and is therefore connected. In order to see that $\psi_i^2(v)$ is connected for an internal node, we first show that $\psi_i^1(v)$ is a subtree. Define $X = \hat{K}(v) \cap K$ for some maximal clique / node K of $\psi_i^1(v)$. Note that X is a clique with two or more vertices. $\mathcal{T}_X$ and $\mathcal{T}_{\hat{K}(v)}$ are non-empty subtrees of $\mathcal{T}$ by Lemma 1, and further, $\mathcal{T}_{\hat{K}(v)} \subseteq \mathcal{T}_X$ because $X \subseteq \hat{K}(v)$. By definition, $\psi_i^1(v)$ is the union of $\mathcal{T}_X$ for all such cliques X. Since $\psi_i^1(v)$ is a union of subtrees that all contain $\mathcal{T}_{\hat{K}(v)}$, $\psi_1^i(v)$ is connected. Last, adding adjacent paths to $\psi_i^1(v)$ to create $\psi_i^2(v)$ preserves connectivity. Therefore ψ_i^2 maps nodes of t_i to subtrees of T.

(**Disjoint Subtrees**). To show disjointness, first assume for the sake of contradiction that $\psi_i^1(u)$ and $\psi_i^1(v)$ intersect. If they intersect, neither can be leaves of t_i due to the label preservation property. Let $K \in \psi_i^1(u) \cap \psi_i^1(v)$. By definition of ψ_i^1 for internal nodes, K shares two or more vertices with both $\hat{K}(u)$ and $\hat{K}(v)$, contradicting Lemma 6. Therefore ψ_i^1 has the disjoint subtrees property. Before discussing ψ_i^2, we prove the following.

(Observation) Let $(u,v) \in E(t_i)$ and $K \in \overline{P}(u,v)$. Then $K \cap V(L^*(t_i)) = \{\{u,v\}\}$.

First we show that $\overline{P}(u,v) \subseteq T_{\{u,v\}}$ by case analysis. Suppose u and v are internal nodes. $\psi_i^1(u)$ intersects $T_{\{u,v\}}$ because there is a maximal clique of $L^*(\mathcal{P})_F$ that is a superset of $\hat{K}(u)$ which is a node of both $\psi_i^1(u)$ and $T_{\{u,v\}}$. Similarly, $\psi_i^2(u)$ and $T_{\{u,v\}}$ intersect. If $\overline{P}(u,v) \nsubseteq T_{\{u,v\}}$ then $\overline{P}(u,v)$ and $T_{\{u,v\}}$ form distinct paths between two disjoint subtrees of a tree (i.e. $\psi_i^1(u)$ and $\psi_i^1(v)$) which is impossible. Hence if u and v are internal then $\overline{P}(u,v) \subseteq T_{\{u,v\}}$. Otherwise, without loss of generality, u is a leaf of t_i labeled by some $a \in \mathcal{L}(t_i)$. Recall $K(a)$ can be either a leaf or an internal node of $\mathcal{T}$. In both cases, $\overline{P}(u,v)$ either contains $K(a)$ or is adjacent to $K(a)$ (by $\mathcal{L}(\mathcal{P})$-tree construction). Now, $K(a)$ is a node of $T_{\{a,v\}}$, and as before it must be that $\overline{P}(u,v) \subseteq T_{\{u,v\}}$.

Now we finish proving the observation. We have shown that $\overline{P}(u,v) \subseteq T_{\{u,v\}}$ so it must be that $\{u,v\} \in K \cap V(L^*(t_i))$. If $K \cap V(L^*(t_i))$ contains another vertex, say $\{u',v'\}$, either without loss of generality $u = u'$ and K is a node of $\psi_i^1(u)$, contradicting the definition of $\overline{P}(u,v)$, or $\{u,v\} \cap \{u',v'\} = \emptyset$ and $(\{u,v\},\{u',v'\})$ is an invalid fill edge, contradicting the assumption that $L^*(\mathcal{P})_F$ is a restricted triangulation. This completes the proof of the observation.

To finish showing disjointness, it suffices to show that for any distinct (u, v), (u', v') in $E(t_i)$ and $w \in V(t_i)$ we have the following.

- $\overline{P}(u, v)$ and $\overline{P}(u', v')$ are disjoint.
- $\overline{P}(u, v)$ and $\psi_i^1(w)$ are disjoint.

Through our observation, if $K \in \overline{P}(u, v) \cap \overline{P}(u', v'))$ then $K \cap V(L^*(t_i))$ is exactly the single vertex $\{u, v\} = \{u', v'\}$. Finally, if $K \in \overline{P}(u, v)$ it contains a single vertex from $L^*(t_i)$ and K is not a node of $\psi_i^1(w)$ by definition of ψ_i^1. Therefore ψ_i^2 satisfies the disjoint subtrees property.

(**Adjacency**). The construction of ψ_i^2 ensures that if $(u, v) \in E(t_i)$ then $\psi_i^2(u)$ and $\psi_i^2(v)$ satisfy the adjacency property of embedding functions. Therefore ψ_i^2 embeds t_i in T.

Now let ψ_i be the map obtained from ψ_i^2 and T when nodes of degree two are suppressed to obtain the $\mathcal{L}(\mathcal{P})$-tree of $\mathcal{T}$. All of the properties of embedding functions hold, but we must consider if an internal node v of t_i maps solely to a node of degree two in T. This cannot occur because v has three or more neighbors, and by the adjacency property of ψ_i^2 we require $\psi_i^2(v)$ to map to more than just this node. Hence ψ_i embeds t_i in the $\mathcal{L}(\mathcal{P})$-tree of $\mathcal{T}$, and by Lemma 4, the theorem is proved.   □

We may simplify $L^*(\mathcal{P})$ and the triangulation condition to obtain a result analogous to Theorem 4. The graph $L(\mathcal{P}) = \text{int}(E(\mathcal{P}))$ is obtained from $L^*(\mathcal{P})$ by removing vertices of the form $\{a\}$ for a species $a \in \mathcal{L}(\mathcal{P})$. A *restricted triangulation* of $L(\mathcal{P})$ is obtained using the same fill edges for restricted triangulations of $L^*(\mathcal{P})$. The interpretation of valid fill edges for $L(\mathcal{P})$ is simpler than that of $L^*(\mathcal{P})$ because we only require that a fill edge be adjacent to vertices from different trees. Readers familiar with the display graph defined in [3,21] will observe that $L(\mathcal{P})$ is the line graph of the display graph of $\mathcal{P}$ (justifying the notation). Finally, we note that without the vertices corresponding to the species $\mathcal{L}(\mathcal{P})$ the clique tree of a restricted triangulation $L(\mathcal{P})_F$ will not correspond to a unique supertree T as is the case for restricted triangulations of $L^*(\mathcal{P})$. Due to space considerations, we omit the details of T's construction.

Theorem 7. *A profile $\mathcal{P}$ is compatible iff $L(\mathcal{P})$ has a restricted triangulation.*

Proof. If the input trees in $\mathcal{P}$ are compatible, then there is a restricted triangulation $L^*(\mathcal{P})_F$ of $L^*(\mathcal{P})$ by Theorem 4. The inheritance property of chordal graphs guarantees that an induced subgraph of a chordal graph is chordal [17], because any chordless cycle in the induced subgraph must also be present in the original graph. $L(\mathcal{P})_F$ is an induced subgraph of $L^*(\mathcal{P})_F$, so it is chordal. Because valid fill edges are the same for both graphs, $L(\mathcal{P})_F$ is a restricted triangulation of $L(\mathcal{P})$.

To show the reverse direction we use the property that $\hat{K}(a)$ is a clique containing the neighborhood of $\{a\}$ in $L^*(\mathcal{P})$. Let $L(\mathcal{P})_F$ be a restricted triangulation of $L(\mathcal{P})$. We will show that the supergraph $L^*(\mathcal{P})_F$, obtained by adding F to $L^*(\mathcal{P})$, is a restricted triangulation of $L^*(\mathcal{P})$. Any cycle C on four or more vertices in $L^*(\mathcal{P})_F$ not having $\{a\} \in C$ for any $a \in \mathcal{L}(\mathcal{P})$ is also a cycle of $L(\mathcal{P})_F$ and therefore has a chord. The only other possibility is that $\{a\} \in C$ for some $a \in \mathcal{L}(\mathcal{P})$ in which case the vertices occurring before and after $\{a\}$ in C are adjacent to $\{a\}$ in $L^*(\mathcal{P})$. Hence these vertices are contained in $\hat{K}(a)$ and are adjacent so they form a chord of C. Therefore every cycle on four or more vertices in $L^*(\mathcal{P})_F$ has a chord and $L^*(\mathcal{P})_F$ is chordal. Finally, the set of valid fill edges for $L(\mathcal{P})$ and $L^*(\mathcal{P})$ are the same, so $L^*(\mathcal{P})_F$ is a restricted triangulation of $L^*(\mathcal{P})$ and $\mathcal{P}$ is compatible by Theorem 4.   □

It is often useful to frame results in terms of minimal triangulations due to its volume of study [13]. A triangulation G_F of G is a *minimal triangulation* if $G_{F'}$ is not chordal for any $F' \subset F$. Removing valid fill edges preserves restricted triangulations, so if H is a restricted triangulation of $L(\mathcal{P})$ there is a minimal triangulation $H' \subseteq H$ that is also restricted.

Theorem 8. *$\mathcal{P}$ is compatible if and only if $L(\mathcal{P})$ and $L^*(\mathcal{P})$ have restricted minimal triangulations.*

4   Maximum Compatibility

In this section, we consider when the profile is not compatible, a situation which frequently arises in practice with real data sets. The *maximum compatibility problem* asks for the largest set of unrooted trees that are compatible. Analogously, it finds a supertree that displays the largest number of trees in a profile. A *subprofile* $\mathcal{P}^* \subseteq \mathcal{P}$ is an *optimal solution* when it is the largest subset of $\mathcal{P}$ that is compatible. We will prove that unrestricted triangulations of $L(\mathcal{P})$ are related to finding maximum compatible subprofiles. A fill edge f of $L(\mathcal{P})$ is an *invalid fill edge* if it is incident to vertices corresponding to edges from the same tree t_i and we say that f *invalidates* t_i. The *valid trees* $\mathrm{vt}(H)$ of a triangulation H are the trees of $\mathcal{P}$ that are not invalidated by any fill edge of H.

Lemma 7. *Let $\mathcal{P}$ be a profile and let $\mathcal{P}' \subseteq \mathcal{P}$ be a compatible set of unrooted trees. Then there is a triangulation H of $L(\mathcal{P})$ such that $\mathcal{P}' \subseteq \mathrm{vt}(H)$.*

Proof. If $\mathcal{P}'$ is compatible, then by Theorem 7 there is a restricted triangulation H' of $L(\mathcal{P}')$. Construct a supergraph H of $L(\mathcal{P})$ beginning with $H = L(\mathcal{P})$ and adding edges as follows. Add the restricted fill edges of H' to H. Now for each nonadjacent $u \in V(L(\mathcal{P})) - V(L(\mathcal{P}'))$ and $v \in V(L(\mathcal{P}))$ add the fill edge (u, v) to H. An invalid fill edge of H can only invalidate trees in $\mathcal{P} - \mathcal{P}'$ so $\mathcal{P}' \subseteq \mathrm{vt}(H)$. To see that H is chordal, consider any cycle $C = x_1, x_2, \ldots, x_k$ of length four or more in H. We proceed by cases to show that C always has a chord. If $x_i \in V(L(\mathcal{P}'))$ for $i = 1, \ldots, k$ then C is a cycle of H' and therefore has a chord. Otherwise there is some $x_j \in V(L(\mathcal{P})) - V(L(\mathcal{P}'))$. Let x_l be any vertex of C that is not x_j and does not occur directly before or after x_j in the cycle (recall C has four or more vertices). By the construction of H, x_l and x_j are adjacent so they form a chord in C. Therefore H is chordal.   □

Lemma 8. *If H is a triangulation of $L(\mathcal{P})$ then $\mathrm{vt}(H) \subseteq \mathcal{P}$ is compatible.*

Proof. Let H be a triangulation of $L(\mathcal{P})$, $\mathcal{P}' = \mathrm{vt}(H)$, and $X = V(L(\mathcal{P}'))$. We will show that $H[X]$ is a restricted triangulation of $L(\mathcal{P}')$. Chordality is inherited [17], so $H[X]$ is chordal. To see that $H[X]$ is a triangulation of $L(\mathcal{P}')$, next we prove that each edge of $L(\mathcal{P}')$ is an edge of $H[X]$. Now, $L(\mathcal{P}')$ is an induced subgraph of $L(\mathcal{P}) \subseteq H$ so an edge e of $L(\mathcal{P}')$ is an edge of H. The vertices incident to e are both in X, so e is an edge of $H[X]$. Hence $H[X]$ is a triangulation of $L(\mathcal{P}')$. This triangulation is restricted, because if not, there is an invalid fill edge incident to two vertices in X from a tree in $\mathcal{P}'$, contradicting the assumption that $\mathcal{P}' = \mathrm{vt}(H)$. Hence $H[X]$ is a restricted triangulation of $L(\mathcal{P}')$ and by Theorem 7 $\mathrm{vt}(H)$ is compatible. □

We now prove the main result of this section, which implies that it is enough to consider the minimal triangulations of $L(\mathcal{P})$ to solve maximum compatibility.

Theorem 9. *Suppose $\mathcal{P}$ is not compatible. Then $\mathcal{P}^* \subseteq \mathcal{P}$ is an optimal solution to the maximum compatibility problem iff there is a minimal triangulation H of $L(\mathcal{P})$ such that $\mathrm{vt}(H) = \mathcal{P}^*$ and for any other minimal triangulation H' of $L(\mathcal{P})$, $|\mathrm{vt}(H')| \leq |\mathrm{vt}(H)|$.*

Proof. ($\Rightarrow$) : Let $\mathcal{P}^*$ be an optimal solution to the maximum compatibility problem. From Lemma 7 there is a triangulation H^* of $L(\mathcal{P})$ such that every invalid fill edge invalidates only trees from $\mathcal{P} - \mathcal{P}^*$. Let $H \subseteq H^*$ be a minimal triangulation of $L(\mathcal{P})$. We have only removed edges to obtain H, so $\mathcal{P}^* \subseteq \mathrm{vt}(H)$. $\mathrm{vt}(H)$ is compatible by Lemma 8, so $\mathcal{P}^* = \mathrm{vt}(H)$ because $\mathcal{P}^*$ is an optimal solution. If H' is a minimal triangulation satisfying $|\mathcal{P}^*| < |\mathrm{vt}(H')|$ then $\mathcal{P}^*$ would not be optimal, and therefore $|\mathrm{vt}(H')| \leq |\mathrm{vt}(H)|$.

($\Leftarrow$) : Suppose $\mathcal{P}^* \subseteq \mathcal{P}$ satisfies the converse with minimal triangulation H but $\mathcal{P}^*$ is not an optimal solution. Then there is a $\mathcal{P}' \subseteq \mathcal{P}$ with more trees than $\mathcal{P}^*$ that is compatible. By Lemma 7 there is a triangulation of $L(\mathcal{P})_F$ of $L(\mathcal{P})$ that does not invalidate any tree of $\mathcal{P}'$. Remove edges from this graph until we obtain a minimal triangulation H'. Since removing edges can only increase the number of valid trees we have $|\mathrm{vt}(H)| = |\mathcal{P}^*| < |\mathcal{P}'| \leq |\mathrm{vt}(H')|$. This contradicts the assumption that $|\mathrm{vt}(H')| \leq |\mathrm{vt}(H)|$, hence $\mathcal{P}^*$ must be optimal. □

5   Discussion and Further Work

Theorems in [10] concerning minimal separators and legal triangulations of the partition intersection graph can be generalized and applied to restricted triangulations of $L(\mathcal{P})$. These methods have been effective for perfect phylogeny problems, and this algorithmic approach should be studied for $L(\mathcal{P})$. Theorem 7 suggests that the intersection graphs of this paper and the display graph of a profile are related. It would be interesting to investigate if and how legal triangulations of the display graph correspond to restricted triangulations of $L(\mathcal{P})$.

Acknowledgments. This research was partially supported by NSF grants CCF-0515378, IIS-0803564, and CCF-1017580 and NIH R01-HG002942.

References

1. Bernstein, P.A., Goodman, N.: Power of natural semijoins. SIAM Journal on Computing 10, 751–771 (1981)
2. Bodlaender, H.L.: Discovering Treewidth. In: Vojtáš, P., Bieliková, M., Charron-Bost, B., Sýkora, O. (eds.) SOFSEM 2005. LNCS, vol. 3381, pp. 1–16. Springer, Heidelberg (2005)
3. Bryant, D., Lagergren, J.: Compatibility of unrooted phylogenetic trees is fpt. Theoretical Computer Science 351(3), 296–302 (2006)
4. Buneman, P.: A characterization of rigid circuit graphs. Discrete Mathematics 9, 205–212 (1974)
5. Chandrasekaran, R., Tamir, A.: Polynomially bounded algorithms for locating p-centers on a tree. Mathematical Programming 22(3), 304–315 (1982)
6. Fomin, F.V., Villanger, Y.: Subexponential parameterized algorithm for minimum fill-in. In: SODA 2012 Proceedings, pp. 1737–1746 (2012)
7. Gavril, F.: The intersection graphs of subtrees in trees are exactly the chordal graphs. Journal of Combinatorial Theory 16(1), 47–56 (1974)
8. Gavril, F.: Generating the maximum spanning trees of a weighted graph. Journal of Algorithms 8, 592–597 (1987)
9. Golumbic, M.C.: Algorithmic Graph Theory and Perfect Graphs. Annals of Discrete Mathematics, vol. 57. Elsevier, Amsterdam (2004)
10. Gusfield, D.: The multi-state perfect phylogeny problem with missing and removable data. Journal of Computational Biology, 383–399 (2010)
11. Gysel, R., Gusfield, D.: Extensions and improvements to the chordal graph approach to the multistate perfect phylogeny problem. IEEE/ACM Transactions on Computational Biology and Bioinformatics 8(4), 912–917 (2011)
12. Gysel, R., Lam, F., Gusfield, D.: Constructing perfect phylogenies and proper triangulations for three-state characters. In: Przytycka and Sagot [10], pp. 104–115
13. Heggernes, P.: Minimal triangulation of graphs: a survey. Discrete Mathematics 306(3), 297–317 (2006)
14. Heggernes, P., Mancini, F., Nederlof, J., Villanger, Y.: A Parameterized Algorithm for CHORDAL SANDWICH. In: Calamoneri, T., Diaz, J. (eds.) CIAC 2010. LNCS, vol. 6078, pp. 120–130. Springer, Heidelberg (2010)
15. Hemminger, R.L., Beineke, L.W.: Line graphs and line digraphs. Academic Press Inc. (1978)
16. Przytycka, T.M., Sagot, M.-F. (eds.): WABI 2011. LNCS, vol. 6833. Springer, Heidelberg (2011)
17. Rose, D.J.: Triangulated graphs and the elimination process. Journal of Mathematical Analysis and Applications 32(3), 597–609 (1970)
18. Semple, C., Steel, M.: Phylogenetics. Oxford Lecture Series in Mathematics and Its Applications. Oxford University Press, Oxford (2003)
19. Steel, M.: The complexity of reconstructing trees from qualitative characters and subtrees. Journal of Classification 9(1), 91–116 (1992)
20. Stevens, K., Kirkpatrick, B.: Efficiently solvable perfect phylogeny problems on binary and k-state data with missing values. In: Przytycka and Sagot [16], pp. 282–297
21. Vakati, S., Fernández-Baca, D.: Graph triangulations and the compatibility of unrooted phylogenetic trees. Applied Mathematics Letters 24(5), 719–723 (2011)
22. Walter, J.R.: Representations of chordal graphs as subtrees of a tree. Journal of Graph Theory 2, 265–267 (1978)

An Optimal Reconciliation Algorithm
for Gene Trees with Polytomies

Manuel Lafond[1], Krister M. Swenson[2], and Nadia El-Mabrouk[3]

[1] DIRO, Université de Montréal, H3C 3J7, Canada
lafonman@iro.umontreal.ca
[2] DIRO
swensonk@iro.umontreal.ca
[3] DIRO
mabrouk@iro.umontreal.ca

Abstract. Reconciliation is a method widely used to infer the evolutionary relationship between the members of a gene family. It consists of comparing a gene tree with a species tree, and interpreting the incongruence between the two trees as evidence of duplication and loss. In the case of binary rooted trees, linear-time algorithms have been developed for the duplication, loss, and mutation (duplication + loss) costs. However, a strict prerequisite to reconciliation is to have a gene tree free from error, as few misplaced edges may lead to a completely different result in terms of the number and position of inferred duplications and losses. How should the weak edges be handled? One reasonable answer is to transform the binary gene tree into a non-binary tree by removing each weak edge and collapsing its two incident vertices into one. The created polytomies are "apparent" as they do not reflect a true simultaneous divergence of many copies from a common ancestor, but rather a lack of resolution. In this paper, we consider the problem of reconciling a non-binary rooted gene tree G with a binary rooted species tree S, were polytomies of G are assumed to be apparent. We give a linear-time algorithm that infers a reconciliation of minimum mutation cost between a binary refinement of a polytomy and S, improving on the best known result, which is cubic. This implies a straightforward generalization to a gene tree G with nodes of arbitrary degree, that runs in time $O(|S||G|)$, which is shown to be an optimal algorithm.

1   Introduction

The evolutionary history of a gene family is determined by a combination of microevolutionary events at the sequence level, and macroevolutionary events (duplications, losses, horizontal gene transfer) affecting the number and distribution of genes among genomes [11]. While sequence similarity can be considered as a footprint of microevolution and used to construct a gene tree G for the gene family, macroevolution is harder to predict as it is not explicitly reflected by the gene tree. Having a clear picture of the speciation, duplication and loss mechanisms that have shaped a gene family is however crucial to the study of

B. Raphael and J. Tang (Eds.): WABI 2012, LNBI 7534, pp. 106–122, 2012.

gene function. Indeed, following a duplication, the most common occurrence is for only one of the two gene copies to maintain the parental function, while the other becomes non-functional (pseudogenization) or acquires a new function (neofunctionalization) [19].

Reconciliation, first introduced by Goodman in 1979 [13] — and since widely studied and implemented in comparative genomics software [12] — is a method that compares the gene tree G with a phylogeny S of the considered species (species tree), and interprets the incongruence between the two trees as evidence describing evolution of the gene family through duplication and loss. A reconciliation $R(G, S)$ is a tree obtained from G by inserting "lost" branches so that the obtained tree is in agreement with the phylogeny S. As there can be several reconciliations for a given tree pair, a natural approach is then to select one, or a subset, that optimize some probabilistic [1,2] or combinatorial [16] criterion such as the number of duplications (duplication cost), losses (loss cost) or both combined (mutation cost). Reconciliation of binary rooted trees is a well-studied problem, and linear-time algorithms based on the so called lowest common ancestor (LCA) mapping have been developed for the duplication, loss and mutation costs [6,20,22]. Generalizations of reconciliation accounting for horizontal gene tranfers have also been considered [9]. In particular, minimizing the number of duplications, losses and transfers has been shown to be computationally hard [18], but feasible in polynomial time if the input species tree is dated [10].

The fundamental hypothesis behind reconciliation is that the gene tree reflects the true phylogeny of the gene family. Therefore, a strict prerequisite is to have both gene tree and species tree free from error [8,15]. Unfortunately gene trees are not always well-supported, and frequently many equally-supported trees are obtained as the output of a phylogenetic method. Typically bootstrap values are used as a measure of confidence in each edge of a phylogeny. How should the weak edges of a gene tree be handled? One strategy adopted in [6] is to explore the space of gene trees obtained from the original tree G by performing Nearest Neighbor Interchanges around weakly-supported edges. Another reasonable answer is to transform the binary gene tree into a non-binary tree by removing each weak edge and collapsing its two incident vertices into one. A polytomy (node with more than two children) in a gene tree is called *true* (or *hard*) if it reflects a true simultaneous divergence of its children from a common ancestor, and it is called *apparent* (or *soft*) otherwise [17]. Implicitly, polytomies of a gene tree obtained by the method of collapsing short or poorly supported internal branches are apparent polytomies, reflecting a lack of resolution.

In this paper, we consider the problem of reconciling a non-binary rooted gene tree G with a binary rooted species tree S, where polytomies of G are assumed to be apparent. More precisely, we seek out a reconciliation of minimum mutation cost between a binary refinement of G and S. Chang and Eulenstein were the first to consider this problem [3]. They showed that each polytomy P can be treated independently in $O(|S| \times |P|^2)$ time, implying an $O(|S| \times |G|^2)$ algorithm for the entire tree. In a recent paper [21], a linear-time algorithm is developed for reconciling a non-binary gene tree G with a binary species tree S, but for the duplication cost. The output is a reconciliation with optimal loss cost over all the reconciliations with the optimal duplication cost, which does not necessarily minimize the mutation cost. Here, we describe an algorithm that infers the minimum mutation cost reconciliation between P and S in $O(max(|P|, |S|))$ time, implying an $O(|G| \times |S|)$ algorithm over the entire gene tree. This algorithm is optimal, since there exists a family of instances leading to a most parsimonious reconciliation of size $\Omega(|G| \times |S|)$.

2   Preliminary Notation

In this paper, all the trees are considered rooted (we ommit to mention it each time). Given a tree T, we denote by T_x the subtree of T rooted at x, and by $L(T_x)$ (or simply $L(x)$ if unambiguous) the set of leaves of T_x. We also denote by $root(T)$ the root of T, by $V(T)$ the set of nodes of T and by $|T|$ the number of nodes $|V(T)|$ of T. The *degree* of an internal node x in a tree T is the number of children of x. If T is binary, we denote by x_l and x_r the two children of x in T.

A *phylogeny* over a set L is a tree with internal nodes of degree 2 or more, uniquely leaf-labeled by L. A *polytomy* (or star tree) over a set of L is a phylogeny with a single internal node, which is of degree $|L|$, adjacent to each leaf of L. For example, the tree G in Figure 2 is a polytomy.

A *species tree* S is a phylogeny over a set of species Σ, which represents the evolutionary relationship between these species. Similarly, we can consider the evolutionary relationship between a family of genes Γ, that appear in the genomes of Σ: a *gene tree* G for Γ is a phylogeny accompanied by a function $g : \Gamma \to \Sigma$ indicating the species where each gene is found. See Figure 1 for an example. Given a gene tree G, we denote by $\mathcal{S}(G_x)$ the subset of Σ corresponding to $L(G_x)$ (i.e. $\mathcal{S}(G_x) = \{g(l) \mid l \in L(G_x)\}$).

In this paper, we assume a binary species tree S and a non-binary gene tree G. As stated in the introduction, the polytomies of G are considered apparent (i.e. reflecting non resolved parts of the tree). The goal is then to find a "binary refinement" of G. For any internal node x of G with children $\{x_1, x_2, \ldots, x_n\}$, any rooted binary tree on the set of leaves $\{G_{x_1}, G_{x_2}, \ldots, G_{x_n}\}$ is a *refinement* of the polytomy G_x. The following definition generalizes this fact.

Definition 1 (binary refinement). *A binary refinement $B(G)$ of a gene tree G is defined as follows.*

- *If r is a leaf then $B(G) = G$;*
- *Otherwise, $B(G)$ is a rooted binary tree on the set $\{B(G_1), B(G_2) \ldots, B(G_n)\}$, where G_i is the tree rooted at the ith child of $root(G)$ (for some ordering of the children), and $B(G_i)$ is a binary refinement of G_i.*

2.1   Histories and Reconciliation

We study the evolution of a family of genes Γ taken from genomes Σ through duplication and loss. Conceptually, a *duplication/loss/speciation history* (or simply *history*) is a tree H reflecting the evolution from a single ancestral gene to

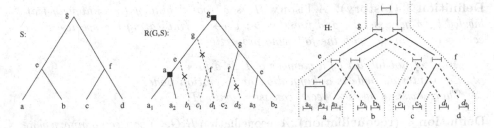

Fig. 1. S is a species tree over $\Sigma = \{a, b, c, d\}$; $R(G, S)$ is a reconciliation between S and the gene tree G represented by plain lines. Here $\Gamma = \{a_1, a_2, a_3, b_2, c_2\}$, and for each $x_i \in \Gamma$, $g(x_i) = x$. Internal nodes of $R(G, S)$ are labeled according to the LCA mapping. Artificial genes $\{b_1, c_1, d_1, d_2\}$ are added to illustrate lost branches. Duplication nodes are indicated by bold squares, and loss leaves are represented by crosses. This reconciliation has cost 5: 2 duplications and 3 losses; H illustrates the history that has led to the gene family Γ. H is the same tree as $R(G, H)$, but represented differently (embedded in the species tree).

a set of genes through duplication, loss, and speciation events. Given a binary gene tree G for the gene family and a species tree S for Σ, a reconciliation is a history obtained from G, in "agreement" with the phylogeny S. In this section we formally define history and reconciliation, as well as presenting tools for working with them. All these concepts are illustrated in Figure 1.

The most popular method for finding a parsimonious reconciliation is based on the "LCA mapping". The *LCA mapping* between G and S, denoted by $\mu()$, maps every node x of G to the *lowest common ancestor* of $S(G_x)$ in S, which is the common ancestor of $S(G_x)$ in S that is farthest from the root. We call $\mu(x)$ the *label* of x. A node x of G is considered a *duplication* with respect to S if and only if $\mu(x_\ell) = \mu(x)$ and/or $\mu(x_r) = \mu(x)$. Any node of G that is not a duplication node, is a *speciation* node.

Take a binary tree T, labeled by the LCA mapping, where there exists exactly one leaf labeled by each gene in Γ, and a function $g : \Gamma \mapsto \Sigma$ indicating the species where each gene is found. A *duplication-free restriction* $D(T)$ of a tree T is obtained by removing either T_{x_ℓ} or T_{x_r} for each duplication node x, along with x, and if x is not the root, joining the parent of x and the remaining child by a new edge. Each duplication-free restriction $D(T)$ can be considered to be a copy of a species tree S, in which case each loss leaf u corresponds subtree S_u of the species tree that is missing in $D(T)$.

A duplication-free restriction $D(T)$ *agrees* with a species tree S iff relabeling each leaf l of $D(T)$ by $g(l)$, and replacing each loss leaf u in $D(T)$ with the subtree S_u, results in a tree isomorphic to S.

Definition 2 (consistent). *Take a species tree S and a rooted binary tree T where there exists exactly one leaf labeled by each gene in Γ, and all other leaves are labeled as losses. T is said to be* consistent *with S iff every duplication-free restriction of T agrees with S.*

Definition 3 (history). *A* history *H* *is a rooted binary tree uniquely leaf-labeled by a gene set Γ, and function $g : \Gamma \mapsto \Sigma$ (indicating the species where each gene is found) with the following properties:*

1. *Any leaf not labeled by a member of Γ is a loss.*
2. *Each internal node is a duplication or speciation node.*
3. *There exists a species tree S consistent with H.*

Definition 4 (reconciliation). *A* reconciliation *$R(G, S)$ between a binary gene tree G and a species tree S is a history that can be obtained from G by inserting loss leaves and labeling internal nodes as speciations or duplications so that it is consistent with S.*

The parsimony criteria used to choose among the large set of possible reconciliations are the number of duplications (*duplication cost*), the number of losses (*loss cost*) or both combined (*mutation cost*). The LCA mapping induces a reconciliation $R(G, S)$ between G and S, where an internal node x of G leads to a duplication node in R if and only if x is a duplication node of G with respect to S. Moreover, $R(G, S)$ is a reconciliation that minimizes the duplication, loss, and mutation costs [5,14].

In the rest of this paper, the *cost* of a reconciliation refers to its mutation cost.

2.2   Problem Statement

Given a binary species tree S and a non-binary gene tree G, we seek out a full resolution of G leading to a reconciliation of minimum mutation cost. We formally define the notion of a resolution of G as being a reconciled refinement of G.

Definition 5 (Resolution). *A tree $R(G, S)$ is a* resolution *of G with respect to S if and only if $R(G, S)$ is a reconciliation between a binary refinement $B(G)$ of G, and S.*

We are now ready to state our optimization problem.

Minimum Resolution:
Input: A binary species tree S and a non-binary gene tree G.
Output: A *Minimum Resolution* of G with respect to S (or simply a *Minimum Resolution of G* if there is no ambiguity on S), e.g. a resolution of G with respect to S of minimum mutation cost.

We first show that each polytomy of G can be resolved independently.

Theorem 1. *Let $\{G_{x_i}$, for $1 \leq i \leq p\}$ be the set of subtrees of G rooted at the p children $\{x_i$, for $1 \leq i \leq p\}$ of the root of G. Let $R_{min}(G_{x_i}, S)$ be a minimum resolution of G_{x_i} w.r.t. S. Let G' be the tree obtained from G by replacing each G_{x_i} by $R_{min}(G_{x_i}, S)$. Then a minimum resolution of G' is a minimum resolution of G.*

Proof. This statement was proved by Chang and Eulenstein in [4], which led them to a dynamic programming algorithm with running-time complexity $O(|S| \times |G|^2)$, for the MINIMUM RESOLUTION problem. The reader interested in this proof might refer to Chang's MSc. Thesis written in 2006.

It follows from Theorem 1 that a minimum resolution of G can be obtained by a depth-first procedure that solves each polytomy G_x iteratively, for each internal node x of G. At each step, whether the children of the polytomy G_x are internal nodes or leaves of G, they are treated as leaves of the polytomy and we refer to each leaf l by its label $\mu(l)$.

In the next section, we consider G as a polytomy whose leaves are labeled (not uniquely) by nodes of S. Furthermore, as the subtrees S_x of S such that $V(S_x) \setminus \{x\}$ has an empty intersection with $\mathcal{S}(G)$, will never be considered in the resolution of G, we can ignore them. We say that S is a *species tree linked to the polytomy* G if and only if any internal node of S has a descendant included in $\mathcal{S}(G)$ and the root of S is the lowest common ancestor of $\mathcal{S}(G)$. For example, in Figure 2, S is a species tree linked to the polytomy G.

3   Method

In this section, we consider G to be a polytomy whose leaves are labeled (not uniquely) by nodes of a species tree S. Notice that a leaf labeled x actually represents a whole subtree of the considered gene tree, which has already been resolved, and thus is consistent with $S_{\mu(x)}$. We assume that S is a species tree linked to G. We describe an approach for computing a minimum resolution $R(G, S)$ of G based on the observation that for any node x in $R(G, S)$, all nodes on a path from x to a leaf in $R(G, S)$ will map to a node that is on a path from $\mu(x)$ to a leaf of S. Thus, we decompose the computation of a minimum resolution of G according to a depth-first traversal of the nodes of S; for each node s of S we consider the cost of having k maximal subtrees of $R(G, S)$ whose roots map to s. For example, Figures 2c and 2d represent two such partial resolutions where there are three maximal subtrees whose roots map to e. Given, for all k, the minimum cost of a so-called "k-partial resolution" corresponding to a node s, we show how to compute the cost of a partial resolution corresponding to the parent of s. Clearly a solution of the Minimum Resolution problem is a minimum 1-partial resolution of G at the root of S.

3.1   Partial Resolutions

Let s be a node of S. The restriction of G by node s, denoted $G_{/s}$, is the tree obtained from G by removing the set of leaves $\mathcal{L}_s$ whose labels are not in S_s.

Definition 6 (partial resolution). *Let s be a node of S. A partial resolution $P(G, S, s)$ of G at s is a polytomy on a set $\mathcal{F}_s \cup \mathcal{L}_s$, where $\mathcal{F}_s$ is obtained from a resolution $R(G_{/s}, S)$ as follows: $\mathcal{F}_s$ is a forest of subtrees of $G_{/s}$, rooted at nodes labeled s, partitioning the set of leaves of $G_{/s}$ (i.e. each leaf of $G_{/s}$ is in a unique tree of the forest).*

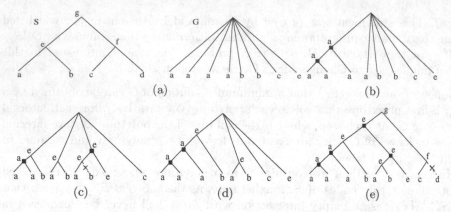

Fig. 2. (a) A species tree S and a polytomy G; (b) A 2-partial resolution of G at a of cost 2 (2 duplications); (c) A 3-partial resolution of G at e of cost 3 (2 duplications and one loss); (d) A 3-partial resolution of G at e of cost 2 (2 duplications); (e) A full resolution of G with minimum cost (4 duplications, 1 loss).

The cost of a partial resolution $P(G, S, s)$ on a set $\mathcal{F}_s \cup \mathcal{L}_s$ is the sum of the cost of all the trees (reconciliations) of $\mathcal{F}_s$. See Figure 2 for an example.

Definition 7 (k-partial resolution). *Let s be a node of S. A k-partial resolution $P^k(G, S, s)$ of G at s is a partial resolution of G at s on a set $\mathcal{F}_s \cup \mathcal{L}_s$ with exactly k trees in $\mathcal{F}_s$.*

For example, the tree G in Figure 2 is itself a 4-partial resolution of G at a, whereas the tree (b) is a 2-partial resolution of G at a, and (c) and (d) are two different 3-partial resolutions of G at e.

Notation 1. *For any integer $k \geq 1$, we denote by $M_{s,k}$ the minimum cost of a k-partial resolution of G at s. We also denote by M_s the vector $(M_{s,k})_{k \geq 1}$.*

A solution for the Minimum Resolution problem is a resolution of G with cost $M_{root(S),1}$. In this section, we describe an algorithm that computes $M_{root(S),1}$ based on the costs $M_{s,k}$ of all partial resolutions of G over all k and s. Before giving a recursive formulation of $M_{s,k}$, we need to introduce a subset of k-partial resolutions, leading to an intermediate cost $C_{s,k}$ for internal nodes s, which can be computed directly from k-partial resolutions corresponding to the children of s.

Definition 8 (k-speciation resolution). *Let s be an internal node of S. A k-partial resolution $P^k(G, S, s)$ of G at node s is a k-speciation resolution of G at s if and only if each node of $P^k(G, S, s)$ labeled s is a speciation or a leaf.*

A k-speciation resolution at s contains no duplication node nor loss leaf labeled s. In Figure 2, the tree G is a 4-speciation resolution of G at node a, while the tree (d) is a 3-speciation resolution at e. Neither the 2-partial resolution of G

at a (b) nor the 3-partial resolutions of G at e (c) is a speciation resolution, as in the first case the left-most child of the root labeled a is a duplication node, while in the second case the right-most child of the root labeled e is a duplication node.

Notation 2. *For any node s of S, we denote by $nb(s)$ the number of leaves of G labeled s.*

Note that there is no k-speciation resolution for $k \leq nb(s)$. Indeed, as G has $nb(s)$ leaves labeled s, any speciation resolution of G has at least $nb(s)$ speciation nodes labeles s, and thus $k \geq nb(s)$. Moreover, as S is a species tree linked to G, at least one descendant of the internal node s of S should be a leaf of G, and thus any partial resolution at s needs to have at least one additional speciation node labeled s.

Notation 3. *For any internal node s of S and any integer $k > nb(s)$, we denote by $C_{s,k}$ the cost of a minimum k-speciation resolution of G at s. For technical reasons, we set $C_{s,k} = \infty$ for $1 \leq k \leq nb(s)$. We denote by C_s the vector $(C_{s,k})_{k \geq 1}$.*

3.2 A Recursive Formulation

There is an infinite range of values of k for which $M_{s,k}$ and $C_{s,k}$ correspond to valid resolutions. However, the following remark is easy to validate and implies that, for some input, we need only consider a fixed-size table of values.

Remark 1. There exists a value $n \in \mathbb{N}$ such that $M_{s,k} < M_{s,n}$, for any node s of S and any integer $0 < k < n$.

The intuition behind this remark is that when k is large enough a k-partial resolution would contain too many losses, so could never be part of an optimal solution.

The following lemma exhibits a relationship between two entries of vector M_s.

Lemma 1. *For any node s of S and any integers $k, i \geq 1$, we have $M_{s,k} \leq M_{s,i} + |k - i|$.*

Proof. Let $P^i(G, S, s)$ be an i-partial resolution at s of cost $M_{s,i}$. If $i < k$, inserting $k - i$ losses of s at the root of $P^i(G, S, s)$ gives us a k-partial resolution of cost $M_{s,i} + k - i$. If $i > k$, joining $i - k + 1$ subtrees rooted at s in $P^i(G, S, s)$ (by duplication nodes) gives us a k-partial resolution of cost $M_{s,i} + i - k$. Both cases imply that $M_{s,k} \leq M_{s,i} + |k - i|$. □

We use Lemma 1 to prove the main recurrence defining $C_{s,k}$ and $M_{s,k}$.

Theorem 2. *Let s be a node of S and $1 \leq k \leq n$.*

1. If s is a leaf of S, then $M_{s,k} = |k - nb(s)|$;

2. *Otherwise, let s_l and s_r be the two children of s in S. Then,*
   (a) $C_{s,k} = M_{s_\ell,k-nb(s)} + M_{s_r,k-nb(s)}$ *if $k > nb(s)$ (∞ otherwise);*

   (b) $M_{s,k} = \min \left(C_{s,k}, \min\limits_{1 \leq i \leq n} \left(M_{s,i} + |k - i| \right) \right)$

Proof. Let $P^k(G, S, s)$ be a minimum k-partial resolution of G at node s of cost $M_{s,k}$. Suppose $P^k(G, S, s)$ is defined on the set of nodes $\mathcal{F}_s \cup \mathcal{L}_s$.

1. Suppose s is a leaf of S. If $k = nb(s)$, then each tree in $\mathcal{F}_s$ is a single node labeled s, with reconciliation cost 0, and thus $M_{s,k} = 0 = |k - nb(s)|$. If $k < nb(s)$ (respectively $k > nb(s)$), then at least $nb(s) - k$ duplication nodes (respec. $k - nb(s)$ losses) should be present in the trees of $\mathcal{F}_s$. As the trees of $\mathcal{F}_s$ are part of an optimal k-partial resolution, the number of duplications (losses) should be exactly $nb(s) - k$ $(k - nb(s))$, and thus $M_{s,k} = |k - nb(s)|$.

2. Otherwise, s is an internal node of S.
2(a) Let $P^k(G, S, s)$ be a k-speciation resolution of G at node s of minimum cost $C_{s,k}$. Since none of the trees in $\mathcal{F}_g$ are rooted at a duplication node nor labeled as a loss, there must be exactly $k - nb(s)$ trees in $\mathcal{F}_s$ that are rooted at speciation nodes labeled by s. Any such node must have one child labeled s_ℓ and one child labeled s_r. Since we are going through each node of S in a depth-first manner, we assume that the values of M_{s_ℓ} and M_{s_r} have been computed. The result follows from the fact that $M_{s_\ell,k-nb(s)}$ (resp. $M_{s_r,k-nb(s)}$) gives the optimal configuration yielding $k - nb(s)$ trees rooted at nodes labeled s_ℓ (resp. s_r).
2(b) If $P^k(G, S, s)$ is a k-speciation resolution of G at s, then clearly $M_{s,k} = C_{s,k}$. Otherwise, let k' be the number of trees rooted at duplication nodes of $\mathcal{F}_s$. For positive k', if each of the two children of those duplication nodes were taken as the roots of two new trees, then we would have a forest of $i' = k + k'$ trees. This gives us $M_{s,k} \geq M_{s,i'} + k' = M_{s,i'} + |k - i'|$. If $k' = 0$ (and $P^k(G, S, s)$ is not a k-speciation resolution), then $\mathcal{F}_s$ must have, say k'', trees corresponding to losses. Consider the $i'' = k - k''$ trees of $\mathcal{F}_s$ that are not losses. This gives us $M_{s,k} \geq M_{s,i''} + k'' = M_{s,i''} + |k - i''|$ as well. Therefore $M_{s,k} \geq \min_i(M_{s,i} + |k - i|)$. On the other hand, Lemma 1 gives us $M_{s,k} \leq \min_i(M_{s,i} + |k - i|)$. □

3.3   A Dynamic Programming Approach

The recurrence 2.b in Theorem 2 induces a circular argument for computing the entries of M_s, as for two different constants k and i, $M_{s,k}$ may be computed from $M_{s,i}$, which in turn may be computed from $M_{s,k}$. In other words, the recurrences of Theorem 2 cannot be used directly in a dynamic programming algorithm for the computation of $C_{s,k}$ and $M_{s,k}$. The rest of this section focuses on reformulating recurrence 2.b. We start by giving two important properties relating M_s to C_s.

Lemma 2. *For an internal node s of S, there exists at least one k such that $M_{s,k} = C_{s,k}$.*

Proof. Let $P^k(G, S, s)$ be a k-partial resolution of G at node s of cost $M_{s,k}$, defined on the set $\mathcal{F}_s \cup \mathcal{L}_g$. Assume that $M_{s,k} \neq C_{s,k}$ for all k. This implies $M_{s,k} < C_{s,k}$ for all k. Consider the subforest $\mathcal{F}'_s$ consisting of the j maximal subtrees of $\mathcal{F}_s$ that are rooted at speciation nodes; $\mathcal{F}'_s$ defines a j-speciation resolution with cost $C' < M_{s,k}$. Since $C_{s,j} \leq C'$, we have $C_{s,j} < M_{s,k}$. But $M_{s,j} < C_{s,j}$, so in general, for any k there exists a j such that $M_{s,j} < M_{s,k}$, a contradiction since the minimum value in M_s must occur in within a finite range (by Remark 1). $\qquad\square$

Theorem 2 shows that for an internal node s of S, $M_{s,k}$ can be computed from $C_{s,k}$, or from some other value in M_s. However, we have not characterized how to easily discern which case will be used, and we have no information about which i gives $M_{s,i} = C_{s,i}$. The following lemma addresses this matter by narrowing the possibilities.

Lemma 3. *For some internal node s of S and integer $k \geq 1$, if $M_{s,k} \neq C_{s,k}$ then there exists an i such that $M_{s,i} = C_{s,i}$ and $M_{s,k} = M_{s,i} + |k - i|$.*

Proof. By the recurrence 2(b) of Theorem 2, if $M_{s,k} \neq C_{s,k}$, then we should have $M_{s,k} = M_{s,i} + |k - i|$ for some i. If $M_{s,i} = C_{s,i}$, then the lemma is verified. Otherwise, $M_{s,i} = M_{s,h} + |i - h|$ for some h. This gives $M_{s,k} = M_{s,h} + |k - i| + |i - h| \geq M_{s,h} + |k - h|$. By lemma 2 we know that there must be some value α for which $M_{s,k_\alpha} = C_{s,k_\alpha}$, so in general, for some integers k_0 and α we have

$$M_{s,k_0} = M_{s,k_\alpha} + |k_\alpha - k_{\alpha-1}| + |k_{\alpha-1} - k_{\alpha-2}| + \cdots + |k_2 - k_1| + |k_1 - k_0|$$

$$= M_{s,k_\alpha} + \sum_{i=1}^{\alpha} |k_i - k_{i-1}|$$

$$\geq M_{s,k_\alpha} + |k_\alpha - k_0| = C_{g,k_\alpha} + |k_\alpha - k_0|.$$

Lemma 1 gives the complementary bound $M_{s,k_0} \leq C_{s,k_\alpha} + |k_\alpha - k_0|$, so equality holds. $\qquad\square$

Lemma 3 allows us to rewrite the recurrence 2(b) of Theorem 2 as follows:

$$M_{s,k} = \min_{nb(s) < i \leq n} C_{s,i} + |k - i| \qquad\qquad (Eq.1)$$

With this new formulation of recurrence 2(b), Theorem 2 leads to a cubic-time dynamic programming algorithm for the computation of the cost of a solution of the Minimum Resolution problem. Indeed, let the height of a node s of S be the maximum number of nodes in a path from s to a leaf of S. Consider an ordering $s_1, s_2 \cdots s_p$ of the nodes of S by increasing height, where $p = |S|$. In other words, leaves are listed before nodes of height 1, etc. In particular $s_p = root(S)$. Consider the tables M and C of $|S|$ lines, where each line i of M and C corresponds respectively to the vectors M_{s_i} and C_{s_i}. The table C is defined only for lines $i > L(S)$. We first compute the $L(S)$ first lines of M in $O(n)$ steps using recurrence (1) of Theorem 2. Then, for each line i, we successively compute C_{s_i} and M_{s_i} for increasing values of i, by using the recurrences (2) a. and b. of Theorem 2. Each line representing C_{s_i} is computed in time $O(n)$, while each line representing M_{s_i} is computed in $O(n^2)$ steps, leading to an $O(n^2|S|)$ algorithm for filling the two tables. The final result (cost of a solution of the Minimum Resolution problem) is just $M_{s_p,1}$. An example is given in Figure 3.

3.4   A Linear-Time Approach

We show in this section that the recurrence (2)b of Theorem 2 can further be simplified in a way leading to a constant time update for each M_s vector according to the M_{s_l} and M_{s_r} vectors. This implies a linear time algorithm for the computation of $M_{root(S),k}$.

We first show that $M_{s,k} = C_{s,k}$ when $C_{s,k}$ is the minimum value among the entries of C_s.

Lemma 4. *For k such that $C_{s,k} = \min\limits_{nb(s)<i\leq n} C_{s,i}$, we have $M_{s,k} = C_{s,k}$.*

Proof. If $M_{s,k} \neq C_{s,k}$, then by Lemma 3, there must exist an i such that $M_{s,k} = M_{s,i} + |k - i|$ and $M_{s,i} = C_{s,i}$. This implies that $C_{s,i} < C_{s,k}$, contradicting the minimality of $C_{s,k}$.   □

The key observation allowing a constant-time computation of any entry in C_s and M_s is that these vectors can be seen as two functions with a cup shape. The following definition formally introduces the notion of a "cup function".

Definition 9 (cup function). *A cup function is a convex piecewise linear function $m()$ which, for a minimum value $\gamma_m \in \mathbb{Z}$ and two breakpoints $m_1, m_2 \in \mathbb{N}$, is strictly decreasing linearly for $x < m_1$, equal to γ_m when $m_1 \leq x \leq m_2$, and strictly increasing linearly when $x > m_2$. It can be written as*

$$m(x) = \begin{cases} \gamma_m + m_1 - x + \mathcal{P}(x) & \text{if } x < m_1 \\ \gamma_m & \text{if } m_1 \leq x \leq m_2 \\ \gamma_m + x - m_2 + \mathcal{Q}(x) & \text{if } x > m_2 \end{cases}$$

where $\mathcal{P} : \mathbb{N} \to \mathbb{Z}$ is non-increasing and $\mathcal{Q} : \mathbb{N} \to \mathbb{Z}$ is non-decreasing.

We say the function $m(x)$ is a simple cup function iff $\mathcal{P}(x) = \mathcal{Q}(x) = 0$ for all x. Roughly speaking, a simple cup function viewed from left to right has a slope of -1, a plateau of minimum values, and a slope of 1.

Assume for now that M_s can be associated with a simple cup function $m()$ such that $M_{s,k} = m(k)$ for $1 \leq k \leq n$. Recall that the values of C_s are obtained by adding the values of M_{s_l} and M_{s_r} (recurrence 2(a) of Theorem 2). We show that C_s can be associated with a cup function by proving that the addition of two simple cup functions yields a cup function.

Lemma 5. *If $\ell()$ and $r()$ are two simple cup functions, then the function $m()$ defined by $m(k) = \ell(k) + r(k)$, is a cup function.*

Furthermore, if ℓ_1, ℓ_2 and r_1, r_2 respectively denote the breakpoints of $\ell()$ and $r()$, and γ_ℓ, γ_r respectively denote the minimum values of $\ell()$ and $r()$, then the breakpoints m_1, m_2 and minimum value γ_m of $m()$ are computed according to the following table:

Condition	γ_m	m_1	m_2
If $\ell_1 < r_1, \ell_2 < r_1$	$\gamma_\ell + \gamma_r + r_1 - \ell_2$	ℓ_2	r_1
If $\ell_1 < r_1, r_1 \leq \ell_2 \leq r_2$	$\gamma_\ell + \gamma_r$	r_1	ℓ_2
If $\ell_1 < r_1, \ell_2 > r_2$	$\gamma_\ell + \gamma_r$	r_1	r_2
If $r_1 \leq \ell_1 \leq r_2, r_1 \leq \ell_1 \leq r_2$	$\gamma_\ell + \gamma_r$	ℓ_1	ℓ_2
If $r_1 \leq \ell_1 \leq r_2, \ell_2 > r_2$	$\gamma_\ell + \gamma_r$	ℓ_2	r_2
If $\ell_1 > r_2, \ell_2 > r_2$	$\gamma_\ell + \gamma_r + \ell_1 - r_2$	r_2	ℓ_1

Proof. The complete proof of this lemma is given in Appendix A. Moreover, a more general version of this lemma has already been proven by Csűrös in [7], where it is shown that the sum of an arbitrary number of cup functions yields a cup function   □

The following theorem states that M_s and C_s can be associated with two cup functions with the same breakpoints m_1, m_2 and minimum value γ_m. Moreover, M_s is associated with a simple cup function. For example, each line of the table of Figure 3 can be rewritten as a cup function, with the breakpoint and minimum values indicated in vectors m_1, m_2 and γ_m.

Theorem 3. *For any node s of S, there exists a simple cup function $m()$ with breakpoints $m_1, m_2 \gtrsim 1$ and minimum value γ_m, such that $M_{s,k} - m(k)$ for $1 \leq k \leq n$.*

Furthermore, if s is an internal node of S, there exists a cup function $c()$ with the same breakpoints $m_1, m_2 > nb(s)$ and same minimum value γ_m, such that $C_{s,k} = c(k)$ for $nb(s) < k \leq n$.

Proof. We prove the theorem by induction over the nodes visited in a postorder traversal of S.

Base Case: If s is a leaf and $nb(s) > 0$, let $m(k) = |k - nb(s)|$. It is clear that $m(k)$ is a simple cup function with breakpoints $m_1 = m_2 = nb(s)$ and $\gamma_m = 0$. If $nb(s) = 0$, let $m(k)$ be a simple cup function with breakpoints $m_1 = m_2 = 1$ and minimum value $\gamma_m = 1$. We have $m(k) = |k - nb(s)|$ for $k \geq 1$. Then from Theorem 2.1, both cases give $M_{s,k} = m(k)$ for $1 \leq k \leq n$.

Induction Step: If s is an internal node, then from the inductive hypothesis, there exist two simple cup functions $\ell(), r()$ such that $M_{s_\ell,k} = \ell(k)$ and $M_{s_r,k} = r(k)$ for $1 \leq k \leq n$. Let $f(k) = \ell(k - nb(s)) + r(k - nb(s))$. By Lemma 5, $f()$ is a cup function. Let f_1, f_2 be the breakpoints of $f()$ and γ_f its minimum value. From part *(2a)* of Theorem 2 we have $C_{s,k} = M_{s_\ell,k-nb(s)} + M_{s_r,k-nb(s)}$, implying that $C_{s,k} = f(k)$ for $nb(s) < k \leq n$. However, it is possible that $f_1 \leq nb(s)$ or $f_2 \leq nb(s)$, making the theorem statement invalid.

Let $c()$ be another cup function with breakpoints $m_1 = \max(f_1, nb(s) + 1)$, $m_2 = \max(f_2, nb(s) + 1)$ and minimum value $\gamma_m = f(m_2) = \min_{nb(s) < k \leq n} C_{s,k}$. It can be verified that $c(k) = f(k)$ when $nb(s) < k \leq n$ and thus, $C_{s,k} = c(k)$ for $nb(s) < k \leq n$. From this, we have

$$
C_{g,k} = \begin{cases}
\infty & \text{if } k \leq nb(s) \\
\gamma_m + m_1 - k + \mathcal{P}(k) & \text{if } nb(s) < k < m_1 \\
\gamma_m & \text{if } m_1 \leq k \leq m_2 \\
\gamma_m + k - m_2 + \mathcal{Q}(k) & \text{if } k > m_2
\end{cases}
$$

for $1 \leq k \leq n$.

By Lemma 4, we know that $M_{s,k} = C_{s,k} = \gamma_m$ for $m_1 \leq k \leq m_2$.

If $k < m_1$, we show that $M_{s,k} = \gamma_m + m_1 - k$. From the equation (Eq.1), we have $M_{s,k} = \min_j(C_{s,j} + |k - j|)$. If $j > m_1$, then by the definition of $C_{s,j}$ given above, we have $C_{s,j} \geq \gamma_m$ and thus $C_{s,j} + j - k \geq \gamma_m + m_1 - k$. If $j < m_1$, then $C_{s,j} + |k - j| \geq \gamma_m + m_1 - j + |k - j| \geq \gamma_m + m_1 - k$ since we can reformulate this inequality as $|k - j| \geq j - k$. Therefore, $C_{s,j} + |k - j|$ has the minimum value when $j = m_1$ and it follows that $M_{s,k} = C_{s,m_1} + m_1 - k = \gamma_m + m_1 - k$ when $k < m_1$.

If $k > m_2$, we show that $M_{s,k} = \gamma_m + k - m_2$. From the (Eq.1), we have $M_{s,k} = \min_j(C_{s,j} + |k - j|)$. If $j < m_2$, we have $C_{s,j} \geq \gamma_m$ and thus $C_{s,j} + k - j \geq \gamma_m + k - m_2$. If $j > m_2$, then $C_{s,j} + |k - j| \geq \gamma_m + j - m_2 + |k - j| \geq \gamma_m + k - m_2$ since we can reformulate this inequality as $|k - j| \geq k - j$. Therefore, $C_{s,j} + |k - j|$ is minimum when $j = m_2$ and it follows that $M_{s,k} = C_{s,m_2} + k - m_2 = \gamma_m + k - m_2$ when $k > m_2$.

The three cases verify that M_s can be associated with a cup function. □

Denote by $m_{1,s}, m_{2,s}$ and γ_s the breakpoints and minimum value of the cup function associated with M_s (and thus from Theorem 3 also with C_s). Theorem 2.(1) allows us to compute $m_{1,s}$, $m_{2,s}$ and γ_s for any leaf s of S. Finally, Theorem 2.(2a) and Lemma 5 allow us to compute, in constant time, the breakpoints $m_{1,s}$, $m_{2,s}$ and minimum value γ_s associated with C_s and M_s, given those associated with M_{s_l} and M_{s_r}.

Stated differently, let $s_1, s_2, \cdots s_p$ be the ordering of the nodes of S defined at the end of Section 3.3, and consider the three vectors $m_1 = (m_{1,s_i})_{1 \leq i \leq s}$, $m_2 = (m_{2,s_i})_{1 \leq i \leq s}$ and $\gamma = (\gamma_{s_i})_{1 \leq i \leq s}$. Then each entry of each of these vectors can be computed in constant-time. Theorem 3 ensures that these vectors allow us to completely define the simple cup function M_s. This leads to an $O(max(|G|, |S|))$ algorithm for computing any value $M_{s,k}$, and in particular the cost $M_{s_p,1}$ of a solution of the minimum Resolution problem.

In Algorithm CupValues(s), we detail the steps required to compute $m_{1,s}, m_{2,s}$ and γ_s for a given node s. Lines 1 to 6 follow from the Theorem 3's base case proof. Line 9 follows from Theorem 2.(2a) and Lemma 5 (discussion above). Line 10 is a correction needed because the table in Lemma 5 gives the result for the addition of two simple cup functions for the same value of k, e.g. $c(k) = \ell(k) + r(k)$. Since $C_{s,k} = M_{s_\ell, k - nb(s)} + M_{s_r, k - nb(s)}$, we need to shift the obtained breakpoints by adding $nb(s)$ to them. Lines 11 and 12 ensure that $m_{1,s}, m_{2,s} > nb(s)$ and that γ_s is the minimum entry in C_s, as stated in Theorem 3.

Algorithm 1. $CupValues(s)$

1: **if** s is a leaf **then**
2:     **if** $nb(s) > 0$ **then**
3:         $m_{1,s} := nb(s); m_{2,s} := nb(s); \gamma_s := 0;$
4:     **else**
5:         $m_{1,s} := 1; m_{2,s} := 1; \gamma_s := 1;$
6:     **end if**
7: **else**
8:     Let s_ℓ, s_r be the two children of s:
9:     Compute $m_{1,s}, m_{2,s}$ and γ_s using   the   children   values
        $m_{1,s_\ell}, m_{2,s_\ell}, m_{1,s_r}, m_{2,s_r}, \gamma_\ell, \gamma_r$ and the table given in Lemma 5;
10:    $m_{1,s} := m_{1,s} + nb(s); m_{2,s} := m_{2,s} + nb(s);$
11:    If $m_{1,s} \leq nb(s)$ then $m_{1,s} := nb(s) + 1;$
12:    If $m_{2,s} \leq nb(s)$ then $m_{2,s} := nb(s) + 1$ and $\gamma_s := M_{s_\ell,1} + M_{s_r,1};$
13: **end if**

	1	2	3	4	5	6	m_1	m_2	γ_m
M_a	3	2	1	0	1	2	4	4	0
M_b	1	0	1	2	3	4	2	2	0
M_c	0	1	2	3	4	5	1	1	0
M_d	1	2	3	4	5	6	1	1	1
C_e	∞	4	2	2	2	4	3	5	2
M_e	4	3	2	2	2	3	3	5	2
C_f	1	3	5	7	9	11	1	1	1
M_f	1	2	3	4	5	6	1	1	1
C_g	5	5	5	6	7	9	1	3	5
M_g	5	5	5	6	7	8	1	3	5

Fig. 3. An illustration of the algorithms for the gene tree G and species tree S of Figure 2(a). The cost of a most parsimonious resolution of G is $M_{g,1} = 5$. The gray cells are those considered by Algorithm DupLoss for computing the Dup and $Loss$ vectors, the values in bold being the first ones evaluated by the algorithm on a given row. The obtained values are $Dup(e) = 2, Dup(a) = 2, Loss(d) = 1$ and $Dup(s) = Loss(s) = 0$ for any other node s. The corresponding resolution is given in Figure 2(e).

3.5   Constructing an Optimal Resolution

Starting with $s = root(S)$ and $k = 1$, we recursively compute the number of losses and duplications required for each node s of S in an optimal reconciliation, based on partial resolutions at s_l and s_r, for s_l and s_r being the children of s. The algorithm presented in this section is based on the following result, which is a corollary of Theorem 3.

Corollary 1 *Let s be a node of S with children s_ℓ and s_r, and $P^k(G, S, s)$ be a minimum k-partial resolution at s defined on the set $\mathcal{F}_g \cup \mathcal{L}_g$ for $1 \leq k \leq n$.*

1. *If $M_{s,k} = M_{s_\ell, k-nb(s)} + M_{s_r, k-nb(s)}$, then the k roots of $\mathcal{F}_s$ are all speciation nodes. Otherwise,*
2. *either $M_{s,k} = \gamma_s + k - m_{2,s}$, in which case $P^k(G, S, s)$ has $k - m_{2,s}$ loss leaves labeled s,*
3. *or $M_{s,k} = \gamma_s + m_{1,s} - k$, in which case $P^k(G, S, s)$ has $m_{1,s} - k$ duplications labeled s.*

Proof. In the first case, $M_{s,k} = C_{s,k}$, indicating that the optimal k-partial resolution at s is a k-speciation resolution. case, taking the $m_{2,s}$-speciation resolution at s of cost γ_s and adding $k - m_{2,s}$ loss leaves labeled s yields a k-partial resolution at s with minimum score $\gamma_s + k - m_{2,s}$. In the third case, taking the $m_{1,s}$-speciation resolution at s of cost γ_s and creating $m_{1,s} - k$ duplications labeled s yields a k-partial resolution at s with minimum score $\gamma_s + m_{1,s} - k$. $\square$

Algorithm DupLoss(s, k) computes the values $Dup(s)$ and $Loss(s)$, being respectively the number of duplications labeled s and the number of loss leaves labeled s in a minimum resolution of G. Starting with a call to Algorithm DupLoss$(root(S), 1)$, the output is a pair $(Dup(s), Loss(s))$ for each node s of S. From these values, it is easy to reconstruct the corresponding solution $R(G, S)$ of the Minimum Resolution problem.

Algorithm 2. *DupLoss(s, k)*

```
if s is a leaf and k ≥ nb(s) then
    Dup(s) := 0; Loss(s) := k − nb(s);
else if s is a leaf and k < nb(s) then
    Dup(s) := nb(s) − k; Loss(s) := 0;
else if k − nb(s) > 0 and M_{s,k} = M_{s_ℓ,k−nb(s)} + M_{s_r,k−nb(s)} then
    Dup(s) := 0; Loss(s) := 0;
    DupLoss(s_ℓ, k − nb(s)); DupLoss(s_r, k − nb(s));
else if k < m_{1,s} then
    Dup(s) := m_{1,s} − k; Loss(s) := 0;
    DupLoss(s_ℓ, m_{1,s} − nb(s)); DupLoss(s_r, m_{1,s} − nb(s));
else if k > m_{2,s} then
    Dup(s) := 0; Loss(s) := k − m_{2,s};
    DupLoss(s_ℓ, m_{2,s} − nb(s)); DupLoss(s_r, m_{2,s} − nb(s));
end if
```

Once the vectors Dup and $Loss$ have been computed, an optimal resolution of G can be constructed easily, knowing $nb(g)$ for each node, and knowing how many of these nodes are joined under duplication or speciation and how many are inserted as losses.

4 Discussion

We have developed an algorithm for constructing the most parsimonious reconciliation, in term of number of duplications + losses, of a polytomy G with a binary species tree S, running in time $O(|S|)$. It naturally leads to an $O(|G| \times |S|)$ algorithm for the reconciliation of a gene tree with an arbitrary number of polytomies. Indeed, it is sufficient to traverse the tree in a depth-first manner, and resolve each polytomy at a time. Interestingly, we can find an example of trees G and S leading to a reconciliation of size $|G| \times |S|$. Indeed, let $\Sigma = \{1, 2, \cdots s\}$, and consider the species tree S over Σ to be a caterpilar tree $(1, 2, \cdots s)$ with leaves ordered from 1 to s, where $s = |S|$. Consider the gene tree G to be the caterpilar tree $((l_1, r_1), \cdots (l_g, r_g))$ composed of g cherries (l_1, r_s), where the l leaves are labeled 1, and the r leaves are labeled s. We have $g = |G|$. Then clearly a most parsimonious reconciliation of G and S is one with $s - 2$ leaves inserted in each cherry of G, which leads to a tree of size $|G| \times |S|$. Therefore, the algorithm is optimal for our considered Minimum Resolution problem. It is likely however that finding the mutation cost of an optimal reconciliation, without displaying the actual reconciliation, can be done in linear-time.

References

1. Akerborg, O., Sennblad, B., Arvestad, L., Lagergren, J.: Simultaneous bayesian gene tree reconstruction and reconciliation analysis. Proceedings of the National Academy of Sciences USA 106(14), 5714–5719 (2009)
2. Arvestad, L., Berglung, A.-C., Lagergren, J., Sennblad, B.: Gene tree reconstruction and orthology analysis based on an integrated model for duplications and sequence evolution. In: Gusfield, D. (ed.) RECOMB 2004: Proceedings of the Eighth Annual International Conference on Research in Computational Molecular Biology, pp. 326–335. ACM, New York (2004)
3. Chang, W.-C., Eulenstein, O.: Reconciling Gene Trees with Apparent Polytomies. In: Chen, D.Z., Lee, D.T. (eds.) COCOON 2006. LNCS, vol. 4112, pp. 235–244. Springer, Heidelberg (2006)
4. Chang, W.C., Eulenstein, O.: Reconciling gene trees with apparent polytomies, technical report. Department of Computer Science, Iowa State University (2006)
5. Chauve, C., El-Mabrouk, N.: New Perspectives on Gene Family Evolution: Losses in Reconciliation and a Link with Supertrees. In: Batzoglou, S. (ed.) RECOMB 2009. LNCS, vol. 5541, pp. 46–58. Springer, Heidelberg (2009)
6. Chen, K., Durand, D., Farach-Colton, M.: Notung: Dating gene duplications using gene family trees. Journal of Computational Biology 7, 429–447 (2000)
7. Csűrös, M.: Ancestral reconstruction by asymmetric wagner parsimony over continuous characteand squared parsimony over distributions. In: Sixth RECOMB Satellite Workshop on Comparative Genomics, pp. 72–86 (2008)
8. Doroftei, A., El-Mabrouk, N.: Removing Noise from Gene Trees. In: Przytycka, T.M., Sagot, M.-F. (eds.) WABI 2011. LNCS (LNBI), vol. 6833, pp. 76–91. Springer, Heidelberg (2011)
9. Doyon, J.-P., Ranwez, V., Daubin, V., Berry, V.: Models, algorithms and programs for phylogeny reconciliation. Brief Bioinform. 12, 392–400 (2011)

10. Doyon, J.-P., Scornavacca, C., Gorbunov, K.Y., Szöllősi, G.J., Ranwez, V., Berry, V.: An Efficient Algorithm for Gene/Species Trees Parsimonious Reconciliation with Losses, Duplications and Transfers. In: Tannier, E. (ed.) RECOMB-CG 2010. LNCS, vol. 6398, pp. 93–108. Springer, Heidelberg (2010)

11. Durand, D., Haldórsson, B.V., Vernot, B.: A hybrid micro-macroevolutionary approach to gene tree reconstruction. Journal of Computational Biology 13, 320–335 (2006)

12. Fang, G., Bhardwaj, N., Robilotto, R., Gerstein, M.B.: Getting started in gene orthology and functional analysis. PLoS Comput. Biol. 6(3), e1000703 (2010)

13. Goodman, M., Czelusniak, J., Moore, G.W., Romero-Herrera, A.E., Matsuda, G.: Fitting the gene lineage into its species lineage, a parsimony strategy illustrated by cladograms constructed from globin sequences. Systematic Zoology 28, 132–163 (1979)

14. Gorecki, P., Tiuryn, J.: DLS-trees: a model of evolutionary scenarios. Theoretical Computer Science 359, 378–399 (2006)

15. Hahn, M.W.: Bias in phylogenetic tree reconciliation methods: implications for vertebrate genome evolution. Genome Biology 8(R141) (2007)

16. Ma, B., Li, M., Zhang, L.: From gene trees to species trees. SIAM J. on Comput. 30, 729–752 (2000)

17. Slowinski, J.B.: Molecular polytomies. Molecular Phylogenetics and Evolution 19(1), 114–120 (2001)

18. Tofigh, A., Hallett, M., Lagergren, J.: Simultaneous identification of duplications and lateral gene transfers. IEEE/ACM Trans. Comput. Biol. Bioinform. 8, 517–535 (2011)

19. Zhang, J.: Evolution by gene duplication: an update. Trends in Ecology and Evolution 18(6), 292–298 (2003)

20. Zhang, L.X.: On Mirkin-Muchnik-Smith conjecture for comparing molecular phylogenies. Journal of Computational Biology 4, 177–188 (1997)

21. Zheng, Y., Wu, T., Zhang, L.: Reconciliation of gene and species trees with polytomies, eprint arXiv:1201.3995 (2012)

22. Zmasek, C.M., Eddy, S.R.: A simple algorithm to infer gene duplication and speciation events on a gene tree. Bioinformatics 17, 821–828 (2001)

Accounting for Gene Tree Uncertainties Improves Gene Trees and Reconciliation Inference

Thi Hau Nguyen[1,2], Jean-Philippe Doyon[1], Stéphanie Pointet[1],
Anne-Muriel Arigon Chifolleau[1], Vincent Ranwez[2], and Vincent Berry[1]

[1] LIRMM, Université Montpellier 2 - CNRS, France
[2] Montpellier SupAgro (UMR AGAP)

Abstract. We propose a reconciliation heuristic accounting for gene duplications, losses and horizontal transfers that specifically takes into account the uncertainties in the gene tree. Rearrangements are tried for gene tree edges that are weakly supported, and are accepted whenever they improve the reconciliation cost. We prove useful properties on the dynamic programming matrix used to compute reconciliations, which allows to speed-up the tree space exploration when rearrangements are generated by Nearest Neighbor Interchanges (NNI) edit operations. Experimental results on simulated and real data confirm that running times are greatly reduced when considering the above-mentioned optimization in comparison to the naïve rearrangement procedure. Results also show that gene trees modified by such NNI rearrangements are closer to the correct (simulated) trees and lead to more correct event predictions on average. The program is available at http://www.atgc-montpellier.fr/
Mowgli/

1 Introduction

A phylogenetic tree or *phylogeny* is a tree depicting evolutionary relationships among biological entities that are believed to have a common ancestor. A gene family is a group of genes descended from a common ancestor that retains similar sequences and often similar functions. A species tree depicts the evolutionary history of a group of species, whereas a gene tree depicts the evolutionary history of a gene family. Gene trees and species tree are often inconsistent due to family-specific evolutionary events such as gene duplications, gene losses, horizontal gene transfers. By comparing gene trees with a species tree, reconciliation methods try to recover those major evolutionary events. Reconciliation is indeed the process of constructing a mapping between a gene tree and a species tree to explain their differences and similitudes with evolutionary events such as speciation ($\mathbb{S}$), duplication ($\mathbb{D}$), loss ($\mathbb{L}$), and horizontal gene transfer ($\mathbb{T}$) events. Reconciliations are most often inferred on the basis of a parsimony criterion: a cost is given to each event type, the cost of a reconciliation is the sum of the costs of the individual events it uses, and a reconciliation of minimum total cost is sought for. This computational problem is often called *Most Parsimonious Reconciliation*, or MPR in short, and many works have been devoted to it recently [1,2,3,4,5,6,7].

B. Raphael and J. Tang (Eds.): WABI 2012, LNBI 7534, pp. 123–134, 2012.
© Springer-Verlag Berlin Heidelberg 2012

The first proposed models focused on reconciliations involving only duplications and losses (the DL model) [8,9,10] or only horizontal transfers and losses [11]. Probabilistic methods have also been developed for the DL model, such as that of Arvestad et al. [12] (see Doyon et al. [13] for a review). Most recent works using a parsimony approach have been devoted to models incorporating duplications, losses and transfers all together (the DTL model) [1,2,4,6], which is necessary to handle prokaryotes. When accounting for transfer events, the history proposed by a reconciliation is consistent if, for any transfer, the donor and receiver species co-exist. Ensuring such a time consistency is difficult and leads to an NP-hard problem in the general case [14,5]. However, in the case divergence dates are available for nodes of the species tree, the problem becomes amenable [15,4]. The difficulty to handle transfers has led to a split within proposed DTL methods, namely those that ensure time-consistency [15,4] and those that don't [1,5,7]. The fastest algorithm for the later category runs in $O(mn \log n)$ where m and n are the sizes of the gene and species trees respectively [7], while the fastest time-consistent algorithm runs in $O(mn^2)$ [4].

A major problem, when applying reconciliation methods, is that parts of the gene trees can be incorrect. This leads reconciliation methods to overestimate $(\mathbb{S})$, $(\mathbb{D})$, $(\mathbb{L})$ and $(\mathbb{T})$ events [16,17]. Errors within the gene tree can be due to sequence alignment problems or phylogenetic reconstruction artifacts such as long branch attraction. Such errors are well-known in phylogenetics and several support measures, such as bootstrap values or bayesian posterior probabilities, have been proposed to detect unreliable edges in a gene tree. Up to now, very few works have solved the reconciliation problem in the presence of unsupported edges, and most of them consider only the DL model [18,19,16,20,21]. Durand et al. proposed an exponential exact algorithm to find the best rearrangement of a gene tree while preserving its strongly supported edges [16]. Some approaches collapse unsupported edges, leading to the creation of nodes with more than two children, called *polytomies* [18,19,20]. They then rely on a generalization of the least common ancestor mapping (LCA) to avoid the need for examining all possible binary rearrangements of the polytomies. In this way, Chang et al. [19], resp. Vernot et al. [20], proposed polynomial time algorithms when considering non-binary gene trees, resp. species trees. Berglund et al. proved that when dealing with species and gene tree that are both non-binary, the problem becomes NP-complete [18]. They thus proposed a heuristic approach tackling a variant of the MPR problem where duplications and losses are optimized separately. Durand et al. used Nearest Neighbor Interchange (NNI) edit operations to rearrange the local topology of the gene tree in the regions of low supports but, unlike Berglund et al., they optimized simultaneously duplications and losses. Chaudhary et al. investigated Subtree Prune and Regraft (SPR) and Tree Bisection and reconnection (TBR) edit operations to search for the gene tree rearrangement that minimizes the number of duplications, regardless of losses and transfers [21].

Due to transfer events, the LCA mapping can not be transposed to the DTL model and it seems hard to have an exact polynomial time algorithm for the MPR problem under this model even when the polytomies are present only in the gene tree or in the species tree. Following the works of Berglund et al. and Durand et al. for the DL model, we propose a heuristic method relying on NNI edit operations to search for a gene tree rearrangement that minimizes the cost of reconciliation to a fixed binary species tree, but in the context of the more complex DTL model. The resulting dynamic program, called MowgliNNI, is a generalization of Mowgli [4] a program initially developed for fixed binary gene trees. Experiments on simulated data show that MowgliNNI provides better $\mathbb{D}$, $\mathbb{T}$, $\mathbb{L}$, $\mathbb{S}$ prediction while improving gene tree inference, i.e. the modified gene tree is closer to the true evolutionary history of the gene family. Experiments on real data show a significant decrease in number of events and in the number of most parsimonious reconciliations.

2   Preliminaries

Trees considered in this paper are rooted and only labeled at their leaves, each leaf being labeled with the name of a studied species. Given a tree T, its nodes, edges, leaves and root are resp. denoted $V(T)$, $E(T)$, $L(T)$ and $r(T)$. The label of a leaf u of T is denoted by $\mathcal{L}(u)$ and the set of labels of leaves of T is denoted by $\mathcal{L}(T)$. When a node u has two children, they are denoted u_1 and u_2. Given two nodes u and v of T, $u \leq_T v$ (resp. $u <_T v$) if and only if v is on the unique path from u to $r(T)$ (resp. and $u \neq v$); if neither $u <_T v$ nor $v <_T u$ then u and v are said to be *incomparable*. As we consider rooted trees T only, we adopt the convention than an edge denoted (u, v) means that $v <_T u$. For a node u of T, T_u denotes the subtree of T rooted at u, u_p the parent node of u, while (u_p, u) is the *parent edge* of u. A tree T' is a *refinement* of a tree T if T can be obtained from T' by collapsing some edges in T', i.e. by merging the two extremities of these edges [22].

A *species tree* is a rooted binary tree depicting the evolutionary relationships of ancestral (internal nodes) species leading to a set of extant (leaf) species. A species tree S is considered here to be *dated*, that is associated to a time function $\theta : V(S) \to \mathbb{R}^+$ such that $y <_S x$ implies that $\theta(y) < \theta(x)$. Such times are usually estimated on the basis of molecular sequences [23] and fossil records. Note that to ensure the time consistency of inferred transfers, absolute dates are not required, the important information being the ordering of the nodes of S induced by the dating. Given a dated binary species tree S, the reconciliation model we rely on considers a variant of S called a *subdivision* and denoted S' (as done also in [24,3,4]). The subdivision S' is constructed from S as follows: for each node $x \in V(S) \setminus L(S)$ and each edge $(y_p, y) \in E(S)$ s.t. $\theta_S(y_p) > \theta_S(x) > \theta_S(y)$, an *artificial* node w is inserted along the edge (y_p, y) in S', with $\theta'_{S'}(w) = \theta_S(x)$.

A *gene tree* is a rooted binary tree explaining the evolutionary history of a gene family, that lead to a set of homologous sequences observed in current organisms. Each leaf of a species tree has a unique label, corresponding to a specific extant sequence of the gene. Though, several leaves of a gene tree can be associated to a same species due to duplication and transfert events. We denote by $s(u)$ the species associated to leaf $u \in V(G)$. Each edge (u, v) of $E(G)$ can be uniquely identified by the subset $\mathcal{L}(T_v) \subseteq \mathcal{L}(G)$. A gene tree G *with supports* is a gene tree whose *internal* edges have a support value. Let $wk_t(G) \subseteq E(G)$ be the set of edges having a support value weaker than threshold t and let $str_t(G)$ be $E(G) - wk_t(G)$, that is the edges having a support equal or stronger than t.

Finding the Most Parsimonious Reconciliation. Reconciling a (binary) gene tree G with a species tree S means building a mapping α that associates each gene $u \in V(G)$ to a sequence $\alpha(u)$ of nodes in the subdivision S'. The $\alpha(u)$ sequence models the evolution of gene u along S' with the following atomic events: $(\mathbb{C})$ contemporary gene, $(\mathbb{S})$ speciation, $(\mathbb{D})$ duplication, $(\mathbb{T})$ transfer, $(\mathbb{SL})$ speciation followed by a loss, $(\mathbb{TL})$ transfer followed by a loss for the donor, and $(\varnothing)$ going from an artificial node of S' to its only child. Observe that each loss is coupled with either a speciation $(\mathbb{SL})$ or a transfer $(\mathbb{TL})$. Indeed, any most parsimonious reconciliation only needs to use a loss when it meets a speciation node of S' where G goes into only one descending edge, or when leaving an edge due to a transfer, with no part of G remaining in the donor edge. For more details, we refer the reader to Definition 3 of Doyon et al. [4] that we follow for reconciliation, except that the mapping α considered here concerns nodes instead of branches.

The *cost* of a reconciliation α is denoted $cost(\alpha) = d\delta + t\tau + l\lambda$, where δ, τ, and λ respectively denote the cost of $\mathbb{D}$, $\mathbb{T}$, and $\mathbb{L}$ events, and d, t, and l denote the number of the corresponding events in α. Moreover, a $\mathbb{TL}$ event is atomic and costs $(\tau + \lambda)$ and a $\mathbb{SL}$ costs λ. The *optimal reconciliation cost* is denoted $C(G, S') = \min\{cost(\alpha) : \alpha$ is a reconciliation between G and $S'\}$.

The optimal cost for mapping a node u of G on a vertex x of S' is defined according to the minimal cost among the events $\mathbb{C}$, $\mathbb{S}$, $\mathbb{D}$, $\mathbb{T}$, $\varnothing$, and $\mathbb{SL}$, together with the cost of a $\mathbb{TL}$ event, which are denoted $c_{\overline{\mathbb{TL}}}(u, x)$ and $c_{\mathbb{TL}}(u, x)$, respectively (see Definition 1 below). This directly follows from the dynamic programming algorithm that computes the optimal cost $C(G, S')$, where the computation of the cost for a $\mathbb{TL}$ event follows that of the other six atomic events, since a $\mathbb{TL}$ event is followed by a $\mathbb{C}$, $\mathbb{S}$, $\mathbb{D}$, $\mathbb{T}$, $\varnothing$, or $\mathbb{SL}$ event [4].

To ensure time consistency of $\mathbb{T}$ and $\mathbb{TL}$ events, the donor $x \in V(S')$ and the receiver $y \in V(S')$ have to be located at the same *time slice* $h(x) = h(y)$ of S' (the term time slice of a vertex refers to its height in S').

These intricated notions are formally detailed in definitions 1 and 2. Though, these definitions depend on one another, there is no circularity as either we progress in the gene tree or we switch from a $\mathbb{TL}$ event to a non-$\mathbb{TL}$ event.

Definition 1 (Reconciliation cost matrix). *Consider a gene tree G and the subdivision S' of a species tree S. Let $c : V(G) \times V(S') \to \mathbb{R}^+$ denote the cost matrix recursively defined as follows for a node u of G and a vertex x of S': $c_{\overline{\mathbb{TL}}}(u, x) = \min\{c_{\mathbb{E}}(u, x) : \mathbb{E} \in \{\mathbb{C}, \mathbb{S}, \mathbb{D}, \mathbb{T}, \varnothing, \mathbb{SL}\}\}$ and $c(u, x) = \min\{c_{\mathbb{TL}}(u, x), c_{\overline{\mathbb{TL}}}(u, x)\}$, where the costs $c_{\mathbb{E}}(u, x)$ for $\mathbb{E} \in \{\mathbb{C}, \mathbb{S}, \mathbb{D}, \mathbb{T}, \varnothing, \mathbb{SL}, \mathbb{TL}\}$ are defined below.*

- $c_{\mathbb{C}}(u, x) = 0$, *if* $u \in L(G)$, $x \in L(S')$ *and* $\mathcal{L}(x) = s(u)$.
- $c_{\mathbb{S}}(u, x) = \min\{c(u_1, x_1) + c(u_2, x_2), c(u_1, x_2) + c(u_2, x_1)\}$
  if $u \notin L(G)$ *and* $x \notin L(S')$.

- $c_{\mathbb{D}}(u, x) = c(u_1, x) + c(u_2, x) + \delta$, if $u \notin L(G)$.
- $c_{\mathbb{T}}(u, x) = \min\{c(u_1, x) + c(u_2, z), c(u_1, y) + c(u_2, x)\} + \tau$
  if u has two children and where z (resp. y) denotes $BR_{\mathbb{T}}(u_2, x)$ (resp. $BR_{\mathbb{T}}(u_1, x)$).
- $c_{\varnothing}(u, x) = c(u, x_1)$, if x has a single child.)
- $c_{\mathbb{SL}}(u, x) = \min\{c(u, x_1), c(u, x_2)\} + \lambda$, if x has two children.
- $c_{\mathbb{TL}}(u, x) = c_{\overline{\mathbb{TL}}}(u, y) + \tau + \lambda$, where y denotes $BR_{\mathbb{TL}}(u, x)$.

If the above constraints for an event $\mathbb{E} \in \{\mathbb{C}, \mathbb{S}, \mathbb{D}, \mathbb{T}, \varnothing, \mathbb{SL}, \mathbb{TL}\}$ on node u and vertex x are not respected, the corresponding cost $c_{\mathbb{E}}(u, x)$ is set to ∞.

Definition 2 (Best receiver). *Consider a node u of G and a vertex x of S'. Let $BR_{\mathbb{T}}(u, x)$ denote a vertex y of S' that minimizes $c(u, y) = \min\{c(u, z) : z \in V_{h(x)}(S')$ and $z \neq x\}$. Similarly, let $BR_{\mathbb{TL}}(u, x)$ denote a vertex y of S' that minimizes $c_{\overline{\mathbb{TL}}}(u, y) = \min\{c_{\overline{\mathbb{TL}}}(u, z) : z \in V_{h(x)}(S')$ and $z \neq x\}$*

The value $c(u, x)$ is the optimal cost when mapping gene node u to node x in S' or on the edge above it. The optimal cost for reconciling G with S', denoted $C(G, S')$, is then $\min_{x \in V(S')}(c(r(G), x)$. The algorithm of Doyon et al.[4], called *Mowgli*, fills the dynamic programming cost matrix $c : V(G) \times V(S') \rightarrow \mathbb{R}^+$ by two embedded loops that visit all slices of S' in backward order and nodes of G in postorder. Due to an optimization in precomputing the best receiver edge for transfer events of nodes u in a given time slice, this algorithm runs in $O(|S|^2.|G|)$ time and space.

The problem considered in this paper is the following:

MOST PARSIMONIOUS RECONCILIATION GENE TREE (MPR-GT)
INPUT: a dated species tree S with a time function θ_S, a gene tree G with supports on the same set of species, costs δ, τ, resp. λ for $\mathbb{D}$, $\mathbb{T}$, resp. $\mathbb{L}$, and a threshold t.
OUTPUT: a tree G' s.t. $str_t(G) \subseteq E(G')$ and $C(G', S')$ is minimum among all such trees.

3   Methods

We describe here a heuristic for the MPR-GT problem that relies on a hill-climbing strategy to seek a (rooted) gene tree G of minimum reconciliation cost (see Def. 1) using NNI edit operations [25].

Performing an NNI operation around an *internal* edge (w, v) means swapping the position of one of the two subtrees connected to v with that of the subtree connected to the sibling of v. Given an initial gene tree G and an edge (w, v) of G, two "alternative" trees can be obtained from G by performing an NNI operation around (w, v) (see Fig. 1 for an example). The hill-climbing proceeds as follows: (1) select a weak edge of G; (2) compute the reconciliation cost for the two alternative gene trees obtained by NNI on that edge; (3) if none of these trees decreases the reconciliation cost, then try another weak edge; if none of the weak edges allows to progress, then G is a local minimum and the hill climbing stops; (4) otherwise one of the alternative gene trees leads to a decrease in reconciliation cost, and the above process continues with the alternative tree of minimum reconciliation cost. MowgliNNI outputs the final binary rearrangement along with its most parsimonious reconciliation. In the worst cases, MowgliNNI examines all unreliable edges and does not find any better binary rearrangement of the given gene tree G since the topology G is already (locally) optimal. Consider now the time complexity of *MowgliNNI*. Identifying the weak edges is done in $O(|G|)$ and generating the two alternative gene trees for an NNI operation is done in constant time. Hence, the complexity bottleneck of *MowgliNNI* is the number of times (denoted N) the $\Theta(|S|^2 \cdot |G|)$ *Mowgli* algorithm is called. Overall, the time complexity of *MowgliNNI* is $\Theta(|S|^2 \cdot |G| \cdot N)$. The next section describes how we can avoid recomputing large parts of the cost matrix, and hence greatly reduce the running time of *MowgliNNI*.

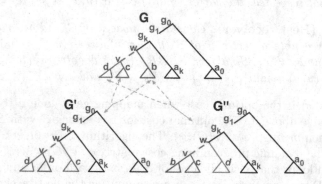

Fig. 1. A gene tree G with a weak edge (w, v) selected for an NNI. v is connected to two subtrees G_c and G_d, while w is connected to v and to the subtree G_b. Performing an NNI operation around (w, v) means exchanging subtree G_b with either G_c or G_d, leading to trees G' and G'' respectively.

Combinatorial Optimization. We now present results that take advantage of the way the dynamic programming matrix is computed (Def. 1) to avoid recomputing from scratch the cost matrix associated to a gene tree G' obtained by an NNI edit operation from a gene tree G. Consider the gene tree G of Figure 1, the NNI operation applied on edge (w, v) that swaps the two subtrees G_b and G_c, and the resulting gene tree denoted G'. We can observe that despite the global architecture of G and G' differs, the local architectures of subtrees $G_b, G_c, G_d, G_{a_0}, \ldots G_{a_k}$ remain unchanged. Hence, any cost that differs between the matrices $c : V(G) \times V(S') \to \mathbb{R}^+$ and $c' : V(G') \times V(S') \to \mathbb{R}^+$ (see Definition 1) is located in a column (i.e. node of the gene tree) associated to an ancestor of v (including v itself). For each of those nodes, there is two cases: (i) the node belongs to the NNI edge and its two children have subtree that have been modified (e.g. nodes w and v); (ii) the node is a strict ancestor of the NNI edge (w, v) and has exactly one child with a subtree that has been modified (e.g. $g_k, \ldots, g_0$). Lemma 1 below indicates which columns of the cost matrix don't need to be recomputed.

Lemma 1. *Consider a gene tree G, the subdivision S' of a species tree S, an edge (w, v) of G, and G' obtained from G by an NNI operation on (w, v). For each node z of G that is incomparable to v and for each vertex x of S', $c(z, x) = c'(z, x)$ holds.*

Unfortunately, there is no extension of Lemma 1 to ensure that when an edge has already been unsuccessful tried for an NNI it is useless to reconsider it later, even if it is a descendant in G of the edge leading to the last successful NNI.

Algorithm 1. $MowgliNNI(G, c)$: seek a gene tree G' of minimum reconciliation cost, starting from a gene tree G and the precomputed matrix reconciliation cost $c : V(G) \times V(S') \to \mathbb{R}^+$, where S' is the subdivided species tree.

1: **for all** edges $(w, v) \in wk_t(G)$ **do**
2:     For each node s of G that is not an ancestor of v, set the column $c'(s, \cdot)$ to $c(s, \cdot)$.

3:     For each vertex x of S', recompute the cost $c'(v, x)$ according to Def. 1.
4:     **for all** strict ancestors s of v according to a bottom-up traversal of G **do**
5:         For each vertex x of S', recompute the cost $c'(s, x)$ according to Def. 1.
6:         If $c(s, x) \leq c'(s, x)$ holds for each vertex x of S', then examine the next edge of loop at line 1 {the NNI rearrangement tree G' is refused}.
7:     **end for**
8:     Return $MowgliNNI(G', c')$ {The rearranged tree G' is accepted}.
9: **end for**
10: Return G {No successful rearrangement of G}

Theorem 1. *Consider a gene tree G, the subdivision S' of a species tree S, an edge (w, v) of G, a gene tree G' obtained by an NNI operation on (w, v), and any strict ancestor u of w in G where the unique child of u that is an ancestor of w is u_1 w.l.o.g. (i.e. $w \leq u_1$ in both G and G'). If $c(u_1, x) \leq c'(u_1, x)$ holds for all $x \in V(S')$, then (1) $c(u, x) \leq c'(u, x)$ holds for all $x \in V(S')$; and (2) $C(G, S') \leq C(G', S')$.*

Computing the cost matrix $c' : V(G') \to V(S')$ given $c : V(G) \to V(S')$ is then achieved in worst-case time $O(|S'| \cdot h(G))$, where $h(G)$ is the height of G.

Theorem 2. *MowgliNNI has worst case running time $O(|S|^2 \cdot |G| + |S|^2 \cdot h(G) \cdot N)$.*

Indeed the steps of Algorithm 1 can be described as follows: initializing the reconciliation matrix for the initial gene tree is done in $O(|S|^2 \cdot |G|)$ time; updating the matrix for each NNI now only costs $O(|S'| \cdot h(G)) = O(|S|^2 \cdot h(G))$.

In *MowgliNNI*'s naïve implementation each rearrangement requires to recompute the cost associated to each and every node of the gene tree. In contrast, in the optimized version, an NNI around edge (w, v) is examined after updating only those costs associated to ancestral nodes of w. This has no impact on the worst case complexity (when the gene tree is a caterpillar $h(G)$ is in $O(|G|)$) but significantly reduces the running times in practice since in most cases the number of nodes in G is much larger than their average height. For some random tree models the average height of a node in an n-leaf tree is indeed proportional to $log(n)$ [26,27].

4   Experimental Evaluation

4.1   Experiments on Simulated Datasets

The phylogeny of 37 proteobacteria proposed by David and Alm [6] was used as a reference species tree. Along this tree, we simulated the evolutionary histories (denoted R_{True}) of 985 gene families (G_{True}), containing 10 to 100 genes, according to the birth and death process [28]. The *initial* gene trees (G_{ML}) were inferred from the simulated molecular sequences of length 1500 - 3000 bp by RAxML under GTR model [29]. *Mowgli*

[4] and Ranger-dtl-D [7] were used to infer the most parsimonious evolutionary history (R_{ML}) between the initial gene tree and the reference species tree. Then, *MowgliNNI* was used to search for an alternative gene tree topology (G_{NNI}) of lower reconciliation cost, along with its most parsimonious evolutionary history (R_{NNI}). The cost of each $\mathbb{D}$, $\mathbb{T}$, $\mathbb{L}$ event considered in reconciliations was computed as follows:

$$Cost_{\mathbb{E}} = \begin{cases} \log(\frac{|\mathbb{D}_{R_{True}}|+|\mathbb{T}_{R_{True}}|+|\mathbb{L}_{R_{True}}|}{|\mathbb{E}_{R_{True}}|}) & \text{if } |\mathbb{E}_{R_{True}}| \neq 0 \\ \log(\frac{|\mathbb{D}_{R_{True}}|+|\mathbb{T}_{R_{True}}|+|\mathbb{L}_{R_{True}}|}{0.1}) & \text{otherwise} \end{cases} \quad (1)$$

where $\mathbb{E}_{R_{True}}$ (with $\mathbb{E}$ beeing $\mathbb{D}$, $\mathbb{T}$ or $\mathbb{L}$) stands for the true events of this type.

We explored the ability of *MowgliNNI* to improve the set of G_{ML} trees using six different bootstrap values as threshold for defining weak edges, i.e. 20, 40, 60, 80, 90, and 95. The G_{ML} trees were inferred from relatively long sequences, they thus contained a large proportion of high bootstrap values, *e.g.* more than 65% edges had a bootstrap value ≥ 80. Though this left only a moderate number of edges in each gene tree to be considered by *MowgliNNI* for rearrangement, the method was still able to improve their quality (see below).

Mowgli and Ranger-dtl-D showed a similar accuracy in inferring duplications and transfers (Fig. 2(a)), though Ranger-dtl-D proposed reconciliations with higher costs in 13% of the cases. As moreover the Ranger-dtl-D software does not provide the mapping of loss events, below we mainly compare the events inferred by *MowgliNNI* with those inferred by *Mowgli* and those of R_{True}.

Table 1 reports the accuracy of the G_{ML} and G_{NNI} trees. The number of families considered for improvement logically increases as the threshold for identifying an edge as weak increases (row 2 of Table 1). Even though bootstrap values in the initial gene trees are high on average, a large number of the 985 processed families are still subject to possible improvements, as for threshold 20, already 43% of the families are concerned, and this goes up to 99% of the families at threshold 95. The first conclusion is that

Table 1. Quality of the gene trees and reconciliations inferred by *MowgliNNI*. For each tested threshold value, the second row indicates the number of gene families containing some weak edges (among the 985 simulated gene families). Third row indicates the percentage of these families where *MowgliNNI* proposes a modified tree of lower reconciliation cost. The last six rows provide the percentage of the former families where *MowgliNNI* provides modified gene trees (resp. reconciliations) that are closer, equally far or farther from the true gene trees (resp. the true evolutions). $RF(G_{True}, G_X)$ denotes the Robinson Fould distance between G_{True} and G_X, $ED(R_{True}, R_X) = |R_{True} - R_X| + |R_X - R_{True}|$, where X stands for NNI or ML.

Threshold	20	40	60	80	90	95
Number of gene families containing weak edges	422	708	911	965	979	981
% of cases where $Cost(S, G_{NNI}) < Cost(S, G_{ML})$	79	81	85	88	88	89
% of cases where $RF(G_{True}, G_{NNI}) < RF(G_{True}, G_{ML})$	38	55	66	71	70	69
% of cases where $RF(G_{True}, G_{NNI}) = RF(G_{True}, G_{ML})$	59	40	28	21	21	21
% of cases where $RF(G_{True}, G_{NNI}) > RF(G_{True}, G_{ML})$	3	5	6	8	9	10
% of cases where $ED(R_{True}, R_{NNI}) < ED(R_{True}, R_{ML})$	59	70	75	79	79	78
% of cases where $ED(R_{True}, R_{NNI}) = ED(R_{True}, R_{ML})$	27	23	19	15	16	15
% of cases where $ED(R_{True}, R_{NNI}) > ED(R_{True}, R_{ML})$	14	8	6	6	6	6

there is a large number of cases where *MowgliNNI* can propose a modified gene tree. The percentage of cases where it actually did is provided in row 3 of Table 1, showing *e.g.* that at threshold 80, *MowgliNNI* proposed a new gene tree in 88% of the cases (853 cases over 965). Even for the lowest considered threshold of 20, a new gene tree is obtained for 79% of the families having weak edges, representing more than a third of the initial 985 families. These modified gene trees (G_{NNI}) represent an improvement over the initial trees (G_{ML}) since they are in most cases closer to the true gene trees (rows 4, 5, 6) and allow to obtain better reconciliations (rows 7, 8, 9). For instance at threshold 80, G_{NNI} is better in 71% of the cases, and worse in only 8%. Similarly, the reconciliation is better in 79% of the cases, and worse in only 6%. The higher the threshold value, the more edges are considered for NNI moves by *MowgliNNI*. Up to a certain point, broadening the search space of *MowgliNNI* allows to improve both the gene trees and the reconciliations. Yet, for threshold greater than 80, *MowgliNNI*'s performance starts to decrease due to the fact that the sequence signal is no longer sufficiently taken into account.

MowgliNNI progressively reduced the number of predicted duplications, transfers and losses as the threshold increased. At threshold 0 (where *MowgliNNI* = *Mowgli*), 5403 duplications, 2460 transfers and 12007 losses were predicted on the whole dataset; going to threshold 80, these numbers drop to 4510 duplications, 1160 transfers and 8016 losses, *i.e.* values that are much closer to the 4443 duplications and 8142 losses contained in the true reconciliations.

The average number of false positive events (FP) of the R_{NNI} reconciliations decreases as the threshold increases (Fig. 2(b)). However, as in Doyon et al. [4], the average number of FP transfers is quite high compared to that of duplications and losses. This can be explained by several reasons. First, a transfer is judged incorrect as soon as i) it does not depart or end in the same edges of the species tree as the

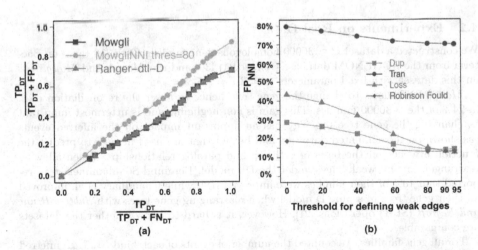

Fig. 2. (a) The accuracy of *Mowgli*, Ranger-dtl-D and *MowgliNNI* (threshold=80) in inferring duplications and transfers, where TP_{DT} (resp. FP_{DT}, FN_{DT}) denotes the true positive (resp. false positive, false negative) of duplications and transfers predicted. (b) Average false positive (FP) of the NNI trees – note that FP values at threshold 0 correspond to *Mowgli* results.

corresponding true transfer, or ii) it does not concern the same edge in the gene tree. Overall, there is an additional constraint w.r.t. duplications and loss events, leading on average to more incorrect events. This point is all the more sensitive that several most parsimonious reconciliations (MPR) are obtained in a number of cases, while we just accounted for one of them for each gene family. Hence, event error rates we report are pessimistic (note that this does not affect RF distance results). Last, incorrect gene trees lead to incorrect event inferences, but the latter are very sensitive to only small errors in gene trees. The event FP error grows exponentially when the RF distance between the initial and the true tree increases from 0 to 10% (data not shown).

Inferring G_{ML} trees from shorter sequences (400 bp), led to a decrease in their quality both in terms of RF distance to G_{True} and in event distance between inferred and true reconciliation. Starting from a less accurate gene tree, the accuracy of the NNI trees and of their evolutionary histories is also lowered, though the relative improvement provided by *MowgliNNI* over *Mowgli* is higher than with long sequences (data not shown).

In order to measure the dependance of *MowgliNNI* on the precise costs used for each kind of event, we ran the method on G_{ML} trees with costs varying up to 10%, 20%, then 50% w.r.t. those computed from Formula (1). The paired t-test for RF distances shows that the G_{NNI} trees obtained with the new costs are not significantly different from those obtained with the former costs (p-value=0.296, 0.2723, 0.2028 respectively). The accuracy of inferred events also does not change much. Transfers have the highest variation with 3.6% (resp. 3%) increase in FP (resp. FN) when the event costs vary up to 50%. Thus, *MowgliNNI* is quite robust to changes in the event costs.

In summary, *MowgliNNI* successfully uses the reconciliation cost as additional information to resolve the uncertain parts of gene trees inferred from sequences only. Though the gene tree resolutions are partly guided by reconciliations with the species tree, they are not attracted away from the true gene trees, but are closer to them than the initial gene trees. As a result, *MowgliNNI* infers gene events more accurately, which is of prior importance to distinguish orthologs from paralogs and xenologs [13].

4.2 Experiments on Real Data

We constructed a dataset of $\approx$ 30000 homologous gene families (3 to 312 taxa) on Bacteria from the HOGENOM database (release 04) [30] and ran *Mowgli* and *MowgliNNI* on this dataset with fixed parameters ($\tau = 3, \delta = 3.5, \lambda = 1$).

MowgliNNI allows to change the gene tree, hence to lower the reconciliation cost, in 24% of the $\approx$ 30000 families. This gain is non-negligible and is uttermost important as changing the gene tree topology has an important impact on the inferred events (as shown on the simulated data sets and below) that are used in turn to predict the function of genes on the basis of ortholog and paralog relationships. When allowing rearrangements on weak edges under the DL model, Berglund-Sonnhammer et al. reported that 10% of their families were improved [18], while Chaudhary el al. improved all their gene trees in a pure D model when rearranging gene trees with *Subtree Prune and Regraft* (SPR) operations [21]. However, it is hard to know whether the datasets are comparable.

For all gene families, we counted the number of events of each kind ($\mathbb{D}$, $\mathbb{T}$, $\mathbb{L}$) inferred by *Mowgli* and *MowgliNNI*. As a rule, *MowgliNNI* led to a decrease in the number of events in inferred evolutionary histories, the reduction being considerable for transfers and losses (88.3% and 59.9% resp.) but quite small for duplications (5.2%). These results obtained in the DTL model are consistent with those of Durand et al. reporting that in the DL model gene tree rearrangements substantially reduce the number of events needed to explain the data [16]. The differences in reductions we observed depending on the kind of events can be explained by the fact that given the costs we used for the events ($\tau = 3, \delta = 3.5, \lambda = 1$), it is usually more parsimonious to explain the conflicts between gene and species tree by a combination of $\mathbb{T}$ and $\mathbb{L}$ rather than a combination of $\mathbb{D}$, and $\mathbb{L}$. Thus, when *MowgliNNI* infers a gene tree closer to the species tree, it mostly removes the need for artificial transfers (and losses to a lesser extent), while not altering that much the number of duplications.

In addition to reductions in errors and number of events, the new gene tree proposed by *MowgliNNI* usually reduced the number of alternative MPRs, i.e. histories. On a random sample of two dozens new gene trees, the number of MPRs is reduced in 63% of the cases (by a factor of 18 in the best case), and increased in 21% (by a factor 3 at worst). This echoes similar findings of Durand et al. in a DL model [16].

We measured the running time of *Mowgli* and of both the non-optimized and optimized versions of *MowgliNNI* (see Methods) on a random sample of 100 families having from 10 to 80 taxa. The results show that the optimized version of *MowgliNNI* is 20 (resp. 50 and 80) times faster than the non-optimized one, when facing 1-20 (resp. 20-40 and 40-60) weak edges. The increase in accuracy due to *MowgliNNI* is obtained at the price of a small computation time overcost.

Acknowledgement. We thank Gergely J. Szöllősi for his help in determining the event costs of the real dataset and the referees that helped strengthen the experimental validation. This work was funded by the *Languedoc-Roussillon Chercheur d'Avenir* program and by the french *Agence Nationale de la Recherche, programmes Domaines Emergents* (ANR-08-EMER-011 *Phylariane*), *6ème Extinction* (ANR-09-PEXT-000 *PhyloSpace*) and *Investissements d'avenir / Bioinformatique* (ANR-10-BINF-01-02, *Ancestrome*).

References

1. Hallett, M., Lagergren, J., Tofigh, A.: Simultaneous identification of duplications and lateral transfers. In: RECOMB 2004, pp. 347–356. ACM, New York (2004)
2. Górecki, P.: Reconciliation problems for duplication, loss and horizontal gene transfer. In: Bourne, P.E., Gusfield, D. (eds.) RECOMB, pp. 316–325. ACM (2004)
3. Conow, C., Fielder, D., Ovadia, Y., Libeskind-Hadas, R.: Jane: a new tool for the cophylogeny reconstruction problem. Algorithms Mol. Biol. 5, 16 (2010)
4. Doyon, J.-P., Scornavacca, C., Gorbunov, K.Y., Szöllősi, G.J., Ranwez, V., Berry, V.: An Efficient Algorithm for Gene/Species Trees Parsimonious Reconciliation with Losses, Duplications and Transfers. In: Tannier, E. (ed.) RECOMB-CG 2010. LNCS, vol. 6398, pp. 93–108. Springer, Heidelberg (2010)
5. Tofigh, A., Hallett, M., Lagergren, J.: Simultaneous identification of duplications and lateral gene transfers. IEEE/ACM TCBB 8(2), 517–535 (2011)
6. David, L.A., Alm, E.J.: Rapid evolutionary innovation during an archaean genetic expansion. Nature 469(7328), 93–96 (2011)
7. Bansal, M.S., Alm, E.J., Kellis, M.: Efficient algorithms for the reconciliation problem with gene duplication, horizontal transfer, and loss. In: Proceeding ISBM 2012 (2012)
8. Goodman, M., Czelusniak, J., Moore, G.W., Herrera, R.A., Matsuda, G.: Fitting the gene lineage into its species lineage, a parsimony strategy illustrated by cladograms constructed from globin sequences. Syst. Zool. 28, 132–163 (1979)

9. Page, R.D.: Extracting species trees from complex gene trees: reconciled trees and vertebrate phylogeny. Mol. Phylogenet. Evol. 14, 89–106 (2000)

10. Ma, B., Li, M., Zhang, L.: From gene trees to species trees. SIAM Journal on Computing 30(3), 729–752 (2001)

11. Nakhleh, L., Warnow, T., Linder, C.R.: Reconstructing reticulate evolution in species: theory and practice. In: Proceedings of the Eighth Annual International Conference on Research in Computational Molecular Biology, RECOMB 2004, pp. 337–346. ACM, New York (2004)

12. Arvestad, L., Lagergren, J., Sennblad, B.: The gene evolution model and computing its associated probabilities. J. ACM 56(2) (2009)

13. Doyon, J.-P., Ranwez, V., Daubin, V., Berry, V.: Models, algorithms and programs for phylogeny reconciliation. Brief Bioinform. 12(5), 392–400 (2011)

14. Ovadia, Y., Fielder, D., Conow, C., Libeskind-Hadas, R.: The co phylogeny reconstruction problem is NP-complete. J. Comput. Biol. 18(1), 59–65 (2011)

15. Libeskind-Hadas, R., Charleston, M.A.: On the computational complexity of the reticulate cophylogeny reconstruction problem. JCB 16(1), 105–117 (2009)

16. Durand, D., Halldorsson, B.V., Vernot, B.: A hybrid micro-macroevolutionary approach to gene tree reconstruction. J. Comput. Biol. 13(2), 320–335 (2006)

17. Hahn, M.W.: Bias in phylogenetic tree reconciliation methods: implications for vertebrate genome evolution. Genome Biology 8(R141) (2007)

18. Berglund-Sonnhammer, A.C., Steffansson, P., Betts, M.J., Liberles, D.A.: Optimal gene trees from sequences and species trees using a soft interpretation of parsimony. J. Mol. Evol. 63(2), 240–250 (2006)

19. Chang, W.-C., Eulenstein, O.: Reconciling Gene Trees with Apparent Polytomies. In: Chen, D.Z., Lee, D.T. (eds.) COCOON 2006. LNCS, vol. 4112, pp. 235–244. Springer, Heidelberg (2006)

20. Vernot, B., Stolzer, M., Goldman, A., Durand, D.: Reconciliation with non-binary species trees. J. Comput. Biol. 15, 981–1006 (2008)

21. Chaudhary, R., Burleigh, J.G., Eulenstein, O.: Algorithms for Rapid Error Correction for the Gene Duplication Problem. In: Chen, J., Wang, J., Zelikovsky, A. (eds.) ISBRA 2011. LNCS, vol. 6674, pp. 227–239. Springer, Heidelberg (2011)

22. Semple, C., Steel, M.A.: Phylogenetics. Oxford Lecture Series in Mathematics and its Applications, vol. 24. Oxford University Press (2003)

23. Sanderson, M.J.: Inferring absolute rates of evolution and divergence times in the absence of a molecular clock. Bioinformatics 19, 301–302 (2003)

24. Tofigh, A.: Using Trees to Capture Reticulate Evolution, Lateral Gene Transfers and Cancer Progression. PhD thesis, KTH Royal Institute of Technology, Sweden (2009)

25. Felsenstein, J.: Inferring phylogenies. Sinauer Associates, Inc. (2004)

26. Knuth, D.E.: The Art of Computer Programmingm, Sorting and Searching, vol. 3. Addison-Wesley (1973)

27. Devroye, L.: A note on the height of binary search trees. Journal of the ACM 33(3), 489–498 (1986)

28. Kendall, D.G.: On the generalized birth-and-death process. Ann. Math. Stat. 19, 1–15 (1948)

29. Stamatakis, A.: Raxml-vi-hpc: maximum likelihood-based phylogenetic analyses with thousands of taxa and mixed models. Bioinformatics 22(21), 2688–2690 (2006)

30. Penel, S., Arigon, A.M., Dufayard, J.F., Sertier, A.S., Daubin, V., Duret, L., Gouy, M., Perriere, G.: Databases of homologous gene families for comparative genomics. BMC Bioinformatics 10(suppl. 6), 3 (2009)

RNA Tree Comparisons via Unrooted Unordered Alignments

Nimrod Milo[1,*], Shay Zakov[2,*], Erez Katzenelson[1], Eitan Bachmat[1],
Yefim Dinitz[1], and Michal Ziv-Ukelson[1,**]

[1] Dept. of Computer Science, Ben-Gurion University of the Negev, Israel
{milon,erezkatz,ebachmat,dinitz,michaluz}@cs.bgu.ac.il
[2] Dept. of Computer Science and Engineering, UC San Diego, La Jolla, CA, USA
szakov@eng.ucsd.edu

Abstract. We generalize some current approaches for RNA tree alignment, which are traditionally confined to *ordered rooted* mappings, to also consider *unordered unrooted* mappings. We define the *Homeomorphic Subtree Alignment* problem, and present a new algorithm which applies to several modes, including global or local, ordered or unordered, and rooted or unrooted tree alignments. Our algorithm generalizes previous algorithms that either solved the problem in an asymmetric manner, or were restricted to the rooted and/or ordered cases. Focusing here on the most general unrooted unordered case, we show that our algorithm has an $O(n_T n_S \min(d_T, d_S))$ time complexity, where n_T and n_S are the number of nodes and d_T and d_S are the maximum node degrees in the input trees T and S, respectively. This maintains (and slightly improves) the time complexity of previous, less general algorithms for the problem. Supplemental materials, source code, and web-interface for our tool are found in http://www.cs.bgu.ac.il/~negevcb/FRUUT.

1   Introduction

Secondary structure of RNA molecules serves important functions in many noncoding RNAs [2]. Functional constraints lead to evolutionary structural conservation that in many cases exceeds the level of sequence conservation. Thus, detecting similarity between RNA secondary structures is of major importance in functional RNA research [3,4]. A mainstream approach for (pseudoknot free) RNA secondary structure comparison represents them as trees, and applies tree comparison algorithms [5–7]. Several variants of tree edit distance and alignment problems were previously studied. These variants differ in the type of trees they examine (ordered/unordered, rooted/unrooted), and in the type of edit operations or alignment restrictions they apply [8].

Currently available bioinformatic softwares for alignment of RNA trees usually assume *rooted ordered tree alignment* [3,9–11]. However, there are known evolutionary phenomena, such as segment insertions, translocations and reversals,

---

* These authors contributed equally to the paper.
** Corresponding author.

B. Raphael and J. Tang (Eds.): WABI 2012, LNBI 7534, pp. 135–148, 2012.
© Springer-Verlag Berlin Heidelberg 2012

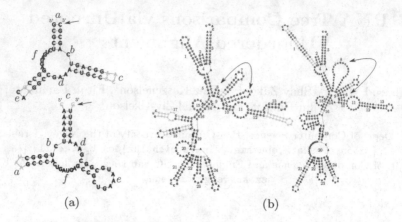

(a)   (b)

Fig. 1. Unrooted and unordered RNA similarities. Nodes of the RNA trees are clustered to motifs marked by letters or numbers (stems, loops, and unpaired nucleotide intervals), where aligned motifs share the same annotation, and unaligned nodes are in gray. Nomenclature is according to [1]. (a) An unrooted alignment between Hammerhead RNAs: *PDB_00693* (Type I, top) and *RFA_00388* (Type III, bottom). Arrows mark the external loops, which are usually chosen as roots by rooted RNA alignment tools. The unrooted mode of FRUUT identifies the high similarity between the molecules, not being restricted to align external loops to each other. (b) An unordered alignment between RNAse P RNAs: *ASE_00047* (left) and *ASE_00334* (right). In the unordered mode of FRUUT, the aligned motifs marked by 6 and 8 do not preserve order. In both molecules, pseudoknots occur between intervals annotated by 8 and 2 (see Figure S8), asserting the validity of the alignment.

which may result in a reordering or re-rooting of RNA structural elements [12]. These events can yield two similarly structured motifs, which are rooted differently (with respect to the standard "external loop" corresponding roots) [13], or permuted with respect to branching order. There are known examples of such unrooted/unordered RNA structural conservations [14,15] (Figure 1a), therefore, it is possible that searching for unordered and unrooted structural similarity may reveal new relations between RNA molecules that were previously undetected.

The general unordered tree edit distance problem is MAX-SNP hard [16], promoting the study of constrained variants. The *Subtree Isomorphism* problem [17] is, given a pattern tree S and a text tree T, to find if there is some subtree T' of T which is isomorphic to S. The *Subtree Homeomorphism* problem [18] is a variant of the former problem, where degree-2 nodes may be deleted from the selected subtree T' of the text. Pinter et al. [19] efficiently solved the Subtree Homeomorphism problem, under the unrooted unordered settings. In addition, their algorithm assigns costs for alignments and finds an alignment of minimum cost, thus solving a weighted variant of the problem. The running time of the algorithm of [19] is $O(n_S^2 n_T + n_S n_T \log n_T)$, where n_T and n_S are the number of nodes in T and S, respectively (improved time complexities under some scoring scheme restrictions were also shown in [19]). The *Constrained Edit Distance Between Unordered Labeled Trees* problem, presented by Zhang in [20], is a restricted version of rooted unordered tree edit distance, which allows the edit operations of node relabeling, subtree pruning, and deletions of degree-2 nodes (where in the general edit distance variant nodes of arbitrary degrees may be deleted). Zhang gave an $O(n_T n_S(d_T + d_S)log(d_T + d_S))$ time algorithm for this variant, where d_T and d_S are the maximum node degrees in T and S respectively. In this sense, the algorithm of [20] can be viewed as a symmetric (allowing deletions from both input trees), yet rooted variant of the algorithm of [19].

Our Contribution. We propose an efficient algorithm for comparing unordered unrooted trees. Specifically, we define the *Homeomorphic Subtree Alignment* problem, for which we give an $O(n_T n_S min(d_T, d_S))$ running time algorithm. Our approach can be viewed as a generalization of the two previous works of [19] and [20], which relaxes the asymmetric "text-pattern" restriction of [19], as well as the rooting restriction of [20]. Both algorithms in [19] and [20], as well as the algorithm presented here, make use of subroutines for solving the *Minimum Cost Bipartite Matching* problem which dictate their time complexities. Here, we define the *All Pairs Cavity Bipartite Matching* problem, a generalization of the *All Cavity Bipartite Matching* problem [21], design an efficient algorithm for it and show how to integrate it into our tree alignment algorithm. This modification allows our algorithm to match, and even improve, the running times of the previous (less general) algorithms of [19] and [20]. We apply our new publicly available tool to the comparison of RNA trees, and report on some results from a preliminary experiment, revealing similarities which may not be detected by the traditional rooted ordered alignment approaches.

Due to space constraints figures, pseudocode, and proofs for all theorems and lemmas are deferred to an online supplemental materials found in our website.

2 Preliminaries

2.1 Tree Notations

A *tree* $T = (V, E)$ is a connected, undirected and acyclic graph. For a node $v \in V$, denote by $N(v)$ the set of *neighbors* of v: $N(v) = \{u \in V : (v, u) \in E\}$. Denote by $d_v = |N(v)|$ the *degree* of v. A node v for which $d_v \leq 1$ is called a *leaf* in T. For simplicity, we henceforth use the notation $v \in T$ and $(v, u) \in T$ to imply that v is a node and (v, u) is an edge in a tree T. We use the notation $(v \rightarrow u)$ to indicate that the generally undirected edge (v, u) is being considered with respect to the specific direction from v to u. Denote by n_T and d_T the number of nodes and the maximum degree of a node in T, respectively.

A *rooted tree* is a tree in which one of the nodes is selected as its *root*. Denote by T^v the tree T when rooted upon the node $v \in T$. An *ordered* tree is a tree T in which for each node $v \in T$, the elements in $N(v)$ are ordered. In this work we consider *unrooted unordered* trees, *rooted unordered* trees, *unrooted ordered* trees, and *rooted ordered* trees. If no indication is given, we assume that the mentioned trees are unrooted and unordered.

Let $T = (V, E)$ be a tree. A *smoothing of a node* v of degree 2 in T is obtained by removing v from T and connecting its two neighbors by an edge. A *smoothing* of T is a tree obtained by smoothing zero or more nodes in T. A *subtree* of T is a connected subgraph of T. For an edge $(v \to u) \in T$, denote by T_u^v the *rooted subtree* of T induced by v as a root, and all nodes x in T such that the path between v and x in T starts with $(v \to u)$. Since a tree T with n nodes contains $n-1$ undirected edges, the total number of directed edges, and hence the number of rooted subtrees of the form T_u^v, is $2(n-1)$.

A *pruning* of a tree T with respect to an edge $(v \to u)$ is the removal from T of all nodes in T_u^v, except for v. Observe that every nonempty subtree of T is obtained by pruning T with respect to zero or more edges.

2.2 Homeomorphic Subtree Alignment

An *isomorphic alignment* between two trees $T = (V, E)$ and $S = (V', E')$ is a bijection $A : V \to V'$, such that for every pair of nodes $v, u \in V$ we have that $(v, u) \in E \Leftrightarrow (A(v), A(u)) \in E'$. A *homeomorphic alignment* between T and S is an isomorphic alignment between some smoothing T' of T and some smoothing S' of S, and a *homeomorphic subtree alignment* (HSA) between T and S is a homeomorphic alignment between some subtree T' of T and some subtree S' of S (Figure S1). For short, we write $(v, v') \in A$ to indicate that $A(v) = v'$.

Let T' and S' be the subtrees of T and S, and let T'' and S'' be the smoothings of T' and S', respectively, such that A is an isomorphic alignment between T'' and S''. Say that a node $v \in T$ is *aligned* by A if $v \in T''$, and that v is *smoothed* by A if $v \in T'$ and $v \notin T''$. Say that a subtree T_u^v is *pruned* by A if $v \in T'$ and $u \notin T'$. Let $prune(T_u^v)$ be a cost associated with pruning the subtree T_u^v from T, $smooth(v)$ be a cost associated with smoothing node v, and $align(v, v')$ be a cost associated with aligning node v against some node v'. Definitions for S are similar. Denote by $\pi(A)$ the set of pruned subtrees and by $\delta(A)$ the set of smoothed nodes of T and S, with respect to A. Define the *alignment cost*:

$$w(T, S, A) = \sum_{(v,v') \in A} align(v, v') + \sum_{T' \in \pi(A)} prune(T') + \sum_{v \in \delta(A)} smooth(v) \quad (1)$$

Denote by $HSA(T, S)$ the minimum alignment cost of an HSA between T and S, and call an HSA A *optimal* with respect to T and S if $w(T, S, A) = HSA(T, S)$. The *Min-Cost HSA* problem is, given a pair of trees T and S, to compute $HSA(T, S)$.

Rooted and Ordered Alignments. In addition to the general Min-Cost HSA problem, we also consider special cases of the problem in which the two input trees are rooted and/or ordered, and the alignment is required to satisfy certain restrictions with respect to these additional properties. For two rooted trees T^v and $S^{v'}$, say that A is a *rooted HSA* between T^v and $S^{v'}$ if A is an HSA between T^v and $S^{v'}$, and $(v, v') \in A$. The definition of *ordered HSA* requires some additional formalism, related to *bipartite matchings*.

Let X and Y be two sets. A *bipartite matching* M between X and Y is a set of pairs $M \subseteq X \times Y$, such that each element in $X \cup Y$ participates in at most one pair in M. If some element $z \in X \cup Y$ does not participate in any pair in M, we say that z is *unmatched* by M and denote $z \notin M$. When $X = \langle x_1, x_2, \ldots, x_n \rangle$ and $Y = \langle y_1, y_2, \ldots, y_m \rangle$ are ordered sets, say that M *preserves linear order* if for every $(x_i, y_j), (x_{i'}, y_{j'}) \in M$, $i \le i' \Leftrightarrow j \le j'$. Say that M *preserves cyclic order* if M preserves linear order with respect to X and some rotation $Y_k = \langle y_k, y_{k+1}, \ldots, y_m, y_1, \ldots, y_{k-1} \rangle$ of Y.

Let A be an HSA between the trees T and S. For an edge $(v \to u) \in T$, say that u is a *relevant neighbor* of v (with respect to A) if there is some $x \in T_u^v, x \neq v$, which is aligned by A. Define relevant neighbors in S similarly.

Lemma 1. *Let $(v, v') \in A$, and let u be a relevant neighbor of v. Then, there is a unique relevant neighbor u' of v' such that for every $(y, y') \in A$, $y \in T_u^v \Leftrightarrow y' \in S_{u'}^{v'}$.*

Lemma 1 implies that for $(v, v') \in A$, the alignment induces a bipartite matching $M_{v,v'}^A$ between $N(v)$ and $N(v')$ in which the matched elements are exactly those relevant neighbors of v and v' (Figure S2). Say that A is *ordered* if T and S are ordered trees, and for every $(v, v') \in A$, the corresponding bipartite matching $M_{v,v'}^A$ is cyclically ordered. Now, we can define three additional variants of the HSA problem. Let T^v and $S^{v'}$ be rooted and ordered trees. Denote by $Ordered\text{-}HSA(T, S)$, $Rooted\text{-}HSA(T^v, S^{v'})$ and $Ordered\text{-}Rooted\text{-}HSA(T^v, S^{v'})$ the minimum costs of an ordered HSA, a rooted HSA, and an ordered and rooted HSA between T^v and $S^{v'}$, respectively. Define the corresponding variants of the Min-Cost HSA problem whose goals are to compute these values.

2.3 Min-cost Bipartite Matching

Similarly to previous tree alignment and edit distance algorithms [19–21], the algorithm presented here makes use of min-cost bipartite matching algorithms as subroutines. We next define an extended variant of the matching problem, in which the score incorporates penalties for unmatched elements, in addition to standard element matching scoring terms.

Let X and Y be two sets. A (generalized) *matching cost function* w for X and Y assigns costs $w(x, y)$ for every $(x, y) \in X \times Y$ and costs $w(z)$ for every $z \in X \cup Y$. The *cost* of a bipartite matching M between X and Y with respect to w is given by $w(M) = \sum_{(x,y) \in M} w(x, y) + \sum_{z \in X \cup Y, z \notin M} w(z).$

A *matching input instance* is a triplet (X, Y, w), where X and Y are two sets, and w is a matching cost function for X and Y. The *Min-Cost Bipartite Matching* problem (MCM) is, given a matching instance (X, Y, w), to find the minimum cost of a matching between X and Y with respect to w. Denote by $MCM(X, Y, w)$ the *solution* of the MCM problem for the instance (X, Y, w), and call a matching whose cost equals to the solution *optimal*.

Numerous works study and suggest algorithms for the MCM problem, usually when no unmatched element costs are taken into account [22–25]. A standard approach is to reduce *MCM* to the *Min-Cost Max-Flow* problem, which may be solved in a cubic time with respect to the size of the input sets. In [20], an adapted reduction was presented which also generalizes the problem definition to incorporate unmatched element costs and runs in $O(nm(n+m))$ time, where n and m are the sizes of the two sets in the matching instance. In the supporting materials we give a refined reduction for the generalized variant, which obtains the running time of $O(nm\min(n,m))$.

In addition to the definition above, the tree alignment definitions and algorithms we present in this work make use of the following derivatives of the *MCM* problem. The *Linear Ordered* MCM problem (*Linear-MCM*) and the *Cyclic Ordered* MCM problem (*Cyclic-MCM*) are defined similarly to *MCM*, with the restrictions that the considered matchings have to preserve linear or cyclic order, respectively.

3   A Basic Algorithm for Homeomorphic Subtree Alignment

In this section we describe a basic algorithm for HSA for its unordered unrooted variant (though it is adequate for the other variants as well).

Let A be an HSA between T and S. Let $(v,v') \in A$, and $M^A_{v,v'}$ the corresponding bipartite matching between $N(v)$ and $N(v')$, as defined in Section 2.2. Note that A can be viewed as a rooted alignment between T^v and $S^{v'}$, which is the union of a set of rooted sub-alignments A^v_u between rooted subtree pairs of the form T^v_u and $S^{v'}_{u'}$, where $(u,u') \in M^A_{v,v'}$ (Figure S2). The alignment cost can therefore be obtained by summing the costs of these sub-alignments, which cover all scoring terms implied by matching nodes, smoothing nodes, and pruning subtrees by the corresponding sub-alignments, and the additional pruning costs of pruned subtrees of the forms T^v_u and $S^{v'}_{u'}$ (where u,u' are unmatched by $M^A_{v,v'}$). Note that the pair (v,v') belongs by definition to each of the sub-alignments A^v_u. In order to avoid multiple additions of the term $align(v,v')$ when summing sub-alignment costs, define $w^{-r}(T^v, S^{v'}, A) = w(T^v, S^{v'}, A) - align(v,v')$. The cost $w(T^v, S^{v'}, A)$ can then be written as follows:

$$w(T^v, S^{v'}, A) = w^{-r}(T^v, S^{v'}, A) + align(v,v') \tag{2}$$

$$w^{-r}(T^v, S^{v'}, A) = \sum_{\substack{u \in N(v), \\ u \notin M^A_{v,v'}}} prune(T^v_u) + \sum_{\substack{u' \in N(v'), \\ u' \notin M^A_{v,v'}}} prune(S^{v'}_{u'}) + \sum_{(u,u') \in M^A_{v,v'}} w^{-r}(T^v_u, S^{v'}_{u'}, A^v_u) \tag{3}$$

Call a rooted alignment *non-trivial* if it aligns at least one additional pair of nodes besides the roots. Note that every rooted sub-alignment A^v_u is non-trivial (since u and u' are relevant neighbors of v and v'). Denote by $Rooted\text{-}HSA^{-r}\left(T^v_u, S^{v'}_{u'}\right)$ the minimum w^{-r} cost of a non-trivial rooted alignment between T^v_u and $S^{v'}_{u'}$. Clearly, if A is an optimal rooted HSA between T^v and $S^{v'}$, then for each $(u,u') \in M^A_{v,v'}$ $w^{-r}\left(T^v_u, S^{v'}_{u'}, A^v_u\right) = Rooted\text{-}HSA^{-r}\left(T^v_u, S^{v'}_{u'}\right)$ (otherwise, it is possible to produce a rooted alignment with a better cost than A for

T^v and $S^{v'}$). Define the bipartite matching instance $(N(v), N(v'), w_{v,v'})$, where for $u \in N(v)$, $u' \in N(v')$, set $w_{v,v'}(u, u') = Rooted\text{-}HSA^{\text{-}r}(T_u^v, S_{u'}^{v'})$, $w_{v,v'}(u) = prune(T_u^v)$, and $w_{v,v'}(u') = prune(S_{u'}^{v'})$. Observe that the right-hand side of Equation 3 equals to the cost of the bipartite matching $M_{v,v'}^A$ for the matching instance $(N(v), N(v'), w_{v,v'})$ (see Fig. S2b). In addition, every bipartite matching between $N(v)$ and $N(v')$ corresponds to some valid rooted HSA between T^v and $S^{v'}$, so that the matching and alignment costs are equal. Therefore, a minimum cost bipartite matching induces a minimum cost alignment, and we get that

$$Rooted\text{-}HSA(T^v, S^{v'}) = align(v, v') + MCM\left(N(v), N(v'), w_{v,v'}\right) \qquad (4)$$

Assuming the non-degenerate case where an optimal HSA between T and S contains at least one pair (v, v'), we can compute the cost of an optimal HSA by considering all possible such pairs for all the sub-instances:

$$HSA(T, S) = \min_{v \in T, v' \in S} Rooted\text{-}HSA(T^v, S^{v'}) \qquad (5)$$

In order to obtain cost functions of the form $w_{v,v'}$ for the computation of Equation 4, we need to compute solutions of the form $Rooted\text{-}HSA^{\text{-}r}(T_u^v, S_{u'}^{v'})$ for sub-instances of the input. When u and u' are leaves, the only non-trivial rooted alignment between T_u^v and $S_{u'}^{v'}$ contains both pairs (v, v') and (u, u'), and therefore we get that $Rooted\text{-}HSA^{\text{-}r}\left(T_u^v, S_{u'}^{v'}\right) = align(u, u')$. Otherwise, Equation 6, whose correction is shown in the supporting materials, computes $Rooted\text{-}HSA^{\text{-}r}\left(T_u^v, S_{u'}^{v'}\right)$ recursively (see Figure S3), where $w_{u,u'}^{v'',v''}$ is defined similarly to $w_{u,u'}$ with respect to the sets $N(u) \setminus \{v\}$ and $N(u') \setminus \{v'\}$.

$$Rooted\text{-}HSA^{\text{-}r}\left(T_u^v, S_{u'}^{v'}\right) =$$

$$\min \begin{cases} I.\ smooth(u) + \min_{x \in N(u) \setminus \{v\}} \left(Rooted\text{-}HSA^{\text{-}r}\left(T_x^u, S_{u'}^{v'}\right) + \sum_{y \in N(u) \setminus \{v,x\}} prune(T_y^u) \right) \\ II.\ smooth(u') + \min_{x' \in N(u') \setminus \{v'\}} \left(Rooted\text{-}HSA^{\text{-}r}\left(T_u^v, S_{x'}^{u'}\right) + \sum_{y' \in N(u') \setminus \{v',x'\}} prune(S_{y'}^{u'}) \right) \\ III.\ align(u, u') + MCM\left(N(u) \setminus \{v\}, N(u') \setminus \{v'\}, w_{u,u'}^{v,v'}\right) \end{cases}$$

$$(6)$$

The computation of $w_{u,u'}^{v,v'}$ requires the computation of scores of the form $Rooted\text{-}HSA^{\text{-}r}\left(T_x^u, S_{x'}^{u'}\right)$ for all $x \in N(u) \setminus \{v\}$ and all $x' \in N(u') \setminus \{v'\}$. It can be shown that all $Rooted\text{-}HSA^{\text{-}r}$ solutions required for the computation of the right-hand side of the equation are for strictly smaller sub-instances than the sub-instance appearing in the left-hand side, thus the termination of the recursive computation is guaranteed. Equation 6 can be efficiently computed using Dynamic Programming (DP), as summarized by Algorithm 1 below.

Algorithm 1. $HSA(T, S)$

---

1 Construct a DP matrix H of size $2(n_T - 1) \times 2(n_S - 1)$, where n_T and n_S are the numbers of nodes in T and S, respectively. The rows of H correspond to subtrees T_u^v and the columns correspond to subtrees $S_{u'}^{v'}$, ordered with nondecreasing sizes (see Figure S4c);

2 Fill the entries of H row by row with increasing row indices, each row column by column, with increasing column indices. Each entry is filled according to Equation 6 with respect to the pair of subtrees corresponding to its row and column (all solutions for relevant instances in the right-hand side of the equation are already computed and stored in H, due to the order-by-size organization of the subtrees along the rows/columns of the matrix);

3 Compute Equation 4 for every T^v, $S^{v'}$ such that $v \in T, v' \in S$;

4 Compute Equation 5 to return $HSA(T, S)$;

In the supporting materials we show that a straightforward implementation of Algorithm 1 obtains the running time of $O(n_T n_S d_T d_S \min(d_T, d_S))$. For some trees with $O(n)$ nodes and a maximum node degree of $O(n)$ (e.g. "star" trees), this implies an $O(n^5)$ running time. In the following section, we show how to improve this time bound and obtain a cubic time algorithm for the problem.

We note that Algorithm 1 generalizes to also solve the ordered unrooted, unordered rooted, and ordered rooted variants of *Min-Cost HSA* in polynomial time. In case a rooted alignment is sought, the algorithm can compute Equation 4 in line 3 only with respect to the two roots, and avoid the computation of Equation 5. In case an ordered alignment is sought, the *MCM* application in Equations 4 can be replaced by *Cyclic-MCM*, and in Equation 6 *MCM* can be replaced by *Linear-MCM*, similarly to [20]. Traditionally, ordered matchings are implemented via reduction to sequence alignment [8,19,26]. Similarly to the algorithm given in the next section for unordered tree alignments, fast incremental/decremental versions of ordered matchings [27–29] can be integrated into the ordered variant of our algorithm to improve its time complexity. However, the detailed description of these techniques are beyond the scope of this paper.

3.1   Improving the Algorithm's Time Complexity

Next, we show how to improve the time complexity of Algorithm 1 by incorporating new *cavity matching* algorithms. The *All-Cavity-MCM* problem [21] is, given a matching instance (X, Y, w), to compute $MCM(X, Y \setminus \{y\}, w)$ for all $y \in Y$. The *All-Pairs-Cavity-MCM* problem is, given a matching instance (X, Y, w), to compute $MCM(X \setminus \{x\}, Y \setminus \{y\}, w)$ for all $x \in X$ and $y \in Y$. Clearly, algorithms for both *All-Cavity-MCM* and *All-Pairs-Cavity-MCM* problems can be implemented by repeatedly running an algorithm for *MCM* on all required inputs. In [21], an algorithm for *All-Cavity-MCM* was proposed which is more efficient than the naïve algorithm, and retains the same cubic running time as the standard algorithm for *MCM*. To the best of our knowledge, no algorithm for *All-Pairs-Cavity-MCM* which improves upon the naïve algorithm was previously described. Algorithm 2 in the supplemental materials efficiently solves the *All-Pairs-Cavity-MCM* problem in the same time complexity as the standard *MCM* algorithm. Intuitively, this algorithm obtains a cubic running time by exploiting an observation that a solution for the sub-instance $(X \setminus \{x\}, Y \setminus \{y\}, w)$ is obtained by summing the solutions for the complete instance (X, Y, w) with the minimum cost of a path from y to x in the corresponding residual flow network. The running times of all three unordered bipartite matching problems are summarized in the following theorem.

Theorem 1. *Let (X, Y, w) be a matching instance. Then, each one of the problems* MCM, All-Cavity-MCM, *and* All-Pairs-Cavity-MCM *over the instance (X, Y, w) can be solved in $O(|X||Y| \min(|X|, |Y|))$ running time.*

The following lemma identifies *MCM* computations as the time consuming bottleneck of Algorithm 1.

Lemma 2. *It is possible to implement Algorithm 1 so that all operations, besides computation of solutions to the* MCM *problem, require $O(n_T n_S)$ running time.*

It thus remains to analyze the time required for executing *MCM* computations. Such computations are applied when computing term *III* of Equation 6 in line 2, and when computing Equation 4 in line 3. In what follows, we explain how cavity-matching algorithms can be used to speed-up these computations.

Let $index(T_u^v)$ and $index(S_{u'}^{v'})$ denote the row and column indices of T_u^v and $S_{u'}^{v'}$ in H, respectively. Let $u \in T$ and $u' \in S$ be a pair of nodes. The set of subtrees T_u'' for $v \in N(u)$ corresponds to a subset of rows in H. Similarly, the set of subtrees $S_{u'}^v$ for $v' \in N(u')$ corresponds to a subset of columns in H, and thus all solutions of the form *Rooted-HSA*$^{-r}\left(T_u^v, S_{u'}^{v'}\right)$ are stored in a sub-matrix of H of size $d_u \times d_{u'}$ (Figure S5). Let $H_{u,u'}$ denote this sub-matrix, and let v_i and v_j' denote nodes in $N(u)$ and $N(u')$ such that $T_u^{v_i}$ and $S_{u'}^{v_j'}$ correspond to the i-th row and j-th column in $H_{u,u'}$, respectively (i.e. $index(T_u^{v_1})$ is the first row in $H_{u,u'}$, $index(T_u^{v_2})$ is the second row, etc.).

The following observation identifies special properties of the second column and second row in $H_{u,u'}$, which are exploited later on for the incorporation of cavity matching subroutines within the DP framework of Algorithm 1. Observe that for every $1 < i \le d_u$, $T_{v_i}^u$ is a subtree of $T_u^{v_1}$, and therefore $index(T_{v_i}^u) < index(T_u^{v_1})$. Also, $T_{v_1}^u$ is a subtree of $T_u^{v_2}$, and therefore $index(T_{v_1}^u) < index(T_u^{v_2})$. Since $index(T_u^{v_1}) < index(T_u^{v_2})$, we get the following observation (Figure S4):

Observation 1. *For every $1 \le i \le d_u$, $index(T_{v_i}^u) < index(T_u^{v_2})$. Similarly, for every $1 \le j \le d_{u'}$, $index(S_{v_j'}^u) < index(S_{u'}^{v_2'})$.*

Consider the computation of term *III* of Equation 6 for instances in the first row in $H_{u,u'}$. Note that for the entries in this row, the first group in the bipartite matching instance is fixed and equals to $N(u) \setminus \{v_1\}$, whereas for each column j, the second group in the matching instance is $N(u') \setminus \{v'_j\}$. The first entry in this row is computed by solving the MCM problem directly for the matching instance $(N(u) \setminus \{v_1\}, N(u') \setminus \{v'_1\}, w_{u,u'}^{v_1,v'_1})$ (Figures S5a, S5b). Based on Observation 1, upon reaching the second entry in this row, all solutions $Rooted\text{-}HSA^{-r}\left(T_{v_i}^u, S_{v'_j}^{u'}\right)$ for $i > 1$ and $j \geq 1$ are computed and stored in H. Therefore, the $All\text{-}Cavity\text{-}MCM$ problem can be solved for the matching instance $(N(u) \setminus \{v_1\}, N(u'), w_{u,u'}^{v_1})$, where $w_{u,u'}^{v_1}$ is defined similarly as $w_{u,u'}$ with respect to the sets $N(u) \setminus \{v_1\}$ and $N(u')$. This allows to compute term III for each one of the remaining entries in this row of $H_{u,u'}$ in $O(1)$ time (Figures S5c, S5d).

The first entry of the second row in $H_{u,u'}$ is again computed directly by solving the MCM problem for the matching instance $(N(u) \setminus \{v_2\}, N(u') \setminus \{v'_1\}, w_{u,u'}^{v_2,v'_1})$. Upon reaching the second entry of the second row of $H_{u,u'}$, Observation 1 implies that all solutions $Rooted\text{-}HSA^{-r}\left(T_{v_i}^u, S_{v'_j}^{u'}\right)$ for $i, j \geq 1$ are already computed and stored in H. Therefore, the $All\text{-}Pairs\text{-}Cavity\text{-}MCM$ problem can be solved for the matching instance $(N(u), N(u'), w_{u,u'})$, allowing to compute term III for each one of the remaining entries in $H_{u,u'}$ in $O(1)$ time (Figures S5e, S5f). Thus, computing term III for all entries in $H_{u,u'}$ is done by solving the MCM problem twice, solving the $All\text{-}Cavity\text{-}MCM$ problem once, and solving the $All\text{-}Pairs\text{-}Cavity\text{-}MCM$ problem once, where the sizes of the two groups in the matching instances for these problems are at most d_u and $d_{u'}$. Based on Theorem 1, this whole computation takes $O(d_u d_{u'} \min(d_u, d_{u'}))$ time. As all computations of the MCM problem conducted by the algorithm are done either in the computation of term III for some entry in a submatrix $H_{u,u'}$, or for some pair $v \in T$, $v' \in S$ when computing Equation 4 in line 3, we can argue that in total, for each pair $u \in T$, $u' \in S$, the algorithm spends $O(d_u d_{u'} \min(d_u, d_{u'}))$ in computations of the MCM problem. As $\sum_{u \in T} \sum_{u' \in S} d_u d_{u'} \min(d_u, d_{u'}) \leq \min(d_T, d_S) \sum_{u \in T} d_u \sum_{u' \in S} d_{u'} = O(\min(d_T, d_S) n_T n_S)$, the total running time for solving the MCM problem along the entire run of the algorithm is $O(n_T n_S \min(d_T, d_S))$. Based on Lemma 2, all other operations require $O(n_T n_S)$ time, and we get the following theorem:

Theorem 2. *Algorithm 1 can be implemented with an $O(n_T n_S \min(d_T, d_S))$ time complexity.*

In the supplemental materials we give a refined time complexity analysis showing that the algorithm's time is in fact $O(n_T n_S + L_T L_S \min(d_T, d_S))$, where L_T and L_S are the number of leaves in T and S, respectively. In addition, we show that for RNA trees T in our specific application it is expected that $L_T \ll n_T$.

4   Results

Our algorithm is implemented (in Java) as a tool called FRUUT (Fast RNA Unordered Unrooted Tree mapper). The RNA tree representation and scoring scheme employed by FRUUT are described in detail in Section S5. FRUUT allows the user to select any alignment mode (rooted/unrooted, ordered/unordered, local/global) and to compute optimal pairwise alignments of RNA trees. We also provide an interactive PHP web-server for running FRUUT in our website (RNA plots are are generated by the Vienna Package [30]).

RNase P is the endoribonuclease responsible for the 5' maturation of tRNA precursors [15]. Secondary structures of bacterial RNase P RNAs have been studied in detail, primarily using comparative methods [31], and were shown to share a common core of primary and secondary structure. In bacteria, synthetic minimal RNase P RNAs consisting only of these core sequences and structures were shown to be catalytically proficient. Sequences encoding RNase P RNAs of various genomes have been determined and a huge database established [32], which consists of a compilation of ribonuclease P RNA sequences, sequence alignments, secondary structures, three-dimensional models and accessory information.

We conducted a preliminary experiment, intended to identify examples of pairs of RNA trees for which an RNA structural comparison approach supporting unrooting and branch shuffling may detect (otherwise hidden) structural similarity. To achieve this, we ran a benchmark of all-against-all pairwise alignments of bacterial RNAse P RNA secondary structure trees, using our tool's different tree-alignment modes and comparing the differences between the obtained alignment costs. The alignment cost functions and parameters used in our experiment are given in Section S5.2.

Our benchmark was based on 470 RNAse P structures, ranging across various organisms, taken from the RNAse P database [32] (molecule naming conventions are according to [33]). After filtering out partial and environmental sequences, 170 distinct structures remained, yielding 14,365 distinct pairs of trees. The sizes of the trees in this dataset ranged from 82 to 230 nodes, averaging at 141.99. The total running time of the benchmark was approximately 33 minutes on a single Xeon X5690 using around 300Mb of memory.

Each pair of trees T, S was compared in two modes to obtain the corresponding scores and alignments: rooted-ordered (RO) and rooted-unordered (RU). Within each mode, we used a relative score formula described by Höchsmann et al. [34] to assess the similarity of two trees, normalizing the alignment cost by the average of the self-alignment costs of the compared trees. Let $HSA_m(T, S)$ denote the optimal alignment cost of trees T and S in alignment mode m, where $m \in \{RO, RU\}$. Let $RelScore_m(T, S)$ denote the relative score of T and S in alignment mode m, given by $RelScore_m(T, S) = \frac{2HSA_m(T,S)}{HSA_m(T,T)+HSA_m(S,S)}$. The scoring scheme we have satisfies that for every tree T, $HSA_m(T,T) < 0$, and for every pair of trees T, S, $HSA_m(T,T), HSA_m(S,S) \leq HSA_m(T,S)$. Under these conditions, the relative score for any pair of trees is upper bounded by 1, and the similarity of the trees increases as the score approaches 1.

Our goal in this experiment was to identify evolutionary events that can be explained by unordered alignments. Thus, we sought pairs of RNAseP RNAs that are highly conserved, and yet their alignment can still benefit from unordered mappings. To achieve this, we removed from the set pairs of trees for which

$RelScore_{RU}(T, S) < 0.5$. We sorted the remaining pairs of trees according to $RelScore_{RU}(T, S) - RelScore_{RO}(T, S)$.

When examining the top 50 alignments carefully, two distinct types of mapping patterns were observed among them, where each of the top 50 pairs belongs (with slight variations) to one of these two types (33 to Type 1 and 17 to Type 2). In Section S6, we exemplify the highest ranking alignment of each of the two types (the first type is shown in Figure 1b). As mentioned before, the input for FRUUT alignments consisted only of sequence and secondary structure information. The tertiary structure (pseudoknot annotations) for the top-ranking alignments were only considered later, during the alignment interpretations.

Another type of homology detected by our tool is exemplified in the HammerHead family, which is characterized by two distinct transcript types yielding the same functional RNA (Figure 1a). Our tool can detect the similarity between these two HammerHead types by applying the unrooted tree alignment mode. Additional details regarding the experiment for detecting such unroooted alignments are given in the supplemental materials.

Acknowledgments. The authors are grateful to the anonymous WABI referees for their helpful comments, and to Naama Amir for pointing out challenges in extending previous algorithms to allow deletions from both trees. This research was partially supported by ISF grant 478/10 and by the Frankel Center for Computer Science at Ben Gurion University of the Negev.

References

1. Andronescu, M., Bereg, V., Hoos, H., Condon, A.: Rna strand: the rna secondary structure and statistical analysis database. BMC Bioinformatics 9, 340 (2008)
2. Agmon, I., Auerbach, T., Baram, D., Bartels, H., Bashan, A., Berisio, R., Fucini, P., Hansen, H., Harms, J., Kessler, M., et al.: On peptide bond formation, translocation, nascent protein progression and the regulatory properties of ribosomes. European Journal of Biochemistry 270, 2543–2556 (2003)
3. Hofacker, I., Fontana, W., Stadler, P., Bonhoeffer, L., Tacker, M., Schuster, P.: Fast folding and comparison of RNA secondary structures. Monatshefte fur Chemie/Chemical Monthly 125, 167–188 (1994)
4. Steffen, P., Voss, B., Rehmsmeier, M., Reeder, J., Giegerich, R.: RNAshapes: an integrated RNA analysis package based on abstract shapes (2006)
5. Hochsmann, M., Toller, T., Giegerich, R., Kurtz, S.: Local similarity in RNA secondary structures. In: Proceedings of the 2003 IEEE Bioinformatics Conference, CSB 2003, pp. 159–168. IEEE (2003)
6. Jiang, T., Lin, G., Ma, B., Zhang, K.: A general edit distance between RNA structures. Journal of Computational Biology 9, 371–388 (2002)
7. Zhang, K., Wang, L., Ma, B.: Computing Similarity between RNA Structures. In: Crochemore, M., Paterson, M. (eds.) CPM 1999. LNCS, vol. 1645, pp. 281–293. Springer, Heidelberg (1999)
8. Bille, P.: A survey on tree edit distance and related problems. Theoretical Computer Science 337, 217–239 (2005)

9. Schirmer, S., Giegerich, R.: Forest Alignment with Affine Gaps and Anchors. In: Giancarlo, R., Manzini, G. (eds.) CPM 2011. LNCS, vol. 6661, pp. 104–117. Springer, Heidelberg (2011)

10. Allali, J., Sagot, M.-F.: A Multiple Graph Layers Model with Application to RNA Secondary Structures Comparison. In: Consens, M.P., Navarro, G. (eds.) SPIRE 2005. LNCS, vol. 3772, pp. 348–359. Springer, Heidelberg (2005)

11. Blin, G., Denise, A., Dulucq, S., Herrbach, C., Touzet, H.: Alignments of RNA structures. IEEE/ACM Transactions on Computational Biology and Bioinformatics 7, 309–322 (2010)

12. Jan, E.: Divergent ires elements in invertebrates. Virus Research 119, 16–28 (2006)

13. Perreault, J., Weinberg, Z., Roth, A., Popescu, O., Chartrand, P., Ferbeyre, G., Breaker, R.: Identification of hammerhead ribozymes in all domains of life reveals novel structural variations. PLoS Computational Biology 7, e1002031 (2011)

14. Birikh, K., Heaton, P., Eckstein, F.: The structure, function and application of the hammerhead ribozyme. European Journal of Biochemistry 245, 1–16 (1997)

15. Haas, E., Brown, J.: Evolutionary variation in bacterial RNase P RNAs. Nucleic Acids Research 26, 4093–4099 (1998)

16. Zhang, K., Jiang, T.: Some MAX SNP-hard results concerning unordered labeled trees. Information Processing Letters 49, 249–254 (1994)

17. Shamir, R., Tsur, D.: Faster subtree isomorphism. J. of Algorithms 33, 267–280 (1999)

18. Chung, M.: O (n2. 5) time algorithms for the subgraph homeomorphism problem on trees. Journal of Algorithms 8, 106 112 (1987)

19. Pinter, R.Y., Rokhlenko, O., Tsur, D., Ziv-Ukelson, M.: Approximate labelled subtree homeomorphism. Journal of Discrete Algorithms 6, 480–496 (2008)

20. Zhang, K.: A constrained edit distance between unordered labeled trees. Algorithmica 15, 205–222 (1996)

21. Kao, M., Lam, T., Sung, W., Ting, H.: Cavity matchings, label compressions, and unrooted evolutionary trees. Arxiv preprint cs/0101031 (2001)

22. Edmonds, J., Karp, R.: Theoretical improvements in algorithmic efficiency for network flow problems. Journal of the ACM (JACM) 19, 248–264 (1972)

23. Fredman, M., Tarjan, R.: Fibonacci heaps and their uses in improved network optimization algorithms. Journal of the ACM (JACM) 34, 596–615 (1987)

24. Gabow, H., Tarjan, R.: Faster scaling algorithms for network problems. SIAM Journal on Computing 18, 1013 (1989)

25. Orlin, J., Ahuja, R.: New scaling algorithms for the assignment and minimum mean cycle problems. Mathematical Programming 54, 41–56 (1992)

26. Zhang, K.: Algorithms for the constrained editing distance between ordered labeled trees and related problems. Pattern Recognition 28, 463–474 (1995)

27. Maes, M.: On a cyclic string-to-string correction problem. Information Processing Letters 35, 73–78 (1990)

28. Schmidt, J.P.: All highest scoring paths in weighted grid graphs and their application to finding all approximate repeats in strings. SIAM J. of Computing 27, 972–992 (1998)

29. Tiskin, A.: Semi-local string comparison: Algorithmic techniques and applications. Mathematics in Computer Science 1, 571–603 (2008)

30. Hofacker, I.: Vienna RNA secondary structure server. Nucleic Acids Research 31, 3429 (2003)

31. Pace, N.R., Brown, J.W.: Evolutionary perspective on the structure and function of ribonuclease P, a ribozyme. J. Bacteriol. 177, 1919–1928 (1995)
32. Brown, J.: The ribonuclease p database. Nucleic acids research 27, 314 (1999)
33. Andronescu, M., Bereg, V., Hoos, H.H., Condon, A.: RNA STRAND: the RNA secondary structure and statistical analysis database. BMC Bioinformatics 9, 340 (2008)
34. Höchsmann, M.: The tree alignment model: algorithms, implementations and applications for the analysis of RNA secondary structures. PhD thesis, Universitätsbibliothek Bielefeld (2005)

Tree Decomposition and Parameterized Algorithms for RNA Structure-Sequence Alignment Including Tertiary Interactions and Pseudoknots

(Extended Abstract)

Philippe Rinaudo[1,2], Yann Ponty[3], Dominique Barth[2], and Alain Denise[1,4]

[1] LRI, Univ Paris-Sud, CNRS UMR8623 and INRIA AMIB, Orsay, F91405
[2] PRISM, Univ Versailles Saint-Quentin and CNRS UMR8144 Versailles, F78000
[3] LIX, Ecole Polytechnique, CNRS UMR7161 and INRIA AMIB, Palaiseau, F91128
[4] IGM, Univ Paris-Sud and CNRS UMR8621, Orsay, F91405

Abstract. We present a general setting for structure-sequence comparison in a large class of RNA structures, that unifies and generalizes a number of recent works on specific families of structures. Our approach is based on a *tree decomposition* of structures, and gives rise to a general parameterized algorithm having complexity in $\mathcal{O}(N \cdot m^t)$, where N (resp. m) is the structure (resp. sequence) length, and the exponent t depends on the family of structures. For each family considered by previous approaches, our contribution specializes into an algorithm whose complexity either matches or outperforms previous solutions.

1 Introduction

The problem of RNA structure-sequence comparison arises in two main kinds of contexts. It allows the search for a given structured RNA within a long sequence, or even a set of sequences, and helps in the task of three-dimensional modeling by homology. Jiang *et al* [6] addressed the problem of the pairwise comparison of RNA structures in its full generality. They defined the edit distance problem on RNA structures, represented as graphs, using a set of atomic edit operations. Notably, the authors gave a dynamic programming algorithm in $O(n \cdot m^3)$ time complexity for comparing a sequence to a *nested* structure, where n is the length of the sequence with known structure, and m the length of the sequence of unknown structure. They also established that the problem of computing such an edit distance is Max-SNP-hard if the structure is allowed to contain pseudoknots.

Meanwhile, considering all known interactions in RNAs, including non-canonical ones [8] and pseudoknots, is crucial to perform a precise structure-sequence alignment. This motivates the design of efficient algorithm for restricted instances of the problem. Jiang *et al* [6] devised a polynomial-time algorithm for pseudoknotted structures, by enforcing constraints on the costs of the edit operations. Other attempts are based on the observation that the so-called *H-type* and *kissing-hairpin* pseudoknots represent more than 80% of the pseudoknots in

B. Raphael and J. Tang (Eds.): WABI 2012, LNBI 7534, pp. 149–164, 2012.
© Springer-Verlag Berlin Heidelberg 2012

Table 1. Summary of existing algorithms for structure-sequence alignment. Our approach unifies all these algorithms and, for each class of structures captured by pre-existing works, specializes into time complexities that matches previous efforts: the tree structure represents an inclusion relation. Hence the root class RCS includes all other classes. Notation: k is the degree of the pseudoknot/(simple) standard structure.

Class of Structures	Time comp.	Multiple interactions	Ref.
RCS – Recursive Classical Structures	$O(n \cdot m^{k+2})$	✓	–
└─ PKF – Secondary Structures (Pseudoknot-free)	$O(n \cdot m^3)$		[6]
└─ ESP – Embedded Standard Pseudoknots.....................	$O(n \cdot m^{k+1})$		[5]
└─ SST – Standard Structures...................................	$O(n \cdot m^k)$	✓	–
└─ SPK – Standard Pseudoknots............................	$O(n \cdot m^k)$		[5]
└─ 2RP – 2-Level Recursive Simple Non-Standard Pseudoknots...	$O(n \cdot m^{k+2})$		[12]
└─ SNS – Simple Non-Standard Structures......................	$O(n \cdot m^{k+1})$	✓	–
└─ SNP – Simple Non-Standard Pseudoknots.................	$O(n \cdot m^{k+1})$		[12]
└─ ETH – Extended Triple Helices	$O(n \cdot m^3)$	✓	–
└─ STH – Triple Helices	$O(n \cdot m^3)$	✓	[11]

known structures [10]. For H-type pseudoknots, the alignment problem can be solved in $O(n \cdot m^3)$ [5,12]. Moreover, Han *et al* [5] introduced the class of *standard pseudoknots*, which contains both H-type and kissing hairpin pseudoknots. They gave $O(n \cdot m^k)$ and $O(n \cdot m^{k+1})$ algorithms, where k is the *degree* of the pseudoknot, respectively for single standard pseudoknots, and standard pseudoknots that are embedded into a nested structure. More recently, the general class of simple non-standard pseudoknots was defined, and an algorithm was given in $O(n \cdot m^{k+1})$ to align a single such structure to a sequence, or in $O(n \cdot m^{k+2})$ if the structure is embedded into a simple structure [12].

In the present paper, we give a general setting for sequence-structure comparison in a large class of RNA structures that unifies and generalizes all the above families of structures. Notably, our scheme supports structures whose nucleotides may pair with multiple partners, thus capturing non-canonical interactions [8], and RNA structural motifs [9]. Our approach is based on a *tree decomposition* of the input structure, and gives rises to a general parameterized algorithm, where the exponential part of the complexity depends on the family of structures. For each of the previously studied families, our algorithm has the same complexity as the specific algorithms that had been given before. Table 1 gives a summary of the previous works that are generalized by our approaches, and the time complexity of our algorithm for each of the classes.

2   Sequence-Structure Alignment

At first let us state some definitions, starting with the concept of arc-annotated sequence, which will be used as an abstract representation for RNA structure.

Definition 1 (Arc-annotated sequence). *An arc-annotated sequence is a pair (S, P), where S is a sequence over an alphabet Σ and P is a set of unordered pairs of positions in S.*

For RNA structures, S is the nucleotide sequence and P the set of the interactions over S, as illustrated by Figure 1 (Upper part). Accordingly, $\Sigma = \{A, U, G, C\}$, and any $(i, j) \in P$ represents an interaction between the nucleotides at position i and j. Nucleotides are numbered from 1 to n in the usual 5'→3' order, and $S[i]$ denotes the nucleotide at position i. Unlike classic definitions, any position i may be involved in multiple interactions. In the following, we sometimes refer to such an arc-annotated sequence as an **RNA graph interaction structure** or, for short, an **RNA graph**.

There exist several equivalent ways to define a structure-sequence alignment, that is an alignment between an arc-annotated sequence and a (plain) sequence. We represent an alignment as a partial mapping between positions in the arc-annotated sequence and positions in the plain sequence, as illustrated by Figure 1.

Definition 2 (Structure-Sequence Alignment). *A structure-sequence alignment between an arc-annotated sequence $A = (S_A, P_A)$ of size n and a plain sequence $B = (S_B, \varnothing)$ of size m is a partial mapping μ from $[1, n]$ to $[1, m + 1]$ such that:*

- *μ is injective.*
- *μ preserves the order: $\mu(i) < \mu(j) \Rightarrow i < j$.*

Let us denote by $\mathcal{F}(A, B)$ (or $\mathcal{F}$ when there is no ambiguity) the set of all possible alignments between A and B.

Remark that some positions in the arc-annotated sequence may have no corresponding position in the plain sequence. We note $\mu(i) = \bot$ if position i does not have an image by μ, and qualify it as **unmatched**. Consecutive sequences of unmatched positions in the structure ($\mu(i) = \bot$) or in the sequence ($\mu^{-1}(i) = \bot$) are usually grouped and scored together. A **(composite) gap** is then the maximum set of consecutive positions $(i, \ldots, j)$ that are either unmatched or do no have an antecedent through μ. The **length** $|g|$ of a gap g is the number of positions it contains. By grouping unmatched positions within gaps, one may handle affine penalties for gaps in the cost function.

Let us now define the cost of an alignment, which captures the similarity of two RNAs.

Definition 3 (Cost of a Sequence-Structure Alignment). *The cost of a structure-sequence alignment μ between an arc-annotated sequence $A = (S_A, P_A)$ and a plain sequence $B = (S_B, \varnothing)$ is defined by:*

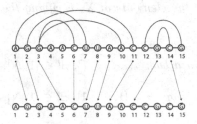

Fig. 1. A representation of a partial mapping between an arc-annotated sequence (upper part) and a (plain) sequence (lower part).

$$\text{Cost}(\mu) = \sum_{i \in [1,n], \mu(i) \neq \perp} \gamma(i, \mu(i)) + \sum_{gap \ g \subset A} \lambda_A(|g|) + \sum_{gap \ g \subset B} \lambda_B(|g|) + \sum_{(i,j) \in P_A} \varphi(i, j, \mu(i), \mu(j))$$

(1)

where

- $\gamma(i, \mu(i))$ *is the cost of a **base substitution** between position i in A and position $\mu(i)$ in B.*
- $\lambda_Y(x) = \alpha_Y \cdot x + \beta_Y$, *is the **affine cost penalty** for a gap of length x in a sequence Y.*
- $\varphi(i, j, \mu(i), \mu(j))$ *is the cost of an **arc removing** ($\mu(i) = \mu(j) = \perp$), **arc altering** ($\mu(i)$ or $\mu(j) = \perp$) or **arc substitution** involving paired positions i and j in A.*

Note that, unlike previously proposed scoring schemes [5,12], our scheme considers arc-alterations and arc-breakings as atomic operations [6], allowing for general cost schemes.

Definition 4 (Structure-Sequence Alignment Problem). *Given an arc-annotated sequence A and a plain sequence $B = (S_B, \varnothing)$, the structure-sequence alignment problem consists in finding a minimal-cost alignment between A and B.*

As stated in [13,2] the problem is NP-Hard when CROSSING interactions are considered (*i.e.* pseudoknots, at most one partner). Therefore, we will adopt a parameterized approach to handle general structures, including unrestricted crossing interactions and multiple interactions per position, assuming that the total number of interactions per position is bounded by a constant.

3   Tree Decomposition and Alignment Algorithm

Our approach relies on a tree decomposition of the input arc-annotated sequence. Tree decompositions are usually defined on graphs rather than on arc-annotated sequences. Here we give a straightforward adaptation, that preserves all the properties of the standard tree decompositions.

3.1   Definitions

Definition 5 (Tree Decomposition of an arc-annotated sequence). *Given an arc-annotated sequence $A = (S, P)$ of length n, a tree decomposition of A is a pair (X, T) where $X = \{X_1, \ldots, X_N\}$ is a set of **bags**, i.e. subsets of positions $\{i, i \in [1, n]\}$, and T is a tree whose nodes are the elements of X, verifying the following conditions:*

1. *Each position belongs to a bag: $\bigcup_{l \in [1,N]} X_l = [1, n]$.*
2. *Both ends of any interaction are present in a bag: $\forall (i, j) \in P, \exists l \in [1, N], \{i, j\} \subset X_l$.*
3. *Consecutive positions are both present in a bag: $\forall i \in [1, n-1], \exists l \in [1, N], \{i, i+1\} \subset X_l$.*

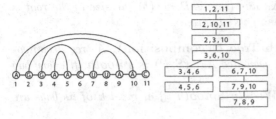

Fig. 2. An arc-annotated sequence and an associated tree decomposition. Each bag contains three positions, so the tree width of this tree decomposition is 2. As there is only one bag which does not respect the smooth property (the one with two children), the tree decomposition is 1-weakly-smooth.

4. For every X_l and X_s, $l, s \in [1, N]$, $X_l \cap X_s \subset X_r$ for all X_r on the path between X_l and X_s.

Definition 6 (Treewidth). *The width of a tree decomposition (X, T) is the size of its largest set X_l minus one. The **treewidth** $t_w(A)$ of an arc-annotated sequence A is the minimum width among all possible tree decompositions of A.*

The tree decomposition of an arc-annotated sequence is illustrated by Figure 2. In general, tree decompositions are not rooted. Nevertheless, for the sake of clarity, we will arbitrarily choose a root X_0 in our decompositions. Additionally, $X_{l,r}$ will be used as a shorthand for $X_l \cap X_r$, where X_l and X_r are two bags.

Our scoring scheme includes affine gaps penalties, therefore we must design a dynamic programming algorithm that supports such functions. This can be done through explicit control over the gap-length (i.e. a dedicated loop), leading to an increase, by an order of magnitude, of the time-complexity. However, under certain conditions, one may adapt a trick underlying Gotoh's algorithm [4] for the pairwise sequence alignment under affine gap penalties.

This trick relies on two matrices of equal dimensions, which correspond to gap opening (constant cost) and extensions (linear cost) respectively. The search for the optimal gap length is broken into unit steps, upon which the mapped position of an adjacent sequence element is lost for subsequent calls. While this is not error-prone for pairwise sequence alignment, our structure-sequence alignment requires additional care, since the precise position of the adjacent sequence element may be used deeper in the tree (e.g. if the position is involved in an interaction). Therefore we define a notion of smoothness of a bag, which enforces the presence of a single unmatched element, adjacent to a matched element whose position will no longer be used (and can be safely lost) while aligning further elements of the tree.

Definition 7 (Smooth Bag of a Tree Decomposition). *Let $X_l \in X$ be a bag in a tree decomposition (X, T) for an arc-annotated sequence $A = (S, P)$. If $X_l \neq X_0$, then let X_r be its father. X_l is then **smooth** iff there exist two consecutive positions i and j such that $i \in X_l - X_r$, $j \in X_{l,r}$, j is not in one of the children of X_l, and there is no $i' \in X_l - X_r$ such that $(i', j) \in P$ and no other position in $X_l - X_r$ consecutive to j. The root X_0 is **smooth** iff one of the two following conditions holds: (a) there exist two consecutive positions $i, j \in X_0$*

*such that j is not in any child of X_0 and $(i,j) \notin P$, or (b) the size of the root is
strictly smaller than the size of one of its children.*

Definition 8 ((Weakly-)Smooth Tree Decomposition). *A tree decomposition (X,T) for an arc-annotated sequence $A = (S,P)$ is **smooth** iff every bag
$X_l \in X$ is smooth.*
*A tree decomposition (X,T) is k-**Weakly-Smooth** iff at most k of its bags are
not smooth.*

As stated in [3], any tree decomposition can be changed into a binary tree in
linear time. Moreover, this transformation can be done without breaking the
smoothness of the tree decomposition. Therefore, we will limit, without loss of
generality, the scope of our algorithm and analysis to binary tree decompositions.
At last, if the tree decomposition is smooth and root X_0 is smooth by condition
(a) of Definition 7, then the tree can also be made smooth by condition (b): it
suffices to create a new root composed of the positions of the old one except the
position i (with i as in the definition above) and add it as the father of X_0. For
this reason, we will only consider case (b) for the alignment algorithms in the
rest of the paper.

3.2   An Alignment Algorithm Based on Tree Decomposition

Let us describe an algorithm that computes the minimum cost alignment of
an arc-annotated sequence A and a (flat) sequence B ($P_B = \varnothing$). This algorithm
implements a dynamic programming strategy, based on a k-Weakly-Smooth tree
decomposition of A.

An Alternative Internal Representation for Alignments. The recursive
step in our scheme consists in extending a partial alignment, assigning positions
that are **proper** to the bag (*i.e.* not present in its father). One of the main
challenge is to preserve the sequential order. Indeed, only consecutive positions
$(i, i+1)$ are guaranteed to be simultaneously present in a bag, allowing for
a direct control over the sequential ordering ($\mu(i) < \mu(i+1)$) at the time of
an assignment. Intuitively, this property may extend transitively over μ, since
$\mu(i) < \mu(i+1)$ and $\mu(i+1) < \mu(i+2)$ implies $\mu(i) < \mu(i+2)$. However,
this property no longer holds when a position is unmatched, as $\mu(i+1)$ is then
undefined and cannot serve as a reference point for the relative positioning of
$\mu(i)$ and $\mu(i+2)$.

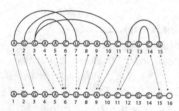

Fig. 3. Internal representation of alignments as a pair
$f = (\mu, \delta)$, used within our dynamic programming algorithm. A function μ defines a complete mapping,
while δ discriminates matched positions (solid stroke)
from unmatched ones (dashed stroke). Additionally,
unmatched positions are forced to aggregate to their
nearest matched neighbor to their right, and we create
an additional virtual position (16 here) to provide an
image to trailing unmatched positions in the structure.

To work around this issue, we identify, in the description of our algorithm, an alignment with a pair $f = (\mu, \delta)$, where $\mu : [1, n] \to [1, m + 1]$ is a full ordered mapping $(\mu(\cdot) \neq \perp)$ and $\delta : [1, n] \to [0, 1]$ distinguishes between matched $(\delta(i) = 1)$ and unmatched positions $(\delta(i) = 0)$. As illustrated by Figure 3, this new representation aggregates consecutive unmatched positions in A to their nearest rightward position that is matched in the alignment. A *virtual position* $m + 1$ is then added to B to serve as an image for the unmatched positions appearing at the end of A.

Dynamic Programming Recursion. Our dynamic programming algorithm assigns positions in B to the elements of a bag, proceeding to a recursive call that assigns the positions found further down in the tree decomposition. This requires a few additional notions and definitions.

Let (S, P) be an arc-annotated sequence, and S' be a subset of S. The **arc-annotated subsequence induced by** S' is the pair (S', P') such that P' is the subset of arcs in P whose both extremities are in S'. Let X_l be a bag in the tree-decomposition of A. The **descending subsequence** of X_l is the subset of positions appearing in the descendants of X_l in the tree. The **descending arc-annotated subsequence** of X_l is the arc-annotated subsequence induced by its descending subsequence. The notion of alignment between A and B naturally extends to alignments involving a sub-arc-annotated structure of A, and we denote by $\mathcal{F}|_{S'}$ the set of all possible alignments between the arc-annotated subsequence induced by S', and B.

Let X_l be a bag having father X_r ($X_r := X_l$ when $l = 0$), let $f \in \mathcal{F}|_{X_{l,r}}$ be an alignment for the common positions of X_l and X_r to B. Let us denote by C_f^l the cost of the best alignment between the descending arc-annotated subsequence of X_l and B, which matches f on $X_{l,r}$. It can be shown that C_f^l obeys

$$C_f^l = \min_{\substack{f' = (\mu', \delta') \in \mathcal{F}|_{X_l} \\ f'(i) = f(i), \forall i \in X_{l,r}}} \left\{ \mathsf{LCost}(X_l, f') + \sum_{s \in sons(l)} C_{f'|_{X_{s,l}}}^s \right\}, \qquad (2)$$

where $\mathsf{LCost}(X_l, f)$ stands for the local contribution of a bag X_l to an alignment $f = (\mu, \delta)$. It depends on the alignments of the individual bases and interactions in the bag, as well as on gaps:

$$\mathsf{LCost}(X_l, f) = \sum_{\substack{i \in X_l - X_r \\ \delta(i) = 1}} \gamma(i, \mu(i)) + \sum_{\substack{i, j \in X_l \text{ s.t.} \\ i \text{ or } j \in X_l - X_r \\ \text{and } (i, j) \in P_A}} \varphi(i, j, f(i), f(j))$$

$$\text{// Bases and interactions}$$

$$+ \sum_{\substack{i, i+1 \in X_l \text{ s.t.} \\ i \text{ or } i+1 \in X_l - X_r \\ \text{and } \mu(i+1) > \mu(i)}} \alpha_B \cdot (\mu(i + 1) - \mu(i)) + \beta_B,$$

$$\text{// Gaps in sequence } B$$

$$+ \sum_{\substack{i,i+1\in X_l \text{ s.t.} \\ i \text{ or } i+1\in X_l-X_r \\ \delta(i)=1 \text{ and } \delta(i+1)=0}} \beta_A + \sum_{\substack{i\in X_l-X_r \\ \text{s.t. } \delta(i)=0}} \alpha_A, \qquad // \text{ Gaps in sequence } A$$

assuming a gentle abuse of notation, in which $f(i) = \{\mu(i) \text{ if } \delta(i) = 1, \text{ or } \perp$ otherwise$\}$.

A dynamic programming algorithm follows from this general recurrence equation, in a standard way. The cost of the best alignment is given by $min\{C_f^0 \mid f \in \mathcal{F}|_{X_0}\}$. A simple backtrack procedure gives the best alignment between A and B. The following theorem gives the worst-case complexity of the algorithm.

Theorem 1. *Let A and B be two arc-annotated sequences ($P_B = \varnothing$), and let (X,T) be a tree decomposition of A. The structure-sequence alignment of A and B can be computed in $\Theta(N \cdot m^{t+1})$ time and $\Theta(N \cdot m^t)$ in space, where $N = |X|$, t is the tree-width of (X,T), and $m = |B|$.*

This complexity can be further improved when the tree decomposition is smooth (or even only k-weakly smooth), taking advantage of the affine nature of gap penalty functions. As previously described in Section 3.1, we use the general principle underlying Gotoh's algorithm [4], and introduce a secondary matrix to distinguish gap openings from gap-extensions. We obtain the dynamic programming equations summarized in Figure 4

Theorem 2. *If the tree decomposition of A is k-weakly smooth, then the sequence-structure alignment of A and B, can be computed in $\Theta(k \cdot m^{t+1} + (N-k) \cdot m^t)$ time.*

Corollary 1. *Let A and B be as before. If the tree decomposition (X,T) of A is smooth and has width t, then time complexity of the structure-sequence alignment algorithm is in $\Theta(N \cdot m^t)$.*

4   Tree Decomposition and Sequence Structure Alignment of RNA Structures

In its full generality, the problem of computing a tree decomposition of minimum width for an arc-annotated sequence is NP-Hard [1]. However, by restricting the problem to some specific RNA structure families, one can obtain a tree decomposition of small width in reasonable time. The key idea relies on a total ordering of the positions in the arc-annotated sequence, as shown in the following. For the sake of clarity, we suppose at first that there is no unpaired position in the structure.

Definition 9. *A **wave embedding** W of an arc-annotated sequence $A = (S, P)$ is defined by an increasing sequence of **pivot** positions $\mathbf{y} = \{y_i\}_{i=0}^k$, such that $y_0 := 1$ (Figure 5). The **degree** of a Wave Embedding is its number of pivots.*

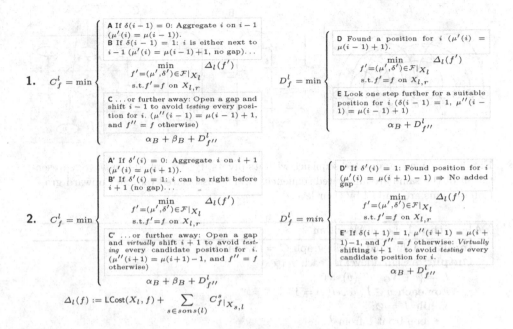

Fig. 4. Dynamic programming equation for aligning a smooth bag X_l, whose father X_r has previously been assigned. Case **1.** and **2.** above apply respectively to smooth bags such that $i - 1 \in X_r$ or $i + 1 \in X_r$ (note that these two cases are mutually exclusive from the definition of smoothness).

Now, a total ordering on the interactions can be inferred from a wave embedding, in which case the wave embedding is said to be **ordering**. Let us now give a sufficient condition for a given embedding to be ordering. To that purpose, we introduce the **upward graph**, and show that its acyclicity is a sufficient condition for the embedding to be ordering. Given a wave embedding W of an arc-annotated sequence $A = (S, P)$, we call **intervals of** W the half open intervals: $I_t = [y_t, y_{t+1}[$ for $t \in [1, k - 1]$ and the interval $I_k = [y_k, n]$. Now, let us start by defining a partial order $\cdot \prec \cdot$ on the positions of a single interval, as follows: for any $i, j \in I_t$, one has $i \prec j$ if either $i < j$ and t is odd, or $i > j$ and t is even. This relation can be used to define the position **directly below** i, i.e. the closest position i^- to i such that $i^- \prec j$. In other words, one has $i^- = i - 1$ if $i, i - 1$ belong to an odd interval, and $i - 1 \in I_t$ or $i^- = i + 1$ if $i, i + 1$ belongs to an even interval. In the absence of such a position, we set $i^- = 0$. The highest position of an interval I_t is the position $i \in I_t$ such that there is no position j in I_t with $i \prec j$. Now we can define the upward graph:

Definition 10. *Given a wave embedding W of an arc-annotated sequence $A = (S, P)$, the **upward graph** of A associated to W is the directed graph $G = (V_G, A_G)$ such that $V_G = P$, and A_G is the set of arcs $((i, j) \mapsto (i', j'))$ such that i or j is directly below i' or j' in W.*

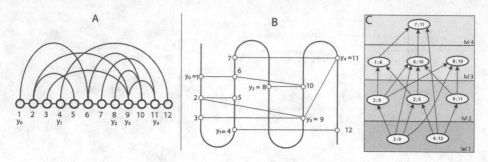

Fig. 5. A: An arc-annotated sequence with its pivots. B: An embedding wave representation of the same arc-annotated sequence. C: The associated (acyclic) upward graph (interactions are ranked by their level).

Algorithm 1. Level Algorithm

Input : a directed acyclic graph $G = (V_G, A_G)$
Output: Assign a level for each vertex
- $L = \{v \in V_G,\ d^-(v) = 0\}$
- **For each** $v \in L$, $level(v) = 1$
- **While** $L \neq \varnothing$:
  - pop front v from L.
  - **For each** v' such that $(v, v') \in A_G$:
    * remove (v, v') from A_G
    * **If** $d^-(v') = 0$:
      · push back v' into L
      · $level(v') = level(v) + 1$

---

*A wave embedding of an arc-annotated sequence $A = (S, P)$ is **ordering** if its upward graph is acyclic.*

Algorithm 1 takes as input an upward graph, supposed to be acyclic, and assigns a level for each vertex, thus a level for each interaction of the associated arc-annotated sequence . This algorithm is a straightforward modification of Kahn's topological ordering algorithm [7], illustrated in Figure 6.

Now we define the level of any position as the minimum level of all interactions in which the position is implicated, as illustrated in Figure 6.

Definition 11 (Level of a position). *Given an ordering wave embedding W of an arc-annotated sequence $A = (S, P)$, the **level of a position** i defined by:* $level(i) = \min_{(i,j) \in P} (level((i, j)))$. *We define a **total order** $\preccurlyeq$ on S through: $i \preccurlyeq j$ iff either $level(i) < level(j)$, or $level(i) = level(j)$ and $i < j$.*

Then, we introduce Algorithm 2 which, starting from an ordering wave embedding, decomposes any arc-annotated sequence. The key idea is to create a root which contains the highest position in each interval. The successor of a bag is then obtained by changing the highest level position into the position directly below it (Figure 6).

Algorithm 2. Chaining Algorithm

Input : an arc-annotated sequence A and an ordering wave embedding of
          degree k
Output: A tree decomposition of A
  – Assign a level for each interaction using Algorithm 1, and map level to each
    position
  – $X = \varnothing$ and T is an empty tree
  – Create a node X_0 composed of the highest position of each interval and set
    X_0 as the root of T
  – $l = 0$
  – **While** there is a position $i \in X_l$ such that $i^- \neq 0$
    • Search the position $p \in X_l$ with the highest level and such that $p^- \neq 0$
    • Add p^- to X_l
    • Add X_l to X
    • If $l > 0$, set X_l as the son of X_{l-1}
    • Set $l = l + 1$ and $X_l = X_{l-1} - \{p\}$
  – **Return** (X, T)

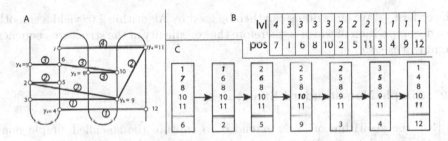

Fig. 6. A: Arc-annotated sequence and its pivots. B: Ranking of the positions by decreasing for the level. C: Tree decomposition obtained with Algorithm 2. The highlighted position in each bag is the position denoted as position p within Algorithm 2. The last position in each bag is the position p^-.

Theorem 3. *Given an ordering wave embedding of degree k for an arc-annotated sequence A, then a tree decomposition of A having width k can always be computed in time $O(k \cdot n)$.*

Corollary 2. *Let A and B be two arc-annotated sequences with $P_B = \varnothing$. Given an ordering wave embedding of degree k of A, the structure-sequence alignment of A and B can be computed in $O(n \cdot m^k)$.*

5   Application to Three General Classes of Structures

In this section, we define three new structure classes, which respectively generalize the standard pseudoknots [5], the simple non-standard pseudoknots [12] and the standard triple helices [11]. For each of them, a *natural* ordering wave embedding can be found, such that our general alignment algorithm has the same complexity as its, previously introduced, *ad hoc* alternatives. In these new classes, each base can interact with several other bases. This notably allows to consider multiple non-canonical interactions (*e.g.* base triples), that occur in most important RNA structural motifs as kink-turns, reverse kink-turns, sarcin-like motifs, C-loop etc. [9].

5.1   Standard Structures

Here we define and describe the **standard structures** (see Figure 7.A), a natural generalization of the standard pseudoknots defined by Han *et al* [5], but where multiple interactions are allowed.

Definition 12 (Standard Structure). *An arc-annotated sequence $A = (S, P)$ is a **standard structure** if there exists an ordering wave embedding, based on a pivot list $\mathbf{y} = \{y_i\}_{i=0}^k$, $k > 1$, such that the extremities of any interaction $(i, j) \in P$ are separated by exactly one pivot.*

The ordering wave embedding can then be used by Algorithm 2 to yield a smooth tree decomposition of width k, therefore the complexity of the structure-sequence alignment is $O(n \cdot m^k)$.

5.2   Simple Non-standard Structures

In [12], the algorithm of [5] is extended to capture the so-called simple non-standard pseudoknots. Briefly, a simple non-standard pseudoknot contains a standard pseudoknot, and defines a special region from which interactions may initiate, possibly crossing interactions in the standard pseudoknot. We extend this class in order to capture multiple interactions, as illustrated by Figure 7.B.

Definition 13 (Simple Non-Standard Structure). *An arc-annotated sequence $A = (S, P)$ is a **simple non-standard structure** (Type I) if there exist an ordering wave embedding, based on a pivot list $\mathbf{y} = \{y_i\}_{i=0}^k$, $k > 1$ and $\tau \in [1, k - k' - 1]$, $k' \in \{0, 1\}$, such that the extremities of any interaction $(i, j) \in P$ with $j < y_{k-k'}$ are separated by exactly one pivot and the others interactions $(i, j) \in P$ with $y_{k-k'} \leq j$ are such that $y_{\tau-1} \leq i < y_\tau$.*

As in [12], Type II simple non-standard structures are symmetric to Type I: the special region lies on the beginning of the sequence. To be coherent with the definition of simple non-standard pseudoknots given in [12], we define the **degree** of a standard structure as its number of pivots in its ordering wave embedding. Therefore, the treewidth of a simple non-standard structure of degree $k + 1$ is at most $k + 1$ and, given its pivots sequence, we can build a smooth tree decomposition of width $k + 1$. Hence the complexity of the structure-sequence alignment is $O(n \cdot m^{k+1})$.

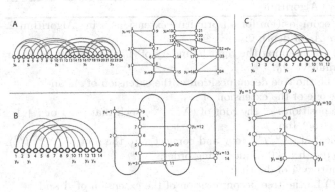

Fig. 7. A: A standard structure and one of its ordering wave embedding representation. B: A simple Non-Standard structure and one of its ordering wave embedding representation ($k' = 2$). C: An extended triple helix and one of its ordering wave embedding representation.

5.3   Extended Standard Triple Helices

To our knowledge, the standard triple helices [11] constitute the first attempt to handle base triples in sequence/structure alignments. A standard triple helix is a kind of standard pseudoknot of degree 3 where some positions are allowed to be involved in multiple base pairs.

We define the **extended standard triple helix** as the structures admitting an ordering wave embedding of degree 3 (Figure 7.C). This new class strictly includes standard triple helices. Furthermore, each such structure can be represented by a tree-decomposition which is smooth and has width at most 3. This gives an algorithm in $O(n \cdot m^3)$ for the structure-sequence alignment.

6   Recursive Structures

Now we consider much more general RNA structures, where different kinds of pseudoknots can occur anywhere. As will be seen below, such structures can be decomposed into **primitives**, and from the tree-decomposition of each primitive a global tree-decomposition of the structure can be built. The set of **primitive sub-arc-annotated subsequences** (**primitives** for short) of an arc-annotated sequence is the set of all sub-arc-annotated sequences induced by the connected components of its **conflict graph**, which is defined as follows. The conflict graph $G = (V, E)$ of an arc-annotated sequence $A = (S, P)$ is the graph such that:

- $V = P$ (the nodes of G are the interactions of A).
- $(v_1, v_2) \in E$ with $v_1 = (i_1, j_1)$ and $v_2 = (i_2, j_2)$ ($i_1 < i_2$) iff $i_1 < i_2 < j_1 < j_2$ (interactions cross).

The **boundaries** of a primitive are its left-most and right-most positions.

Let A and A' be two primitives of an arc-annotated sequence, and let i' and j' be the boundaries of A' ($i' < j'$). We say that A is **encapsulated** in A' iff for any position $i \in A$, one has $i' \leq i \leq j'$, and there exists at least one position $j \in A$

Algorithm 3. Recursive Algorithm

- Compute a tree decomposition for all primitive extensions using Algorithm 2.
- For each primitive A_0 of depth 0:
  - Create a bag χ containing the right boundary of A_0 and the left boundary of its next primitive A_0'.
  - Let (X^0, T^0) be the tree decomposition of the extension of A_0 and (X'^0, T'^0) the one of the extension of A_0'.
  - Add to χ the position i of the root of (X'^0, T'^0) such that $i - 1$ or $i + 1$ belongs to the root too.
  - Connect the leaf of (X^0, T^0) to χ and connect χ to the root of (X'^0, T'^0).
- For each possible depth $i > 0$ in increasing order:
  - For each primitive A of depth i:
    * Let (X, T) be the tree decomposition of the extension of A and (X', T') the one of the extension of the arc-annotated sequence A' in which A is encapsulated.
    * Find a bag χ in (X', T') such that χ contains the boundaries of A.
    * Connect (X, T) to (X', T') by connecting the root of (X, T) to χ.
    * Add the right-most boundary of A in (X, T) to all bags from the root to the first bag containing it.

---

such that $i' < j < j'$. The **depth** of a primitive of an arc-annotated sequence is the number of primitives which encapsulate it. We say that A is **directly** encapsulated in A' if A is encapsulated in A' and $depth(A) = depth(A') + 1$.

The **extension** of a primitive A of depth i is the arc-annotated subsequence consisting of: the primitive A; the boundaries of any primitive that is directly encapsulated in A; and the unpaired positions that are directly encapsulated in A.

Given a primitive A_0 of depth 0 of an arc-annotated sequence A, the **next primitive** of A_0 is the following primitive of level 0 in the sequential order (note that they can share a boundary).

Theorem 4. *Let A be an arc-annotated sequence of size n. If there exist an ordering wave embeddings of degree at most k for each extension of its primitives, then the treewidth of A is at most $k+1$, and a κ-weakly smooth tree decomposition of A can be built in $O(k \cdot n)$ time, where κ is the number of primitives of odd degree, whose depth is greater or equal to 1.*

Corollary 3. *Let A and B be two arc-annotated sequences with $P_B = \varnothing$. Given an ordering wave embedding of degree k (at most) for each extension of the primitives of A, the structure-sequence alignment of A and B can be computed in $O(\kappa \cdot m^{k+1} + n \cdot m^k)$ (with κ defined in Theorem 4).*

7   Conclusion

We have given a general parameterized dynamic programming scheme for sequence-structure comparison in a large class of RNA structures, which unifies and generalizes several families of structures that have been independently

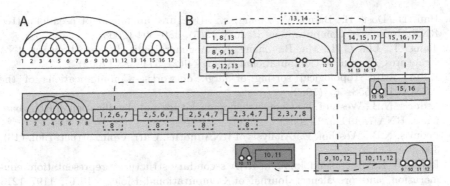

Fig. 8. (A) An arc-annotated sequence and (B) a representation of the tree decomposition given by Algorithm 3. Each box correspond to an extension of a primitive and its associated tree decomposition. Dashed links correspond to the connections made by Algorithm 3. The Dashed bag (top of B) correspond to the bag χ added to connect two consecutive primitive of level 0. The position 8 in a dashed box illustrates the case where the right boundary of a primitive need to be added in its tree decomposition.

considered by previous works. Notably, we can handle structures where each nucleotide can be paired to any number of other nucleotides, thus capturing any type of non-canonical interactions. Our approach relies on a tree decomposition approach of arc-annotated sequences represented as wave embeddings, and the treewidth of the decomposition is then equal to the degree of the wave embedding. Computing a wave embedding of small degree is easy for all classes of pseudoknotted structures considered in this paper. However, the problem of finding a minimum degree wave embedding for any kind of pseudoknotted structure remains open.

Acknowledgments. We warmly thank Robert Giegerich for helpful comments and discussions. This work was supported by the DIGITEO RNAOmics project (AD, PR and YP), the ANR AMIS ARN project (ANR-09-BLAN-0160, DB, AD and PR) and the ANR MAGNUM project (ANR-2010-BLAN-0204, YP).

References

1. Arnborg, S., Corneil, D., Proskurowski, A.: Complexity of finding embeddings in a k-tree. SIAM J. Alg. Disc. Meth. 8(2), 277–284 (1987)
2. Blin, G., Denise, A., Dulucq, S., Herrbach, C., Touzet, H.: Alignment of RNA structures. IEEE/ACM Transactions on Computational Biology and Bioinformatics 7(2), 309–322 (2010)
3. Bodlaender, H.L.: Treewidth: Algorithmic Techniques and Results. In: Privara, I., Ružička, P. (eds.) MFCS 1997. LNCS, vol. 1295, pp. 19–36. Springer, Heidelberg (1997)
4. Gotoh, O.: An improved algorithm for matching biological sequences. J. Mol. Biol. 162(3), 705–708 (1982)

5. Han, B., Dost, B., Bafna, V., Zhang, S.: Structural alignment of pseudoknotted RNA. Journal of Computational Biology 15(5), 489–504 (2008)
6. Jiang, T., Lin, G.H., Ma, B., Zhang, K.: A general edit distance between RNA structures. Journal of Computational Biology 9(2), 371–388 (2002)
7. Kahn, A.B.: Topological sorting of large networks. Communications of the ACM 5(11), 558–562 (1962)
8. Leontis, N.B., Westhof, E.: Geometric nomenclature and classification of RNA base pairs. RNA 7, 499–512 (2001)
9. Leontis, N.B., Westhof, E.: Analysis of RNA motifs. Curr. Opin. Struct. Biol. (13), 300–308 (2003)
10. Rødland, E.A.A.: Pseudoknots in RNA secondary structures: representation, enumeration, and prevalence. Journal of Computational Biology 13(6), 1197–1213 (2006)
11. Wong, T.K., Yiu, S.M.: Structural alignment of RNA with triple helix structure. Journal of Computational Biology 19(4), 365–378 (2012)
12. Wong, T., Lam, T., Sung, W., Cheung, B., Yiu, S.: Structural alignment of RNA with complex pseudoknot structure. Journal of Computational Biology 18(1) (2011)
13. Zhang, K., Wang, L., Ma, B.: Computing Similarity between RNA Structures. In: Crochemore, M., Paterson, M. (eds.) CPM 1999. LNCS, vol. 1645, pp. 281–293. Springer, Heidelberg (1999)

δ-TRIMAX: Extracting Triclusters and Analysing Coregulation in Time Series Gene Expression Data

Anirban Bhar[1], Martin Haubrock[1], Anirban Mukhopadhyay[2], Ujjwal Maulik[3],
Sanghamitra Bandyopadhyay[4], and Edgar Wingender[1,*]

[1] Department of Bioinformatics, Medical School, Georg August University of
Goettingen, Goldschmidtstrasse 1, D-37077 Goettingen, Germany
{anirban.bhar,martin.haubrock,edgar.wingender}
@bioinf.med.uni-goettingen.de
[2] Department of Computer Science and Engineering, University of Kalyani,
Kalyani-741235, India
anirban@klyuniv.ac.in
[3] Department of Computer Science and Engineering, Jadavpur University,
Kolkata-700032, India
umaulik@cse.jdvu.ac.in
[4] Machine Intelligence Unit, Indian Statistical Institute, Kolkata-700108, India
sanghami@isical.ac.in

Abstract. In an attempt to analyse coexpression in a time series microarray gene expression dataset, we introduce here a novel, fast triclustering algorithm δ-TRIMAX that aims to find a group of genes that are coexpressed over a subset of samples across a subset of time-points. Here we defined a novel mean-squared residue score for such 3D dataset. At first it uses a greedy approach to find triclusters that have a mean-squared residue score below a threshold δ by deleting nodes from the dataset and then in the next step adds some nodes, keeping the mean squared residue score of the resultant tricluster below δ. So, the goal of our algorithm is to find large and coherent triclusters from the 3D gene expression dataset. Additionally, we have defined an affirmation score to measure the performance of our triclustering algorithm for an artificial dataset. To show biological significance of the triclusters we have conducted GO enrichment analysis. We have also performed enrichment analysis of transcription factor binding sites to establish coregulation of a group of coexpressed genes.

Keywords: Time series gene expression data, Tricluster, Mean-squared residue, Affirmation score, Gene ontology, KEGG Pathway, TRANSFAC.

1 Introduction

In the context of genomics research, the functional approach is based on the ability to analyze genome-wide patterns of gene expression and the mechanisms

---

* Corresponding author.

B. Raphael and J. Tang (Eds.): WABI 2012, LNBI 7534, pp. 165–177, 2012.

by which gene expression is coordinated. Microarray technology and other high-throughput methods are used to measure expression values of thousands of genes over different samples/experimental conditions. In recent years the microarray technology has been used to measure in a single experiment expression values of thousands of genes under a huge variety of experimental conditions across different timepoints. This kind of dataset can be referred to as time series microarray dataset. Because of the large data volume, computational methods are used to analyze such datasets. Clustering is one of the most common methods for identifying coexpressed genes [6]. This kind of analysis is facilitative for constructing gene regulatory networks in which single or groups of genes interact with other genes. Besides this, coexpression analysis also reveals information about some unknown genes that form a cluster with some known genes.

A clustering algorithm is used to group genes that are coexpressed over all conditions/samples or to group experimental conditions over all genes based on some similarity/ dissimilarity metric. However clustering may fail to find the group of genes that are similarly expressed over a subset of samples/experimental conditions i.e. clustering algorithms are unable to find such local patterns in the gene expression dataset. To deal with that problem, biclustering algorithms are used. A bicluster can be defined as a subset of genes that are coexpressed over a subset of samples/experimental conditions. The first biclustering algorithm that was used to analyse gene expression dataset was proposed by Cheng and Church and they used a greedy search heuristic approach to retrieve largest possible bicluster having mean squared residue (MSR) under a predefined threshold value δ (δ-bicluster) [4]. But nowadays, biologists are eager to analyze 3D microarray dataset to answer the question: "*Which genes are coexpressed under which subset of experimental conditions/samples across which subset of time-points*?" Biclustering is not a proper method to answer this question. So, in this case we need some other clustering technique that can deal with the problem. Hence the term *Triclustering* has been defined and a tricluster can be delineated as a subset of genes that are similarly expressed across a subset of experimental conditions/ samples over a subset of time-points. Zhao and Zaki proposed a triclustering algorithm *TRICLUSTER* that is based on graph-based approach. They defined coherence of a tricluster as $\frac{max(e_{ib}/e_{ia}, e_{jb}/e_{ja})}{min(e_{ib}/e_{ia}, e_{jb}/e_{ja})} - 1$, where e_{ia}, e_{ib} denote the expression values of two columns a and b respectively for a row i. A tricluster is valid if it has a ratio below a maximum ratio threshold ϵ [23].

Here we propose an efficient triclustering algorithm δ-*TRIMAX* that aims to cope with noisy 3D gene expression dataset and is less sensitive to input parameters. The normalization method does not influence the performance of our algorithm, as it produces the same results for both normalized and raw datasets. Here we propose a novel extension of MSR [4] for 3D gene expression data and use a greedy search heuristic approach to retrieve triclusters, having MSR values below a threshold δ. Hence the triclusters can be defined as δ-tricluster. The performance of the proposed algorithm is demonstrated on a synthetic dataset as well as a real-life dataset.

2   Methods

2.1   Definitions

Definition 1 (Time Series Microarray Gene Expression Dataset). *We can model a* time series microarray gene expression dataset *(D) as a $G \times C \times T$ matrix and each element of D (d_{ijk}) corresponds to the expression value of gene i over jth sample/experimental condition across time-point k, where $i \in (g_1, g_2, ..., g_G)$, $j \in (c_1, c_2, ..., c_C)$ and $k \in (t_1, t_2, ..., t_T)$.*

Definition 2 (Tricluster). *A* tricluster *is defined as a submatrix $M(I,J,K) = [m_{ijk}]$, where $i \in I$, $j \in J$ and $k \in K$. The submatrix M represents a subset of genes (I) that are coexpressed over a subset of conditions (J) across a subset of time-points (K).*

Definition 3 (Perfect Shifting Tricluster). *A Tricluster $M(I,J,K) = m_{ijk}$, where $i \in I$, $j \in J$ and $k \in K$, is called a* perfect shifting tricluster *if each element of the submatrix M is represented as: $m_{ijk} = \Gamma + \alpha_i + \beta_j + \eta_k$, where Γ is a constant value for the tricluster, α_i, β_j and η_k are shifting factors of ith gene, jth samples/experimental condition and kth time-point, respectively. As noise is present in microarray datasets, the deviation from actual value and expected value of each element in the dataset also exists. For this deviation, every tricluster is not a perfect one.*

Cheng and Church proposed an algorithm for retrieving large and maximal biclusters that have mean squared residue score (MSR) below a threshold δ in 2D microarray gene expression dataset. They also showed that MSR of a perfect δ-bicluster and perfect shifting bicluster is zero ($\mathbf{S} = \delta = 0$) [4,6]. Now extending this idea, here we present a novel definition of Mean Squared Residue score for 3D microarray gene expression datasets. The MSR ($\mathbf{S}$) of a perfect shifting tricluster becomes also zero, where each element $m_{ijk} = \Gamma + \alpha_i + \beta_j + \eta_k$. For delineating new MSR score ($\mathbf{S}$), at first we need to define the residue score:

Let the mean of ith gene (m_{iJK}): $m_{iJK} = \frac{1}{|J||K|} \sum_{j \in J, k \in K} m_{ijk}$, the mean of jth sample/experimental condition (m_{IjK}): $m_{IjK} = \frac{1}{|I||K|} \sum_{i \in I, k \in K} m_{ijk}$, the mean of kth time-point (m_{IJk}): $m_{IJk} = \frac{1}{|I||J|} \sum_{i \in I, j \in J} m_{ijk}$, and the mean of tricluster (m_{IJK}): $m_{IJK} = \frac{1}{|I||J||K|} \sum_{i \in I, j \in J, k \in K} m_{ijk}$. Now the mean of the tricluster can be considered as the value of constant i.e. $\Gamma = m_{IJK}$. We can define the shifting factor for the ith gene (α_i) as the difference between m_{iJK} and m_{IJK} i.e. $\alpha_i = m_{iJK} - m_{IJK}$. Similarly, we can define shifting factor for the jth condition (β_j) as $\beta_j = m_{IjK} - m_{IJK}$ and shifting factor for the kth time-point (η_k) can be defined as $\eta_k = m_{IJk} - m_{IJK}$. Hence we can define each element of a perfect shifting tricluster as $m_{ijk} = \Gamma + \alpha_i + \beta_j + \eta_k = m_{IJK} + (m_{iJK} - m_{IJK}) + (m_{IjK} - m_{IJK}) + (m_{IJk} - m_{IJK}) = (m_{iJK} + m_{IjK} + m_{IJk} - 2m_{IJK})$. But usually noise is evident in microarray gene expression dataset. Therefore to evaluate the difference between the actual value of an element (m_{ijk}) and its expected value, obtained from above equation, the term *"residue"* can be used [6]. Thus the residue of a tricluster (r_{ijk}) can be defined as follows: $r_{ijk} = m_{ijk} - (m_{iJK} + m_{IjK} + m_{IJk} - 2m_{IJK}) = (m_{ijk} - m_{iJK} - m_{IjK} - m_{IJk} + 2m_{IJK})$.

Definition 4 (Mean Squared Residue). *We define the term Mean Squared Residue MSR(I,J,K) or* **S** *of a tricluster M(I,J,K) to estimate the quality of a tricluster i.e. the level of coherence among the elements of a tricluster as follows:*

$$S = \frac{1}{|I||J||K|} \sum_{i \in I, j \in J, k \in K} r_{ijk}^2$$
$$= \frac{1}{|I||J||K|} \sum_{i \in I, j \in J, k \in K} (m_{ijk} - m_{iJK} - m_{IjK} - m_{IJk} + 2m_{IJK})^2. \quad (1)$$

Lower residue score represents larger coherence and better quality of a tricluster.

2.2   Proposed Method

The method proposed here (δ-*TRIMAX*) aims to find largest and maximal triclusters in a 3D microarray gene expression dataset. It is an extension of Cheng and Church biclustering algorithm [4] that deals with 2-D microarray datasets. In contrast, our algorithm is capable to mine 3D gene expression dataset. There is always a submatrix in an expression dataset that has a perfect MSR(I,J,K) or **S** score i.e. **S** = 0 and this submatrix is each element of the dataset. But as mentioned above, our algorithm finds maximal triclusters having **S** score under a threshold δ, hence we have used a greedy heuristic approach to find triclusters. Our algorithm therefore starts with the entire dataset containing all genes, all samples/experimental conditions and all time-points.

Algorithm I (δ-TRIMAX):
Input. D, a matrix that represents 3D microarray gene expression dataset, λ > 1, an input parameter for multiple node deletion algorithm, $\delta \geq 0$, maximum allowable MSR score.
Output. All possible δ-triclusters.
Initialization. Missing elements in D $\leftarrow$ random numbers, D' $\leftarrow$ D
repeat
   a. D'$_1$ $\leftarrow$ Results of Algorithm II on D' using δ and λ. If the no. of genes (conditions/samples and/or no. of time-points) is 50 (This value can be chosen experimentally. Large value increases the execution time of the algorithm as it then executes more number of iterations), then do not apply Algorithm II on genes (conditions/samples and/or time-points).
   b. D'$_2$ $\leftarrow$ Results of Algorithm III on D'$_1$ using δ.
   c. D'$_3$ $\leftarrow$ Results of Algorithm IV on D'$_2$.
   d. Return D'$_3$ and replace the elements that exist in D' and D'$_3$ with random numbers.
until (there is no gene in δ-tricluster)

Initially, our algorithm removes genes or conditions or time-points from the dataset to accomplish largest diminishing of score **S**; this step is described in the following section in which a node corresponds to a gene or experimental condition or time-point in the 3D microarray gene expression dataset.

Algorithm II (Multiple Node Deletion):

Input. D, a matrix of real numbers that represents 3D microarray gene expression dataset; $\delta \geq 0$, maximum allowable MSR threshold, $\lambda > 1$, threshold for multiple node deletion. The value of λ is set experimentally to optimize the speed and performance (to avoid falling into local optimum) of the algorithm.

Output. M_{IJK}, a δ-tricluster, consisting of a subset(I) of genes, a subset(J) of samples/ experimental conditions and a subset of time-points, having MSR score (**S**) less than or equal to δ.

Initialization. I ← {set of all genes}, J ← {set of all experimental conditions/ samples} and K ← {set of all time-points} and to M(I,J,K) ← D

repeat

   Calculate m_{iJK}, $\forall$ i ∈ I; m_{IjK}, $\forall$ j ∈ J; m_{IJk}, $\forall$ k ∈ K; m_{IJK} and **S**.

   if S ≤ δ then

   return M(I,J,K)

   else

   Delete genes i ∈ I that satisfy the following inequality

$$\frac{1}{|J||K|} \Sigma_{j \subset J, k \in K}(m_{ijk} - m_{iJK} - m_{IjK} - m_{IJk} + 2m_{IJK})^2 > \lambda S$$

   Recalculate m_{iJK}, $\forall$ i ∈ I; m_{IiK}, $\forall$ j ∈ J; m_{IJk}, $\forall$ k ∈ K; m_{IJK} and **S**
   Delete samples/experimental conditions j ∈ J that satisfy the following inequality

$$\frac{1}{|I||K|} \Sigma_{i \in I, k \in K}(m_{ijk} - m_{iJK} - m_{IjK} - m_{IJk} + 2m_{IJK})^2 > \lambda S$$

   Recalculate m_{iJK}, $\forall$ i ∈ I; m_{IjK}, $\forall$ j ∈ J; m_{IJk}, $\forall$ k ∈ K; m_{IJK} and **S**
   Delete time-points k ∈ K that satisfy the following inequality

$$\frac{1}{|I||J|} \Sigma_{i \in I, j \in J}(m_{ijk} - m_{iJK} - m_{IjK} - m_{IJk} + 2m_{IJK})^2 > \lambda S$$

   end if

until (There is no change in I, J and/or K)

The complexity of this algorithm is O(max(m,n,p)) where m, n and p are the number of genes, samples and time-points in the 3D microarray dataset.

In the second step, we delete one node at each iteration from the resultant submatrix, produced by Algorithm II, until the score **S** of the resultant submatrix is less than or equal to δ. This step results in a δ-tricluster.

Algorithm III (Single Node Deletion):

Input. D, a matrix of real numbers that represents 3D microarray gene expression dataset; $\delta \geq 0$, maximum allowable MSR threshold.

Output. M_{IJK}, a δ-tricluster, consisting of a subset(I) of genes, a subset(J) of samples/experimental conditions and a subset of time-points, having MSR score (**S**) less than or equal to δ.

Initialization. I $\leftarrow$ {set of all genes in D}, J $\leftarrow$ {set of experimental conditions/samples in D} and K $\leftarrow$ {set of time-points in D} and to M(I,J,K) $\leftarrow$ D

Calculate m_{iJK}, $\forall$ i $\in$ I; m_{IjK}, $\forall$ j $\in$ J; m_{IJk}, $\forall$ k $\in$ K; m_{IJK} and **S**.

while S $>$ δ do

Detect gene i $\in$ I that has the highest score

$$\mu(i) = \frac{1}{|J||K|} \Sigma_{j \in J, k \in K} (m_{ijk} - m_{iJK} - m_{IjK} - m_{IJk} + 2m_{IJK})^2$$

Detect sample/experimental condition j $\in$ J that has the highest score

$$\mu(j) = \frac{1}{|I||K|} \Sigma_{i \in I, k \in K} (m_{ijk} - m_{iJK} - m_{IjK} - m_{IJk} + 2m_{IJK})^2$$

Detect time-point k $\in$ K that has the highest score

$$\mu(k) = \frac{1}{|I||J|} \Sigma_{i \in I, j \in J} (m_{ijk} - m_{iJK} - m_{IjK} - m_{IJk} + 2m_{IJK})^2$$

Delete gene or sample/experimental condition or time-point that has highest μ score and modify I or J or K.

Recalculate m_{iJK}, $\forall$ i $\in$ I; m_{IjK}, $\forall$ j $\in$ J; m_{IJk}, $\forall$ k $\in$ K; m_{IJK} and **S**.

end while

Return M(I,J,K)

The complexity of first and second steps is O(mnp) as those will iterate (m+n+p) times. The complexity of selection of best genes, samples and time-points is O(log m + log n + log p). So it is suggested to use algorithm II before algorithm III.

As the goal of our algorithm is to find maximal triclusters, having MSR score (**S**) below the threshold δ, the resultant tricluster M(I,J,K) may not be the largest one. That means some genes and/or experimental conditions/samples and/or time-points may be added to the resultant tricluster T produced by node deletion algorithm, so that the MSR score of new tricluster T' produced after node addition does not exceed the MSR score of T. Now the third step of our algorithm is described below.

Algorithm IV (Node Addition):

Input. D, a matrix of real numbers that represents δ-tricluster, having a subset of genes (I), a subset of experimental conditions/samples (J) and a subset of time-points (K).

Output. $M_{I'J'K'}$, a δ-tricluster, consisting of a subset of genes (I') , a subset of samples/experimental conditions (J') and a subset of time-points (K'), such that $I \subset I'$, $J \subset J'$, $K \subset K'$ and MSR(I',J',K') $\leq$ MSR of D.

Initialization. $M(I,J,K) \leftarrow D$

repeat

Calculate m_{iJK}, $\forall$ i; m_{IjK}, $\forall$ j; m_{IJk}, $\forall$ k; m_{IJK} and **S**.

Add genes $i \notin I$ that satisfy the following inequality

$$\frac{1}{|J||K|} \Sigma_{j \in J, k \in K} (m_{ijk} - m_{iJK} - m_{IjK} - m_{IJk} + 2m_{IJK})^2 \leq \mathbf{S}$$

Recalculate m_{IjK}, $\forall$ j; m_{IJk}, $\forall$; m_{IJK} and **S**

Add samples/experimental conditions $j \notin J$ that satisfy the following inequality

$$\frac{1}{|I||K|} \Sigma_{i \in I, k \in K} (m_{ijk} - m_{iJK} - m_{IjK} - m_{IJk} + 2m_{IJK})^2 \leq \mathbf{S}$$

Recalculate m_{iJK}, $\forall$ i; m_{IJk}, $\forall$ k; m_{IJK} and **S**

Add time-points $k \notin K$ that satisfy the following inequality

$$\frac{1}{|I||J|} \Sigma_{i \in I, j \in J} (m_{ijk} - m_{iJK} - m_{IjK} - m_{IJk} + 2m_{IJK})^2 \leq \mathbf{S}$$

until (There is no change in I, J and/or K)

I' $\leftarrow$ I, J' $\leftarrow$ J, K' $\leftarrow$ K

Return I', J', K'

The complexity of this algorithm is O(mnp) as each step iterates (m+n+p) times.

3   Results and Discussion

3.1   Results on Simulated Dataset

We have produced one simulated dataset SMD of size 2000 × 30 × 30. At first we have implanted three perfect shifting triclusters of size 100 × 6 × 6, 80 × 6 × 6 and 60 × 5 × 5 into the dataset SMD and then implanted three noisy shifting triclusters of the same size mentioned before into it. To estimate the degree of similarity between the implanted and obtained triclusters, we define *affirmation score* in the same way as Prelic et. al. defined for two sets of biclusters [6,7]. So, overall average affirmation score of T_1 with respect to T_2 is as follows, where $(SM^*{}_G(T_1, T_2))$ is the average gene affirmation score, $(SM^*{}_C(T_1, T_2))$ is the average sample affirmation score and $(SM^*{}_K(T_1, T_2))$ is the average time-point affirmation score of T_1 with respect to T_2:

$$SM^*(T_1, T_2) = \sqrt{(SM^*_G(T_1, T_2) \times SM^*_C(T_1, T_2) \times SM^*_T(T_1, T_2))} \quad (2)$$

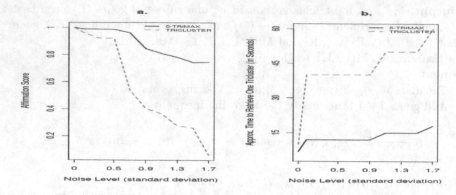

Fig. 1. a. Comparison of Affirmation scores produced by δ-TRIMAX and *TRICLUS-TER* algorithm. b. Comparison of running time of δ-TRIMAX and *TRICLUSTER* algorithm on the synthetic dataset.

Suppose, we have two sets of triclusters T_{im} and T_{res} where T_{im} represents the set of implanted triclusters and T_{res} corresponds to the set of triclusters retrieved by any triclustering algorithm. Hence SM*(T_{im}, T_{res}) denotes how well the triclustering algorithm finds the true triclusters that have been implanted into the dataset. This score varies from 0 to 1 (if $T_{im} = T_{res}$).

In this case we have assigned 0.35 and 1.0005 to the parameters δ and λ, respectively. The value of δ varies from one dataset to another dataset. Here we have first implanted three perfect shifting triclusters into the dataset SMD and then added noise to those triclusters with different standard deviations (σ = 0.1, 0.3, 0.5, 0.7, 0.9, 1.1, 1.3, 1.5, 1.7). To have an idea about the δ value, we have first clustered the genes over all time-points and then the time-points over the subset of genes for each gene cluster in each sample plane using the K-means algorithm. Then we have computed the MSR value(S) of the submatrix, considering a randomly selected sample plane, gene and time-pont cluster for 100 times. Then we have taken the lowest value as the value of δ. For these noisy datasets, we have assigned 3.75 and 1.004 to the parameters δ and λ, respectively. In Figure 1 we have compared the performance of our algorithm with that of the *TRICLUSTER* algorithm [23] in terms of affirmation score using the artificial dataset. Our δ-TRIMAX algorithm performs better than *TRICLUSTER* algorithm for the noisy dataset. For perfect additive triclusters, performances of both these algorithms are comparable with each other.

3.2   Results on Real-Life Dataset

Datasets for Genome-Wide Analysis of Estrogen Receptor Binding Sites: This dataset contains 54675 affymetrix probe-set ids, 3 biological replicates and 4 time-points. In this experiment MCF7 cells are stimulated with 100 nm estrogen for 0, 3, 6 and 12 hours and the experiments are performed in triplicate. This dataset can be downloaded from the following webpage publicly:

http://www.ncbi.nlm.nih.gov/geo/query/acc.cgi?acc=GSE11324. This has been used for discovering of cis-regulatory sites in previously uninvestigated regions and cooperating transcription factors underlying estrogen signaling in breast cancer [11]. We assign 0.012382 and 1.2 to δ and λ respectively. From Figure 2, we observe that the genes in tricluster 4 have similar expression profiles over all three samples across 0, 6 and 12 hours but not at 3 hour. Our algorithm results in 115 triclusters that cover 96.37% of all genes, 100% of all samples and 100% of all time-points.

Biological Significance. We have used the *GOstats* package [22] in R to perform GO and pathway enrichment analysis for genes belonging to each tricluster. We have adjusted the p-values using Benjamini-Hochberg FDR method [1] and considered those terms as significant ones that have a p-value below a threshold of 0.05. The smaller p-value represents higher significance level. We have found statistically enriched GO terms for genes belonging to each tricluster. Additionally to analyse the potential coregulation of coexpressed genes, we have done transcription factor binding site (TFBS) enrichment analysis using the TRANS-FAC library (version 2009.4) [8] that contains eukaryotic transcription factors, their experimentally-proven binding sites, and regulated genes. Here we used 42,544,964 TFBS predictions that have high affinity scores and are conserved between human, mouse, dog and cow (Haubrock et al., in preparation). Out of

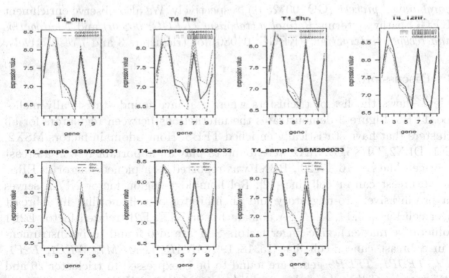

Fig. 2. a.The figures in first row show the expression profiles of genes *ESR1, HOXA11, FAM71A, SPEF2, IFIH1, FPR2, SPAG9, NCF4, ADAM3A, CCNYL1*, respectively of tricluster 4 over all samples; The red-colored time-point (3 hr.) is not a member of this tricluster. b. The figures in second row show the expression profiles of the same genes across 0, 6 and 12 hour.

Table 1. TRANSFAC Matrices for Triclusters, having statistically enriched TFBS for real-life dataset. Transcription factors of red, green, blue, orange, brown and violet colored matrices are Helix-turn-helix, β-Sheet binding to DNA, Zinc-coordinating DNA-binding, basic, all-α-helical DNA-binding and Immunoglobulin fold domain factors, respectively.

Tricluster (no. of genes)	5 most significant TRANSFAC matrices (in ascending order of p-values)	FDR-BY corrected p-value of top-most matrix
Tricluster 3 (875)	VNCX_02, VMSX1_02, V$PAX4_02, V$POU3F2_01, V$TBP_01	4.29e-08
Tricluster 1 (4477)	VNCX_02, VHDX_01, V$BCL6_01, V$ZNF333_01, V$DLX2_01	1.27e-05
Tricluster 26 (3177)	V$E2F_Q2, V$ZF5_01, V$USF2_Q6, V$SP1_Q6_01, V$KID3_01	2.99e-05
Tricluster 4 (3482)	V$BCL6_01, V$HOXA10_01, VSRY_01, VNKX23_01, V$WT1_Q6	9.51e-05
Tricluster 2 (2186)	V$CHCH_01, V$MOVOB_01, VMAZ_Q6, VPAX4_03, V$CACD_01	0.0001
Tricluster 12 (476)	VSRY_02, VNCX_02, V$BCL6_01, V$HB24_01, V$HOXA10_01	0.002
Tricluster 17 (999)	V$CREB_01, V$CREBATF_Q6, V$SP1_Q6_01, V$ATF3_Q6, V$CREBPICJUN_01	0.004
Tricluster 50 (182)	V$ETF_Q6	0.006
Tricluster 18 (260)	V$STAT1STAT1_Q3	0.042
Tricluster 31 (2465)	V$SP1_Q6_01	0.046

these 42 million conserved TFBS we have selected the best 1% for each TRANSFAC matrix individually to identify the most specific regulators (transcription factors). We have used hypergeometric test [9] and Benjamini Yekutieli-FDR method [2] for p-value correction to find over-represented binding sites (p-value ≤ 0.05) in the upstream regions of genes belonging to each tricluster. Here we rank the triclusters in ascending order of corrected p-values of transcription factor matrices. Triclusters 3 and 1 are found to be the top ranked triclusters as per our ranking. Genes belonging to triclusters 3 and 1 are enriched with GOBP terms *cyclic nucleotide catabolic process* (GO: 0009214) and *multicellular organismal process* (GO: 0032501), respectively. We also observe enrichment of KEGG pathway terms *Retinol metabolism (KEGG: 00830)* and *Neuroactive ligand receptor interaction* (KEGG: 04080) for triclusters 3 and 1 respectively.

3.3   Discussion

Table 1 shows the list of triclusters where we have found statistically meliorated TFBS. Figure 3 demonstrates the interactions between regulators for all triclusters that have statistically enriched TFBS. Homeodomain factors MSX2, TLX1, DLX2, DLX5, BARX2 were found to play an important role in a breast cancer cell line [10, 16, 17, 20]. PBX1 was reported as a pioneer factor in ERα-positive breast cancer cell line [12]. Rel homology region factor NFAT serves as a pro-invasive, pro-migratory and an inhibitor of cell motility in a breast cancer cell line [13, 14, 21]. TEA domain factor ETF, E2F related factor E2F1 (proliferative marker), zinc finger factors SP1 are also found to be instrumental in a breast cancer cell line [15, 18, 19]. *ANAPC1, MCM7, WEE1, E2F1, E2F4, TFDP1, TFDP2* genes are found to be coexpressed in tricluster 26 and participate in *Cell cycle* pathway. *ANAPC1, MCM7, WEE1* genes also share conserved TFBS for V$E2F_Q2 matrix and *E2F1, E2F4, TFDP1, TFDP2* are coding genes of the same TRANSFAC matrix. The CREB-related factors CREB1, ATF and Jun-related factors JunB and JunD play an important role in estrogen receptor-mediated regulation in a breast cancer cell line [3, 5]. Fig. 2 in

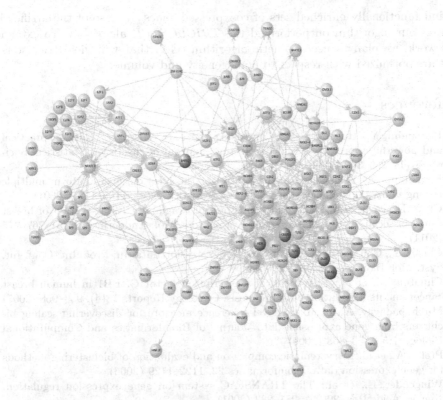

Fig. 3. Gene Regulatory Network for TFBS enriched triclusters

supplementary file (Suppl.pdf) shows the regulatory behavior of transcription factors BARX2, TLX1, DLX2, DLX6, HDX and CREB5 over all time-points. BARX2, member of tricluster 26 have been found to be a regulator of tricluster 3 and 1 and activates HDX and CREB5 at 6 hrs., whereas BARX2, TLX1 and DLX2 jointly act as activators of CREB5 at 6 hrs. We have also observed that a group of coexpressed genes can be regulated by a transcription factor that itself is a member of another tricluster. We can also observe that the expressions of DLX2 and DLX6 are mutually exclusive and it could be the fact that both DLX6 and DLX2 play the role of activator of CREB5, as they belong to the same sub family. The supplementary file is available at http://www.bioinf.med.uni-goettingen.de/services/talks/wabi_2012/.

4   Conclusion

From the above work, we can conclude that the proposed δ-TRIMAX triclustering algorithm is able to retrieve large and coherent groups of genes, having an MSR score below a threshold δ. Genes belonging to each tricluster are coexpressed over a subset of samples/ experimental conditions and across subset of time-points. The results show that the proposed triclustering algorithm is able to find functionally enriched sets of coexpressed genes. In case of the artificial dataset our algorithm outperformed the *TRICLUSTER* algorithm. To extend this work, we plan to use a genetic algorithm (GA) that will yield triclusters that are optimized with respect to homogeneity and volume.

References

1. Benjamini, Y., Hochberg, Y.: Controlling the False Discovery Rate: a practical and powerful approach to multiple testing. Journal of the Royal Statistical Society 57(1), 289–300 (1995)
2. Benjamini, Y., Yekutieli, D.: The control of the false discovery rate in multiple testing under dependency. The Annals of Statistics 29(4), 1165–1188 (2001)
3. Chen, D., et al.: JunD and JunB integrate prostaglandin E2 activation of breast cancer-associated proximal aromatase promoters. Mol. Endocrinol. 25(5), 767–775 (2011)
4. Cheng, Y., Church, G.M.: Biclustering of expression data. In: Proc. Int. Conf. Int. Syst. Mol. Biol., pp. 93–103 (2000)
5. Chhabra, A., et al.: Expression of transcription factor CREB1 in human breast cancer and its correlation with prognosis. Oncology Reports 18(4), 953–958 (2007)
6. Mukhopadhyay, A., et al.: A novel coherence measure for discovering scaling biclusters from gene expression data. Journal of Bioinformatics and Computational Biology 7(5), 853–868 (2009)
7. Prelic, A., et al.: A systematic comparison and evaluation of biclustering methods for gene expression data. Bioinformatics 22, 1122–1129 (2006)
8. Wingender, E., et al.: The TRANSFAC system on gene expression regulation. Nucleic Acids Res. 29(29), 281–283 (2001)
9. Boyle, E.I., et al.: GO::TermFinder-open source software for accessing gene ontology information and finding significantly enriched gene ontology terms associated with a list of genes. Bioinformatics 20(18), 3710–3715 (2004)
10. Lanigan, F., et al.: Homeobox transcription factor muscle segment homeobox 2(Msx2) correlates with good prognosis in breast cancer patients and induces apoptosis in vitro. Breast Cancer Research 12(R59) (2010)
11. Carroll, J.S., et al.: Genome-wide analysis of estrogen receptor binding sites. Nature Genetics 38(11) (November 2006)
12. Magnani, L., et al.: PBX1 genomic pioneer function drives ERα signaling underlying progression in breast cancer. PLOS Genetics 7(11) (November 2011)
13. Fougere, M., et al.: NFAT3 transcription factor inhibits breast cancer cell motility by targeting the Lipocalin 2 gene. Oncogene 29(15), 2292–2301 (2010)
14. Yoeli-Lerner, M., et al.: Akt blocks breast cancer cell motility and invasion through the transcription factor NFAT. Molecular Cell 20(4), 539–550 (2005)
15. Khan, S., et al.: Role of specificity protein transcription factors in estrogeninduced gene expression in mcf-7 breast cancer cells. Journal of Molecular Endocrinology 39, 289–304 (2007)
16. Tommasi, S., et al.: Methylation of homeobox genes is a frequent and early epigenetic event in breast cancer. Breast Cancer Research 11(R14) (2009)
17. Lee, S.Y., et al.: Homeobox gene Dlx-2 is implicated in metabolic stress-induced necrosis. Molecular Cancer 10(113) (2011)
18. Zhang, S.Y., et al.: E2F-1: a proliferative marker of breast neoplasia. Cancer Epidemiology, Biomarkers & Prevention 9, 395–401 (2000)

19. Maeda, T., et al.: TEF-1 transcription factors regulate activity of the mouse mammary tumor virus LTR. Biochemical and Biophysical Research Communications 296(5), 1279–1285 (2002)
20. Stevens, T.A., et al.: BARX2 and estrogen receptor-alpha (ESR1) coordinately regulate the production of alternatively spliced esr1 isoforms and control breast cancer cell growth and invasion. Oncogene 25, 5426–5435 (2006)
21. Jauliac, S., et al.: The role of NFAT transcription factors in integrin-mediated carcinoma invasion. Nature Cell Biology 4(7), 540–544 (2002)
22. Falcon, S., Gentleman, R.: Using GOstats to test gene lists for GO term association. Bioinformatics 23(2), 257–258 (2007)
23. Zhao, L., Zaki, M.J.: TRICLUSTER: An effective algorithm for mining coherent clusters in 3D microarry data. In: SIGMOD (June 2005)

CLIIQ: Accurate Comparative Detection and Quantification of Expressed Isoforms in a Population

Yen-Yi Lin[1,*,**], Phuong Dao[1,**], Faraz Hach[1,**], Marzieh Bakhshi[1],
Fan Mo[2], Anna Lapuk[2], Colin Collins[2], and S. Cenk Sahinalp[1,*]

[1] School of Computing Science, Simon Fraser University, Burnaby, BC, Canada
[2] Vancouver Prostate Centre & Department of Urologic Sciences,
University of British Columbia, Vancouver, BC, Canada
yenyil@sfu.ca, cenk@cs.sfu.ca

Abstract. The recently developed RNA-Seq technology provides a high-throughput and reasonably accurate way to analyze the transcriptomic landscape of a tissue. Unfortunately, from a computational perspective, identification and quantification of a gene's isoforms from RNA-Seq data remains to be a non-trivial problem. We propose CLIIQ, a novel computational method for identification and quantification of expressed isoforms from *multiple samples* in a *population*. Motivated by ideas from compressed sensing literature, CLIIQ is based on an integer linear programming formulation for identifying and quantifying "the most parsimonious" set of isoforms. We show through simulations that, on a single sample, CLIIQ provides better results in isoform identification and quantification to alternative popular tools. More importantly, CLIIQ has an option to jointly analyze multiple samples, which significantly outperforms other tools in both isoform identification and quantification.

Keywords: Isoform Identification, Isoform Quantification, RNA-Seq, Transcriptomics, Integer Linear Programming.

1 Introduction

Recent advances in high throughput sequencing (HTS) technologies have replaced array-based technologies to obtain the mRNA content of a tissue sample. These technologies, commonly known as RNA-Seq have also offered more diverse applications than array based technologies, including the detection of gene fusions, novel isoforms, alternative/differential splicing events and differentially expressed isoforms in multiple samples.

Transcriptome assembly and quantification are challenging problems in current status of RNA-Seq studies. The transcriptome assembly focuses on discovering all the expressed isoforms of a sample. Once the isoforms are determined, the

---

* Corresponding authors.
** These authors contributed equally to this work.

B. Raphael and J. Tang (Eds.): WABI 2012, LNBI 7534, pp. 178–189, 2012.

next step is to report expression levels for all identified isoforms for all genes. Current approaches to transcriptome assembly [1] are (i) *de novo* (reference genome independent) assembly and (ii) genome-guided assembly. Trinity [2] and Trans-ABySS [3] are *de novo* assemblers trying to assemble novel transcripts solely from RNA-Seq reads. It is possible to map the assembled transcripts to discover novel splice junctions as well as isoforms - provided the assembly quality is high. Guided assemblers such as Scripture [4], Cufflinks [5], IsoLasso [6], and SLIDE [7] rely on the mapping of the RNA-Seq reads to reference genome. In these methods, the reads are first mapped onto the genome by TopHat [8], SpliceMap [9], and MapSplice [10]. Although the genome guided assemblers use reference genome sequences and even gene annotations, their performance are still effected by many factors such as sequencing errors, multiple mapping locations and short exons.

Currently available genome-guided transcriptome assemblers typically focus on one of the following objectives (i) sensitivity of identification, (ii) precision of identification, or (iii) quantification accuracy. The objective of Scripture [4] for example, is to optimize the sensitivity of identification. It focuses on reporting as many isoforms expressed in a single sample as possible. Although, Scripture [4] might help to identify transcripts with low expression levels, it may also report many incorrect isoforms, especially when a gene has many number of exons and complicated splicing events. Cufflinks [5] focuses on the precision in identifying isoforms. Its goal is to find the minimum number of isoforms that explain the mapping results. In other words, it aims to make sure that each read is included in at least one reported isoform. Scripture and Cufflinks primarily aim to detect expressed isoforms, but they also provide means to solve the isoform quantification problem in a followup stage, which is based on maximum-likelihood estimation. Unlike these two methods, IsoLasso [6] tries to solve isoform identification and quantification simultaneously by considering the quantification errors in their formulation. More specifically, it uses quadratic programming to formulate the differences between observed and estimated quantification errors, and use the LASSO technique [11] to prevent overfitting. This strategy helps to compromise between the quantification errors and sparsity.

The problem of detecting novel isoforms or transcripts from RNA-Seq data remains a difficult problem. One of the main issues is the lower specificity and sensitivity of available methods. This problem is also of concern in the detection of genomic, and in particular structural variations [12]. However, rather than analyzing each genome independently, joint analyzing multiple related samples have been shown to improve the accuracy in detecting structural variants significantly [12]. In fact, a recent study [13] demonstrates that joint analysis of RNA-Seq data from multiple samples can improve the estimation of expression levels. However, this study does not aim to perform isoform identification - the focus is only quantification. Thus, it is very tempting to design a computational method for detecting novel isoforms by jointly analyzing RNA-Seq data from many related samples (eg. primary tumor v.s. metastasis v.s. normal samples from the same patient, or samples from the same population).

Our Contributions. In this paper, we present CLIIQ, a novel computational approach for Common Loci Isoform Identification and Quantification problem. The objective of CLIIQ is to simultaneously identify isoforms and estimate their expression levels by jointly considering RNA-Seq data from multiple samples. It is based on a two-stage integer linear programming formulation to solve this problem. CLIIQ first determines the minimum number of expressed isoforms of a gene such that estimated expression values of the exons and their junctions are within a user defined error threshold. Then it minimizes the difference between the observed and estimated expression values of the exons and their junctions for a fixed number of isoforms, determined in the earlier step. CLIIQ can be used for isoform identification and quantification for a single sample. We show though simulations that CLIIQ improves popular transcript assembly tools both in terms of isoform identification and quantification for a single transcriptome. More importantly, on multiple samples, CLIIQ significantly outperforms these tools (as well as CLIIQ when it is applied to each sample idependently) with respect to both isoform identification and quantification, especially when samples share many of the expressed isoforms.

2   Methods

In this section, we introduce integer linear programming formulations for isoform detection and quantification from one or more RNA-Seq data samples. In Section 2.1, we show how to determine the number of isoform candidates from a given set of samples. In Section 2.2, we introduce mathematical annotations for exon junctions and isoforms. Finally, in Section 2.3, we describe ILP formulations to model and solve the problem.

2.1   Enumerating Isoforms

One of the challenges for isoform identification problem is the large search space of potential solutions. For a gene containing m exons, there are $O(2^m)$ possible isoforms and thus $O(2^{2^m})$ possible sets of isoforms which can, in theory, form a solution. We can reduce the number of possible isoforms (and hence the solution space) by simple filtration techniques: an isoform will be only considered "expressed" if each of its exons and exon-exon junctions "attracts" more than a user specified number of reads.

2.2   Annotations and Their Relations

From now on, we consider the problem of isoform identification and quantification for a particular gene across many samples. We denote the number of samples as s, the number of exons of a specific gene as m and the number of isoforms as t. Let $E = \{E_1, E_2, \cdots, E_m\}$ be the set of the exons (or exon segments) of the gene considered. A junction $J_{i,j}$ $(1 \leq i < j \leq m)$ is defined as the concatenation of two exons (exon segments) E_i and E_j. We denote by $J = \{J_{i,j} | (1 \leq i < j \leq m)\}$

the set of junctions. Let $I = \{I_j | (1 \leq j \leq t)\}$ stand for the set of isoforms and let $Seg(I_j)$ be the set of exons and junctions in the isoform I_j. For a segment S (an exon or a junction), we denote its length as $len(S)$. Finally we denote the read length with L.

We now establish the relation between the expression level of the isoform I_j and the number of reads mapped its exons and junctions. We denote by the variable $E_l(I_j)$ the estimated expression value of the isoform I_j in the lth sample. Intuitively, this value corresponds to the average number of mapped reads per base. Dohm $et\ al.$ [14] show that there are biases in the number mapped reads among the positions of a segment: the starting and ending positions receive much fewer mapped reads than the middle positions. We denote the observed number of reads that passes through the "middle position" of a segment S in the lth sample as $O_l(S)$. Similarly, we denote by $N_l(S)$, the estimated number of reads that "pass through" the middle position of a segment S in l^{th} sample. $O_l(S)$ is estimated from the experimental data while $N_l(S)$ is estimated through the expression of the isoforms that contain S.

Let $f(S, I_j)$ $(S \in Seg(I_j)$) be the number of starting locations of the mapped reads that cover the middle position of S. In order to define $f(S, I_j)$, let $Pos(S, I_j)$ be the position of the middle point of the segment in the isoform I_j. Then we define $Le(S, I_j) = \max(0, Pos(S, I_j) - L + 1)$ as the leftmost starting position and $Ri(S, I_j) = \min(Pos(S, I_j) + L - 1, len(I_j)) - L + 1$ as the rightmost starting position of a mapped read that cover the middle point of S in the isoform I_j. Now we can define $f(S, I_j)$ as follows:

$$
f(S, I_j)
= \begin{cases}
Ri(S, I_j) - Le(S, I_j) + 1 & \text{if } Le(S, I_j) = 0 \\
& \text{or } len(I_j) \leq Pos(S, I_j) + L - 1 \\
\\
L - 1 & \text{otherwise.}
\end{cases}
$$

In short, $f(S, I_j)$ is not equal to $L - 1$ when the mapping locations of the reads exceed the left or right boundary of an isoform. Now, the estimated number of reads that cover the middle of a segment S $(N_l(S))$ could be written as:

$$
N_l(S) = \sum_{\{j | S \in Seg(I_j)\}} f(S, I_j) \times E_l(I_j)
$$

2.3   An ILP Solution

Since we would like to minimize the number of isoforms while estimating their expression values as close as possible to the observed ones, we will introduce a two-stage formulation. We first determine the minimum number of expressed isoforms such that the estimated expressions of segments are within a user defined fraction ϵ $(0 < \epsilon \leq 1)$ of the observed ones. There could be many feasible optimal solutions. Thus, in the second stage, we try to minimize the difference between the observed and estimated expressions of the exons and junctions among all these optimal solutions.

We will describe the constraints on the expressions of segments (exons, junctions) or isoforms in the following subsections. We describe these constraints for each sample l^{th} $(1 \leq j \leq s)$. Finally, we describe the objective function of the ILPs.

Constraints on Exon/Junction Expression Values. We first estimate the total estimated number of reads that are mapped to exons given the expression of the isoforms:

$$\sum_{\{E_i \in E\}} N_l(E_i)$$

Similarly, the total estimated number of reads that are mapped to junctions could be calculated as follows:

$$\sum_{\{J_{ik} \in J\}} N_l(J_{ik})$$

And we have the total observed numbers of reads that are mapped to exons and junctions: $\sum_{\{E_i \in E\}} O_l(E_i)$ and $\sum_{\{J_{ik} \in J\}} O_l(J_{ik})$. Now we enforce the ratio between the estimated number of mapped reads and the observed number of mapped reads of the exons and junctions should be within the interval $[1-\epsilon, 1+\epsilon]$:

$$(1 - \epsilon) \sum_{\{E_i \in E\}} O_l(E_i) \leq \sum_{\{E_i \in E\}} N_l(E_i) \leq (1 + \epsilon) \sum_{\{E_i \in E\}} O_l(E_i)$$

$$(1 - \epsilon) \sum_{\{J_{ik} \in J\}} O_l(J_{ik}) \leq \sum_{\{J_{ik} \in J\}} N_l(J_{ik}) \leq (1 + \epsilon) \sum_{\{J_{ik} \in J\}} O_l(J_{ik})$$

Finally, we require that for each junction segment J_{ik}, its estimated number of mapped reads should be at least the minimum observed number of mapped reads to any junction. We denote $Low(J)$ as the minimum observed number of mapped reads to any junction and $Low(J) = min\{O_l(J_{ik})|(1 \leq i < k \leq m)\}$. We enforce the following constraint for each junction $J_{ik} \in J$:

$$N_l(J_{ik}) \geq Low(J)$$

Constraints on Isoform Expression Values. The following constraints ensure that the estimated expression of each isoform I_j is no more than $(1 + \epsilon)$ factor of the upper bound on the isoform expression value, which, in the ideal case, would be defined as the minimum expression value of the isoform's exons and junctions. As the minimum expression value of the exons and junctions of an isoform I_j in this ideal case is $min\{O_l(S)/f(S, I_j)|S \in Seg(I_j)\}$, we have the following constraint for each isoform I_j:

$$E_l(I_j) \leq (1 + \epsilon) \times min\{O_l(S)/f(S, I_j)|S \in Seg(I_j)\}$$

In practice, to estimate the upper bound of the expression of an isoform, we may consider to use the median (rather than the minimum) of expression values of its exons and junctions.

Let $Iso(I_j)$ be the indicator variable denoting whether the isoform I_j is expressed in at least one sample. Thus, $Iso(I_j) = 1$ if the isoform I_j is expressed in some sample; 0 otherwise. Thus, the estimated expression $E_l(I_j)$ of an isoform I_j from sample l^{th} is bounded as below:

$$E_l(I_j) \leq B \times Iso(I_j)$$

where B is an upper bound on the expression levels of all the isoforms.

The Objective Functions. In the first stage, with the above constraints, we try to minimize the number of isoforms which are expressed in at least one sample. Here, we give an example: suppose that we have two samples and three isoforms. Sample 1 has expressed isoforms I_1, I_2 and sample 2 has expressed isoforms I_1 and I_3. The number of isoforms expressed in at least one sample can be determined to be 3 (I_1, I_2 and I_3). As a result, the objective function in this stage is to minimize $\sum_{\{I_j \in I\}} Iso(I_j)$.

There could be many feasible optimal solutions by solving the ILP in the first stage. In the second stage, we try to minimize the difference between the observed and estimated expressions of the exons and junctions with the obtained number of expressed isoforms from the first stage. Thus after obtaining the minimum number of isoforms M from the first stage, we add another constraint to the ILP as $\sum_{\{I_j \in I\}} Iso(I_j) \leq M$. Now we try to minimize the absolute error between the observed number of reads and the estimated number of reads:

$$\sum_{1 \leq l \leq s} \sum_{S \in E \cup J} |N_l(S) - O_l(S)|$$

In order achieve this through ILP, for each absolute value $|a$ above, we add the following constraints to the ILP:

$$-e \leq a \leq e$$

$$0 \leq e$$

and also add e in the objective function of the ILP.

3   Experimental Results

We evaluate the performance of CLIIQ based on simulated datasets, and compare results between single sample and multiple sample formulations. We also provide isoform identification and quantification results from two popular tools, namely Cufflinks and IsoLasso, for comparison purposes. In particular, we compare CLIIQ with Cufflinks version 1.3.0 and IsoLasso version 2.5.2 with the following parameters.

- Cufflinks: we use default settings.
- IsoLasso: we set *min-frac* = 0.05 (the minimum fraction of reported isoforms should be at least 0.05 in a sample), and *minexp*= 0.2 (the minimum expression level of isoforms) to filter isoforms with low or even 0 expression level.

The ILP formulations for CLIIQ are solved by IBM ILOG CPLEX (version 12.2). We monitor the simulation process and generate the corresponding mapping results such that every read is mapped to exactly one location.

Simulated Data. We use UCSC hg19 human gene annotations and known isoforms to generate 50-bp single-end reads at coverage 30x. Based on the annotations, the simulator generates reads uniformly at random and randomly assigns expression levels to all isoforms such that the ratio between maximum and minimum expression levels from all isoforms of a gene in single sample is less than 10. Here we only consider perfect matching reads, and evaluate performances of CLIIQ and other tools in this case. Note that we only focused on genes for which the coordinates of the exons have no overlap with that of other genes' exons. Our simulations are based on all 770 genes from chromosomes 1, 2, 3 and 4, which satisfy the above condition and contain two to ten isoforms. These genes have a total of 2151 known isoforms.(It is possible to extend CLIIQ formulation to genes with overlapping exons by partitioning each chromosome to disjoint exonic regions; we leave the exploration of this extension to a later study.)

Experiment Design. We consider two different experiment designs for different samples in a population. For each experiment, we assess the performance of CLIIQ in single sample and multiple sample settings separately. We analyze whether the multiple sample formulation helps to improve performances. In the first experiment, all known isoforms of a gene are expressed in every sample in the population, but with different expression levels. In the second experiment, for each gene we randomly select at most two isoforms such that these isoforms are only expressed in some but not all samples. For example, suppose a gene has three known isoforms I_1, I_2, and I_3. In the first experiment, every sample has these 3 isoforms expressed. In the second experiment, some samples only have I_1 and I_2 expressed, and others contain I_1 and I_3. Although there is a common isoform I_1 for the whole population, the isoforms I_2 and I_3 are not expressed in all the samples. For both experiments, we generate five different samples for each gene. We evaluate the performance of CLIIQ in these two experiments: (1) single sample: we run CLIIQ to identify and quantify isoforms for each sample separately, (2) multiple samples: we formulate all five samples in a population simultaneously.

Running Time. Since the formulation of CLIIQ is based on ILP, it is important to ensure that ILP programs can be solve in reasonable time. We restrict the maximum running time for each ILP program for each gene to 10 mins and there are only 6/914 genes that requires longer running time. Below we report the execution time of all the methods on all the genes selected from chromosome 1, 2, 3 and 4: As seen in the above table, both formulations of CLIIQ on single

Table 1. Execution time of various methods on isoform identification and quantification

	Cufflinks	IsoLasso	CLIIQ (single sample)	CLIIQ (multiple samples)
First Experiment	32 min	12 min	44 min	85 min
Second Experiment	20 min	7 min	34 min	104 min

or multiple samples still have reasonable running time even though it is slower than other methods.

Isoform Identification. We define precision, recall, and F-score values as follows. Suppose a method reports N transcripts for a sample containing M transcripts. If there are totally C correctly identified isoforms, these three values can be calculated by

$$\text{Precision} = \frac{C}{N}$$

$$\text{Recall} = \frac{C}{M}$$

$$\text{F-score} = \frac{2 \times \text{Precision} \times \text{Recall}}{\text{Precision} + \text{Recall}}$$

By correct identification between a reported and an isoform derived from an experiment, we mean (1) these two isoforms contain an identical set of exons, and (2) the same exon between these two transcripts can differ at most 2 bases for the boundaries. The second rule tries to eliminate the biases for the first and last exons for an isoform.

Since the performance of CLIIQ relies on a user defined error threshold ϵ, we determine the value for our experiments by running different ϵ of CLIIQ in randomly selected 100 genes. Based on Table 2, although $\epsilon=0.15$ can provide best available F-score, we select $\epsilon = 0.2$ which achieves highest precision and the second highest F-score.

The isoform identification results of Cufflinks, IsoLasso, CLIIQ with single sample and multiple samples in two different experiments are shown in Table 3.

In single sample formulation, CLIIQ can achieve slightly better identification results than Cufflinks and IsoLasso, and CLIIQ with multiple samples provides more than 10% improvements. More specifically, multiple sample formulation of CLIIQ has similar precision with single sample CLIIQ, but has higher recall rate. It shows that by combining multiple samples, we can retrieve the isoforms

Table 2. Performance of CLIIQ on isoform identification for test data with different ϵ values

	0.1	0.15	0.2	0.25	0.3
ID Precision	0.8208	0.8325	0.8562	0.8440	0.8509
ID F-score	0.7993	0.8025	0.8021	0.7974	0.8011

Table 3. Performance of various methods on isoform identification of $\epsilon=0.2$

	First Experiment				Second Experiment			
	Cufflinks	IsoLasso	CLIIQ (Single Sample)	CLIIQ (Multiple Samples)	Cufflinks	IsoLasso	CLIIQ (Single Sample)	CLIIQ (Multiple Samples)
Precision	0.8011	0.7587	0.8351	0.8831	0.8233	0.7825	0.8588	0.8571
Recall	0.6505	0.6739	0.7353	0.7836	0.7213	0.7381	0.8080	0.8788
F-Score	0.7180	0.7138	0.7820	0.8304	0.7690	0.7596	0.8327	0.8678

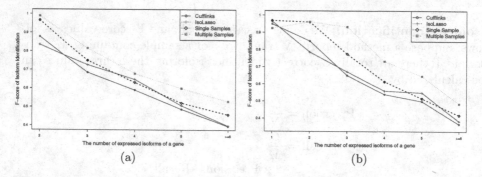

(a)   (b)

Fig. 1. The F-score of isoform identification of the first experiment(left) and the second experiment(right) with respect to isoform number in a gene

effectively. Note that the above results are based on perfect mapping location. For real mapping results using TopHat, multiple sample formulation of CLIIQ also outperforms Cufflinks and IsoLasso as in Supplementary Material [15].

We analyze the performance of isoform identification with respect to the number of expressed isoforms in a gene as in Figure 1. Higher number of expressed isoforms is the result of multiple and complicated alternative splicing events. Thus, it is more challenging in identifying and quantifying these expressed isoforms. From Figure 1, although the performances of all the methods decrease as the number of expressed isoform grows, multiple sample formulation of CLIIQ still outperforms other tools.

Isoform Quantification. We follow the definition of precision, recall, and F-score for isoform identification, but we adjust the definition of correct isoforms as follows. A reported isoform is considered as a correctly quantified one if (1) this is a correct identification, and (2) the relative quantification error δ should be less than a threshold. Suppose a reported set of isoforms contains I_1 with expression level 10 and I_2 with 15, and the true answer is I_1, I_2 with expression levels 12, 10 respectively. With the error threshold $\delta=0.2$, we only consider I_1 as a correct quantification ($\frac{|12-10|}{10} \leq 0.2$). I_2 is an incorrect quantification ($\frac{|15-10|}{15} >$ 0.2). Table 3 show the performance of various methods on isoform identification and quantification. Here we report RPKM values as the abundance estimation results.

Table 4. Performance of various methods on isoform identification and quantification with error 0.1

	First Experiment				Second Experiment			
	Cufflinks	IsoLasso	CLIIQ (Single Sample)	CLIIQ (Multiple Samples)	Cufflinks	IsoLasso	CLIIQ (Single Sample)	CLIIQ (Multiple Samples)
Precision	0.3410	0.4152	0.5810	0.6172	0.4100	0.5697	0.6846	0.6832
Recall	0.2769	0.3688	0.5115	0.5476	0.3592	0.5373	0.4440	0.7004
F-Score	0.3056	0.3907	0.5440	0.5803	0.3829	0.5530	0.6637	0.6917

From Table 4, we find that in both experiment settings, the quantification performances of single sample CLIIQ are better than Cufflinks and IsoLasso in most metrics given relative quantification error threshold $\delta= 0.1$. Once we combine multiple samples from a population, we can enhance the performance of CLIIQ further, especially for the first experiment in which all samples contain an identical set of expressed isoforms. We also perform the same experiment using real mapping results of TopHat and report in Supplementary Material [15]. And the performance of multiple sample formulation of CLIIQ is better than ones of Cufflinks and IsoLasso.

In Figure 2, we show the performance of isoform quantification of all the methods when δ ranges from 0.1 to 0.5. From Figure 2, we find that the curve of CLIIQ and IsoLasso become smooth since $\delta = 0.2$, but Cufflinks has significant improvement when we increase δ. It shows that CLIIQ and IsoLasso provide more accurate estimation values by considering identification and quantification factors simultaneously.

4   Discussion

The improvement provided by CLIIQ can be attributed to two primary aspects. First, performing identification and quantification simultaneously seems to help considerably. Second, when the RNA-Seq data samples considered contain similar sets of isoforms, it is possible to recover individual isoforms missed in an individually analyzed sample (due to coverage issues, possibility of alternative solutions etc.) through the joint analysis of all samples. We observed that, because of its own "brand" of maximum parsimony objective function, Cufflinks

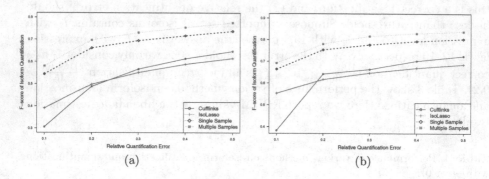

Fig. 2. The F-score of isoform quantification results (with respect to all reported iso-forms) of the first experiment(left) and the second experiment(right) with respect to different error tolerances

typically reports fewer isoforms than those that are present in the data set. Iso-Lasso on the other hand tends to generate several false positives CLIIQ provides a better F-score than the alternative tools because it combines these two factors in a single objective function. Moreover, multiple sample formulation of CLIIQ targets to find the most parsimonious set of isoforms for the entire sample set, and not just for one sample.

References

1. Garber, M., Grabherr, M.G., Guttman, M., Trapnell, C.: Computational methods for transcriptome annotation and quantification using RNA-seq. Nature Method 8(6), 469–477 (2011)
2. Grabherr, M.G., Haas, B.J., Yassour, M., Levin, J.Z., Thompson, D.A., Amit, I., Adiconis, X., Fan, L., Raychowdhury, R., Zeng, Q., Chen, Z., Mauceli, E., Hacohen, N., Gnirke, A., Rhind, N., di Palma, F., Birren, B.W., Nusbaum, C., Lindblad-Toh, K., Friedman, N., Regev, A.: Full-length transcriptome assembly from RNA-seq data without a reference genome. Nature Biotechnology 29(7), 644–652 (2011)
3. Robertson, G., Schein, J., Chiu, R., Corbett, R., Field, M., Jackman, S.D., Mungall, K., Lee, S., Okada, H.M., Qian, J.Q., Griffith, M., Raymond, A., Thiessen, N., Cezard, T., Butterfield, Y.S., Newsome, R., Chan, S.K., She, R., Varhol, R., Kamoh, B., Prabhu, A.L., Tam, A., Zhao, Y., Moore, R.A., Hirst, M., Marra, M.A., Jones, S.J.M., Hoodless, P.A., Birol, I.: De novo assembly and analysis of RNA-seq data. Nat. Meth. 7(11), 909–912 (2010)
4. Guttman, M., Garber, M., Levin, J.Z., Donaghey, J., Robinson, J., Adiconis, X., Fan, L., Koziol, M.J., Gnirke, A., Nusbaum, C., Rinn, J.L., Lander, E.S., Regev, A.: Ab initio reconstruction of cell type-specific transcriptomes in mouse reveals the conserved multi-exonic structure of lincRNAs. Nature Biotechnology 28(5), 503–510 (2010)
5. Trapnell, C., Williams, B.A., Pertea, G., Mortazavi, A., Kwan, G., van Baren, M.J., Salzberg, S.L., Wold, B.J., Pachter, L.: Transcript assembly and quantification by RNA-seq reveals unannotated transcripts and isoform switching during cell differentiation. Nat. Biotech. 28(5), 511–515 (2010)

6. Li, W., Feng, J., Jiang, T.: IsoLasso: A LASSO Regression Approach to RNA-Seq Based Transcriptome Assembly (Extended Abstract). In: Bafna, V., Sahinalp, S.C. (eds.) RECOMB 2011. LNCS, vol. 6577, pp. 168–188. Springer, Heidelberg (2011)
7. Li, J.J., Jiang, C.R., Brown, J.B., Huang, H., Bickel, P.J.: Sparse linear modeling of next-generation mRNA sequencing (RNA-seq) data for isoform discovery and abundance estimation. Proceedings of the National Academy of Sciences 108(50), 19867–19872 (2011)
8. Trapnell, C., Pachter, L., Salzberg, S.L.: TopHat: discovering splice junctions with RNA-seq. Bioinformatics 25(9), 1105–1111 (2009)
9. Au, K.F., Jiang, H., Lin, L., Xing, Y., Wong, W.H.: Detection of splice junctions from paired-end RNA-seq data by SpliceMap. Nucleic Acids Research 38(14), 4570–4578 (2010)
10. Wang, K., Singh, D., Zeng, Z., Coleman, S.J., Huang, Y., Savich, G.L., He, X., Mieczkowski, P., Grimm, S.A., Perou, C.M., MacLeod, J.N., Chiang, D.Y., Prins, J.F., Liu, J.: MapSplice: Accurate mapping of RNA-seq reads for splice junction discovery. Nucleic Acids Research 38(18), e178 (2010)
11. Tibshirani, R.: Regression shrinkage and selection via the lasso. Journal of the Royal Statistical Society. Series B (Methodological) 58(1), 267–288 (1996)
12. Hormozdiari, F., Hajirasouliha, I., McPherson, A., Eichler, E.E., Sahinalp, S.C.: Simultaneous structural variation discovery among multiple paired-end sequenced genomes. Genome Research 21(12), 2203–2212 (2011)
13. Rozov, R., Halperin, E., Shamir, R.: MGMR: leveraging RNA-Seq population data to optimize expression estimation. BMC Bioinformatics 13(suppl. 6), S2 (2012)
14. Dohm, J.C., Lottaz, C., Borodina, T., Himmelbauer, H.: Substantial biases in ultra-short read data sets from high-throughput DNA sequencing. Nucleic Acids Research 36(16), e105 (2008)
15. CLIIQ Supplementary Material (2012), http://compbio.cs.sfu.ca/publications/CLIIQSup.pdf

Improved Lower Bounds on the Compatibility of Quartets, Triplets, and Multi-state Characters*

Brad Shutters, Sudheer Vakati, and David Fernández-Baca

Department of Computer Science, Iowa State University, Ames, IA 50011, USA
{shutters,svakati,fernande}@iastate.edu

Abstract. We study a long standing conjecture on the necessary and sufficient conditions for the compatibility of multi-state characters: There exists a function $f(r)$ such that, for any set C of r-state characters, C is compatible if and only if every subset of $f(r)$ characters of C is compatible. We show that for every $r \geq 2$, there exists an incompatible set C of $\lfloor \frac{r}{2} \rfloor \cdot \lceil \frac{r}{2} \rceil + 1$ r-state characters such that every proper subset of C is compatible. Thus, $f(r) \geq \lfloor \frac{r}{2} \rfloor \cdot \lceil \frac{r}{2} \rceil + 1$ for every $r \geq 2$. This improves the previous lower bound of $f(r) \geq r$ given by Meacham (1983), and generalizes the construction showing that $f(4) \geq 5$ given by Habib and To (2011). We prove our result via a result on quartet compatibility that may be of independent interest: For every integer $n \geq 4$, there exists an incompatible set Q of $\lfloor \frac{n-2}{2} \rfloor \cdot \lceil \frac{n-2}{2} \rceil + 1$ quartets over n labels such that every proper subset of Q is compatible. We contrast this with a result on the compatibility of triplets: For every $n \geq 3$, if R is an incompatible set of more than $n - 1$ triplets over n labels, then some proper subset of R is incompatible. We show this bound is tight by exhibiting, for every $n \geq 3$, a set of $n - 1$ triplets over n taxa such that R is incompatible, but every proper subset of R is compatible.

1 Introduction

The multi-state character compatibility (or perfect phylogeny) problem is a basic question in computational phylogenetics [21]. Given a set C of characters, we are asked whether there exists a phylogenetic tree that displays every character in C; if so, C is said to be compatible, and incompatible otherwise. The problem is known to be NP-complete [3,24], but certain special cases are known to be polynomially-solvable [1,9,15,17,18,19,22]. See [11] for more on the perfect phylogeny problem.

In this paper we study a long standing conjecture on the necessary and sufficient conditions for the compatibility of multi-state characters.

Conjecture 1. *There exists a function $f(r)$ such that, for any set C of r-state characters, C is compatible if and only if every subset of $f(r)$ characters of C is compatible.*

---

* This work was supported in part by the National Science Foundation under grants CCF-1017189 and DEB-0829674.

B. Raphael and J. Tang (Eds.): WABI 2012, LNBI 7534, pp. 190–200, 2012.

If Conjecture 1 is true, it would follow that we can determine if any set C of r-state characters is compatible by testing the compatibility of each subset of $f(r)$ characters of C, and, in case of incompatibility, output a subset of at most $f(r)$ characters of C that is incompatible.

A classic result on binary character compatibility shows that $f(2) = 2$; see [5,10,15,20,21]. In 1975, Fitch [12,13] gave an example of a set C of three 3-state characters such that C is incompatible, but every pair of characters in C is compatible, showing that $f(3) \geq 3$. In 1983, Meacham [20] generalized this example to r-state characters for every $r \geq 3$, demonstrating a lower bound of $f(r) \geq r$ for all r; see also [19]. A recent breakthrough by Lam, Gusfield, and Sridhar [19] showed that $f(3) = 3$. While the previous results could lead one to conjecture that $f(r) = r$ for all r, Habib and To [16] recently disproved this possibility by exhibiting a set C of five 4-state characters such that C is incompatible, but every proper subset of the characters in C are compatible, showing that $f(4) \geq 5$. They conjectured that $f(r) \geq r + 1$ for every $r \geq 4$.

The main result of this paper is to prove the conjecture stated in [16] by giving a quadratic lower bound on $f(r)$. Formally, we show that for every integer $r \geq 2$, there exists a set C of r-state characters such that all of the following conditions hold.

1. C is incompatible.
2. Every proper subset of C is compatible.
3. $|C| = \lfloor \frac{r}{2} \rfloor \cdot \lceil \frac{r}{2} \rceil + 1$.

Therefore, $f(r) \geq \lfloor \frac{r}{2} \rfloor \cdot \lceil \frac{r}{2} \rceil + 1$ for every $r \geq 2$.

Our proof relies on a new result on quartet compatibility we believe is of independent interest. We show that for every integer $n \geq 4$, there exists a set Q of quartets over a set of n labels such that all of the following conditions hold.

1. Q is incompatible.
2. Every proper subset of Q is compatible.
3. $|Q| = \lfloor \frac{n-2}{2} \rfloor \cdot \lceil \frac{n-2}{2} \rceil + 1$.

This represents an improvement over the previous lower bound on the maximum cardinality of such an incompatible set of quartets of $n-2$ given in [24]. We note here that the construction given in [16] showing that $f(4) \geq 5$ can be viewed as a special case of the construction given here when $n = 6$.

We contrast our result on quartet compatibility with a result on the compatibility of triplets: For every $n \geq 3$, if R is an incompatible set of triplets over n labels, and $|R| > n - 1$, then some proper subset of R is incompatible. We show this bound is tight by exhibiting, for every $n \geq 3$, a set of $n - 1$ triplets over n labels such that R is incompatible, but every proper subset of R is compatible.

2   Preliminaries

Given a graph G, we represent the vertices and edges of G by $V(G)$ and $E(G)$ respectively. We use the abbreviated notation uv for an edge $\{u, v\} \in E(G)$. For any $e \subset E(G)$, $G - e$ represents the graph obtained from G by deleting edge e. For an integer i, we use $[i]$ to represent the set $\{1, 2, \cdots, i\}$.

2.1 Unrooted Phylogenetic Trees

An *unrooted phylogenetic tree* (or just *tree*) is a tree T whose leaves are in one to one correspondence with a label set $L(T)$, and has no vertex of degree two. See Fig. 1(a) for an example. For a collection $\mathcal{T}$ of trees, the *label set* of $\mathcal{T}$, denoted $L(\mathcal{T})$, is the union of the label sets of the trees in $\mathcal{T}$. A tree is *binary* if every internal (non-leaf) vertex has degree three. A *quartet* is a binary tree with exactly four leaves. A quartet with label set $\{a, b, c, d\}$ is denoted $ab|cd$ if the path between the leaves labeled a and b does not intersect with the path between the leaves labeled c and d.

For a tree T, and a label set $L \subseteq L(T)$, the *restriction* of T to L, denoted by $T|L$, is the tree obtained from the minimal subtree of T connecting all the leaves with labels in L by suppressing vertices of degree two. See Fig. 1(b) for an example. A tree T *displays* another tree T', if T' can be obtained from $T|L(T')$ by contracting edges. A tree T displays a collection of trees $\mathcal{T}$ if T displays every tree in $\mathcal{T}$. If such a tree T exists, then we say that $\mathcal{T}$ is *compatible*; otherwise, we say that $\mathcal{T}$ is *incompatible*. See Fig. 1(a) for an example. Determining if a collection of unrooted trees is compatible is NP-complete [24].

2.2 Multi-state Characters

There is also a notion of compatibility for sets of partitions of a label set L. A *character* χ on L is a partition of L; the parts of χ are called *states*. If χ has at most r parts, then χ is an r-state character. Given a tree T with $L = L(T)$ and a state s of χ, we denote by $T_s(\chi)$ the minimal subtree of T connecting all leaves with labels having state s for χ. We say that χ is *convex* on T, or equivalently T *displays* χ, if the subtrees $T_i(\chi)$ and $T_j(\chi)$ are vertex disjoint for all states i and j of χ where $i \neq j$. A collection C of characters is *compatible* if there exists a tree T on which every character in C is convex. If no such tree exists, then we say that C is *incompatible*. See Fig. 1(a) for an example. The *perfect phylogeny problem* (or *character compatibility problem*) is to determine whether a given set of characters is compatible.

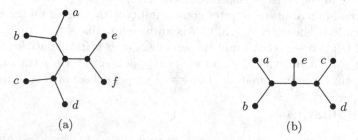

(a)   (b)

Fig. 1. (a) shows a tree T witnessing that the quartets $q_1 = ab|ce$, $q_2 = cd|bf$, and $q_3 = ad|ef$ are compatible; T is also a witness that the characters $\chi_{q_1} = ab|ce|d|f$, $\chi_{q_2} = cd|bf|a|e$, and $\chi_{q_3} = ad|ef|b|c$ are compatible; (b) shows $T|\{a, b, c, d, e\}$

There is a natural correspondence between quartet compatibility and character compatibility that we now describe. Let Q be a set of quartets, $n = |L(Q)|$, and $r = n - 2$. For each $q = ab|cd \in Q$, we define the r-state character corresponding to q, denoted χ_q, as the character where a and b have state 0 for χ_q; c and d have state 1 for χ_q; and, for each $\ell \in L(Q) \setminus \{a, b, c, d\}$, there is a state s of χ_q such that ℓ is the only label with state s for character χ_q (see Example 1). We define the set of r-state characters corresponding to Q by $C_Q = \bigcup_{q \in Q} \{\chi_q\}$.

Example 1. Consider the quartets and characters given in Fig. 1(a): χ_{q_1} is the character corresponding to q_1, χ_{q_2} is the character corresponding to q_2, and χ_{q_3} is the character corresponding to q_3.

The following lemma relating quartet compatibility to character compatibility is well known [23] and is omitted here.

Lemma 1. *A set Q of quartets is compatible if and only if C_Q is compatible.*

2.3 Quartet Closure Rules

We now introduce *quartet (closure) rules* which were originally used in the contexts of psychology [6] and linguistics [7]. The idea is that for a collection Q of quartets, any tree that displays Q may also necessarily display another quartet $q \notin Q$, in which case we write $Q \vdash q$.

Example 2. Let $Q = \{ab|ce, ae|cd\}$. Then the tree of Fig. 1(b) displays Q, and furthermore, it is easy to see that it is the only tree that displays Q. Hence, $Q \vdash ab|de$, $Q \vdash ab|cd$, and $Q \vdash be|cd$.

We will make use of the following quartet rule:

$$\{ab|cd, ab|ce\} \vdash ab|de \tag{r1}$$

which was originally observed in [7]. For the purposes of this paper, we define the *closure* of an arbitrary collection Q of quartets, denoted cl(Q), as the minimal set of quartets that contains Q, and has the property that if for some $q_1, q_2 \in$ cl(Q), $\{q_1, q_2\} \vdash q_3$ using rule (r1), then $q_3 \in$ cl(Q). Clearly, any tree that displays Q must also display cl(Q).

We will make use of the following observation which follows by repeated application of rule (r1).

Observation 1. *Let Q be an arbitrary set of quartets with $\{x, y, a_1, \ldots, a_k\} \subseteq L(Q)$. If*

$$\bigcup_{i=1}^{k-1} \{xy|a_i a_{i+1}\} \subseteq \text{cl}(Q),$$

then $xy|a_1 a_k \in$ cl(Q).

We refer the reader to [21] and [14] for more on quartet rules.

3 Compatibility of Quartets and Multi-state Characters

For every $s, t \geq 2$, we fix a set of labels $L_{s,t} = \{a_1, a_2, \ldots, a_s, b_1, b_2, \ldots, b_t\}$ and define the set

$$Q_{s,t} = \{a_1 b_1 | a_s b_t\} \cup \bigcup_{i=1}^{s-1} \bigcup_{j=1}^{t-1} \{a_i a_{i+1} | b_j b_{j+1}\}$$

of quartets with $L(Q_{s,t}) = L_{s,t}$. We denote the quartet $a_1 b_1 | a_s b_t$ by q_0, and a quartet of the form $a_i a_{i+1} | b_j b_{j+1}$ by $q_{i,j}$.

Observation 2. *For all $s, t \geq 2$, $|Q_{s,t}| = (s-1)(t-1) + 1$.*

Lemma 2. *For all $s, t \geq 2$, $Q_{s,t}$ is incompatible.*

Proof. For each $i \in [s-1]$,

$$\bigcup_{j=1}^{t-1} \{a_i a_{i+1} | b_j b_{j+1}\} \subseteq Q_{s,t} \subseteq \mathrm{cl}(Q_{s,t}).$$

Then, by Observation 1, it follows that for each $i \in [s-1]$, $a_i a_{i+1} | b_1 b_t \in \mathrm{cl}(Q_{s,t})$. So,

$$\bigcup_{i=1}^{s-1} \{b_1 b_t | a_i a_{i+1}\} \subseteq \mathrm{cl}(Q_{s,t}).$$

Then, again by Observation 1, it follows that $b_1 b_t | a_1 a_s \in \mathrm{cl}(Q_{s,t})$. But since $q_0 = a_1 b_1 | a_s b_t \in Q_{s,t}$, and $Q_{s,t} \subseteq \mathrm{cl}(Q_{s,t})$, we have that $\{a_1 b_1 | a_s b_t, b_1 b_t | a_1 a_s\} \subseteq \mathrm{cl}(Q_{s,t})$. It follows that any tree that displays $Q_{s,t}$ must display both $a_1 b_1 | a_s b_t$ and $b_1 b_t | a_1 a_s$. However, no such tree exists. Hence, $Q_{s,t}$ is incompatible.   □

Lemma 3. *For all $s, t \geq 2$, and every $q \in Q_{s,t}$, $Q_{s,t} \setminus \{q\}$ is compatible.*

Proof. Let $q \in Q_{s,t}$. Either $q = q_0$ or $q = q_{x,y}$ for some $1 \leq x < s$ and $1 \leq y < t$. In every case, we present a tree witnessing that $Q_{s,t} \setminus \{q\}$ is compatible.

Case 1. Suppose $q = q_0$. Create the tree T as follows: There is a node for each label in $L_{s,t}$, and two additional nodes a and b. There is an edge ab. For every $a_x \in L_{s,t}$ there is an edge $a_x a$. For every $b_x \in L_{s,t}$, there is an edge $b_x b$. There are no other nodes or edges in T. See Fig. 2(a) for an illustration of T. Consider any quartet $q \in Q_{s,t} \setminus \{q_0\}$. Then $q = a_i a_{i+1} | b_j b_{j+1}$ for some $1 \leq i < s$ and $1 \leq j < t$. Then, the minimal subgraph of T connecting leaves with labels in $\{a_i, a_{i+1}, b_j, b_{j+1}\}$ is the quartet q.

Case 2. Suppose $q = q_{x,y}$ for some $1 \leq x < s$ and $1 \leq y < t$. Create the tree T as follows: There is a node for each label in $L_{s,t}$ and six additional nodes $a_\ell, b_\ell, \ell, h, a_h$, and b_h. There are edges $a_\ell \ell, b_\ell \ell, \ell h, h a_h$, and $h b_h$. For every $a_i \in L_{s,t}$, there is an edge $a_i a_\ell$ if $i \leq x$, and an edge $a_i a_h$ if $i > x$. For every $b_j \in L_{s,t}$ there is an edge $b_j b_\ell$ if $j \leq x$, and an edge $b_j b_h$ if $j > y$. There

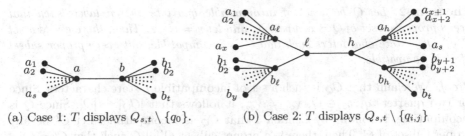

(a) Case 1: T displays $Q_{s,t} \setminus \{q_0\}$.   (b) Case 2: T displays $Q_{s,t} \setminus \{q_{i,j}\}$

Fig. 2. Illustrating the proof of Lemma 3

are no other nodes or edges in T. See Fig. 2(b). Now consider any quartet $q \in Q_{s,t} \setminus \{q_{x,y}\}$. Either $q = q_0$ or $q = q_{i,j}$ where $i \neq x$ or $j \neq y$. If $q = q_0$, then the minimal subgraph of T connecting leaves with labels in $\{a_1, b_1, a_s, b_t\}$ is the subtree of T induced by the nodes in $\{a_1, a_\ell, \ell, b_\ell, b_1, a_s, a_h, h, b_h, b_t\}$. Suppressing all degree two vertices results in a tree that is the same as q_0. So T displays q. So assume that $q = a_i a_{i+1} | b_j b_{j+1}$ where $i \neq x$ or $j \neq y$. We define the following subset of the nodes in T:

$$V = \begin{cases} \{a_i, a_{i+1}, a_\ell, \ell, b_\ell, b_j, b_{j+1}\} & \text{if } i < x \text{ and } j < y, \\ \{a_i, a_{i+1}, a_\ell, \ell, b_y, b_\ell, h, b_h, b_{y+1}\} & \text{if } i < x \text{ and } j = y, \\ \{a_i, a_{i+1}, a_\ell, \ell, h, b_h, b_j, b_{j+1}\} & \text{if } i < x \text{ and } j > y, \\ \{a_x, a_\ell, \ell, h, a_h, a_{x+1}, b_\ell, b_j, b_{j+1}\} & \text{if } i = x \text{ and } j < y, \\ \{a_x, a_\ell, \ell, h, a_h, a_{x+1}, b_h, b_j, b_{j+1}\} & \text{if } i = x \text{ and } j > y, \\ \{a_j, a_{j+1}, a_h, h, \ell, b_\ell, b_j, b_{j+1}\} & \text{if } i > x \text{ and } j < y, \\ \{a_j, a_{j+1}, a_h, h, b_y, b_\ell, \ell, b_h, b_{y+1}\} & \text{if } i > x \text{ and } j = y, \\ \{a_j, a_{j+1}, a_h, h, b_h, b_j, b_{j+1}\} & \text{if } i > x \text{ and } j > y. \end{cases}$$

Now, the subgraph of T induced by the nodes in V is the minimal subgraph of T connecting leaves with labels in q. Suppressing all degree two vertices results in a tree that is the same as q. Hence, T displays q. $\qquad \square$

Theorem 1. *For every integer $n \geq 4$, there exists a set Q of quartets over n taxa such that all of the following conditions hold.*

1. *Q is incompatible.*
2. *Every proper subset of Q is compatible.*
3. *$|Q| = \lfloor \frac{n-2}{2} \rfloor \cdot \lceil \frac{n-2}{2} \rceil + 1$.*

Proof. By letting $s = \lfloor \frac{n}{2} \rfloor$ and $t = \lceil \frac{n}{2} \rceil$, it follows from Observation 2 and Lemmas 2 and 3 that $Q_{s,t}$ is of size $(\lfloor \frac{n}{2} \rfloor - 1) \cdot (\lceil \frac{n}{2} \rceil - 1) + 1 = \lfloor \frac{n-2}{2} \rfloor \cdot \lceil \frac{n-2}{2} \rceil + 1$ and is incompatible, but every proper subset of $Q_{s,t}$ is compatible. $\qquad \square$

The following theorem allows us to use our result on quartet compatibility to establish a lower bound on $f(r)$.

Theorem 2. *Let Q be a set of incompatible quartets over n labels such that every proper subset of Q is compatible, and let $r = n-2$. Then, there exists a set C of $|Q|$ r-state characters such that C is incompatible, but every proper subset of C is compatible.*

Proof. We claim that C_Q is such a set of incompatible r-state characters. Since for two quartets $q_1, q_2 \in Q$, $\chi_{q_1} \neq \chi_{q_2}$, it follows that $|C_Q| = |Q|$. Since Q is incompatible, it follows by Lemma 1 that C_Q is incompatible. Let C' be any proper subset of C. Then, there is a proper subset Q' of Q such that $C' = C_{Q'}$. Then, since Q' is compatible, it follows by Lemma 1 that C' is compatible. $\square$

Theorem 1 together with Theorem 2 gives the following theorem.

Theorem 3. *For every integer $r \geq 2$, there exists a set C of r-state characters such that all of the following hold.*

1. *C is incompatible.*
2. *Every proper subset of C is compatible.*
3. *$|C| = \lfloor \frac{r}{2} \rfloor \cdot \lceil \frac{r}{2} \rceil + 1$.*

Proof. By Theorem 1 and Observation 2, there exists a set Q of $\lfloor \frac{r}{2} \rfloor \cdot \lceil \frac{r}{2} \rceil + 1$ quartets over $r + 2$ labels that that are incompatible, but every proper subset is compatible, namely $Q_{\lfloor \frac{r+2}{2} \rfloor, \lceil \frac{r+2}{2} \rceil}$. The theorem follows from Theorem 2. $\square$

The quadratic lower bound on $f(r)$ follows from Theorem 3.

Corollary 1. *$f(r) \geq \lfloor \frac{r}{2} \rfloor \cdot \lceil \frac{r}{2} \rceil + 1$.*

4   Compatibility of Triplets

A *rooted phylogenetic tree* (or just *rooted tree*) is a tree whose leaves are in one to one correspondence with a label set $L(T)$, has a distinguished vertex called the *root*, and no vertex other than the root has degree two. See Fig. 3(a) for an example. A rooted tree is *binary* if the root vertex has degree two, and every other internal (non-leaf) vertex has degree three. A *triplet* is a rooted binary tree

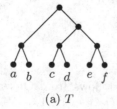

(a) T

(b) $T|\{a, b, c, e\}$

Fig. 3. Example of rooted phylogenetic trees: (a) shows a tree T that is a witness that the triplets $ab|c$, $de|b$, $ef|c$, and $ec|b$ are compatible; (b) shows the tree T restricted to the label set $\{a, b, c, e\}$

with exactly three leaves. A triplet with label set $\{a, b, c\}$ is denoted $ab|c$ if the path between the leaves labeled a and b avoids the path between the leaf labeled c and the root vertex. For a tree T, and a label set $L \subseteq L(T)$, let T' be the minimal subtree of T connecting all the leaves with labels in L. The *restriction* of T to L, denoted by $T|L$, is the rooted tree obtained from T' by distinguishing the vertex closest to the root of T as the root of T', and suppressing every vertex other than the root having degree two. A rooted tree T *displays* another rooted tree T' if T' can be obtained from $T|L(T')$ by contracting edges. A rooted tree T displays a collection of rooted trees $\mathcal{T}$ if T displays every tree in $\mathcal{T}$. If such a tree T exists, then we say that $\mathcal{T}$ is compatible; otherwise, we say that $\mathcal{T}$ is incompatible. Given a collection of rooted trees $\mathcal{T}$, it can be determined in polynomial time if $\mathcal{T}$ is compatible [2,24].

The following theorems follow from the connection between collections of un-rooted trees with at least one common label across all the trees, and collections of rooted trees [24].

Theorem 4. *Let Q be a collection of quartets where every quartet in Q shares a common label ℓ. Let R be the set of triplets such that there exists a triplet $ab|c$ in R if and only if there exists a quartet $ab|c\ell$ in Q. Then, Q is compatible if and only if R is compatible.*

Let R be a collection of triplets. For a subset $S \subseteq L(R)$, we define the graph $[R, S]$ as the graph having a vertex for each label in S, and an edge $\{a, b\}$ if and only if $ab|c \in R$ for some $c \in S$. The following theorem is from [4].

Theorem 5. *A collection R of rooted triplets is compatible if and only if $[R, S]$ is not connected for every $S \subseteq L(R)$ with $|S| \geq 3$.*

Corollary 2. *Let R be a set of rooted triplets such that R is incompatible but every proper subset of R is compatible. Then, $[R, L(R)]$ is connected.*

We now contrast our result on quartet compatibility with a result on triplets.

Theorem 6. *For every $n \geq 3$, if R is an incompatible set of triplets over n labels, and $|R| > n - 1$, then some proper subset of R is incompatible.*

Proof. For sake of contradiction, let R be a set of triplets such that R is incompatible, every proper subset of R is compatible, $|L(R)| = n$, and $|R| > n-1$. The graph $[R, L(R)]$ will contain n vertices and at least n edges. Since each triplet in R is distinct, there will be a cycle C of length at least three in $[R, L(R)]$. Since R is incompatible but every proper subset of R is compatible, by Corollary 2, $[R, L(R)]$ is connected.

Consider any edge e in the cycle C. Let t be the triplet that contributed edge e in $[R, L(R)]$. Let $R' = R \backslash t$. Since the graph $[R, L(R)] - e$ is connected, $[R', L(R')]$ is connected. By Theorem 5, R' is incompatible. But $R' \subset R$, contradicting that every proper subset of R is compatible. $\square$

To show the bound is tight, we first prove a more restricted form of Theorem 1.

Theorem 7. *For every* $n \geq 4$, *there exists a set of quartets* Q *with* $|L(Q)| = n$, *and a label* $\ell \in L(Q)$, *such that all of the following hold.*

1. *Every* $q \in Q$ *contains a leaf labeled by* ℓ.
2. Q *is incompatible.*
3. *Every proper subset of* Q *is compatible.*
4. $|Q| = n - 2$.

Proof. Consider the set of quartets $Q_{2,n-2}$. From Lemmas 2 and 3, $Q_{2,n-2}$ is incompatible but every proper subset of $Q_{2,n-2}$ is compatible. The set $Q_{2,n-2}$ contains exactly $n - 2$ quartets. From the construction, there are two labels in L which are present in all the quartets in $Q_{2,n-2}$. Set one of them to be ℓ. □

The following is a consequence of Theorem 7 and Theorem 4.

Corollary 3. *For every* $n \geq 3$, *there exists a set* R *of triplets with* $|L(R)| = n$ *such that all of the following hold.*

1. R *is incompatible.*
2. *Every proper subset of* R *is compatible.*
3. $|R| = n - 1$.

The generalization of the Fitch-Meacham examples given in [19] can also be expressed in terms of triplets. For any $r \geq 2$, let $L = \{a, b_1, b_2, \cdots, b_r\}$. Let

$$R_r = ab_r|b_1 \cup \bigcup_{i=1}^{r-1} ab_i|b_{i+1}$$

Let $Q = \{ab|c\ell : ab|c \in R_r\}$ for some label $\ell \notin L$. The set C_Q of r-state characters corresponding to the quartet set Q is exactly the set of characters built for r in [19]. In the partition intersection graph of C_Q, (following the terminology in [19]) labels ℓ and a correspond to the end cliques and the rest of the r labels $\{b_1, b_2, \cdots, b_r\}$ correspond to the r tower cliques. From Lemma 1 and Theorem 4, R_r is compatible if and only of Q is compatible.

5   Conclusion

We have shown that for every $r \geq 2$, $f(r) \geq \lfloor \frac{r}{2} \rfloor \cdot \lceil \frac{r}{2} \rceil + 1$, by showing that for every $n \geq 4$, there exists an incompatible set Q of $\lfloor \frac{n-2}{2} \rfloor \cdot \lceil \frac{n-2}{2} \rceil + 1$ quartets over a set of n labels such that every proper subset of Q is compatible. Previous results show that our lower bound on $f(r)$ is tight for $r = 2$ and $r = 3$ [5,10,15,19,20,21]. We give the following conjecture.

Conjecture 2. *For every* $r \geq 2$, $f(r) = \lfloor \frac{r}{2} \rfloor \cdot \lceil \frac{r}{2} \rceil + 1$.

Note that, due to Theorem 2, a proof of Conjecture 2 would also show that the number of incompatible quartets given in the statement of Theorem 1 is also as large as possible.

Another direction for future work is to show an upper bound on the function $f(r)$, which would prove Conjecture 1. For quartets, an upper bound follows from the following argument. A set Q of quartets is *irreducible* if there does not exist a quartet $q \in Q$ such that $Q \setminus \{q\} \vdash q$. If Q is incompatible, but every proper subset of Q is compatible, then it follows that every proper subset Q' of Q with $|Q'| = |Q| - 1$ is irreducible. A recent result by Dietrich, McCartin, and Semple [8] shows that the maximum size of an irreducible set of quartets over a set of n labels has cardinality less than n^3. Thus, the cardinality of a set of quartets over n labels such that every proper subset is compatible is at most n^3. However, an upper bound on the cardinality of a set of r-state characters over an arbitrary number of labels that is both incompatible and has every proper subset compatible remains a central open question.

Acknowledgments. We thank Sylvain Guillemot and Mike Steel for valuable comments.

References

1. Agarwala, R., Fernández-Baca, D.: A polynomial-time algorithm for the perfect phylogeny problem when the number of character states is fixed. SIAM Journal on Computing 23(6), 1216–1224 (1994)
2. Aho, A.V., Sagiv, Y., Szymanski, T.G., Ullman, J.D.: Inferring a tree from lowest common ancestors with an application to the optimization of relational expressions. SIAM Journal on Computing 10(3), 405–421 (1981)
3. Bodlaender, H., Fellows, M., Warnow, T.: Two Strikes against Perfect Phylogeny. In: Kuich, W. (ed.) ICALP 1992. LNCS, vol. 623, pp. 273–283. Springer, Heidelberg (1992)
4. Bryant, D., Steel, M.: Extension operations on sets of leaf-labelled trees. Advances in Applied Mathematics 16, 425–453 (1995)
5. Buneman, P.: The recovery of trees from measurements of dissimilarity. In: Mathematics in the Archeological and Historical Sciences, pp. 387–395. Edinburgh University Press (1971)
6. Colonius, H., Schulze, H.H.: Tree structures for proximity data. British Journal of Mathematical and Statistical Psychology 34(2), 167–180 (1981)
7. Dekker, M.C.H.: Reconstruction Methods for Derivation Trees. Master's thesis, Vrije Universiteit, Amsterdam, Netherlands (1986)
8. Dietrich, M., McCartin, C., Semple, C.: Bounding the maximum size of a minimal definitive set of quartets. Information Processing Letters 112(16), 651–655 (2012)
9. Dress, A., Steel, M.: Convex tree realizations of partitions. Applied Mathematics Letters 5(3), 3–6 (1992)
10. Estabrook, G.F., Johnson, J., McMorris, F.R.: A mathematical foundation for the analysis of cladistic character compatibility. Mathematical Biosciences 29(1-2), 181–187 (1976)
11. Fernández-Baca, D.: The Perfect Phylogeny Problem. In: Steiner Trees in Industry, pp. 203–234. Kluwer (2001)
12. Fitch, W.M.: Toward finding the tree of maximum parsimony. In: Proceedings of the 8th International Conference on Numerical Taxonomy, pp. 189–230 (1975)

13. Fitch, W.M.: On the problem of discovering the most parsimonious tree. The American Naturalist 111(978), 223–257 (1977)
14. Grünewald, S., Huber, K.T.: Identifying and defining trees. In: Gascuel, O., Steel, M. (eds.) Reconstructing Evolution: New Mathematical and Computational Advances. Oxford University Press (2007)
15. Gusfield, D.: Efficient algorithms for inferring evolutionary trees. Networks 21(1), 19–28 (1991)
16. Habib, M., To, T.H.: On a Conjecture about Compatibility of Multi-states Characters. In: Przytycka, T.M., Sagot, M.-F. (eds.) WABI 2011. LNCS, vol. 6833, pp. 116–127. Springer, Heidelberg (2011)
17. Kannan, S., Warnow, T.: Inferring Evolutionary History From DNA Sequences. SIAM Journal on Computing 23(4), 713–737 (1994)
18. Kannan, S., Warnow, T.: A fast algorithm for the computation and enumeration of perfect phylogenies. SIAM Journal on Computing 26(6), 1749–1763 (1997)
19. Lam, F., Gusfield, D., Sridhar, S.: Generalizing the Splits Equivalence Theorem and Four Gamete Condition: Perfect Phylogeny on Three-State Characters. SIAM Journal on Discrete Mathematics 25(3), 1144–1175 (2011)
20. Meacham, C.A.: Theoretical and computational considerations of the compatibility of qualitative taxonomic characters. In: Numerical Taxonomy. Nato ASI Series, vol. G1, Springer (1983)
21. Semple, C., Steel, M.: Phylogenetics. Oxford Lecture Series in Mathematics and its Applications. Oxford University Press (2003)
22. Shutters, B., Fernández-Baca, D.: A simple characterization of the minimal obstruction sets for three-state perfect phylogenies. Applied Mathematics Letters 25(9), 1226–1229 (2012)
23. Steel, M.: Personal communications (2012)
24. Steel, M.: The complexity of reconstructing trees from qualitative characters and subtrees. Journal of Classification 9(1), 91–116 (1992)

Succinct Multibit Tree: Compact Representation of Multibit Trees by Using Succinct Data Structures in Chemical Fingerprint Searches

Yasuo Tabei

ERATO Minato Project, Japan Science and Technology Agency, Sapporo, Japan
tabei.y.aa@m.titech.ac.jp

Abstract. Similarity searches in the databases of chemical fingerprints are a fundamental task in discovering novel drug-like molecules. Multibit trees have a data structure that enables fast similarity searches of chemical fingerprints (Kristensen et al., WABI'09). A standard pointer-based representation of multibit trees consumes a large amount of memory to index large-scale fingerprint databases. To make matters worse, original fingerprint databases need to be stored in memory to filter out false positives. A succinct data structure is compact and enables fast operations. Many succinct data structures have been proposed thus far, and have been applied to many fields such as full text indexing and genome mapping. We present compact representations of both multibit trees and fingerprint databases by applying these data structures. Experiments revealed that memory usage in our representations was much smaller than that of the standard pointer-based representation. Moreover, our representations enabled us to efficiently perform PubChem scale similarity searches.

1 Introduction

Chemically similar molecules tend to have similar molecular functions. This means molecular functions can be predicted by searching for similar molecules in databases. Thus, similarity searches of chemical compounds are an important research topic in chemoinformatics [8]. The number of available molecules has been constantly increasing. For example, more than 30 million molecules are now stored in the National Center for Biotechnology Information (NCBI) PubChem database. Since the size of the whole chemical space has roughly been estimated to be 10^{60} molecules [8], it will certainly continue to grow after this.

Molecules are typically represented by bit strings that summarize information on molecules. Such representations are called molecular fingerprints. Using binary variables, fingerprints enable us to record the presence or absence of particular functional groups or combinatorial features. Jaccard similarity, also called Tanimoto similarity, has commonly been used in chemoinformatics [9] to measure the similarity between fingerprints.

Several approaches have been proposed to improve similarity searches of fingerprints. As far as we know, most of them have employed the bounds for Jaccard

B. Raphael and J. Tang (Eds.): WABI 2012, LNBI 7534, pp. 201–213, 2012.

similarity to reduce the number of calculations of similarity [15,3,10,1]. Although these methods proposed tight upper bounds, they did not present an efficient data structure to use these bounds, which resulted in limited scalability with respect to database size.

Thomas et al. [17,18] solved the efficiency problem by introducing the *pointer-based multibit tree* (MT) which is an efficient tree-based data structure built on the upper bound of [10,1]. By clustering the database and then building a binary tree by recursively splitting fingerprints for each cluster with the upper bound information, MT enables fast searches of a given query by pruning out useless portions of the search space. Although MT has both theoretical [11] and empirical grounds [18,11] for efficiency, it has a serious issue with the memory bottleneck caused by the pointers required for its tree-structured implementation, resulting in limited scalability of memory. Moreover, an original fingerprint database needs to be stored in memory to filter out false positives. Since the number of available molecules is ever increasing, developing algorithms using smaller amounts of memory currently remains a challenge.

We present a *succinct multibit tree* (SMT) in this paper, which is a novel compact representation of a multibit tree and fingerprint databases. SMT leverages the idea behind succinct data structures [7] that achieve space-efficient representations of data structures while preserving the property of efficient operations. While the multibit tree and fingerprint databases themselves are represented by succinct data structures, their auxiliaries, e.g. labels, are not always small. In such cases, memory usage is dominated by the auxiliaries to the data structures. To prevent this, we present a novel succinct variable-length array in which elements are represented by bit strings of different lengths. The main difficulty in designing such an array is how to retain the addressability to any element in $O(1)$ time. We overcame this difficulty with the assistance of fast operations in a succinct data structure.

We applied our SMT successfully in experiments to 30 million chemical compounds from PubChem and demonstrated significantly better memory efficiency than that with the pointer-based representation while performing fast similarity searches. In fact, our SMT reduced memory by a factor of 10 compared to the pointer-based approach, resulting in a requirement for only 2 GB of memory for a database managing 30 million fingerprints. The main drawback with our succinct-based approach could be the increase in search time due to multiple calls of operations. We therefore experimented with trade-offs between memory usage and search speed. Surprisingly, our approach was only a few times slower than MT and remained practical, if about 4 GB of memory was used, which is the memory in commercially available PCs.

The rest of this paper is organized as follows. Section 2 introduces MT for chemical fingerprint searches. We also formulate a similarity search of chemical fingerprints with respect to Jaccard similarity in this same section. Section 3 reviews succinct data structures, and our succinct representations are presented for both the multibit tree and fingerprint databases in Section 4. In addition, our variable-length array is presented. Section 5 reports the experimental results. We conclude the paper in Section 6.

2   Multibit Tree

We introduce the pointer-based multibit tree (MT) [18,11] in this section.

First, we formulate the similarity search problem with a chemical fingerprint database. Fingerprint is a bit string of fixex-length, and has the equivalent expression as set which includes an element i if i-th bit of the fingerprint is 1. We use a set representation of fingerprint in the following. The fingerprint of a query compound is described as Q. We denote a fingerprint database of n compounds with $W_1,...,W_n$. We call each element in Q and W_i a word. We also denote the number of words, i.e., cardinality, in a fingerprint Q with $|Q|$. Given two fingerprints, $W = (w_1, ..., w_s)$ and $W' = (w'_1, ..., w'_t)$, Jaccard similarity is a normalized measure defined as

$$J(W, W') = \frac{|W \cap W'|}{|W \cup W'|}.$$

Then, our search problem is to retrieve the fingerprints within a similarity threshold, ϵ, in terms of Jaccard similarity:

$$I_N = \{i \; ; \; \epsilon \leq J(W_i, Q)\}.$$

MT is an indexing structure for fingerprint databases, and it enables fast searches of a query fingerprint by traversing the structure. Binary trees are built on the notion of two constraints. The first establishes a lower bound and an upper bound for the cardinality of fingerprint W as set for given query Q and threshold ϵ, as follows. If $\epsilon \leq J(W, Q)$, then

$$\epsilon |Q| \leq |W| \leq \frac{|Q|}{\epsilon}. \tag{1}$$

Therefore, if we can solve the following transformed problem of

$$I_1 = \{i \; ; \; \epsilon |Q| \leq |W_i| \leq \frac{|Q|}{\epsilon}\},$$

it contains all solutions, $I_N \subseteq I_1$.

The second constraint gives an upper bound for Jaccard similarity, and results from the intersection inequality in [2] as follows:

$$J(W, Q) \leq \frac{\min(|W| - N_0, |Q| - N_1)}{|W| + |Q| - \min(|W| - N_0, |Q| - N_1)}, \tag{2}$$

where N_0 is the number of words contained in W and not in Q and N_1 is the number of words not contained in W but in Q. If we can also solve the following problem of

$$I_2 = \{i \; ; \; \epsilon \leq \frac{\min(|W_i| - N_0, |Q| - N_1)}{|W_i| + |Q| - \min(|W_i| - N_0, |Q| - N_1)}\},$$

it contains all solutions, $I_N \subseteq I_2$.

The goal of MT is to build multiple binary trees that leverage these constraints. First, to find candidate solutions I_1 efficiently, a fingerprint database, $W_1, ..., W_n$, is clustered into bins consisting of fingerprints of the same cardinality:

$$K_c = \{i \; ; \; |W_i| = c\}.$$

Then, a binary tree is built on each bin K_c to find candidate solutions I_2. The root of a binary tree contains all fingerprints in K_c. Fingerprints are recursively split to the left and right children according to whether or not a word is included (Figure 1). Such a word is chosen by entropy maximization in the current fingerprints. Splitting stops when the number of fingerprints reaches one, i.e., each leaf contains a fingerprint.

Words existing in the current set of fingerprints are stored in set M_v^1 at each node v, and words not existing in the current set of fingerprints are stored in set M_v^0 at each node v. M_v^1 and M_v^0 are respectively used to calculate N_0 and N_1 at each node v.

Given query Q, the search space is limited for binary trees built on bins satisfying the constraint (1) to narrow down candidate solutions in I_1. The similarity search of a binary tree starts from the root and recursively descends to every leaf. The upper bound (2) is calculated at each node v by referring to M_v^1 and M_v^0. The search is pruned if it is less than threshold ϵ. Candidate solutions $I_1 \cap I_2$ are found through the similarity search. Such candidates might not be included in actual solutions I_N, i.e., false positives. To filter out false positives, Jaccard similarities between every element in $I_1 \cap I_2$ and query Q were calculated, and finally the set of solutions I_N was found.

A crucial drawback of MT is that it requires a large amount of memory. It requires $O(\sum_{c=1}^{C} 2(2|K_c| - 1)\log(2|K_c| - 1) + 2MC)$ bits for the total number of bins C for word size M to store binary trees and auxiliaries M_v^1 and M_v^0 at each node v. Moreover, the original fingerprint database needs to be stored in memory to calculate Jaccard similarity. Memory usage prevents practical usage of MT for large-scale chemical databases. In contrast, without drastically slowing down speed, our SMT significantly reduces the memory requirements of both MT and fingerprint databases by using a rank/select dictionary and a succinct ordered tree, i.e., representative succinct data structures, which will be reviewed in the following section.

3   Succinct Data Structures

The *succinct data structure* is memory efficient and it enables fast operations. Precursors were published by Elias [5], Tarjan and Yao [16], and Chazelle [4]. Jacobson had the first paper published on a succinct data structure from his Ph.D thesis in 1989 [7]. Various succinct data structures have been proposed

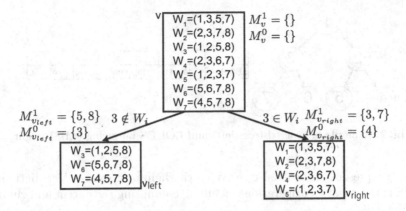

Fig. 1. Example of node of multibit tree. Fingerprints are split into left v_{left} and right v_{right} children according to whether 3 is included in fingerprints or not. v_{left} has W_3, W_6, and W_7, which do not include 3. v_{right} has W_1, W_2, W_4, and W_5, which include 3. Since all fingerprints in v_{left} include $\{8\}$ and not $\{3, 6\}$, $M^1_{v_{left}} = \{8\}$ and $M^0_{v_{left}} = \{3, 6\}$. Since all fingerprints in v_{right} include $\{3, 7\}$ and not $\{5\}$, $M^1_{v_{right}} = \{3, 7\}$ and $M^0_{v_{right}} = \{5\}$.

thus far for bit strings [14,13], sequences [6], trees [7], and graphs [19]. Of these, succinct bit strings called rank/select dictionaries, and they are used as building blocks for other succinct data structures. We review a rank/select dictionary [14,13] and a succinct ordered tree called LOUDS [7] in this section.

3.1 Rank/Select Dictionary

Rank/select dictionaries are data structures for bit string B of length n [14]. They support the rank and select queries as follows:

- $rank_b(B, i)$ returns the number of occurrences of bit $b \in \{0, 1\}$ in $B[1, i]$,
- $select_b(B, i)$ returns the position of the i-th occurrence of bit $b \in \{0, 1\}$ in B.

Naively, they take $O(n)$ time to compute the rank. There are, however, several data structures achieving $n + o(n)$ bit memory and $O(1)$ query time for the rank [13,12]. Most methods compute the select queries by a binary search on a bit string in $O(\log n)$ time. A data structure to compute the select queries in $O(1)$ time has also been proposed [14].

3.2 LOUDS

The *level-order unary degree sequence* (LOUDS) [7] represents an ordered tree as a bit string. Standard pointer-based representations of ordered trees require $O(n \log n)$ bits in memory for the number of nodes n. This memory requirement is inefficient in representing a large tree. LOUDS only uses a bit string of length

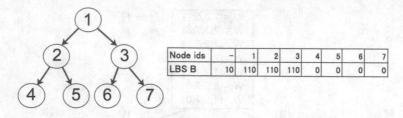

Node ids	–	1	2	3	4	5	6	7
LBS B	10	110	110	110	0	0	0	0

Fig. 2. Example of ordered tree (left) and LOUDS bit string (LBS) (right)

$2n + 1$ to represent an ordered tree. With the help of a rank/select dictionary, LOUDS enables fast tree operations including computing the parent and children for a tree node.

A LOUDS bit string (LBS) is a representation of an ordered tree in LOUDS (Figure 2). LBS B for an ordered tree is a binary string that is built by traversing the tree in a breadth-first manner and placing k 1s followed by 0 for a k-degree node in preorder. For example, a tree node having three children is represented by "1110" in LBS. The first two bits "10" on LBS represent the super root, and are placed for fast tree operations without calculating boundary conditions.

A property of LBS is that for a tree consisting of n nodes there are n 1s and $n+1$ 0s on LBS. Thus, these 1s and 0s except the first 0 on LBS correspond to tree nodes one-by-one. After B is indexed by the rank/select dictionary, the position of the parent and first child for a tree node on LBS B can be calculated by combining the rank/select operations on B. In the following, let $B[1]$ correspond to the root, and let traversing start from $B[1]$.

The position p of the first child for given position i of $B[i] = 1$ on B is calculated with $p = select_0(B, rank_1(B, i)) + 1$, because $q = rank_1(B, i)$ returns the node identifier corresponding to position i on B. Then, $r = select_0(B, q)$ returns the position of $B[r] = 0$ corresponding to the identifier, q, of a tree node. Hence, $p = r + 1$ is the position on B corresponding to the first child of position i. If $B[p] = 0$, the node does not have any children, i.e., the node corresponds to a leaf. For position p for the first child, $p + 1$ computes the position of the next child. If $B[p + 1] = 0$, p is the position of the last child. Hence, any children for a given node can be computed.

The position of the parent for given position i of $B[i] = 1$ on B is calculated with $p = select_1(B, rank_0(B, i))$, because $q = rank_0(B, i)$ returns identifier q of the parent node for position i on B. Then, $p = select_1(B, q)$ returns the position of parent node p.

One can efficiently traverse a tree by using the calculations of the parent and children for a given node.

4   Succinct Multibit Tree

The succinct multibit tree (SMT) consists of succinct representations of multibit trees and fingerprint databases. We present two compact representations with a variable-length array and a succinct trie for fingerprint databases. Our representations are built on a rank/select dictionary and LOUDS.

4.1 Succinct Representation of Multibit Tree

Our basic idea was to represent a binary tree in MT by using LOUDS. Since MT is built on a binary tree, our SMT represents it with LOUDS (Figure 3). Let B_c be the LBS for a binary tree built on bin K_c. Auxiliaries M_v^1 and M_v^0 at each node v are needed to compute an upper bound for Jaccard similarity (2). M_v^1 (respectively M_v^0) can be represented as a set; we can apply efficient compression methods to M_v^1 and M_v^0. More specifically, we use VerByte codes that enable high compression and fast decoding [20].

The main advantages of our approach are that auxiliaries M_v^1 and M_v^0 can be accessed in $O(1)$ time from given position p on LBS B_c. Since each 1 bit in LBS corresponds to a node, node identifier v corresponding to $B_c[p]$ can be computed by using the $rank_1$ operation, $v = rank_1(B_c, p)$.

The identifiers of the original fingerprints are needed at every leaf to refer to these fingerprints for the Jaccard similarity calculations. The main difficulty in our representation is how to relate these identifiers to LBS. We solved this problem by extracting the identifiers from left to right leaves, and we stored them in array IDs. In addition, we prepared bit string L_c to represent whether a position on B_c is a leaf or not, i.e., if $B_c[p]$ corresponds to a leaf $L_c[p] = 1$, otherwise $L_c[p] = 0$. L_c was indexed by the rank/select dictionary. Thus, the identifier of the fingerprint corresponding to a leaf can be obtained by accessing IDs with $rank_1$ on L_c. Given position p of a leaf, i.e., $L_c[p] = 1$, the identifier of the corresponding fingerprint can be obtained with $IDs[rank_1(L_c, p)]$

SMT performs a similarity search by using these representations for a given query with the same method explained in Section 2. While SMT causes a slight slow down with respect to the search time, SMT significantly reduces the memory usage of the multibit tree. This is explained further in Section 5.

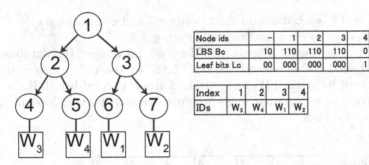

Node ids	–	1	2	3	4	5	6	7
LBS Bc	10	110	110	110	0	0	0	0
Leaf bits Lc	00	000	000	000	1	1	1	1

Index	1	2	3	4
IDs	W_3	W_4	W_1	W_2

Fig. 3. Figure at left represents MT built by splitting process on fingerprints $\{W_1, W_2, W_3, W_4\}$, where node 2 corresponds to splitting of $\{W_3, W_4\}$ and node 3 corresponds to splitting of $\{W_1, W_2\}$. In figure at right, multibit tree on bin K_c is represented by LBS B_c, leaf bits L_c, and array IDs. Leaves contain fingerprint identifiers that are stored in IDs.

4.2   Variable-Length Array for Compactly Representing Fingerprint Databases

We represent a fingerprint by an array sorted in increasing order. To minimize each word in fingerprint W_i as much as possible, we compute the difference between the k-th and $(k-1)$-th words as $W_i[k] - W_i[k-1]$, $k = 1...|W_i|$, and store them in new array A_i. We can recover original words in W_i by cumulatively adding each element in A_i from 1 to $|A_i|$. Each element in A_i is smaller than the original word in W_i. Although most elements in A_i are expected to be small by computing differences between words, some elements can remain large. For maximum difference m, $|A_i| \log m$ bits are basically required to store A_i. When m is large, the size for storing A_i is not different from that for storing original fingerprint W_i.

We present a variable-length array to compactly store A_i. The variable-length array is an array in which elements are represented by bit strings of different lengths. In addition, our variable-length array is directly accessible by any element in $O(1)$ time. We designed such an array with the help of the rank/select dictionary.

Let A be an array converted to a variable-length array. We use two bit strings R and P to represent A as follows (Figure 4).

- R is a bit string whose k-th substring corresponds to the bit representation of $A[k]$.
- P is a bit string whose k-th substring consists of ($\lceil \log A[k] \rceil$ - 1) 0s followed by 1.

Bit string P is indexed by the rank/select dictionary. $A[k]$ is recovered by $select_1$ operation on P as

1. If $k = 1$ s=1, else $s = select_1(P, k-1) + 1$
2. $e = select_1(P, k)$
3. Convert substring $R[s, e]$ to an integer.

Through steps 1 and 2, we can obtain start position s and end position e on R, and $R[s, e]$ corresponds to the substring representing $A[k]$.

We represent A_i as a variable-length array. Here, we call a fingerprint database represented by A_i and variable-length arrays VLA. We also call a fingerprint database represented by original W_i and standard arrays of fixed-length RAW.

We present a trie representation of fingerprint databases with a variable-length array in the next subsection.

Index	1	2	3	4	5	6
Array A	1	3	8	5	3	2
Bit string R	1	11	1000	101	11	10
Bit string P	1	01	0001	001	01	01

Fig. 4. Variable-length array

4.3   Succinct Trie for Representing Fingerprint Databases

We applied a trie, also called a prefix tree, to succinctly represent fingerprint databases. The trie is an ordered tree that is used to store an associative array where the keys are usually strings. Yet, the trie can be applied to fingerprints considered as strings, and still remain effective. Each node in the trie defines the key with which it is associated. All the descendants of a node have a common prefix of the string associated with that node, and the root is associated with the empty string. Values are not associated with every node, only with leaves and some inner nodes that correspond to keys. The values in our case correspond to the identifiers of fingerprints. Figure 5 shows a trie built on a fingerprint database.

The alphabet tends to be small for trie applications in a string. Examples are the DNA alphabet, which typically contains 4 characters, and English which contains 26. A difficulty with our application of a trie is that the size of the words to organize the fingerprints is not always small. Although a tree itself for such applications is compactly represented by LOUDS, the memory usage of a trie is dominated by that of the words attached to each node. We computed the differences between every pair of the word of a node and that of the parent to minimize word sizes. The sizes of resulting words are certainly smaller than those of the original words (Figure 5).

Since a trie is an ordered tree, it can efficiently be represented by LOUDS (Figure 5). Let LBS for the trie built on a fingerprint database be T. The words created by the difference computations are stored in variable-length array D. These words are placed in D in the breath-first order of the trie. We prepared an array, $idconv$, whose indices were the identifiers of fingerprints and elements were the node identifiers of leaves.

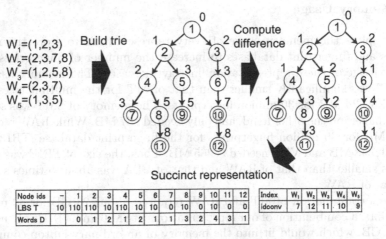

$W_1=(1,2,3)$
$W_2=(2,3,7,8)$
$W_3=(1,2,5,8)$
$W_4=(2,3,7)$
$W_5=(1,3,5)$

Build trie   Compute difference

Succinct representation

Node ids	-	1	2	3	4	5	6	7	8	9	10	11	12
LBS T	10	110	110	10	110	10	10	0	10	0	10	0	0
Words D		0	1	2	1	2	1	1	3	2	4	3	1

Index	W_1	W_2	W_3	W_4	W_5
idconv	7	12	11	10	9

Fig. 5. Succinct representation of fingerprint database. Node identifiers are represented in nodes. Node labels are attached to nodes. Root has dummy label of 0. Nodes containing identifiers of fingerprints are represented by concentric circles.

The decoding of a fingerprint starts from a leaf or an inner node corresponding to it. The identifiers of fingerprints similar to a query are computed by using the similarity search step of the multibit tree. By looking *idconv* up from each identifier, we obtain the corresponding leaf or inner node. When recovering a fingerprint, we first obtain words by ascending from the leaf or the inner node related to the identifier of a fingerprint to the root. Since each element in D is represented by our variable-length array and can be accessed by any element in $O(1)$ time, the recovery process is efficiently performed. The original words in the fingerprint can be recovered by cumulatively adding the words obtained from the root's word to the leaf's or inner node's word.

5   Experiments

We evaluated our succinct representations of multibit trees and fingerprint database by comparing them with other representations. We compared our SMT with MT for multibit trees. We compared TRIE with VLA and RAW for the fingerprint database. We downloaded all $30,129,866$ chemical fingerprints in the PubChem database[1]. Of these, 30 million fingerprints were randomly sampled. Since there are 881 chemical substructures defined in the PubChem database, the dimension of those fingerprints was 881. We used the average search time and memory usage as evaluation measures. All methods were implemented in C++. We used Okanohara's implementation[2] of Navarro and Providel's rank/select dictionary [12] for the rank/select operations. All experiments were carried out with a Linux machine on a quad-core AMD Opteron^{TM} Processor 8393 SE (3.1GHz) with 512 GB of memory.

5.1   Memory Usage

Figure 6 plots the memory usage of every representation to store the multibit trees and fingerprint databases to increase the number of fingerprints. The memory usage of these representation linearly increased. The memory usage of our SMT was significantly smaller than that of MT for the multibit trees. MT required $5,929$ MB for 30 million fingerprints. The memory of SMT was significantly smaller, on the other hand, and only needed 847 MB. While RAW required $16,673$ MB for 30 million fingerprints for the fingerprint database, TRIE only needed $1,348$ MB and VLA needed $3,265$ MB. Thus, the size of TRIE was about 12 times smaller than that of RAW and that of VLA was about 5 times smaller than that of RAW.

While a combination of MT and RAW required about 23 GB for 30 million fingerprints, a combination of our representations of SMT and TRIE only needed about 2 GB, which would fit into the memory of an ordinary laptop computer.

---

[1] http://pubchem.ncbi.nlm.nih.gov/
[2] http://code.google.com/p/rsdic/

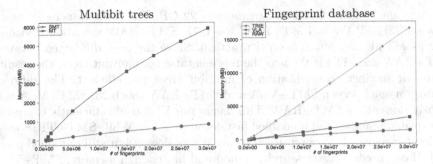

Fig. 6. Memory usage for increasing the number of fingerprints

Even a combination of SMT and VLA only needed about 4 GB. Although minimizing the representations of fingerprint databases is more effective than minimizing those of multibit trees, the representations of multibit trees reduced the total memory from 23 GB with MT and RAW to 18 GB with SMT and RAW, which means that our compact representation of multibit tree is effective for 30 million fingerprints.

5.2   Search Time

We evaluated the average search time per query by using combinations of representations on 30 million fingerprints. We randomly sampled 100 fingerprints from all fingerprints as queries to measure the average search time. We tried all six combinations of representations: SMT+TRIE, SMT+VLA, SMT+RAW, MT+TRIE, MT+VLA, and MT+RAW. We used three similarity thresholds of $\epsilon = 0.98, 0.95$, and 0.9 that respectively output 10, 160, and 1440 per query.

Table 1 summarizes the memory usage, average search time I_1, $I_1 \cap I_2$, and I_N for 30 million fingerprints. MT+RAW was the fastest of these combinations,

Table 1. Average search time in seconds and memory in megabytes on 30 million fingerprints for similarity thresholds of $\epsilon = 0.98, 0.95$, and 0.9

	Time (sec)			Memory (MB)
	$\epsilon = 0.98$	$\epsilon = 0.95$	$\epsilon = 0.9$	
SMT+TRIE	0.021 ± 0.042	0.182 ± 0.384	1.706 ± 2.592	$2,195$
SMT+VLA	0.014 ± 0.015	0.114 ± 0.127	0.724 ± 0.726	$4,112$
SMT+RAW	0.013 ± 0.010	0.090 ± 0.065	0.506 ± 0.351	$17,520$
MT+TRIE	0.009 ± 0.000	0.082 ± 0.047	0.579 ± 2.562	$7,277$
MT+VLA	0.007 ± 0.000	0.059 ± 0.000	0.329 ± 0.124	$9,194$
MT+RAW	0.006 ± 0.000	0.048 ± 0.041	0.294 ± 0.041	$22,602$
$|I_1|$	$1,372,261$	$3,474,693$	$6,984,526$	
$|I_1 \cap I_2|$	103	979	$8,875$	
$|I_N|$	10	160	$1,440$	

but consumed a large amount of memory of 22 GB. The ratio of memory usage between SMT+RAW and MT+RAW was 0.77. SMT+RAW was about twice as slow as MT+RAW, but remained practical. Since the only difference between SMT+RAW and MT+RAW was the representation of multibit trees, the results mean our succinct representation of multibit trees was efficient. The ratio of memory usage between SMT+VLA and SMT+RAW was 0.23. SMT+VLA was slightly slower than SMT+RAW. This means our VLA could efficiently compress fingerprint databases and it enabled fast decompression. While SMT+TRIE used the least memory of these representations, it was slowest but still remained practical. If one needs a faster search method and has enough memory, SMT+VLA is a good alternative to SMT+TRIE.

6   Conclusion

We presented novel representations of multibit trees and a fingerprint database for memory-efficient similarity searches. Our representations of multibit trees and the fingerprint database were called succinct multibit trees, and were built on a rank/select dictionary and a succinct tree. We also designed a variable-length array to compactly represent the auxiliaries of trees. While our variable-length array could represent each element represented by bit strings of different lengths, it enabled us to access any element in $O(1)$ time. We represented fingerprint databases with two representations of a succinct trie and a variable-length array. Our experimental results demonstrated that our representations significantly reduced memory usage and the loss of search time was reasonable for a moderate number of outputs. Although the succinct trie was slower than the variable-length array for the fingerprint databases, it was moderate. If one is concerned about such a loss of search time and has a computer with a large memory, the variable-length array is a better choice to represent fingerprint databases. Although we did not conduct any chemical or pharmaceutical analyses in this research, we think our succinct multibit tree is quite useful. The C++ implementation of our succinct multibit tree can be downloaded and tried from http://code.google.com/p/smbt/.

References

1. Aung, Z., Ng, S.-K.: An Indexing Scheme for Fast and Accurate Chemical Fingerprint Database Searching. In: Gertz, M., Ludäscher, B. (eds.) SSDBM 2010. LNCS, vol. 6187, pp. 288–305. Springer, Heidelberg (2010)
2. Baldi, P., Hirschberg, D.: An Intersection Inequality Sharper than the Tanimoto Triangle Inequality for Efficiently Searching Large Databases. Journal of Chemical Information and Modeling 49, 1866–1870 (2009)
3. Baldi, P., Hirschberg, D., Nasr, R.: Speeding Up Chemical Database Searches Using a Proximity Filter Based on the Logical Exclusive-OR. Journal of Chemical Information and Modeling 48, 1367–1378 (2008)
4. Chazelle, B.: A Functional Approach to Data Structures and its Use in Multidimensional Searching. SIAM Journal on Computing 17 (1988)

5. Elias, P.: Efficient Storage and Retrieval by Content and Address of Static Files. Journal of the ACM 21, 246–260 (1974)
6. Ferragina, P., Manzini, G.: An experimental study of an opportunistic index. In: Proceedings of the Twelfth Annual ACM-SIAM Symposium on Discrete Algorithms, pp. 269–278. Society for Industrial and Applied Mathematics (2001)
7. Jacobson, G.: Space-efficient Static Trees and Graphs. In: Proceedings of the 30th Annual Symposium of Foundations of Computer Science, pp. 549–554 (1989)
8. Keiser, M., Roth, B., Armbruster, B., Ernsberger, P., Irwin, J., Shoichet, B.: Relating protein pharmacology by ligand chemistry. Nature Biotechnology 25(2), 197–206 (2007)
9. Leach, A., Gillet, V.: An introduction to chemoinformatics. Kluwer Academic Publishers, The Netherlands, rev. ed. (2007)
10. Nasr, R., Hirschberg, D., Baldi, P.: Hashing Algorithms and Data Structures for Rapid Searches of Fingerprint Vectors. Journal of Chemical Information and Modeling 50, 1358–1368 (2010)
11. Nasr, R., Kristensen, T., Baldi, P.: Tree and hashing data structures to speed up chemical searches: Analysis and experiments. Molecular Informatics 30, 791–800 (2011)
12. Navarro, G., Providel, E.: Fast, Small, Simple Rank/Select on Bitmaps. In: Proc. SEA, pp. 295–306 (2012)
13. Okanohara, D., Sadakane, K.: Practical Entropy-Compressed Rank/Select Dictionary. In: Workshop on Algorithm Engineering & Experiments (2007)
14. Raman, R., Raman, V., Rao, S.: Succinct Indexable dictionaries with applications to encoding k-ary trees and multisets. In: SODA, pp. 232–242 (2002)
15. Swamidass, S., Baldi, P.: Bounds and Algorithms for Exact Searches of Chemical Fingerprints in Linear and Sublinear time. Journal of Chemical Information and Modeling 47, 302–317 (2007)
16. Tarjan, R.E., Yao, A.C.: Storing a Sparse Table. Communications of the ACM 22, 606–611 (1979)
17. Kristensen, T.G., Nielsen, J., Pedersen, C.N.S.: A Tree Based Method for the Rapid Screening of Chemical Fingerprints. In: Salzberg, S.L., Warnow, T. (eds.) WABI 2009. LNCS, vol. 5724, pp. 194–205. Springer, Heidelberg (2009)
18. Kristensen, T.G., Nielsen, J., Pedersen, C.N.S.: A tree-based method for the rapid screening of chemical fingerprints. Algorithms for Molecular Biology 5 (2010)
19. Turan, G.: Succinct Representation of Graphs. Discrete Applied Math. 8, 289–294 (1984)
20. Williams, H.E., Zobel, J.: Compressing integers for fast file access. Comput. J. 42, 193–201 (1999)

Comparing DNA Sequence Collections by Direct Comparison of Compressed Text Indexes

Anthony J. Cox[1], Tobias Jakobi[2],
Giovanna Rosone[3], and Ole B. Schulz-Trieglaff[1]

[1] Illumina Cambridge Ltd., United Kingdom
{acox,oschulz-trieglaff}@illumina.com
[2] Computational Genomics, CeBiTec, Bielefeld University, Germany
tjakobi@cebitec.uni-bielefeld.de
[3] University of Palermo, Dipartimento di Matematica e Informatica, Italy
giovanna@math.unipa.it

Abstract. Popular sequence alignment tools such as BWA convert a reference genome to an indexing data structure based on the Burrows-Wheeler Transform (BWT), from which matches to individual query sequences can be rapidly determined. However the utility of also indexing the query sequences themselves remains relatively unexplored.

Here we show that an all-against-all comparison of two sequence collections can be computed from the BWT of each collection with the BWTs held entirely in external memory, i.e. on disk and not in RAM. As an application of this technique, we show that BWTs of transcriptomic and genomic reads can be compared to obtain reference-free predictions of splice junctions that have high overlap with results from more standard reference-based methods.

Code to construct and compare the BWT of large genomic data sets is available at http://beetl.github.com/BEETL/ as part of the BEETL library.

1 Introduction

In computer science, a *suffix tree* is the classical example of an *indexing* data structure which, when built from some text T, allows the presence or absence of a query string S in T to be rapidly determined. A suffix tree is several times larger than the text it indexes, but research since 2000 (well summarized in [13]) has led to *compressed full-text indexes* that provide the same functionality as the suffix tree while taking up less space than the text itself.

Of these, the *FM-index* has become central to bioinformatics as the computational heart of popular sequence alignment tools such as BWA [9], Bowtie [8] and SOAP2 [10]. All these programs work in a similar way, with individual query sequences being searched for one-by-one in the index of a reference genome. The FM-index of, say, the latest human reference sequence can be viewed as a constant and precomputed, so the cost of building it is not important for this particular use case.

B. Raphael and J. Tang (Eds.): WABI 2012, LNBI 7534, pp. 214–224, 2012.
© Springer-Verlag Berlin Heidelberg 2012

Constructing the FM-index of T is dominated by the computation of the *Burrows-Wheeler transform*, a permutation of the symbols of T that also has widespread applications in data compression [2]. For large T, this calculation requires either a large amount of RAM or a cumbersome divide-and-conquer strategy. However, in [3,4], two of the present authors demonstrated that if T can be considered to be a large number of independent short patterns then its BWT can be built partially or entirely in external memory (that is, by sequential access to files held on disk). This leads us to the aim of the present work, which is to introduce some of the additional possibilities that arise if the set of query sequences is also indexed.

The search for a single pattern in an FM-index may potentially need access to any part of the BWT, requiring the entire BWT to be held in RAM to guarantee that this can be efficiently achieved. However, building on our previous work, we show that the computations needed for an all-against-all comparison of the sequences in two collections can be arranged so that the BWTs of the collections are both accessed in a series of sequential passes, permitting the comparison to be done efficiently with the BWTs held on disk. In k passes, this procedure traverses all k-mers that are present in one or both of the two indexes. This traversal can be viewed as a template upon which different sequence comparison tasks can be defined by specifying particular sets of behaviours according to whether each of the k-mers encountered is unique to one or the other dataset, or shared by both. To illustrate, we show how this template may be adapted to the task of comparing transcriptomic and genomic reads from an individual eukaryotic organism to deduce exon-exon splice junctions.

In a eukaryotic genome, large tracts of intragenic and intronic DNA will be represented in the genome but not the transcriptome, but the sequences present in the transcriptome alone are far fewer and of more interest: notwithstanding experimental artefacts and the relatively rare phenomenon of RNA editing, these must span splice junctions between exons. By obtaining the genomic and transcriptomic samples from the same individual, we eliminate the possibility that differences between the datasets are due to genetic variation between individuals.

We apply our methods to data from the Tasmanian Devil (*S. Harrissii*). The hypothesis- and reference-free nature of our procedure is advantageous for *de novo* projects where no reference sequence is available or cases where, as here, the reference genome is of draft quality.

2   Methods

2.1   Definitions

Consider a string s comprising k symbols from an alphabet $\Sigma = \{c_1, c_2, \ldots, c_\sigma\}$ whose members satisfy $c_1 < c_2 < \cdots < c_\sigma$. We mark the end of s by appending a special *end marker* symbol \$ that satisfies $\$ < c_1$. We can build $k + 1$ distinct *suffixes* from s by starting at different symbols of the string and continuing rightwards until we reach \$. If we imagine placing these suffixes in alphabetical order, then the *Burrows-Wheeler transform* [5,2] of s can be defined such that the

i-th element of the BWT is the symbol in s that precedes the first symbol of the i-th member of this ordered list of suffixes. Each symbol in the BWT therefore has an *associated suffix* in the string. A simple way (but not the only way - see [11]) to generalize the notion of the BWT to a collection of m strings $S = \{s_1, \ldots, s_m\}$ is to imagine that all members s_i of the collection are terminated by distinct end markers $\$_i$ such that $\$_1 < \cdots < \$_m < c_1$.

The characters of $BWT(S)$ whose associated suffixes start with some string Q form a single contiguous substring of $BWT(S)$. We call this the Q-*interval* of $BWT(S)$ and express it as a pair of coordinates $[b_Q, e_Q)$, where b_Q is the position of the first character of this substring and e_Q is the position of the first character after it. This definition is closely related to the *lcp-interval* introduced in [1]: the Q-interval is an lcp-interval of length $|Q|$. If Q is not a substring of any member of S then a consistent definition of its Q-interval is $[b_Q, b_Q)$, where b_Q is the position Q would take if it and the suffixes of S were to be arranged in alphabetical order.

Each occurrence of some character c in the Q-interval corresponds to an occurrence of the string cQ in S. We call cQ a *backward extension* of Q. If all characters in the Q-interval are the same then Q has a *unique backward extension*, which is equivalent to saying that all occurrences of Q in S are preceded by c. A Q-interval of size 1 is a special case of unique backward extension that corresponds to there being a unique occurrence of Q in S: we call this a *singleton backward extension*. Similarly, we can say that appending a character c to Q to give Qc creates a *forward extension* of Q. Since all suffixes that start with Qc must also start with Q, the Qc-interval is clearly a subinterval of the Q-interval.

2.2 All-Against-All Backward Search

We can use Q-intervals to describe the *backward search* algorithm for querying $BWT(S)$ to compute $occ(P)$, the number of occurrences of some query string $P = p_1 \cdots p_n$ in S. This proceeds in at most n stages. At stage j, let Q be the j-*suffix* (*i.e.* the last j symbols) of P, let c be the character preceding Q in P and assume we know the position of the (non-empty) Q-interval in $BWT(S)$. The number of occurrences of cQ in S is given by the number of occurrences of c in the Q-interval (see [7]). If this is zero, then we know that cQ does not occur in S and so $occ(P)$ must be zero. Otherwise, we observe that the number of occurrences of cQ in S is by definition the size of the cQ-interval. The start of the cQ-interval is given by the count of c characters that precede the start of the Q-interval. These are the two pieces of information we need to specify the cQ-interval that we need for the next iteration. At the last stage, the count of p_1 characters in the $p_2 \cdots p_n$-interval gives $occ(P)$.

Running this procedure to completion for a single query P entails counting symbols in intervals that can potentially lie anywhere in $BWT(S)$, the whole of which must therefore reside in RAM if we are to guarantee this can be done efficiently. However, we show that $occ(P)$ may be computed for all strings P of length k or less in S by making k sequential passes through $BWT(S)$, allowing the processing to be done efficiently with $BWT(S)$ held on disk.

At the start of iteration j, we open an array F of σ write-only files and set each entry of an array Π of σ counters to zero. During the iteration, we apply the processInterval() function described in Figure 1 to the Q-intervals of all j-suffixes of S which, by induction, we assume are available in lexicographic order. With this ordering, the intervals $[b_Q, e_Q)$, $[b_{Q'}, e_{Q'})$ of two consecutive j-suffixes Q, Q' satisfy $e_Q \leq b_{Q'}$, which means that we can update the counters Π and π needed by processInterval() by reading the symbols of BWT(S) consecutively.

These arrays simulate the rank() function used in the FM-index - $\Pi[i]$ holds rank(c_i, b_Q), the number of occurrences of c_i prior to the start of the b_Q, whereas $\pi[i]$ counts the occurrences of c_i in the Q-interval $[b_Q, e_Q)$ itself. The pair $(\Pi[i], \pi[i])$ that is appended to the file $F[i]$ specifies the start position and size of the $c_i Q$ interval. The last act of processInterval() is to update Π to count the occurrences of each symbol up to position e_Q and return the updated array ready to be passed in at the next call to the function.

At the end of the iteration, each file $F[i]$ contains the Q-intervals of all $(j+1)$-suffixes that start with symbols c_i, in lexicographic order. If we consider the files in the order $F[1], F[2], \ldots, F[\sigma]$ and read the contents of each sequentially, we have the lexicographic ordering of the $(j+1)$-suffixes that we need for the next iteration.

---

function PROCESSINTERVAL($[b_Q, e_Q), B, \Pi, F$)
    Update Π if necessary so that each $\Pi[i]$ counts occurrences of c_i in $B[O, b_Q)$
    Create π such that each $\pi[i]$ counts occurrences of c_i in $B[b_Q, e_Q)$
    for $i = 1$) σ **do**
        if $\pi[i] > 0$ **then**
            Write $[\Pi[i], \pi[i])$ to file $F[i]$
        end if
    end for
    $\Pi \leftarrow \Pi + \pi$
    return Π
end function

Fig. 1. Given a Q-interval $[b_Q, e_Q)$ of Q in a BWT string B, the function processInterval() computes the $c_i Q$-intervals for all backward extensions $c_i Q$ of Q that are present in $[b_Q, e_Q)$ and appends them to the appropriate file $F[i]$ ready for processing during the next iteration

2.3 All-Against-All Comparison of Two BWTs

The concept of all-against-all backward search can be extended to compute the union of all suffixes of length k or less present in two collections S_A and S_B by making k passes through their BWTs. Figure 2 describes the logic of a single pass. Conceptually, each pass is a simple merge of the two lists of Q-intervals of a given length, with the ordering of Q-intervals being determined by the lexicographic ordering of their associated suffixes Q. The notable implementation detail is that

```
while (1) do
    while (gotQ == true) do
        gotQ = getNextInterval(bQ, eQ, mQ, BWT(SA))
        if  (gotQ == false) or (mQ == true) then
            break
        end if
        doAOnlyBehaviour()
        processInterval(bQ, eQ, BWT(SA), ΠA, FA)
        for i = 1 → σ do
            if (piA[i] > 0) then
                Write false to file MA[i]
            end if
        end for
    end while
    while (gotR == true) do
        gotR = getNextInterval(bR, eR, mR, BWT(SB))
        if  (gotR == false) or (mR == true) then
            break
        end if
        doBOnlyBehaviour()
        processInterval(bR, eR, BWT(SB), ΠB, FB)
        for i = 1 → σ do
            if (piB[i] > 0) then
                Write false to file MB[i]
            end if
        end for
    end while
    if (gotQ == false) then
        break
    end if
    doSharedBehaviour()
    processInterval(bQ, eQ, BWT(SA), ΠA, FA)
    processInterval(bR, eR, BWT(SB), ΠB, FB)
    for i = 1 → σ do
        if (piA[i] > 0) and (piB[i] > 0) then
            Write true to files MA[i], MB[i]
        else if (piA[i] > 0) then
            Write false to file MA[i]
        else if (piB[i] > 0) then
            Write false to file MB[i]
        end if
    end for
end while
```

Fig. 2. Pseudocode for stage j of the all-against-all comparison of the BWTs of the collections S_A and S_B. Consecutive calls to function `getNextInterval()` are assumed to populate b_Q, e_Q and m_Q with details of the Q-intervals of the j-suffixes of the relevant BWT in lexicographic order, returning **false** once the list of intervals has been exhausted. In practice, these intervals are read sequentially from the sets of files F_A, F_B, M_A and M_B that were generated during the previous execution of this procedure.

associating an additional bit m_Q with each Q-interval avoids the need to store and compare strings when deciding whether Q-intervals from the two collections are associated with the same suffix. As in the previous section, this insight is best understood inductively: if a Q-interval is shared (which we know because m_Q is set to **true**), then any common backward extensions cQ must also be common to both collections (and m_{cQ} must be set to **true** to reflect that).

Each suffix Q we encounter during the execution of the algorithm in Figure 2 is either present in S_A only, present in S_B only, or common to both S_A and S_B and the algorithm calls different functions in the event of these three possibilities. We show how to specify the behaviour of these three functions so as to compute differences between genomic and transcriptomic sequence data from an individual eukaryotic organism.

Figure 3 shows a very simple example of how splicing in the transcriptome might give rise to a read set T containing sequence that is not present in the genomic reads G. Figure 4 shows the BWTs of the two datasets. In the function `doSharedBehaviour()`, we look for Q-intervals shared by $BWT(T)$ and $BWT(G)$ for which the Q-interval in $BWT(G)$ has a unique backward extension but the corresponding Q-interval in $BWT(T)$ exhibits significant evidence of one or more different backward extensions cQ. In our implementation, spurious junction predictions due to sequencing error are guarded against by ignoring any such backward extensions for which the number of occurrences (given by the number of c symbols present in the Q-interval) fails to exceed a threshold. Any T-only cQ-intervals that do pass this test are backward-extended in subsequent intervals by `doAOnlyBehaviour()` until a string CQ is obtained for which the size of the CQ-interval fails to exceed a threshold t, which is equivalent to demanding that CQ must occur at least t times in T. The aim of this extension is to accumulate as much sequence context as possible to the left of the putative exon/exon junction. In a similar way, we could improve specificity by allowing `doBOnlyBehaviour()` to extend G-only intervals and thus accumulate sequence context that reaches into the separating intron, although our current implementation does not do this.

If the sequence context to the right of a predicted junction is a prefix of the sequence that lies to the right of another prediction junction, then the former prediction is subsumed into the latter. For example, in Figure 3 the same splice junction gives rise to reads **TCACA** and **CACAT** with rightward contexts **ACA** and **ACAT**: the former is a prefix of the latter. This aggregation of predictions is conveniently done by making a single pass through the final list of predictions once they have been sorting in lexicographic order of their rightmost context. As well as removing repeated predictions, this acts as a further guard against false positives - we discard any predictions whose contexts cannot be rightward-extended in this way.

Finally, we note that the double-stranded nature of DNA is handled by aggregating the individual chromosomal sequences plus their reverse complements into a sequence collection and building the BWT of that.

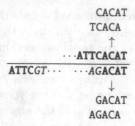

CACAT
TCACA
↑

‧‧‧**ATTCACAT**
―――――――――――――――
ATTCGT‧‧‧ ‧‧‧AG**ACAT**
↓

GACAT
AGACA

Fig. 3. Simple example of genome/transcriptome comparison. In the genome (below the line), the exons ATTC and ACAT (in bold) are separated by an intron (italics). In the transcriptome (above), the splicing together of these exons gives rise to reads $T = \{\text{CACAT}, \text{TCACA}\}$ containing the exon/exon boundary whereas, in the genome, the reads $G = \{\text{AGACA}, \text{GACAT}\}$ extend from the ACAT exon into the intron.

BWT(T)	suffixes	BWT(G)	suffixes
T	$\$_1$	A	$\$_1$
A	$\$_2$	T	$\$_2$
C	A$\$_2$	C	A$\$_1$
C	ACA$\$_2$	G	ACA$\$_2$
C	ACAT$\$_1$	G	ACAT$\$_2$
C	AT$\$_1$	$\$_1$	AGACA$\$_1$
A	CA$\$_2$	C	AT$\$_2$
T	CACA$\$_2$	A	CA$\$_1$
$\$_1$	CACAT$\$_1$	A	CAT$\$_2$
A	CAT$\$_1$	A	GACA$\$_1$
T	T$\$_1$	$\$_2$	GACAT$\$_2$
$\$_2$	TCACA$\$_2$	A	T$\$_2$

Fig. 4. Comparison of the BWTs of T and G from Figure 3. During the second execution of the procedure in Figure 2, we find the AC-interval in BWT(G) has a unique backward extension G, but the corresponding interval is BWT(T) has a different backward extension C. This is corroborated (and the sequence context to the right of the splice junction is extended) at steps 3 and 4 when the ACA- and ACAT-intervals of the two BWTs are compared. The CA-interval also suggests a divergent backward extension, but the lack of a forward extension of CA that is common to both T and G means this observation is not corroborated by subsequent executions of the code in Figure 2 and is therefore discarded.

3   Results

3.1   Reference-Free Detection of Splice Junctions

We tested our approach using data from a recent study [12] during which 1.45 billion genomic reads and 132 million RNA-Seq reads, all 100bp in length, were obtained from an individual Tasmanian devil. The RNA-Seq library was prepared from a mixture of mRNA from 11 different tissues to obtain broad coverage of gene content. The genome of the Tasmanian devil was estimated to be between 2.89 and 3.17 gigabase pairs (Gb) in size and is thus comparable in size to the human genome. *De novo* assembly of the genomic reads yielded 3.17 Gb of sequence with an N50 of 1.85 megabase pairs (Mb). The Ensembl gene annotation pipeline was then applied to the assembled contigs: evidence from alignment of mammalian EST, protein and RNA-Seq sequences was combined and then various gene prediction algorithms were used to refine these alignments and to build gene models (more detail on the annotation procedure is given in the supplement of [12]). We obtained the most recent version (0.67) of the Tasmanian devil gene annotation from the Ensembl FTP site. It contains 20 456 genes which give rise to 187 840 exon junction sites.

We built BWTs of both the genomic and RNA-Seq read sets using the algorithms given in [3] and compared them as described in the previous sections. The sequences to the right and left of each prediction were aligned to the devil assembly using BWA [9] in single-read mode, setting the option to allow up to 10 candidates for each read. Predictions for which the left and right halves aligned to the same contig with appropriate orientation were classified as putative junction sites: we obtained 171 371 of these.

We also predicted gene models and junction sites from the same RNA-Seq reads using version 2.0.0 of Tophat [17], which is a popular tool for this task. Tophat first aligns reads to a reference genome with the Bowtie2 aligner [8] then builds splicing models based on these alignments. The results of our comparison are summarized in Table 5: Tophat predicts 120 010 junction sites, of which 66 587 are not contained in the gene annotation.

Tool	Junctions predicted	True positives	False Negatives	Sensitivity (%)	FDR (%)
BWT	171 371	93 615	94 225	49.84	45.37
Tophat	120 010	66 587	121 253	35.45	44.51

Fig. 5. Comparison of junction site predictions. Our approach predicts 171 371 sites and Tophat predicts 120 010. Treating the Ensembl annotation as a gold standard, we evaluate sensitivity and false discovery rate of each method. The BWT-based approach is competitive with the established software Tophat.

Using the BEDtools software suite [14], we identified junction sites that overlap with sites contained in the Ensembl gene annotation. We used default parameters, apart from requiring a reciprocal overlap of 90% of the feature length. Of the 171 371 sites computed by our approach, 94 225 match known Ensembl predictions. Manual inspection of the remaining sites revealed that many were contained in putative gene annotation derived from EST alignments or from *ab initio* gene recognition algorithms. These putative annotations were not incorporated into the final annotation because of various threshold or partially contradicting evidence. They represent nevertheless likely candidates for coding regions. The EST alignments cover 48.06 Mb in 22 582 alignments and the *ab initio* predictions cover 44.92 Mb in 44 659 regions. Of the 77 756 junction sites detected by our method that did not have a counterpart in the Ensembl prediction, 24 322 did not have a match in the EST alignment data set and 11 168 did not match coding regions predicted by *ab initio* algorithms. Taking these sets together, we found that only 8 755 out of 171 371 (5.11%) did not have any evidence of being in transcribed regions. For Tophat, 53 423 junction sites did not have a match in the Ensembl gene annotation. Of these predictions, 14 668 did not have a match in regions covered by EST alignments and 6 227 did not have a match in *ab initio* gene predictions. In sum, 4 732 Tophat predictions (3.94%) did not match any potentially coding regions.

4   Discussion

In this work, we show that BWTs of transcriptomic and genomic read sets can be compared to obtain reference-free predictions of splice junctions that have high overlap with results from more standard reference-based methods. Our method predicts splice junctions by directly comparing sets of genomic and transcriptomic reads and can therefore provide orthogonal confirmation of gene predictions obtained by comparative genomics approaches. A reference sequence is not required for the prediction process itself (here we map the predicted junctions to the assembly only for comparison purposes), making the method particularly well suited to the analysis of organisms for which no reference genome exists.

When comparing the performance of our method with Tophat we find that, at least on this data, our approach has superior sensitivity and comparable false discovery rate. In order to give a strong proof of principle we deliberately avoided building any sort of prior information about gene structure into our analysis. In contrast, Tophat makes assumptions about the presence of canonical dinucleotide motifs at donor/acceptor sites and the relative abundance of isoforms. This is an entirely reasonable thing to do, but it is conceivable that Tophat's use of prior information might be a disadvantage for this particular dataset as it is not clear to what extent these signals are conserved across species and in particular in the Tasmanian devil.

An obvious piece of prior information needed by both Tophat and the Ensembl annotation pipeline is of course a reference sequence. Although considered to be of 'draft' quality, the Tasmanian devil assembly we used [12] nevertheless required not only both considerable computational and manual effort to generate but also made use of additional sequencing data in the form of long-insert mate pair libraries.

While the Ensembl annotation pipeline is a robust and well-established methodology, we note that our implicit treatment of the Ensembl annotation as absolute truth is an assumption that might be questioned, since the pipeline is being applied here to a draft assembly from a relatively poorly-understood genome. Nevertheless, we believe that our results do demonstrate that direct comparison of BWTs gives results that are biologically credible and that are competitive with existing tools.

The need to sequence the genome as well as the transcriptome means our method is unlikely to supplant methods such as Tophat which can operate on transcriptome data alone. Comparison of transcriptome to exome data might be more practical and have some utility, although it is arguable whether such an approach remains hypothesis-free. However in situations where, as here, both genome and transcriptome data are available our method may provide valuable additional information. Even for a much better characterized genome such as human, our algorithm should provide insight into transcription from regions that are not well-represented in the reference sequence and might also be a useful tool for investigating RNA editing. A further improvement of the method would be to used read-pairing information to link junction sites that are present in the same read pair and hence in the same transcript.

Computing the BWTs of the genomic and transcriptome read sets took around 6 days and 12 hours of wallclock time respectively, although the method employed ran entirely in external memory and so did not require high-end computing hardware. Moreover, our previous work [3] suggests that these computation could be approximately halved by storing the work files on a SSD (flash memory) drive and could be further improved by using a different algorithm that reduces I/O at the expense of moderate RAM usage. Indeed, one could make a case that the cost of BWT computation should not be included in the overall compute time, since it is useful in its own right for lossless compression of the data [6] and for facilitating other analyses such as de novo assembly [15,16].

The comparison of BWTs ran in just under 3 days of wallclock time. Again, all processing was done in external memory and could be sped up by the use of an SSD drive or, alternatively, the sequential nature of the algorithm's I/O access would facilitate cache efficient processing if the BWT files were instead held in RAM on a high-end machine. To put these numbers into context, the analysis using TopHat took 18.6 hours but this obviously does not include the time to assemble and curate the reference genome.

In addition, our implementation is a proof-of-principle with considerable scope for optimization. Future work will focus on such improvements and on exploring further applications of the algorithm described in Figure 2: many important tasks in sequence analysis can be reinterpreted as a comparison between BWTs, not least the comparison of tumour and normal read sets from cancer samples and the comparison of reads to a reference sequence.

Acknowledgement. A.J.C. and O.S.-T. are employees of Illumina Inc., a public company that develops and markets systems for genetic analysis, and receive shares as part of their compensation. Part of T.J.'s contribution was made while on a paid internship at Illumina's offices in Cambridge, UK. We thank Elizabeth Murchison and Zemin Ning for contributing the genomic and RNA-Seq data and the genome assembly of the Tasmanian Devil.

References

1. Abouelhoda, M.I., Kurtz, S., Ohlebusch, E.: Replacing suffix trees with enhanced suffix arrays. Journal of Discrete Algorithms 2(1), 53–86 (2004)
2. Adjeroh, D., Bell, T., Mukherjee, A.: The Burrows-Wheeler Transform: Data Compression, Suffix Arrays, and Pattern Matching, 1st edn. Springer Publishing Company, Incorporated (2008)

3. Bauer, M.J., Cox, A.J., Rosone, G.: Lightweight BWT Construction for Very Large String Collections. In: Giancarlo, R., Manzini, G. (eds.) CPM 2011. LNCS, vol. 6661, pp. 219–231. Springer, Heidelberg (2011)

4. Bauer, M.J., Cox, A.J., Rosone, G.: Lightweight algorithms for constructing and inverting the BWT of string collections. Theoretical Computer Science (2012) (online February 10, 2012)

5. Burrows, M., Wheeler, D.J.: A block sorting data compression algorithm. Technical report, DIGITAL System Research Center (1994)

6. Cox, A.J., Bauer, M.J., Jakobi, T., Rosone, G.: Large-scale compression of genomic sequence databases with the Burrows-Wheeler transform. Bioinformatics 28(11), 1415–1419 (2012)

7. Ferragina, P., Manzini, G.: Opportunistic data structures with applications. In: Proceedings of the 41st Annual Symposium on Foundations of Computer Science, pp. 390–398. IEEE Computer Society, Washington, DC (2000)

8. Langmead, B., Trapnell, C., Pop, M., Salzberg, S.: Ultrafast and memory-efficient alignment of short DNA sequences to the human genome. Genome Biology 10(3), R25+ (2009)

9. Li, H., Durbin, R.: Fast and accurate short read alignment with Burrows-Wheeler transform. Bioinformatics 25(14), 1754–1760 (2009)

10. Li, R., Yu, C., Li, Y., Lam, T.W., Yiu, S.M., Kristiansen, K., Wang, J.: Soap2: an improved ultrafast tool for short read alignment. Bioinformatics 25(15), 1966–1967 (2009)

11. Mantaci, S., Restivo, A., Rosone, G., Sciortino, M.: An extension of the Burrows-Wheeler Transform. Theor. Comput. Sci. 387(3), 298–312 (2007)

12. Murchison, E.P., Schulz-Trieglaff, O.B., Ning, Z., Alexandrov, L.B., Bauer, M.J., Fu, B., Hims, M., Ding, Z., Ivakhno, S., Stewart, C., Ng, B.L., Wong, W., Aken, B., White, S., Alsop, A., Becq, J., Bignell, G.R., Cheetham, R.K., Cheng, W., Connor, T.R., Cox, A.J., Feng, Z., Gu, Y., Grocock, R.J., Harris, S.R., Khrebtukova, I., Kingsbury, Z., Kowarsky, M., Kreiss, A., Luo, S., Marshall, J., McBride, D.J., Murray, L., Pearse, A., Raine, K., Rasolonjatovo, I., Shaw, R., Tedder, P., Tregidgo, C., Vilella, A.J., Wedge, D.C., Woods, G.M., Gormley, N., Humphray, S., Schroth, G., Smith, G., Hall, K., Searle, S.M.J., Carter, N.P., Papenfuss, A.T., Futreal, P.A., Campbell, P.J., Yang, F., Bentley, D.R., Evers, D.J., Stratton, M.R.: Genome sequencing and analysis of the tasmanian devil and its transmissible cancer. Cell 148(4), 780–791 (2012)

13. Navarro, G., Mäkinen, V.: Compressed full-text indexes. ACM Comput. Surv. 39(1) (2007)

14. Quinlan, A.R., Hall, I.M.: Bedtools: a flexible suite of utilities for comparing genomic features. Bioinformatics 26(6), 841–842 (2010)

15. Simpson, J.T., Durbin, R.: Efficient construction of an assembly string graph using the FM-index. Bioinformatics 26(12), i367–i373 (2010)

16. Simpson, J.T., Durbin, R.: Efficient de novo assembly of large genomes using compressed data structures. Genome Research 22(3), 549–556 (2011)

17. Trapnell, C., Pachter, L., Salzberg, S.L.: Tophat: discovering splice junctions with rna-seq. Bioinformatics 25(9), 1105–1111 (2009)

Succinct de Bruijn Graphs

Alexander Bowe[1], Taku Onodera[2], Kunihiko Sadakane[1], and Tetsuo Shibuya[2]

[1] National Institute of Informatics, 2-1-2 Hitotsubashi, Chiyoda-ku,
Tokyo 101-8430, Japan
{alex,sada}@nii.ac.jp
[2] Human Genome Center, Institute of Medical Science,
University of Tokyo 4-6-1 Shirokanedai, Minato-ku, Tokyo 108-8639, Japan
{tk-ono,tshibuya}@hgc.jp

Abstract. We propose a new succinct de Bruijn graph representation. If the de Bruijn graph of k-mers in a DNA sequence of length N has m edges, it can be represented in $4m + o(m)$ bits. This is much smaller than existing ones. The numbers of outgoing and incoming edges of a node are computed in constant time, and the outgoing and incoming edge with given label are found in constant time and $\mathcal{O}(k)$ time, respectively. The data structure is constructed in $\mathcal{O}(Nk \log m / \log \log m)$ time using no additional space.

1 Introduction

Within the last two decades, assembling a genome from enormous amount of reads from various DNA sequencers has been one of the most challenging and important computational problems in molecular biology. Though the problem is proved to be NP-hard [14], many algorithms have been proposed for the problem (see the surveys [10,13,19]). Most of these algorithms follow a so-called Overlap-Layout-Consensus strategy, where an algorithm first finds overlaps between reads, next layouts these reads, and finally finds the consensus genome. These algorithms can be categorized into two types, due to the graph used in the overlap phase.

Most old-time assembly algorithms (especially for the long Sanger reads) first construct a graph called the *overlap graph* after finding the overlapping pairs of reads, where each node represents a read and edges are constructed between nodes *iff* the corresponding two reads have an overlap of enough length [1,9,15]. But this strategy is difficult to apply against the huge data from more recent epoch-making next-generation sequencers (NGSs). The NGS machines can sequence vast amount of genome data. It makes it computationally very hard to compare all the pairs of reads. Moreover, most NGSs cannot read long DNA fragments (*e.g.*, at most 200bp in the case of Illumina HiSeq2000), and their read lengths are not long enough to detect overlaps with enough lengths between reads. To conquer these problems, many recent assembler algorithms utilize a graph called the *de Bruijn graph* in the overlap phase [11,12,18,21,22,24], instead of the overlap graph.

B. Raphael and J. Tang (Eds.): WABI 2012, LNBI 7534, pp. 225–235, 2012.
© Springer-Verlag Berlin Heidelberg 2012

A de Bruijn graph is a graph where each node represents a k-mer (a substring of length k) that exists in the reads, and an edge exists *iff* there is an exact overlap of length $k-1$ between the corresponding k-mers. The de Bruijn graph can be constructed more efficiently than the overlap graph in many cases, but the overlap phase is still the bottleneck of most assembly algorithms based on the de Bruijn graph. This is because storing the de Bruijn graph requires huge amount of memory. Thus we focus on reducing the memory required for the de Bruijn graph in this paper.

There have been proposed only two data structures for reducing the size of memory for the de Bruijn graph. The succinct data structure proposed by Conway and Bromage [5] is a data structure that straightforwardly represents the de Bruijn graph by a bit vector. Its representation should be smaller than a naive ordinary implementation of the de Bruijn graph, but it still requires $O(m \cdot k)$ memory, where k is the k-mer length and m is the number of edges in the de Bruijn graph, which means it would be very large when k is large. The other data structure is by Ye et al. [23], which stores only a subset of nodes of the de Bruijn graph to save memory, but it is not actually the de Bruijn graph.

In this paper, we propose a new succinct representation of a de Bruijn graph which only requires $m(2 + \log \sigma)$ bit to store[1], where σ is the alphabet size (*i.e.*, $\sigma = 4$ in the case of DNA). The size of this representation is not affected by the value of k and is much smaller than either of the two previous methods. Moreover we will present the algorithm to construct the data structure on-line. Our main result is summarized as follows:

Theorem 1. *The k-dimensional de Bruijn graph of M string of total length N on an alphabet of size σ can be stored in $m(2 + \log \sigma) + \mathcal{O}((\sigma + M) \log m) + o(m \log \sigma)$ bits where m is the number of edges in the graph. The numbers of outgoing and incoming edges of a node are computed in $\mathcal{O}(\log \sigma / \log \log m)$ time, and the outgoing and incoming edge with given label are found in $\mathcal{O}(\log \sigma / \log \log m)$ time and $\mathcal{O}(k \log^2 \sigma / \log \log m)$ time, respectively. The node for a given k-mer is found in $\mathcal{O}(k \log \sigma / \log \log m)$ time. If $\sigma = \mathrm{polylog}(m)$, the time complexities become $\mathcal{O}(1)$, $\mathcal{O}(1)$, $\mathcal{O}(k \log \sigma)$, and $\mathcal{O}(k)$ time, respectively.*

Theorem 2. *The k-dimensional de Bruijn graph of a string of length N can be constructed in $\mathcal{O}\left(Nk \cdot \frac{\log m}{\log \log m}(1 + \frac{\log \sigma}{\log \log m})\right)$ time using no additional space. This representation can be converted to the static one in $\mathcal{O}\left(\frac{m \log m}{\log \log m}(1 + \frac{\log \sigma}{\log \log m})\right)$ time.*

For DNA sequences ($\sigma = 4$), the succinct de Bruijn graph can be constructed in $\mathcal{O}(Nk \log m / \log \log m)$ time and its space becomes $4m + o(m)$ bits. This is much smaller than existing ones. For example, the succinct representation of Conway and Bromage [5] uses 40.8GB for storing a de Bruijn graph with $m = 12,292,819,311$ edges and $k = 27$ (28.5 bits per edge). On the other hand, if we use an efficient implementation of *rank/select* data structures [17] for our representation, the estimated size is less than 5 bits per edge. Therefore the above graph is stored in less than 8GB.

---

[1] The base of logarithm is 2.

2 Preliminaries

2.1 de Bruijn Graphs

In the original definition [2], the k-dimensional de Bruijn graph of σ symbols is a directed graph representing overlaps between strings of symbols defined as follows. The graph has σ^k nodes, consisting of all length-k strings of the symbols. A node is denoted by $(u_1, \ldots, u_k)$ where $u_1, \ldots, u_k$ are symbols. For any pair of nodes $u = (u_1, \ldots, u_k)$ and $v = (v_1, \ldots, v_k)$ such that $u_2 = v_1, u_3 = v_2, \ldots, u_k = v_{k-1}$, the graph has a directed edge from u to v labeled with v_k. In this paper we call it the complete k-dimensional de Bruijn graph of σ symbols.

The de Bruijn graphs considered in this paper are subgraphs of the complete de Bruijn graph. We define the k-dimensional de Bruijn graph of a string T as follows. The nodes of the graph correspond to all length-k substrings of T. If the string is of length N, the graph has at most $N - k + 1$ nodes. The edges of the graph are defined in the same way as the complete de Bruijn graph. For convenience, we add k characters $ at the head of the string, and a $ at the end.

We can also store a set of M strings $T_1, \ldots, T_M$ as follows. We append a terminator $_i$ to the tail of each string T_i, and concatenate all the strings. Then we add k characters $_0$ at the head. Figure 1 shows an example.

2.2 Basic Succinct Data Structures

Let $T = T[1]T[2] \cdots T[N]$ be a string of length N on alphabet $\mathcal{A}$, that is, $T[i] \in \mathcal{A}$ for any $i = 1, \ldots, N$. Let $\sigma = |\mathcal{A}|$ denote the alphabet size. We can store T in $N\lceil \log_2 \sigma \rceil$ bits. The space does not depend on the word size of CPU. We can retrieve any character $T[i]$ in constant time using bit operations on words.

The most basic succinct data structure is the one for computing $rank$, $select$, and $access$ values on strings, which are defined as follows. The value $access(T, i)$ returns $T[i]$ for $1 \le i \le N$. The value $rank_c(T, i)$ where $c \in \mathcal{A}$ and $1 \le i \le N$ is the number of c's in $T[1] \cdots T[i]$. For any T and c we define $rank_c(T, 0) = 0$. The value $select_c(T, j)$ where $c \in \mathcal{A}$ and $1 \le j \le rank_c(T, N)$ is the position of j-th c in T. For any T and c we define $select_c(T, 0) = 0$ and for any $j > rank_c(T, N)$ $select_c(T, j) = N + 1$. Let $t_r(N, \sigma)$, $t_s(N, \sigma)$, and $t_a(N, \sigma)$ denote the time complexity for computing $rank$, $select$, and $access$, respectively, on a string of length N and alphabet size σ. For brevity, we assume that for any $N_1 \le N_2$, $t_r(N_1, \sigma) \le t_r(N_2, \sigma)$ and for any $\sigma_1 \le \sigma_2$, $t_r(N, \sigma_1) \le t_r(N, \sigma_2)$. Let $t_b(N, \Sigma)$ denote the maximum of $t_r(N, \sigma), t_s(N, \sigma), t_a(N, \sigma)$.

For convenience, we define $pred_c(T, i) = select_c(T, rank_c(T, i))$ which is the position of the first occurrence of c when we scan T from the position i to the head, and $succ_c(T, i) = select_c(T, rank_c(T, i - 1) + 1)$ which is the position of the first occurrence of c when we scan T from the position i to the end. If $T[i]$ is the first (last) occurrence of c, $pred$ ($succ$) returns 0 ($N + 1$).

There exist many succinct data structures for $rank$ and $select$ on strings. Among them, we use the one by Ferragina et al. [8] for the static case (the case the string does not change). A string of T length n on an alphabet of size σ can

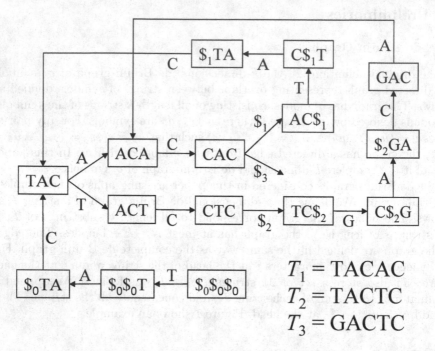

$$T_1 = \text{TACAC}$$
$$T_2 = \text{TACTC}$$
$$T_3 = \text{GACTC}$$

Fig. 1. The 3-dimensional de Bruijn graph of strings 'TACAC', 'TACTC', and 'GACTC'

be stored in $nH_0(T) + \mathcal{O}(\sigma \log n) + o(n \log \sigma)$ bits so that *rank*, *select* and *access* queries take $\mathcal{O}(\log \sigma / \log \log n)$ time, where $H_0(T)$ denotes the order-0 entropy of the string. Note that if the alphabet size σ is polylog(n), the queries are done in constant time. For a binary alphabet case, we can use a simpler data structure that has the same time and space complexities [20].

For the dynamic case where the string is modified by inserting or deleting a character, we use the one by Navarro and Sadakane [16] which stores the string in $nH_0(T) + \mathcal{O}(\sigma \log n) + o(n \log \sigma)$ bits so that *rank*, *select* and *access* queries and insertion and deletion of a character take $\mathcal{O}(\frac{\log n}{\log \log n}(1 + \frac{\log \sigma}{\log \log n}))$ time. For polylog-sized alphabets, the operations are done in optimal $\mathcal{O}(\log n / \log \log n)$ time. The time complexities for insert and delete are denoted by $t_u(n, \sigma)$.

2.3 The XBW Data Structure

The XBW-transform [6] is a method for compressing and indexing labeled trees. It is an extension of the Burrows-Wheeler transform [3] used for compressing and indexing strings. Given a rooted tree with n nodes where each node has a label in the set of size σ, the XBW-transform converts the tree into a representation of $2n + n \log \sigma$ bits. The size of the representation matches the information-theoretic lower bound. We can support tree navigational operations by adding small-size auxiliary indexes.

Because the XBW is for storing a tree, we cannot use it directly for storing de Bruijn graphs, which is a cyclic graph. This paper proposes a new compact representation of de Bruijn graphs of strings.

3   Succinct de Bruijn Graphs

Let G be a k-dimensional de Bruijn graph of a string T of length N on alphabet $\mathcal{A}$. Let n and m be the numbers of nodes and edges of G, respectively. A succinct representation of G supports the following operations:

- *Outdegree*(v) returns the number of outgoing edges from node v.
- *Outgoing*(v, c) returns the node w pointed to by the outgoing edge of node v with edge label c. If no such node exists, it returns -1.
- *Indegree*(v) returns the number of incoming edges to node v.
- *Incoming*(v, c) returns the node $w = (w_1, \ldots, w_k)$ such that there is an edge from w and v and $w_1 = c$. If no such node exists, it returns -1.
- *Index*(s) returns the index i of the node whose label is the string s of length k.

We define $\mathcal{A}^-$ as any set of size $|\mathcal{A}|$ such that $\mathcal{A}^- \cap \mathcal{A} = \emptyset$. Let c^- denote an element of $\mathcal{A}^-$ corresponding to an element $c \in \mathcal{A}$. We also define a function u as $u(c^-) = c$ for any $c^- \in \mathcal{A}^-$ and $u(c) = c$ for any $c \in \mathcal{A}$. We assume that the function is evaluated in constant time.

3.1   The Succinct Representation

The representation consists of the following components:

- a string $W = W[1]W[2] \cdots W[m]$ where each character is from $\mathcal{A} \cup \mathcal{A}^-$.
- a string *last* of length m on the binary alphabet $\{0, 1\}$.
- an array F of length $\sigma = |\mathcal{A}|$.

An example is shown in Figure 2.

The string W is defined as follows. Each character $W[i]$ represents the label of an edge of G. Each edge $u \to v$ of G is associated with the node label of u. Those edge labels are sorted in the lexicographic order of reversals of associated node labels. Ties are broken by edge labels. Let *Node*[i] denote the node label for $W[i]$. This is not explicitly stored.

The string *last* is defined as *last*[i] = 1 if $i = n$ or *Node*[i] is different from *Node*[$i + 1$], or *last*[i] = 0 otherwise. From this definition, all node labels *Node*[i] with *last*[i] = 1 are distinct, and those indices i have one-to-one correspondence with the nodes of G. Therefore we use an index i of the strings such that *last*[i] = 1 to represent a node v. Let n denote the number of nodes.

The array F stores cumulative frequencies of the last characters of node labels. Namely, for any $c \in \mathcal{A}$, $F[c] = |\{i \mid 1 \le i \le m, C(i) < c\}|$ where $C(i)$ denotes the last character of *Node*[i]. Because $F[\$_i] = i$ for $i = 0, 1, \ldots, M$, we need not store them.

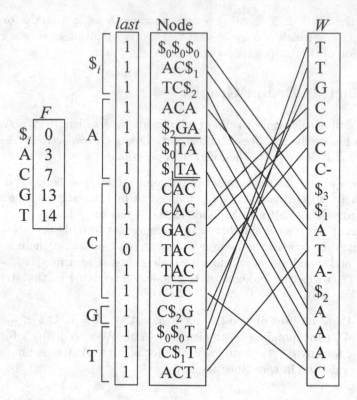

Fig. 2. The succinct representation of the de Bruijn graph in Figure 1. Lines between *Node* and *W* show *fwd* and *bwd* functions.

The array F is represented in $\mathcal{O}(\sigma \log m)$ bits. If F does not change, we can store it as it is using a simple array and $F[c]$ is computed in constant time. In a dynamic case that a new node or edge is inserted to the de Bruijn graph, we have to update F accordingly. By using a balanced binary tree, F can be maintained in $\mathcal{O}(\log \sigma)$ time.

We also use the inverse of F, that is, given i, we need to know the last character c of $Node[i]$. In a static case, this can be computed in constant time using a *rank/select* data structure of $\mathcal{O}(\sigma \log m) + \mathcal{O}(m \log \log m / \log m)$ bits [20]. In a dynamic case, it is done in $\mathcal{O}(\log \sigma)$ time using a balanced binary search tree. It can be improved to $\mathcal{O}(\frac{\log m}{\log \log m}(1 + \frac{\log \sigma}{\log \log m}))$ time using [16]. This data structure uses $\mathcal{O}(\sigma \log n) + \mathcal{O}(m \log \log m / \log m)$ bits. Let t_f denote the largest time complexity of those operations.

A character $W[i]$ is from either $\mathcal{A}$ or $\mathcal{A}^-$. If $W[i]$ is from $\mathcal{A}^-$, it means that there exists $j < i$ such that $W[j] = u(W[i])$ and $Node[j]$ and $Node[i]$ have the identical suffix of length $k - 1$.

We can define a one-to-one mapping between indices i of *last* with $last[i] = 1$ and indices j of W with $W[j] \in \mathcal{A}$. As stated above, the indices i with $last[i] = 1$ have one-to-one correspondence with the nodes of the de Bruijn graph G.

Consider indices j with $W[j] \in \mathcal{A}$. Let $Node'[j]$ denote the concatenation of the length $k-1$ suffix of $Node[j]$ and $W[j]$. For any $Node'[j]$, there exists i such that $Node[i] = Node'[j]$. Because of the definition of W, there are no indices j and j' ($j \neq j'$) such that $Node'[j] = Node'[j']$. Therefore there is a one-to-one mapping. Furthermore, the mapping is represented by $rank$ and $select$ queries on W. Let i, j be indices such that $last[i] = 1$ and $Node[i] = Node'[j]$. Let $c = C(i)$ be the last character of $Node[i]$ and $r = rank_1(last, i) - rank_1(last, F[c])$. Then it holds $j = select_c(W, r)$. From j, i is computed by $c = W[i]$, $r = rank_c(W, j)$ and $i = select_1(last, rank_1(last, F[c]) + r)$. We define $bwd(i) = j$ and $fwd(j) = i$. The time complexities of $bwd(i)$ and $fwd(j)$ are $\mathcal{O}(t_f + t_b(m, 2\sigma))$.

Our data structure is similar to the XBW data structure [6] in the sense that the $last$ array in ours is the same as S_{last} in the XBW. We propose a new encoding scheme for storing labels of a graph.

3.2 The *Outdegree* and *Outgoing* Operations

The *Outdegree(v)* operation is easy to support. We assume that v is the index of $last$ such that $last[v] = 1$ and $Node[v]$ is the label of the node. From the definition of $last$, it is obvious that $Outdegree(v) = v - pred_1(last, v - 1)$. The time complexity is $\mathcal{O}(t_b(m, 2))$.

The *Outgoing(v, c)* operation is done as follows. For any $1 \leq i \leq m$, we define $R(i) = [pred_1(last, i - 1) + 1, succ_1(last, i)]$, which is the range of W and $last$ that for all $j \in R(i)$, $Node[j]$ are identical. The labels of outgoing edges of node v are stored in $W[j]$ for $j \in R(v)$. Let j be the index such that $u(W[j]) = c$. We can find j by $pred_c(W, v)$ and $pred_{c-}(W, v)$. Then $w = Outgoing(v, c)$ can be computed by $r = fwd(j)$.

The time complexity for $Outgoing(v, c)$ is $\mathcal{O}(t_f + t_b(m, 2\sigma))$.

3.3 The *Indegree* and *Incoming* Operations

Consider to compute *Indegree(v)*. Let $d = C(v)$ and $x = bwd(v)$. Then it holds $d = W[x]$ and the first character of $Node[x]$ is the label of an edge pointing to v. Let $y = succ_d(W, x)$. Then all d^- between $W[x]$ and $W[y]$ correspond to parents of v. The number of such d^- is computed by $rank$ on W. The time complexity is $\mathcal{O}(t_f + t_b(m, 2\sigma))$.

To compute *Incoming(v, c)*, we need to obtain the first character of $Node[i]$ such that $x \leq i < y$ and $u(W[i]) = d$. The first character of $Node[i]$ is computed by $C(b^{k-1}(i))$ where b^{k-1} stands for applying $bwd(succ_1(last, i))$ repeatedly $k-1$ times. We perform a binary search to find the index i such that $c = C(bwd^{k-1}(i))$. The time complexity is $\mathcal{O}(k(t_f + t_b(m, 2\sigma)) \log \sigma)$.

3.4 The *Index* operation

Recall that *Index(s)* returns the index i of the node whose label is the string s of length k. Precisely, it returns i such that $last[i] = 1$ and $Node[i] = s$.

The algorithm for $Index(s)$ is similar to [7]. Let $i_1 < i_2 < \cdots < i_w$ be the indices such that $last[i_j] = 1$ and $Node[i_j]$ and s have the same suffix of length d $(1 \le d \le k)$. Let i_0 be the smallest index in $R(i_1)$. Then for any i such that $i_0 \le i \le i_w$, $Node[i]$ and s have the same suffix of length d and for other indices this does not hold. Therefore $Index(s)$ can be done by computing ranges $[i_0, i_w]$ for $d = 1, 2, \ldots, k$. Let c_d denote the d-th character of s $(1 \le d \le k)$. For $d = 1$, the range is $[F[c_1] + 1, F[c_1 + 1]]$. Given the range $[\ell_d, r_d]$ for d, we can compute the range $[\ell_{d+1}, r_{d+1}]$ for $d+1$ as follows. The end of the range r_{d+1} is computed by $r_{d+1} = Outgoing(r_d, c_{d+1})$. The beginning of the range ℓ_d is computed by $pred_1(last, Outgoing(succ_1(last, \ell_d), c_{d+1}) + 1$.

The above algorithm can be simplified. Instead of computing ranges $[i_0, i_w]$, we can use $[i_1, i_w]$. For $d = 1$, the range is $[succ_1(last, F[c_1] + 1), F[c_1 + 1]]$. Given the range $[\ell_d, r_d]$ for d, the range for $d + 1$ is obtained by $r_{d+1} = Outgoing(r_d, c_{d+1})$ and $\ell_{d+1} = Outgoing(\ell_d, c_{d+1})$. The time complexity is $\mathcal{O}(k(t_f + t_b(m, 2\sigma)))$.

3.5   Time and Space Complexities

We implement the above data structure for the static case using known succinct data structures. The array F is stored in $\sigma \log m$ bits. The data structure for computing $C(i)$ uses $\mathcal{O}(\sigma \log m) + \mathcal{O}(m \log \log m / \log m)$ bits. The operation time t_f is constant. The string $last$ is stored in $m + o(m)$ bits so that $rank$, $select$, and $access$ takes constant time [20]. The string W is stored by using [8]. Because the characters of W are from $\mathcal{A} \cup \mathcal{A}^- \cup \{\$_1, \ldots, \$_M\}$, the alphabet size is $2\sigma + M$. We can reduce the alphabet size to $2\sigma + 1$ by unifying the M terminators $\$_1, \ldots, \$_M$ into a character $\$$. We distinguish two terminators, but encode them using the same code.

The string W is stored in $m \log(2\sigma + M) + \mathcal{O}((\sigma + M) \log m) + o(m \log \sigma) = m + m \log \sigma + \mathcal{O}((\sigma + M) \log m) + o(m \log \sigma)$ bits, and the time complexities t_r, t_s, t_a are $\mathcal{O}(\frac{\log \sigma}{\log \log m})$. Therefore the time complexities for $Outdegree$, $Indegree$, $Outgoing$, $Incoming$, and $Index$ are $\mathcal{O}(\frac{\log \sigma}{\log \log n})$, $\mathcal{O}(\frac{\log \sigma}{\log \log n})$, $\mathcal{O}(\frac{\log \sigma}{\log \log n})$, $\mathcal{O}(\frac{k \log^2 \sigma}{\log \log n})$, $\mathcal{O}(\frac{k \log \sigma}{\log \log n})$, respectively.

For polylog-size alphabets, $Outdegree$, $Indegree$ and $Outgoing$ takes constant time, $Incoming$ takes $\mathcal{O}(k \log \sigma)$ time, and $Index$ takes $\mathcal{O}(k)$ time.

4   On-Line Construction

In this section we propose an on-line construction algorithm of the de Bruijn graph of a string. Here on-line means given the succinct de Bruijn graph G of a string $T = T[1] \cdots T[N]$, we change it to the succinct de Bruijn graph G' of the string $T' = T[1] \cdots T[N + 1]$ which is made by appending a character to T.

As stated above, our succinct representation of G assumes that a character $\$$ is appended to the end of T. Let p be the position of $\$$ in W. To construct the succinct representation of G', we first change $W[p]$ from $\$$ to $T[N + 1]$ and modify other parts if necessary, then insert $\$$ to another position of W. The details are as follows.

Let p be the position of $ in W for the string $T = T[1] \cdots T[N]$. If a new character $c = T[N + 1]$ is appended to the end of T, we change $W[p]$ from $ to $T[N + 1]$. We have to maintain the invariant that for all $i \in R(p)$, that is, $Node[i] = Node[p]$, $W[i]$ are distinct. Because before changing $W[p]$ they are distinct, we can check the invariant by finding the character $c = T[N + 1]$ or c^- in $W[i]$ such that $i \in R(p)$. This is done by $rank$ and $select$ on W.

If $T[N + 1]$ already exists in the range, let p' be its position. We delete $W[p]$ and $last[p]$ and we insert $ in W at position $x = fwd(p')$. We also insert 0 in $last[x]$ because $Node[x]$ already exists. We update $p = x$ and the array F accordingly.

If $T[N + 1]$ does not exist in the range, we change $W[p] = $ to either $c = T[N + 1]$ or c^-. To determine c or c^-, we first find the nearest occurrence of c to $W[p]$, namely, its position is $j = pred_c(W, p - 1)$ if it exists $(j > 0)$. We compare $Node[j]$ with $Node[p]$. If they have the same suffix of length $k - 1$, we change $W[p]$ to c^-, and otherwise change $W[p]$ to c. We compare characters of $Node[j]$ and $Node[p]$ one by one using the bwd function. We also compare $Node[j_2]$ with $Node[p]$ where $j_2 = succ_c(W, p + 1)$ if it exists $(j_2 \le m)$. If they share the length $k - 1$ suffix, we change $c_2 = W[j_2]$ to c_2^-. This takes $\mathcal{O}(k(t_f + t_b(m, 2\sigma)))$ time. If the nearest c does not exist $(j = 0)$, let $j = F[c]$. The position x to insert $ is computed by $x = fwd(j)$. We insert 0 to $last[x]$ if $W[p]$ or $W[j_2]$ has a character in $\mathcal{A}^-$, or 1 otherwise. Finally we set $p = x$ and update the array F.

In total, the update operation takes $\mathcal{O}(k(t_f + t_b(m, 2\sigma)))$ time. If we use the dynamic $rank/select$ data structure of [16] for W and $last$, $t_b = \mathcal{O}(\frac{\log m}{\log \log m}(1 + \frac{\log \sigma}{\log \log m}))$ time. We also use [16] for computing $C(i)$. Then $t_f = \mathcal{O}(\frac{\log m}{\log \log m}(1 + \frac{\log \sigma}{\log \log m}))$ and the space is $\mathcal{O}(\sigma \log n) + \mathcal{O}(m \log \log m / \log m)$ bits. Because we repeat this update operation N times for all characters of the input string, the succinct de Bruijn graph can be constructed in $\mathcal{O}\left(Nk \cdot \frac{\log m}{\log \log m}(1 + \frac{\log \sigma}{\log \log m})\right)$ time. For polylog-sized alphabets, it becomes $\mathcal{O}(Nk \cdot \frac{\log m}{\log \log m})$.

It is easy to construct the static data structure from the dynamic one. The strings $last$ and W for the static one are generated by applying $access$ operations to the dynamic one for $i = 1, \ldots, m$ in $\mathcal{O}(mt_b(m, 2\sigma))$ time. After constructing the static strings, the auxiliary data structures for computing $rank/select$ are constructed in $\mathcal{O}(m)$ time.

5   Concluding Remarks

We have proposed a succinct representation of de Bruijn graphs, which can be constructed with efficient time and space complexities, and in an on-line manner. Therefore they are useful for large-scale genome assembly.

The succinct de Bruijn graph can be also used for data compression. The PPM (Prediction by Partial Matching) is a text compression algorithm [4]. In the order-k PPM, a character is compressed using statistical information that it appears after a string of length k based on a given probability distribution. We can easily extend our succinct de Bruijn graph to be used for PPM compression. In addition to the array W, we use another array to store the numbers of times that each edge is traversed. Then we have enough information for compression. The succinct de Bruijn graph is used for natural language processing because it stores all n-grams in a text.

Our future work will be to improve the time complexity for the on-line construction algorithm, and to implement the proposed data structure and apply it for assembling large genomes and PPM data compression. A sample source code is available at http://code.google.com/p/csalib/.

Acknowledgments. KS and TS are supported in part by KAKENHI 23240002.

References

1. Batzoglou, S., Jaffe, D.B., Stanley, K., Butler, J., Gnerre, S., Mauceli, E., Berger, B., Mesirov, J.P., Lander, E.S.: Arachne: a whole-genome shotgun assembler. Genome Research 12, 177–189 (2002)
2. De Bruijn, N.G.: A combinatorial problem. Koninklijke Nederlandse Akademie v. Wetenschappen 49, 758–764 (1946)
3. Burrows, M., Wheeler, D.J.: A Block-sorting Lossless Data Compression Algorithms. Technical Report 124, Digital SRC Research Report (1994)
4. Cleary, J.G., Witten, I.H.: Data Compression Using Adaptive Coding and Partial String Matching. IEEE Trans. on Commun. COM-32(4), 396–402 (1984)
5. Conway, T.C., Bromage, A.J.: Succinct data structures for assembling large genomes. Bioinformatics 27(4), 479–486 (2011)
6. Ferragina, P., Luccio, F., Manzini, G., Muthukrishnan, S.: Compressing and indexing labeled trees, with applications. Journal of the ACM 57(1), 4:1–4:33 (2009)
7. Ferragina, P., Manzini, G.: Indexing compressed texts. Journal of the ACM 52(4), 552–581 (2005)
8. Ferragina, P., Manzini, G., Mäkinen, V., Navarro, G.: Compressed Representations of Sequences and Full-Text Indexes. ACM Transactions on Algorithms 3(2(20)) (2006)
9. Huang, X., Yang, S.P.: Generating a genome assembly with pcap. Current Protocols in Bioinformatics Unit 11.3 (2005)
10. Kasahara, M., Morishita, S.: Large-Scale Genome Sequence Processing. Imperial College Press (2006)
11. Li, R., Zhu, H., Ruan, J., Qjan, W., Fang, X., Shi, Z., Li, Y., Li, S., Shan, G., Kristiansen, K.: H Yang, and J. Wang. De novo assembly of human genomes with massively parallel short read sequencing. Genome Research 20, 265–272 (2009)
12. MacCallum, I., Przybylski, D., Gnerre, S., Burton, J., Shlyakhter, I., Gnirke, A., Malek, J., McKernan, K., Ranade, S., Shea, T.P., Williams, L., Young, S., Nusbaum, C., Jaffe, D.B.: Allpaths 2: small genomes assembled accurately and with high continuity from short paired reads. Genome Biology 10(R103) (2009)
13. Miller, J.R., Koren, S., Sutton, G.: Assembly algorithms for next-generation sequencing data. Genomics 95, 315–327 (2010)
14. Myers, E.W.: Toward simplifying and accurately formulating fragment assembly. Journal of Comutational Biology 2, 275–290 (1995)

15. Myers, E.W., Sutton, G.G., Delcher, A.L., Dew, I.M., Fasulo, D.P., Flanigan, M.J., Kravitz, S.A., Mobarry, C.M., Reinert, K.H.J., Remington, K.A., Anson, E.L., Bolanos, R.A., Chou, H., Jordan, C.M., Halpern, A.L., Lonardi, S., Beasley, E.M., Brandon, R.C., Chen, L., Dunn, P.J., Lai, Z., Liang, Y., Nusskern, D.R., Zhan, M., Zhang, Q., Zheng, X., Rubin, G.M., Adams, M.D., Venter, J.C.: A whole-genome assembly of drosophila. Science 287, 2196–2204 (2000)
16. Navarro, G., Sadakane, K.: Fully-functional static and dynamic succinct trees. Submitted for Journal Publication (2010). A preliminary version appeared In: Proc. ACM-SIAM SODA, pp. 134–149 (2010), http://arxiv.org/abs/0905.0768
17. Okanohara, D., Sadakane, K.: Practical Entropy-Compressed Rank/ Select Dictionary. In: Proc. of Workshop on Algorithm Engineering and Experiments, ALENEX (2007)
18. Pevzner, P.A., Tang, H., Waterman, M.S.: An eulerian path approach to dna fragment assembly. Proceedings of the National Academy of Sciences 98, 9748–9753 (2001)
19. Pop, M.: Genome assembly reborn: recent computational challenges. Briefings in Bioinformatics 10(4), 354–366 (2009)
20. Raman, R., Raman, V., Satti, S.R.: Succinct indexable dictionaries with applications to encoding k-ary trees, prefix sums and multisets. ACM Trans. Algorithms 3(4) (November 2007)
21. Sahli, M., Shibuya, T.: Arapan-s: a fast and highly accurate whole-genome assembly software for viruses and small genomes. BMC Research Notes (in press)
22. Simpson, J.T., Wong, K., Jackman, S.D., Schein, J.E., Jones, S.J.: Abyss: a parallel assembler for short read sequence data. Genome Research 19, 1117–1123 (2009)
23. Ye, C., Ma, Z.S., Cannon, C.H., Pop, M., Yu, D.W.: Exploiting sparseness in de novo genome assembly. BMC Bioinformatics 13(suppl. 6:S1) (2012)
24. Zerbino, D.R., Birney, E.: Velvet: algorithms for de novo short read assembly using de bruijn graphs. Genome Research 18, 821–829 (2008)

Space-Efficient and Exact de Bruijn Graph Representation Based on a Bloom Filter

Rayan Chikhi[1] and Guillaume Rizk[2]

[1] Computer Science department, ENS Cachan/IRISA, 35042 Rennes, France
[2] Algorizk, 75013 Paris, France

Abstract. The de Bruijn graph data structure is widely used in next-generation sequencing (NGS). Many programs, e.g. *de novo* assemblers, rely on in-memory representation of this graph. However, current techniques for representing the de Bruijn graph of a human genome require a large amount of memory (≥ 30 GB).

We propose a new encoding of the de Bruijn graph, which occupies an order of magnitude less space than current representations. The encoding is based on a Bloom filter, with an additional structure to remove critical false positives. An assembly software implementing this structure, Minia, performed a complete *de novo* assembly of human genome short reads using 5.7 GB of memory in 23 hours.

1 Introduction

The de Bruijn graph of a set of DNA or RNA sequences is a data structure which plays an increasingly important role in next-generation sequencing applications. It was first introduced to perform *de novo* assembly of DNA sequences [5]. It has recently been used in a wider set of applications: *de novo* mRNA [4] and metagenome [13] assembly, genomic variants detection [14,6] and *de novo* alternative splicing calling [17]. However, an important practical issue of this structure is its high memory footprint for large organisms. For instance, the straightforward encoding of the de Bruijn graph for the human genome ($n \approx 2.4 \cdot 10^9, k$-mer size $k = 27$) requires 15 GB ($n \cdot k/4$ bytes) of memory to store the nodes sequences alone. Graphs for much larger genomes and metagenomes cannot be constructed on a typical lab cluster, because of the prohibitive memory usage.

Recent research on de Bruijn graphs has been targeted on designing more lightweight data structures. Li *et al.* pioneered minimum-information de Bruijn graphs, by not recording read locations and paired-end information [9]. Simpson *et al.* implemented a distributed de Bruijn graph to reduce the memory usage per node [18]. Conway and Bromage applied sparse bit array structures to store an implicit, immutable graph representation [3]. Targeted methods compute local assemblies around sequences of interest, using negligible memory, with greedy extensions [19] or portions of the de Bruijn graph [15]. Ye *et al.* recently showed that a graph roughly equivalent to the de Bruijn graph can be obtained by storing only one out of g nodes ($10 \leq g \leq 25$) [20].

B. Raphael and J. Tang (Eds.): WABI 2012, LNBI 7534, pp. 236–248, 2012.

Conway and Bromage observed that the self-information of the edges is a lower bound for exactly encoding the de Bruijn graph [3]:

$$\log_2\left(\binom{4^{k+1}}{|E|}\right) \text{ bits,}$$

where $k+1$ is the length of the sequence that uniquely defines an edge, and $|E|$ is the number of edges. In this article, we will consider for simplicity that a de Bruijn graph is fully defined by its nodes. A similar lower bound can then be derived from the self-information of the nodes. For a human genome graph, the self-information of $|N| \approx 2.4 \cdot 10^9$ nodes is $\log_2\left(\binom{4^k}{|N|}\right) \approx 6.8$ GB for $k = 27$, i.e. ≈ 24 bits per node.

A very recent article [12] from Pell *et al.* introduced the *probabilistic de Bruijn graph*, which is a de Bruijn graph stored as a Bloom filter (described in the next section). It is shown that the graph can be encoded with as little as 4 bits per node. An important drawback of this representation is that the Bloom filter introduces false nodes and false branching. However, they observe that the global structure of the graph is approximately preserved, up to a certain false positive rate. Pell *et al.* did not perform assembly directly by traversing the probabilistic graph. Instead, they use the graph to partition the set of reads into smaller sets, which are then assembled in turns using a classical assembler. In the arXiv version of [12] (Dec 2011), it is unclear how much memory is required by the partitioning algorithm.

In this article, we focus on encoding an exact representation of the de Bruijn graph that efficiently implements the following operations:

1. For any node, enumerate its neighbors
2. Sequentially enumerate all the nodes

The first operation requires random access, hence is supported by a structure stored in memory. Specifically, we show in this article that a probabilistic de Bruijn graph can be used to perform the first operation exactly, by recording a set of troublesome false positives. The second operation can be done with sequential access to the list of nodes stored on disk. One highlight of our scheme is that the resulting memory usage is approximated by

$$1.44 \log_2\left(\frac{16k}{2.08}\right) + 2.08 \text{ bits/}k\text{-mer.}$$

For the human genome example above and $k = 27$, the size of the structure is 3.7 GB, i.e. 13.2 bits per node. This is effectively below the self-information of the nodes. While this may appear surprising, this structure does not store the precise set of nodes in memory. In fact, compared to a classical de Bruijn graph, the membership of an arbitrary node cannot be efficiently answered by this representation. However, for the purpose of many applications (e.g. assembly), these membership queries are not needed.

We apply this representation to perform *de novo* assembly by traversing the graph. In our context, we refer by traversal to any algorithm which visits all the nodes of the graph exactly once (e.g. a depth-first search algorithm). Thus, a mechanism is needed to mark which nodes have already been visited. However, nodes of a probabilistic de Bruijn graph cannot store additional information. We show that recording only the visited complex nodes (those with in-degree or out-degree different than one) is a space-efficient solution. The combination of (i) the probabilistic de Bruijn graph along with the set of critical false positives, and (ii) the marking scheme, enables to perform very low-memory *de novo* assembly.

In the first Section, the notions of de Bruijn graphs and Bloom filters are formally defined. Section 3 describes our scheme for exactly encoding the de Bruijn graph using a Bloom filter. Section 4 presents a solution for traversing our representation of the de Bruijn graph. Section 6 presents two experimental results: (i) an evaluation of the usefulness of removing false positives and (ii) an assembly of a real human dataset using an implementation of the structure. A comparison is made with other recent assemblers based on de Bruijn graphs.

2   de Bruijn Graphs and Bloom Filters

The **de Bruijn graph** [5], for a set of strings S, is a directed graph. For simplicity, we adopt a node-centric definition. The nodes are all the k-length substrings (also called k-*mers*) of each string in S. An edge $s_1 \to s_2$ is present if the $(k-1)$-length suffix of s_1 is also a prefix of s_2. Throughout this article, we will indifferently refer to a node and its k-mer sequence as the same object.

A more popular, edge-centric definition of de Bruijn graphs requires that edges reflect consecutive nodes. For k'-mer nodes, an edge $s_1 \to s_2$ is present if there exists a $(k'+1)$-mer in a string of S containing s_1 as a prefix and s_2 as a suffix. The node-centric and edge-centric definitions are essentially equivalent when $k' = k-1$ (although in the former, nodes have length k, and $k-1$ in the latter).

The **Bloom filter** [8] is a space efficient hash-based data structure, designed to test whether an element is in a set. It consists of a bit array of m bits, initialized with zeros, and h hash functions. To insert or test the membership of an element, h hash values are computed, yielding h array positions. The insert operation corresponds to setting all these positions to 1. The membership operation returns *yes* if and only if all of the bits at these positions are 1. A *no* answer means the element is definitely not in the set. A *yes* answer indicates that the element may or may not be in the set. Hence, the Bloom filter has one-sided errors. The probability of false positives increases with the number of elements inserted in the Bloom filter. When considering hash functions that yield equally likely positions in the bit array, and for large enough array size m and number of inserted elements n, the false positive rate $\mathcal{F}$ is [8]:

$$\mathcal{F} \approx \left(1 - e^{-hn/m}\right)^h = \left(1 - e^{-h/r}\right)^h \tag{1}$$

where $r = m/n$ is the number of bits per element. For a fixed ratio r, minimizing Equation 1 yields the optimal number of hash functions $h \approx 0.7r$, for which $\mathcal{F}$ is approximately 0.6185^r. Solving Equation 1 for m, assuming that h is the optimal number of hash function, yields $m \approx 1.44 \log_2(\frac{1}{\mathcal{F}})n$.

The **probabilistic de Bruijn graph** is obtained by inserting all the nodes of a de Bruijn graph (i.e all k-mers) in a Bloom filter [12]. Edges are implicitly deduced by querying the Bloom filter for the membership of all possible extensions of a k-mer. Specifically, an *extension* of a k-mer v is the concatenation of either (i) the $k - 1$ suffix of v with one of the four possible nucleotides, or (ii) one of the four nucleotides with the $k - 1$ prefix of v. The probabilistic de Bruijn graph holds an over-approximation of the original de Bruijn graph. Querying the Bloom filter for the existence of an arbitrary node may return a false positive answer (but never a false negative). This introduces false branching between original and false positive nodes.

3 Removing Critical False Positives

3.1 The cFP Structure

Our contribution is a mechanism that avoids false branching. Specifically, we propose to detect and store false positive elements which are responsible for false branching, in a separate structure. To this end, we introduce the cFP structure of *critical False Positives* k-mers, implemented with a standard set allowing fast membership test. Each query to the Bloom filter is modified such that the *yes* answer is returned if and only if the Bloom filter answers *yes* and the element is not in cFP.

Naturally, if cFP contained all the false positives elements, the benefits of using a Bloom filter for memory efficiency would be lost. The key observation is that the k-mers which will be queried when traversing the graph are not *all* possible k-mers. Let S be the set of true positive nodes, and $\mathcal{E}$ be the set of extensions of nodes from S. Assuming we only traverse the graph by starting from a node in S, false positives that do not belong to $\mathcal{E}$ will never be queried. Therefore, the set cFP will be a subset of $\mathcal{E}$. Let $\mathcal{P}$ be the set of all elements of $\mathcal{E}$ for which the Bloom filter answers *yes*. The **set of critical false positives** cFP is then formally defined as $cFP = \mathcal{P} \setminus S$.

Figure 1 shows a simple graph with the set S of correct nodes in regular circles and cFP in dashed rectangles. The exact representation of the graph is therefore made of two data structures: the Bloom filter, and the set cFP of critical false positives. The set cFP can be constructed using an algorithm that limits its memory usage, e.g. to the size of the Bloom filter. The set $\mathcal{P}$ is created on disk, from which cFP is then gradually constructed by iteratively filtering $\mathcal{P}$ with partitions of S loaded in a hash-table.

3.2 Dimensioning the Bloom Filter for Minimal Memory Usage

The set cFP grows with the number of false positives. To optimize memory usage, a trade-off between the sizes of the Bloom filter and cFP is studied here.

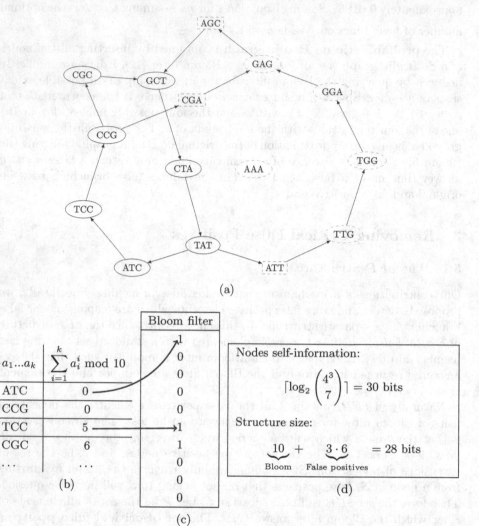

(a)

(b)

$a_1...a_k$	$\sum_{i=1}^{k} a_i^i \bmod 10$
ATC	0
CCG	0
TCC	5
CGC	6
...	...

Bloom filter

Nodes self-information:

$$\lceil \log_2 \binom{4^3}{7} \rceil = 30 \text{ bits}$$

Structure size:

$$\underbrace{10}_{\text{Bloom}} + \underbrace{3 \cdot 6}_{\text{False positives}} = 28 \text{ bits}$$

(d)

(c)

Fig. 1. A complete example of removing false positives in the probabilistic de Bruijn graph. (a) shows $\mathcal{S}$, an example de Bruijn graph (the 7 non-dashed nodes), and $\mathcal{B}$, its probabilistic representation from a Bloom filter (taking the union of all nodes). Dashed rectangular nodes (in red in the electronic version) are immediate neighbors of $\mathcal{S}$ in $\mathcal{B}$. These nodes are the critical false positives. Dashed circular nodes (in green) are all the other nodes of $\mathcal{B}$; (b) shows a sample of the hash values associates to the nodes of $\mathcal{S}$ (a toy hash function is used); (c) shows the complete Bloom filter associated to $\mathcal{S}$; incidentally, the nodes of $\mathcal{B}$ are exactly those to which the Bloom filter answers positively; (d) describes the lower bound for exactly encoding the nodes of $\mathcal{S}$ (self-information) and the space required to encode our structure (Bloom filter, 10 bits, and 3 critical false positives, 6 bits per 3-mer).

Using the same notations as in the definition of the Bloom filter, given that $n = |\mathcal{S}|$, the size of the filter m and the false positive rate $\mathcal{F}$ are related through Equation 1. The expected size of cFP is $8n \cdot \mathcal{F}$, since each node only has eight possible extensions, which might be false positives. In the encoding of cFP, each k-mer occupies $2 \cdot k$ bits. Recall that for a given false positive rate $\mathcal{F}$, the expected optimal Bloom filter size is $1.44n \log_2(\frac{1}{\mathcal{F}})$. The total structure size is thus expected to be

$$\underbrace{1.44n \log_2\left(\frac{1}{\mathcal{F}}\right)}_{\text{Bloom filter}} + \underbrace{(16 \cdot \mathcal{F}nk)}_{cFP} \text{ bits} \tag{2}$$

The size is minimal for $\mathcal{F} \approx (16k/2.08)^{-1}$. Thus, the minimal number of bits required to store the Bloom filter and the set cFP is approximately

$$n \cdot (1.44 \log_2(\frac{16k}{2.08}) + 2.08). \tag{3}$$

For illustration, Figure 2-(a) shows the size of the structure for various Bloom filter sizes and $k = 27$. For this value of k, the optimal size of the Bloom filter is 11.1 bits per k-mer, and the total structure occupies 13.2 bits per k-mer. Figure 2-(b) shows that k has only a modest influence on the optimal structure size. Note that the size of the cFP structure is in fact independent of k.

In comparison, a Bloom filter with virtually no critical false positives would require $\mathcal{F} \cdot 8n < 1$, i.e. $r > 1.44 \log_2(8n)$. For a human genome ($n = 2.4 \cdot 10^9$), r would be greater than 40.2, yielding a Bloom filter of size 13.7 GB.

4   Additional Marking Structure for Graph Traversal

Many NGS applications, e.g. *de novo* assembly of genomes [11] and transcriptomes [4], and *de novo* variant detection [17], rely on (i) simplifying and (ii) traversing the de Bruijn graph. However, the graph as represented in the previous section neither supports (i) simplifications (as it is immutable) nor (ii) traversals (as the Bloom filter cannot store an additional visited bit per node). To address the former issue, we argue that the simplification step can be avoided by designing a slightly more complex traversal procedure [2].

We introduce a novel, lightweight mechanism to record which portions of the graph have already been visited. The idea behind this mechanism is that not every node needs to be marked. Specifically, nodes that are inside simple paths (i.e nodes having an in-degree of 1 and an out-degree of 1) will either be all marked or all unmarked. We will refer to nodes having their in-degree or out-degree different to 1 as *complex* nodes. We propose to store marking information of complex nodes, by explicitly storing complex nodes in a separate hash table. In de Bruijn graphs of genomes, the complete set of nodes dwarfs the set of complex nodes, however the ratio depends on the genome complexity [7].

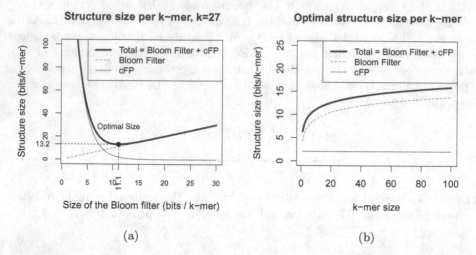

Fig. 2. (a) Structure size (Bloom filter, critical false positives) in function of the number of bits per k-mer allocated to the Bloom filter (also called ratio r) for $k = 32$. The trade-off that optimizes the total size is shown in dashed lines. (b) Optimal size of the structure for different values of k.

The memory usage of the marking structure is $n_c C$, where n_c is the number of complex nodes in the graph and C is the memory usage of each entry in the hash table ($C \approx 2k + 8$).

5   Implementation

The de Bruijn graph structure described in this article was implemented in a new *de novo* assembly software: Minia[1]. An important preliminary step is to retrieve the list of distinct k-mers that appear in the reads, i.e. true graph nodes. To discard likely sequencing errors, only the k-mers which appear at least d times are kept (*solid k-mers*). We experimentally set d to 3. Classical methods that retrieve solid k-mers are based on hash tables [10], and their memory usage scale linearly with the number of distinct k-mers. To avoid using more memory than the whole structure, we implemented a constant-memory k-mer counting procedure (manuscript in preparation). To deal with reverse-complementation, k-mers are identified to their reverse-complements.

We implemented in Minia a graph traversal algorithm that constructs a set of contigs (gap-less sequences). The Bloom filter and the cFP structure are used to determine neighbors of each node. The marking structure records already traversed nodes. A bounded-depth, bounded-breadth BFS algorithm (following Property 2 in [2]) is performed to traverse short, locally complex regions. Specifically, the traversal ignores tips (dead-end paths) shorter than $2k + 1$ nodes. It

<hr />

[1] Source code available at http://minia.genouest.org/

chooses a single path (consistently but arbitrarily), among all possible paths that traverse graph regions of breadth ≤ 20, provided these regions end with a single node of depth ≤ 500. These regions are assumed to be sequencing errors, short variants or short repetitions of length ≤ 500 bp. The breadth limit prevents combinatorial blowup. Note that paired-end reads information is not taken into account in this traversal. In a typical assembly pipeline (e.g. [18]), a separate program (*scaffolder*) can be used to link contigs using pairing information.

6   Results

Throughout the Results section, we will refer to the N50 metric of an assembly as the longest contig size, such that half the assembly is contained in contigs longer than this size.

6.1   On the Usefulness of Removing Critical False Positives

To test whether the combination of the Bloom filter and the cFP structure offers an advantage over a plain probabilistic de Bruijn graph, we compared both structures in terms of memory usage and assembly consistency. We retrieved 20 million *E. coli* short reads from the Short Read Archive (SRX000429), and discarded pairing information. Using this dataset, we constructed the probabilistic de Bruijn graph, the cFP structure, and marking structure, for various Bloom filter sizes (ranging from 5 to 19 bits per k-mer) and $k = 23$ (yielding 4.7 M solid k-mers).

We measured the memory usage of both structures. For each, we performed an assembly using Minia with exactly the same traversal procedure. The assemblies were compared to a reference assembly (using MUMmer), made with an exact graph. The percentage of nucleotides in contigs which aligned to the reference assembly was recorded.

Figure 3 shows that both the probabilistic de Bruijn graph and our structure have the same optimal Bloom filter size (11 bits per k-mer, total structure size of 13.82 bits and 13.62 per k-mer respectively). In the case of the probabilistic de Bruijn graph, the marking structure is prominent. This is because the graph has a significant amount of complex k-mers, most of them are linked to false positive nodes. For the graph equipped with the cFP structure, the marking structure only records the actual complex nodes; it occupies consistently 0.49 bits per k-mer. Both structures have comparable memory usage.

However, Figure 3 shows that the probabilistic de Bruijn graph produces assemblies which strongly depend on the Bloom filter size. Even for large sizes, the probabilistic graph assemblies differ by more than 3 Kbp to the reference assembly. We observed that the majority of these differences were due to missing regions in the probabilistic graph assemblies. This is likely caused by extra branching, which shortens the lengths of some contigs (contigs shorter than 100 bp are discarded).

Below ≈ 9 bits per k-mer, probabilistic graph assemblies significantly deteriorate. This is consistent with another article [12], which observed that when

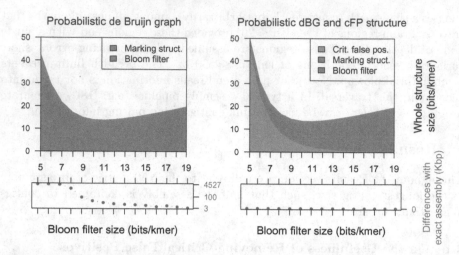

Fig. 3. Whole structures size (Bloom filter, marking structure, and *cFP* if applicable) of the probabilistic de Bruijn graph with (top right) and without the *cFP* structure (top left), for an actual dataset (E. coli, $k = 23$). All plots are in function of the number of bits per k-mer allocated to the Bloom filter. Additionally, the difference is shown (bottom left and bottom right) between a reference assembly made using an exact de Bruijn graph, and an assembly made with each structure.

the false positive rate is over 18% (i.e., the Bloom filter occupies ≤ 4 bits per k-mer), distant nodes in the original graph become connected in the probabilistic de Bruijn graph. To sum up, assemblies produced by the probabilistic de Bruijn graph are prone to randomness, while those produced by our structure are exact.

6.2 *de novo* Assembly

We assembled a complete human genome (NA18507, SRA:SRX016231, 142.3 Gbp of unfiltered reads of length ≈ 100 bp, representing 47x coverage) using Minia. After k-mer counting, 2,712,827,800 solid k-mers ($d = 3$) were inserted in a Bloom filter dimensioned to 11.1 bits per solid k-mer. The *cFP* structure contained 78,762,871 k-mers, which were stored as a sorted list of 64 bits integers, representing 1.86 bits per solid k-mer. A total of 166,649,498 complex k-mers (6% of the solid k-mers) were stored in the marking structure using 4.42 bits per solid k-mer (implementation uses $8\lceil \frac{k}{32} \rceil$ bytes per k-mer). Table 1 shows the time and memory usage required for each step in Minia.

We compared our results with assemblies reported by the authors of ABySS [18], SOAPdenovo [9], and the prototype assembler from Conway and Bromage [3]. Table 2 shows the results for four classical assembly quality metrics, and the time and peak memory usage of the compared programs. We note that Minia has the lowest memory usage (5.7 GB), seconded by the assembler from Conway and Bromage (32 GB). The wall-clock execution time of Minia (23 h) is comparable to the other assemblers; note that it is the only single-threaded assembler. The N50 metric of our assembly (1.2 Kbp) is slightly above that of the other assemblies (seconded by SOAPdenovo, 0.9 Kbp). All the programs except one assembled 2.1 Gbp of sequences.

We furthermore assessed the accuracy of our assembly by aligning the contigs produced by Minia to the GRCh37 human reference using GASSST [16]. Out of the 2,090,828,207 nucleotides assembled, 1,978,520,767 nucleotides (94.6%) were contained in contigs having a full-length alignment to the reference, with at least 98% sequence identity. For comparison, 94.2% of the contigs assembled by ABySS aligned full-length to the reference with 95% identity [18].

To test another recent assembler, SparseAssembler [20], the authors assembled another dataset (NA12878), using much larger effective k values. SparseAssembler stores an approximation of the de Bruijn graph, which can be compared to a classical graph for $k' = k + g$, where g is the sparseness factor. The reported assembly of the NA12878 individual by SparseAssembler ($k + g = 56$) has a N50 value of 2.1 Kbp and was assembled using 26 GB of memory, in a day. As an attempt to perform a fair comparison, we increased the value of k from 27 to 51 for the assembly done in Table 2 ($k = 56$ showed worse contiguity). The N50 obtained by Minia (2.0 Kbp) was computed with respect to the size of SparseAssembler assembly. Minia assembled this dataset using 6.1 GB of memory in 27 h, a 4.2× memory improvement compared to SparseAssembler.

Table 1. Details of steps implemented in Minia, with wall-clock time and memory usage for the human genome assembly. For constant-memory steps, memory usage was automatically set to an estimation of the final memory size. In all steps, only one CPU core was used.

Step	Time (h)	Memory (Gb)
k-mer counting	11.1	Constant (set to 4.0)
Enumerating positive extensions	2.8	3.6 (Bloom filter)
Constructing cFP	2.9	Constant (set to 4.0)
Assembly	6.4	5.7 (Bloom f.+ cFP + mark. struct.)
Overall	23.2	5.7

7 Discussion

This article introduces a new, space-efficient representation of the de Bruijn graph. The graph is implicitly encoded as a Bloom filter. A subset of false positives, those which introduce false branching from true positive nodes, are recorded in a separate structure. A new marking structure is introduced, in order for any traversal algorithm to mark which nodes have already been visited. The marking structure is also space-efficient, as it only stores information for a

Table 2. *de novo* human genome (NA18507) assemblies reported by our assembler (Minia), Conway and Bromage assembler [3], ABySS [18], and SOAPdenovo [9]. Contigs shorter than 100 bp were discarded. Assemblies were made without any pairing information.

Method	Minia	C. & B.	ABySS	SOAPdenovo
Value of k chosen	27	27	27	25
Number of contigs (M)	3.49	7.69	4.35	-
Longest contig (Kbp)	18.6	22.0	15.9	-
Contig N50 (bp)	1156	250	870	886
Sum (Gbp)	2.09	1.72	2.10	2.08
Nb of nodes/cores	1/1	1/8	21/168	1/16
Time (wall-clock, h)	23	50	15	33
Memory (sum of nodes, GB)	**5.7**	**32**	**336**	**140**

subset of k-mers. Combining the Bloom filter, the critical false positives structure and the marking structure, we implemented a new memory-efficient method for *de novo* assembly (Minia).

To the best of our knowledge, Minia is the first method that can create contigs for a complete human genome on a desktop computer. Our method improves the memory usage of de Bruijn graphs by two orders of magnitude compared to ABySS and SOAPdenovo, and by roughly one order of magnitude compared to succinct and sparse de Bruijn graph constructions. Furthermore, the current implementation completes the assembly in 1 day using a single thread.

De Bruijn graphs have more NGS applications than just *de novo* assembly. We plan to port our structure to replace the more expensive graph representations in two pipelines for reference-free alternative splicing detection, and SNP detection [14,17]. We wish to highlight three directions for improvement. First, some steps of Minia could be implemented in parallel, e.g. graph traversal. Second, a more succinct structure can be used to mark complex k-mers. Two candidates are Bloomier filters [1] and minimal perfect hashing.

Third, the set of critical false positives could be reduced, by exploiting the nature of the traversal algorithm used in Minia. The traversal ignores short tips, and in general, graph regions that are eventually unconnected. One could then define n-th order critical false positives (n-cFP) as follows. An extension of a true positive graph node is a n-cFP if and only if a breadth-first search from the true positive node, in the direction of the extension, has at least one node of depth $n + 1$. In other words, false positive neighbors of the original graph which are part of tips, and generally local dead-end graph structures, will not be flagged as critical false positives. This is an extension of the method presented in this article which, in this notation, only detects 0-th order critical false positives.

Acknowledgments. The authors are grateful to Dominique Lavenier for helpful discussions and advice, and Aurélien Rizk for proof-reading the manuscript. This work benefited from the ANR grant associated with the MAPPI project (2010-2014).

References

1. Chazelle, B., Kilian, J., Rubinfeld, R., Tal, A.: The bloomier filter: an efficient data structure for static support lookup tables. In: Proceedings of the Fifteenth Annual ACM-SIAM Symposium on Discrete Algorithms, pp. 30–39. SIAM (2004)
2. Chikhi, R., Lavenier, D.: Localized Genome Assembly from Reads to Scaffolds: Practical Traversal of the Paired String Graph. In: Przytycka, T.M., Sagot, M.-F. (eds.) WABI 2011. LNCS, vol. 6833, pp. 39–48. Springer, Heidelberg (2011)
3. Conway, T.C., Bromage, A.J.: Succinct data structures for assembling large genomes. Bioinformatics 27(4), 479 (2011)
4. Grabherr, M.G.: Full-length transcriptome assembly from RNA-Seq data without a reference genome. Nat. Biotech. 29(7), 644–652 (2011)
5. Idury, R.M., Waterman, M.S.: A new algorithm for DNA sequence assembly. Journal of Computational Biology 2(2), 291–306 (1995)
6. Iqbal, Z., Caccamo, M., Turner, I., Flicek, P., McVean, G.: De novo assembly and genotyping of variants using colored de bruijn graphs. Nature Genetics (2012)
7. Kingsford, C., Schatz, M.C., Pop, M.: Assembly complexity of prokaryotic genomes using short reads. BMC Bioinformatics 11(1), 21 (2010)
8. Kirsch, A., Mitzenmacher, M.: Less Hashing, Same Performance: Building a Better Bloom Filter. In: Azar, Y., Erlebach, T. (eds.) ESA 2006. LNCS, vol. 4168, pp. 456–467. Springer, Heidelberg (2006)
9. Li, R., Zhu, H., Ruan, J., Qian, W., Fang, X., Shi, Z., Li, Y., Li, S., Shan, G., Kristiansen, K.: De novo assembly of human genomes with massively parallel short read sequencing. Genome Research 20(2), 265 (2010)
10. Marais, G., Kingsford, C.: A fast, lock-free approach for efficient parallel counting of occurrences of k-mers. Bioinformatics 27(6), 764–770 (2011)
11. Miller, J.R., Koren, S., Sutton, G.: Assembly algorithms for next-generation sequencing data. Genomics 95(6), 315–327 (2010)
12. Pell, J., Hintze, A., Canino-Koning, R., Howe, A., Tiedje, J.M., Brown, C.T.: Scaling metagenome sequence assembly with probabilistic de bruijn graphs. Arxiv preprint arXiv:1112.4193 (2011)
13. Peng, Y., Leung, H.C.M., Yiu, S.M., Chin, F.Y.L.: Meta-IDBA: a de novo assembler for metagenomic data. Bioinformatics 27(13), i94–i101 (2011)
14. Peterlongo, P., Schnel, N., Pisanti, N., Sagot, M.-F., Lacroix, V.: Identifying SNPs without a Reference Genome by Comparing Raw Reads. In: Chavez, E., Lonardi, S. (eds.) SPIRE 2010. LNCS, vol. 6393, pp. 147–158. Springer, Heidelberg (2010)
15. Peterlongo, P., Chikhi, R.: Mapsembler, targeted and micro assembly of large NGS datasets on a desktop computer. BMC Bioinformatics (1), 48 (2012)
16. Rizk, G., Lavenier, D.: GASSST: global alignment short sequence search tool. Bioinformatics 26(20), 2534 (2010)
17. Sacomoto, G., Kielbassa, J., Chikhi, R., Uricaru, R., Antoniou, P., Sagot, M., Peterlongo, P., Lacroix, V.: KISSPLICE: de-novo calling alternative splicing events from RNA-seq data. BMC Bioinformatics 13(suppl. 6), S5 (2012)

18. Simpson, J.T., Wong, K., Jackman, S.D., Schein, J.E., Jones, S.J.M., Birol, N.: ABySS: a parallel assembler for short read sequence data. Genome Research 19(6), 1117–1123 (2009)
19. Warren, R.L., Holt, R.A.: Targeted assembly of short sequence reads. PloS One 6(5), e19816 (2011)
20. Ye, C., Ma, Z., Cannon, C., Pop, M., Yu, D.: Exploiting sparseness in de novo genome assembly. BMC Bioinformatics 13(suppl. 6), S1 (2012)

From de Bruijn Graphs to Rectangle Graphs for Genome Assembly

Nikolay Vyahhi[1], Alex Pyshkin[1], Son Pham[2,*], and Pavel A. Pevzner[1,2]

[1] Algorithmic Biology Laboratory, St. Petersburg Academic University, Russia
[2] Department of Computer Science and Engineering, UCSD, La Jolla, CA, USA
kspham@cs.ucsd.edu

Abstract. Jigsaw puzzles were originally constructed by painting a picture on a rectangular piece of wood and further cutting it into smaller pieces with a jigsaw. The *Jigsaw Puzzle Problem* is to find an arrangement of these pieces that fills up the rectangle in such a way that neighboring pieces have "matching" boundaries with respect to color and texture. While the general Jigsaw Puzzle Problem is NP-complete [6], we discuss its simpler version (called *Rectangle Puzzle Problem*) and study the *rectangle graphs*, recently introduced by Bankevich et al., 2012 [3], for assembling such puzzles. We establish the connection between Rectangle Puzzle Problem and the problem of assembling genomes from read-pairs, and further extend the analysis in [3] to real challenges encountered in applications of rectangle graphs in genome assembly. We demonstrate that addressing these challenges results in an assembler SPAdes+ that improves on existing assembly algorithms in the case of bacterial genomes (including particularly difficult case of genome assemblies from single cells).

SPAdes+ is freely available from http://bioinf.spbau.ru/spades.

1 Introduction

The recent proliferation of next generation sequencing technologies has enabled new experimental opportunities and, at the same time, raised formidable computational challenges. When the length of a repeat in the genome exceeds the read length, it becomes difficult to "span" the flanking regions of this repeat in the assembly. To alleviate this problem, sequencing technologies were extended to produce *read-pairs*, pairs of reads separated by an estimated *insert length*. Because insert length is longer than the read length, read-pairs span longer repeats and could potentially result in better assemblies. However, while assembling *single* reads can be elegantly modeled by *de Bruijn graphs* [7], equally elegant models for assembling *read-pairs* remain unknown [6].

Pevzner and Tang, 2001 [11] addressed this challenge by constructing the de Bruijn graph and further checking if a path between two reads within a read-pair satisfies the constraint imposed by the insert length. If only one such path exists,

---

* Corresponding author.

B. Raphael and J. Tang (Eds.): WABI 2012, LNBI 7534, pp. 249–261, 2012.
© Springer-Verlag Berlin Heidelberg 2012

the read-pair is transformed into a virtual long read where the gap between reads is filled in with the nucleotide sequence representing the found path.

While this and similar methods [9, 13] had a large impact on genome assembly, they fail in repeat-rich regions, where there are multiple paths between the reads within read-pairs. Recently, Medvedev et al., 2011 [10] introduced *paired de Bruijn graphs* that directly incorporate read-pairs into the graph structure and bypass the problem of multiple paths in previous approaches. However, the paired de Bruijn graph concept was introduced as a theoretical framework and is mainly aimed at an unrealistic case when the distance between reads within read-pairs is exactly d for all read-pairs. To address this bottleneck, Bankevich et al., 2012 [3] introduced the *rectangle graph* by generalizing the problem of a string reconstruction from its paired substrings to a variation of a jigsaw puzzle problem. While fragment assembly is usually modeled as a 1-dimensional *overlapping* puzzle (pieces correspond to individual reads), Bankevich et al. [3] modeled assembly as a 2-dimensional *non-overlapping* puzzle (pieces correspond to pairs of paths in the de Bruijn graph). However, while Bankevich et al. [3] sketched the rectangle graph idea, the various questions arising in applications of rectangle graphs to fragment assembly remained unaddressed.

The jigsaw puzzles were originally constructed by painting a picture on a rectangular piece of wood and further cutting it into smaller pieces with a jigsaw. The *Jigsaw Puzzle Problem* is to find an arrangement of these pieces that fills up the rectangle in such a way that neighboring pieces have "matching" boundaries with respect to color and texture. This paper extends the previous algorithmic studies of the Jigsaw Puzzle Problem (that were motivated by the restoration of archaeological artifacts [8]) to the problem of genome assembly from read-pairs. In section 2, we describe a class of simple jigsaw puzzles (called *rectangle puzzles*) and define the rectangle graph to assemble such puzzles. In section 3, we establish the relation between the rectangle puzzle and genome assembly from read-pairs and address the algorithmic challenges of "missing rectangles" (that often arise in genome assembly) in the rectangle puzzle problem. In Section 4, we apply the rectangle graph to bacterial genome assembly for both standard (multicell) and more difficult single cell datasets.

2   Rectangle Puzzles

Consider $n + 1$ points $x_0 = 0 < x_1 < \ldots < x_n$ on x-axis and $m + 1$ points $y_0 = 0 < y_1 < \ldots < y_m$ on y-axis. Points (x_i, y_j) form a 2-dimensional grid consisting of $n \cdot m$ rectangles filling up the grid with corners $(0, 0), (0, y_m), (x_n, 0)$ and (x_n, y_m). By analogy with the jigsaw puzzle assembly, we assume that the grid is "painted" and the goal is to assemble small rectangles into the painted grid in such a way that rectangles fully fill up the grid and that the colors at the sides of neighboring rectangles match (*valid assembly*). To simplify the matters, we will assume that the "orientation" of each rectangle is known.

Assembling the Ha Long Bay puzzle in Fig. 1a is trivial (for every rectangle, there exists an unambiguous choice of neighboring rectangles). Fig. 1b shows 9 rectangles from this puzzle with 12 blue dotted lines connecting the *unique* matching sides of these rectangles. Fig. 1c shows a more difficult "frogs and butterflies" puzzle (for every rectangle, there are *multiple* choices of neighboring rectangles). Fig. 1d shows 9 rectangles from this puzzle with 6 dotted connections showing all rectangles that match the upper side of the lower-left rectangle. Even a seasoned puzzle enthusiast may have difficult time assembling this puzzle and may end up with a wrong assembly shown in Fig. 1e [1]. Below we introduce a simpler type of puzzles (shown in Fig. 1f and called *rectangle puzzles with traversing curves*) and discuss algorithms for their assembly.

Consider a continuous non-self-intersecting curve from $(0,0)$ to (x_n, y_m) in the grid that crosses the sides of the rectangles at points $p_0 = (0,0), p_1, \ldots, p_N = (x_n, y_m)$ (in their order along the curve). For convenience, we assume that points on this curve are painted "red" and no other points in the puzzle is painted red. The curve is called *traversing* [2] if p_i and p_{i+1} belong to different sides of a rectangle (for $0 \leq i \leq N - 1$) and if there are exactly two red points on the sides of each rectangle. For simplicity we assume that the direction of the traversing curve within each rectangle is known and that the curve does not pass through the corners of rectangles except for the points $p_0 = (0,0)$ and $p_N = (x_n, y_m)$.

In the Rectangle Puzzle Problem we assume that red points in the grid form a traversing curve and the goal is to assemble the grid from rectangles. More precisely, we want to generate all valid assemblies of the rectangles into the grid. Fig. 1g shows 9 rectangles from this puzzle with blue dotted lines connecting all matching sides of these rectangles and illustrates that the number of matching sides is reduced as compared to Fig 1d.

The red curve enters and leaves each rectangle R through sides that we call $source(R)$ and $sink(R)$ correspondingly. We assume that every side S of a rectangle is assigned a label $label(S)$ (that encodes the painting of this side) and that differently painted sides are assigned different labels. We represent a rectangle R by a directed edge $edge(R)$ from vertex $label(source(R))$ to vertex $label(sink(R))$. For convenience, we assign identical and unique labels to the sides of rectangles containing the first and the last points $(0,0)$ and (x_n, y_m) on the red curve. It corresponds to closing the traversing curve as shown in Fig. 1g.

The concept of the traversing curve turns out to be useful for bringing the de Bruijn graph concept in the domain of puzzle assembly. Below, we describe an application of this concept for assembling rectangle puzzles.

Rectangle Graphs. The concept of rectangle graphs was first described in [3]. Given a rectangle puzzle, its *rectangle graph* is constructed as follows:

- Define a graph G with $n \cdot m$ isolated edges (on $2 \cdot n \cdot m$ vertices) by introducing a directed edge $edge(R)$ for each rectangle R. Starting and ending vertices of $edge(R)$ are labeled as $label(source(R))$ and $label(sink(R))$, correspondingly.
- The rectangle graph is formed by gluing identically labeled vertices in G (Fig. 1h).

---

[1] Polynomial algorithms for assembling such puzzles remain unknown.

[2] Intuitively, a traversing curve is a curve that "visits" every rectangle exactly once.

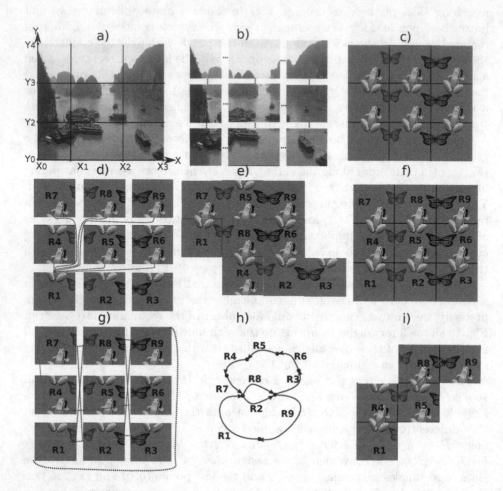

Fig. 1. Rectangle puzzles and rectangle graphs. a) A simple puzzle on the background image of Ha Long Bay. Points $x_0, \ldots, x_3$ on x-axis and $y_0, \ldots, y_3$ on y-axis form a 3×3 grid. (b) Nine rectangles with twelve matching sides in the Ha Long bay puzzle (shown by blue dotted connections) illustrate that there exists an unambiguous choice of matching sides. (c) A more difficult "frogs and butterflies" puzzle with multiple ambiguous choices of matching sides. (d) Nine rectangles with multiple matching sides (only some of them are shown) illustrate ambiguities in the selection of matching sides. (e) Failed attempt at the "frogs and butterflies" puzzle assembly. (f) The traversing curve in the "frogs and butterflies" puzzle makes the assembly easier. (g) Gluing sides of the rectangles in the "frogs and butterflies" puzzle. (h) The rectangle graph with 2 Eulerian cycles: $R_1 R_2 R_3 R_6 R_5 R_4 R_7 R_8 R_9$ and $R_1 R_8 R_3 R_6 R_5 R_4 R_7 R_2 R_9$ where only the first represents a valid solution. (i) Traversing line and subgrid assembly. The red traversing curve is replaced by a line. We are interested in assembling the subgrid formed by all rectangles crossed by the red line.

Obviously, the rectangle graph is the de Bruijn graph on strings of length 2 in the alphabet of labels (each label encodes a side of a rectangle). Each rectangle assembly corresponds to an Eulerian cycle in the rectangle graph. All Eulerian cycles can be generated using the BEST [1] theorem thus reducing the rectangle puzzle assembly to enumerating Eulerian cycles in the rectangle graph [2]. However, not every Eulerian cycle corresponds to a valid solution of the rectangle puzzle since some solutions may correspond to: (i) an assembly where rectangles overlap, (ii) an assembly that does not form a rectangular grid, (iii) an assembly where some sides of rectangles do not match. While the number of Eulerian cycles may be large, it is easy to check if a given Eulerian cycle corresponds to a valid rectangle puzzle assembly in linear time.

Below we limit attention to traversing *lines* (rather than *curves*) and relax the condition of *visiting* all rectangles (Fig. 1i): The traversing line $y = x + d$ visits *some* (not necessary all) rectangles. In this case we are only interested in assembling rectangles into a *subgrid* formed by rectangles crossed by the red line, rather than assembling all rectangles into the full grid. For $d \neq 0$, the traversing line does not necessarily starts at $(0,0)$ or ends at (x_n, y_m). In this case, every Eulerian cycle corresponds to a valid subgrid assembly. It is easy to see that in the case of the traversing *line* (in difference from the traversing *curve*) no additional checks are needed to verify that the assembly (given by an Eulerian cycle) is valid. Below we continue using the term "rectangle puzzle" while referring to the case of traversing *lines* (rather than traversing *curves*).

3 Rectangle Puzzles and Genome Assembly

Generating a Rectangle Puzzle from a Genome

We represent a genome as a circular string over the alphabet of nucleotides $\{A, T, C, G\}$. A *k-mer* is a string of length k in the alphabet of nucleotides.

Given a k-mer $s = s_1 \ldots s_k$, we define $prefix(s) = s_1 \ldots s_{k-1}$ and $suffix(s) = s_2 \ldots s_k$. Given a *Genome*, the de Bruijn graph $DB(Genome, k)$ is defined on the set of vertices representing all $(k-1)$-mers from *Genome* and has a directed edge $(prefix(s), suffix(s))$ for each k-mer s appearing in *Genome*. It is easy to see that *Genome* defines an Eulerian cycle in its de Bruijn graph.

A vertex v in a graph precedes (follows) a vertex w if there exists an edge from v to w (from w to v). The indegree (outdegree) of a vertex is the number of vertices preceding (following) it. A vertex is called a branching vertex if either its indegree or its outdegree is larger than 1. A path in a graph is called a *non-branching path* if all vertices in this path (with exception of the first and the last ones) have indegree and outdegree both equal to 1.

The de Bruijn graph $DB(Genome, k)$ partitions $(k-1)$-mers from *Genome* into branching (if they correspond to branching vertices in $DB(Genome, k)$) and non-branching. Similarly, all positions in *Genome* are partitioned into branching (if the $(k-1)$-mer starting at this position is branching) and non-branching. For example, ACG, CGT, GTT, and TCT are the branching 3-mers in *Genome* (shown as red points in Fig. 2a) while 0, 1, 7, 9, 13, 14, 19, 21, 24 are branching positions (shown as red points in Fig. 2b). For convenience, we assume that the circular genome "starts" at a branching position 0 and "ends" at the branching position N.[3]

We denote the branching positions in *Genome* as $x_0 = 0 < x_1 < \ldots x_n = N$ and define the grid consisting of $n \cdot n$ rectangles formed by points $x_0 = 0 < x_1 < \ldots < x_n = N$ on x axis and the same list of points on y-axis. The segment of *Genome* between positions x_i and x_{i+1} corresponds to a non-branching path in the de Bruijn graph. Thus, every rectangle corresponds to a pair of non-branching paths. The red line is defined by the equation $y = x + d$ [4]. Given an integer d, we define $Puzzle(Genome, k, d)$ as a set of rectangles crossed by the red line. Each rectangle in this set is uniquely defined by a pair of non-branching paths and the position of a red line segment within the rectangle.

Given a position x in *Genome* we define $(\bar{x})$ as the $(k-1)$-mer starting at this position. Thus, each integer 2D coordinate (x, y) defines a *paired* $(k-1)$-*mer* $(a|b)$, where $a = (\bar{x})$ and $b = (\bar{y})$. The label ("paint") of position (x, y) in the grid is defined as $((\bar{x}), (\bar{y}), color)$. where *color* is "red" or "white" depending on whether (x, y) is located on the red line or not. The label of a side of a rectangle is defined as an ordered list of all labels of (integer) points located on this side. It is easy to see that there exists an alternative simpler representation of this label, i.e., by a paired $(k-1)$-mer corresponding to the red position on the corresponding side [5]. A rectangle formed by a pair of non-branching paths (p, p') together with the red line segment on it can be represented as a triple (p, p', t) where t is the position of the red line in the rectangle relative to the low left corner of the rectangle. See Fig. 2b for an example of rectangle puzzle constructed from a genome.

We mention that since the labels along the red line completely define the genome, assembling the red line from the rectangles results in assembling the genome. However, this puzzle may appear useless for genome assembly tasks since the puzzle itself was originally created from the genome that we are trying to assemble in the first place! Below we show that the rectangle puzzle can be created from read-pairs without knowing the genome.

Generating a Rectangle Puzzle from Exact-Distance Read-Pairs

A (k, d)-mer is a pair of k-mers separated by d in the genome. If two reads $r'_1 \ldots r'_n$ and $r''_1 \ldots r''_n$ within a read-pair are separated by an exact distance d, one can extract (k, d)-mers $(r'_i \ldots r'_{i+k-1}|r''_i \ldots r''_{i+k-1})$ from them (for $1 \leq i \leq n - k + 1$). Iterating over all read-pairs results in a large set of (k, d)-mers. For simplicity, we unrealistically assume that the resulting set contains all and only (k, d)-mers from the genome. Before showing that the $Puzzle(Genome, k, d)$ can be constructed only from the (k, d)-mers set without knowing the genome, we

---

[3] Since the genome is circular, these two positions represent the same site in the genome and the same vertex in the de Bruijn graph.

[4] Below we will define d as the median distance between reads within a read-pair.

[5] Since it uniquely defines the label of all other points on the side.

introduce *multirectangle* — a different but equivalent representation of rectangle pieces in the rectangle puzzle [6] (Fig. 3a).

Given r rectangles: $(p, p', t_1), \ldots, (p, p', t_r)$ formed by the same pair of non-branching paths (p, p') but having different positions of their red line segments, we define a multirectangle R^* as a rectangle that is formed by the same pair of non-branching paths (p, p') (with horizontal edges p and vertical edges p') but with r red line segments and represent the multirectangle as $(p, p', \{t_1, \ldots, t_r\})$.

Let $Puzzle^*(Genome, k, d)$ denote a set of multirectangles by transforming rectangles in $Puzzle(Genome, k, d)$ that are formed by the same pairs of non-branching paths into single multirectangles. Within each multirectangle $R^* = (p, p', \{t_1, \ldots t_r\}) \in Puzzle^*(Genome, k, d)$, the red (integer) points on these r line segments represent *all* $(k-1, d)$-mers $(a|b)$ of the genomes such that $a \in p$ and $b \in p'$ [7]. Additionally, each $(k-1, d)$-mer $(a|b)$ such that $a \in p$ and $b \in p'$, corresponds to a *unique* position in the multirectangle. These two observations lead to a simple approach for constructing the $Puzzle^*(Genome, k, d)$ from the (k, d)-mers set: (1) Construct the de Bruijn graph $DB(ReadPairs, k)$ from individual reads in read-pairs; (2) For each pair of non-branching paths (p, p') in the de Bruijn graph that are *connected* by (k, d)-mers (i.e., there exists at least one $(k-1, d)$-mer $(a|b)$ such that $a \in p$ and $b \in p'$), we draw a multirectangle with horizontal edges p and vertical edges p', together with red points corresponding to $(k-1, d)$-mer that connect p and p'. These points form a single or multiple red line segments within the multirectangle.

From the set of multirectangles, we further transform it into the rectangle puzzle by replacing each multirectangle by separated rectangles, each containing a single rod line segment (see Fig. 3a).

Generating a Rectangle Puzzle from Inexact-Distance Read-Pairs

We now show how to construct the rectangle puzzle in a more realistic case of read-pairs with inexact distances between reads. Given integers d and Δ, [8] a pair of k-mers $(a|b)$ is called a (k, d, Δ)-mer in *Genome* if it is a (k, d_0)-mer of *Genome* for some $d_0 \in [d - \Delta, d + \Delta]$. While the set of all $(k-1, d)$-mer of *Genome* forms a line $(d) : y = x + d$ in the 2D grid, a set of $(k-1, d, \Delta)$-mers fills up a band of width Δ around (d), called a Δ-*cloud*. In this case, the rectangle puzzle needs to be redefined since: (1) red line segments in rectangles are substituted by red Δ-clouds, making it difficult to infer the position of the red line segments within the rectangles; (2) some new rectangles with red points are added into the original set of rectangles crossed by the red line (false rectangles); (3) some rectangles crossed by the red line are now missing (missing rectangles). Below we address these complications.

---

[6] While introducing the multirectangle concept does not have any analogy to the jigsaw puzzle assembly, it simplifies the proof that the $Puzzle(Genome, k, d)$ can be constructed only from the set of all (k, d)-mers of the unknown genome.

[7] With a minor exception for points that lie on the edges of the multirectangles.

[8] Integer d refers to the median distance between reads within a read-pair while integer Δ refers to the maximum deviation of the distance between the reads within a read-pair.

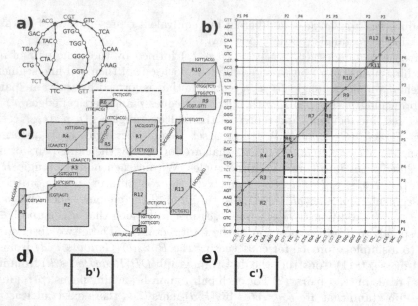

Fig. 2. The rectangle puzzle and the rectangle graph of the genome *Genome* = *ACGTCAAGTTCTGACGTGGGTTCT*. (a) De Bruijn graph *DB*(*Genome, k*) with *k* = 4 can be constructed from *Genome* or individual reads generated from *Genome*. The graph has 4 branching vertices (ACG, CGT, GTT, TCT) colored red that correspond to 8 branching positions in *Genome*. (b) Generating rectangle puzzle when *Genome* is known. *Genome* is represented as a sequence of 3-mers in both vertical and horizontal axes. The 3-mers corresponding to the branching vertices in the de Bruijn graph are colored red. The set of all (3, 5)-mers (pair of 3-mers separated by 5 nucleotides in the genome) forms a line (*d*) : *y* = *x* + 5 on the grid. (c) Rectangle graph is obtained by gluing sides of rectangle with the same labels. (d) The same as figure (b) but with rectangles in the dash box (*R*₅, *R*₆, *R*₇) removed. R_5, R_6, R_7 represent 3 missing rectangles. (e) The same as figure (c) but with rectangles in the dash box (R_5, R_6, R_7) missing. This results in two dead-ends vertices (sides of R_4 and R_8) in the rectangle graph.

For each pair of non-branching path (p, p') in the de Bruijn graph that is connected by $(k-1, d, \Delta)$-mers, we form a *multirectangle* with horizontal edge p and vertical edge p' together with red points corresponding to the $(k-1, d, \Delta)$-mers (a, b) where $a \in p$ and $b \in p'$. While in the case of the exact distance and perfect coverage, red points define a collection of line segments in the multirectangle, in the case of inexact distance these points *fall into* a band of width Δ around these (unknown) red line segments. Thus, red points in each multirectangle should be somehow transformed into the red line segments, a difficult task. Below, we introduce the notion of $(k-1, d)$-tuple, which enables us to draw all possible positions of the red line segments, and later, using the red points in the multirectangle to classify these segments into correct/incorrect red line segments.

Given the de Bruijn graph DB, we define a $(k-1, d)$-*tuple* as a pair of $(k-1)$-mers $(a|b)$ such that there exists a path of length d between vertex a and vertex

b in DB. Obviously, every $(k-1, d)$-mer in $Genome$ corresponds to a $(k-1, d)$-tuple in DB (but not vice versa).

Given a multirectangle formed by a pair of paths (p, p') and a collection of red points within it, we generate all possible [9] $(k-1, d)$-tuples $(a|b)$ such that $a \in p$ and $b \in p'$. These $(k-1, d)$-tuples define 45 degree line segments within this multirectangle. Those $(k-1, d)$-tuples that are also (k, d)-mers, form correct red line segments, while tuples that are not (k, d)-mers, form incorrect line segments. However, such a classification is unknown and we attempt to infer the correct/incorrect red line segments by the red points (corresponding to the $(k-1, d, \Delta$-mers) in the multirectangle.

Intuitively, correct line segments usually lie close to the "center" of red Δ-clouds, while the incorrect ones have few red points surrounding them. However, correctly classifying these segment into correct/incorrect segments still remains a difficult problem[10], since in the case of closely located red line segments, it is difficult to rule out which of them is correct (or whether they both are correct) and often forces us to combine such segments into a a *cluster* (see Fig. 3b) within an assumption that at least one of red line segments in the cluster represents a correct red segment. Below we describe the rectangle graph approach in the case when we deal with clusters of red segments.

In this case, we still represent each rectangle R as a single edge $edge(R)$ but use *multiple* labels for its starting and ending vertices (in the past we labeled these vertices by a single label). Specifically, we label its starting (ending) vertex by a *multiset* of all starting (ending) points of red segments. The *multilabeled rectangle graph* is defined as follows:

- Form a directed edge $edge(R)$ for each rectangle R. Starting (ending) vertices of $edge(R)$ are labeled by a *set* of labels of all starting (ending) points of the red segment within this rectangle.
- The rectangle graph is formed by gluing vertices in G if their sets of labels overlap.

Given a multirectangle R, SPAdes+ identifies T *clusters* of line segments that are supported by $ReadPairs$ using a variation of the approach from [3]. It further generates T rectangles (each rectangle with a single cluster of red segments as in Fig. 3b) and applies the multilabeled rectangle graph to assemble the resulted rectangles.

Missing Rectangles. We now consider the case when some rectangles are missing and ask whether the missing rectangles can be somehow reconstructed to complete the puzzle. Fig. 2d illustrates the case of 3 missing rectangles (R5, R6, and R7) resulting in a "gap" in the rectangle graph between vertices $(GTT|GAG)$ and $(ACG|GGT)$ in Fig. 2e. These *dead-ends* vertices (i.e., vertices with indegree or outdegree zero) provide a clue that some rectangles are missing and, as we show below, often allow one to recover the missing rectangles.

---

[9] If no such $(k-1, d)$-tuple exists, we remove the multirectangle.
[10] A similar problem was addressed in [3, 12].

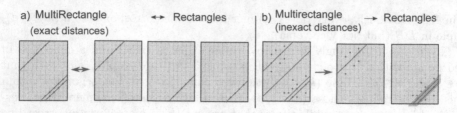

a) MultiRectangle   ↔ Rectangles   b) Multirectangle   → Rectangles
(exact distances)   (inexact distances)

Fig. 3. Rectangles and red line segments. a) Multirectangle: an equivalent representation of multiple rectangles that are formed by the same non-branching paths but have different positions of the red line segments. b) A multirectangle is transformed into 2 rectangles (one of them represents a cluster of two closely positioned red line segments). Note that one segment (the longest) in the multirectangle was classified as incorrect (there is no red point around this segment) and further being removed.

Consider two points with integer coordinates $(x, x + d)$ and $(x + t, x + d + t)$ in the grid located at the intersection of the red line with the sides of the rectangles. We refer to labels of these points as $(a|b)$ and $(a'|b')$, correspondingly. For example, points $(7,11)$ and $(13,17)$ in Fig 2d correspond to paired $(k-1)$-mers $(GTT|GAG)$ and $(ACG|GGT)$. Given *Genome*, we define $Rectangles((x, y) \to (x', y'))$ as the set of all rectangles crossed by the segment of the red line between points $(x, x + d)$ and $(x + t, x + d + t)$. For example, $Rectangles((7,11) \to (13,17)) = \{R5, R6, R7\}$.

Fig. 2c presents an idealized case when *Genome* as well as the points $(7,11)$ and $(13,17)$ (that contain vertices from some missing rectangles) are known. In reality, this information is not available in genome assembly projects. However, one knows the de Bruijn graph and can guess the labels of the points $(7,11)$ and $(13,17)$ (as labels of the dead-ends vertices in the rectangle graph in Fig. 2e). This raises the question whether the missing rectangles $Rectangles((7, 11) \to (13, 17))$ can be inferred from the paired $(k - 1)$-mers $(GTT|GAG)$ and $(ACG|GGT)$ (that represent labels of points $(7,11)$ and $(13,17)$) *without* knowing the coordinates of these points. Given paired $(k - 1)$-mers $(a|b)$ and $(a'|b')$, below we define the set of rectangles $Rectangles((a|b) \to (a'|b'))$ that often approximates $Rectangles((x, y) \to (x', y'))$ well.

Given an integer t, paired $(k-1)$-mers $(a|b)$ and $(a'|b')$ are called t-*tied* if there exist instances of a, b, a', b' located, respectively, at positions $x, y, x + t, y + t$ in *Genome*. Labels of every two red points in the grid represent t-tied paired $(k-1)$-mers. Below we relax the definition of t-tied paired $(k-1)$-mers for the case when *Genome* is unknown and only the de Bruijn graph of *Genome* is given.

Given an integer t and a de Bruijn graph DB, paired $(k - 1)$-tuples $(a|b)$ and $(a'|b')$ are called t-*linked* if there exists a path $p = p_0 \ldots p_t$ of length t between a and a' and a path $q = q_0 \ldots q_t$ of the same length between b and b' in the de Bruijn graph DB. Obviously, every t-tied paired k-mers is also t-linked, but not vice versa. Paths p and q define $t + 1$ paired $(k - 1)$-tuples $(p_i|q_i)$ that may potentially belong to the red line (since the notion of "t-linked" is a relaxation of the notion of "t-tied"). We define $Rectangles_{p,q}((a|b) \to (a'|b'))$ as the set of all rectangles that contain at least one point $(p_i|q_i)$ (for $0 < i < t$). We will often refer to $Rectangles_{p,q}((a|b) \rightarrow (a'|b'))$ as simply $Rectangles((a|b) \rightarrow (a'|b'))$ when it does not cause a confusion.

For example, $(GTT|GAG)$ and $(ACG|GGT)$ are 6-linked since there exists a path p (q) of length 6 from GTT to ACG (from GAG to GGT) in the de Bruijn graph in Fig. 2a. The vertices of the path $p(q)$ are located on 2 (3) non-branching paths in the de Bruijn graph thus contributing to $2 \times 3 = 6$ rectangles. Only 3 of these 6 rectangles ($R5, R6$, and $R7$) contain red points implying that $Rectangles((GTT|GAG) \rightarrow (ACG|GGT)) = \{R5, R6, R7\}$.

This example illustrates that one can close the gap between the dead-ends vertices $(a|b)$ and $(a'|b')$ in the rectangle graph by simply finding t-linked dead-ends in the rectangle graph (for small values of t), generating the set of missing rectangles $Rectangles((a|b) \rightarrow (a'|b'))$, and adding these missing rectangles to the pool of previously generated rectangles. Finally, one can construct the rectangle graph from the resulting enlarged set of rectangles.

4   Results

Assembly Datasets. To evaluate rectangle graph algorithm for genome assembly, we assembled two paired-end datasets from [5]. All these datasets are Illumina short reads with 100bp read length, 600× coverage. The first dataset is

Table 1. Comparison of assemblies for single-cell (ECOLI-SC) and standard (ECOLI-MC) datasets

Assembler	# contigs	NG50 (bp)	Largest (bp)	Total (bp)	Covered (%)	Misassemblies	Mismatches (per 100 kbp)	Complete genes
Single-cell *E. coli* (ECOLI-SC)								
EULER-SR	1344	26662	126616	4369634	87.8	21	11.0	3457
SOAPdenovo	1240	18468	87533	4237595	82.5	13	99.5	3059
Velvet	**428**	22648	132865	3533351	75.8	2	**1.9**	3117
Velvet SC	872	19791	121367	4589603	93.8	2	**1.9**	3654
E+V-SC	501	32051	132865	4570583	93.8	2	6.7	3809
SPAdes-single	1164	42492	166117	4781576	**96.1**	1	6.2	3888
SPAdes	1024	49623	177944	4790509	**96.1**	1	5.2	3911
SPAdes+	509	**56842**	**209690**	4550761	95.5	0	3.6	**3975**
Normal multicell sample of *E. coli* (ECOLI-MC)								
EULER-SR	295	**110153**	221409	4598020	99.5	10	5.2	4232
SOAPdenovo	192	62512	172567	4529677	97.7	1	26.1	4141
Velvet	198	78602	196677	4570131	99.9	4	1.2	4223
Velvet-SC	350	52522	166115	4571760	99.9	0	1.3	4165
E+V-SC	339	54856	166115	4571406	99.9	0	2.9	4172
SPAdes-single	445	59666	166117	4578486	99.9	0	**0.7**	4246
SPAdes	195	86590	**222950**	4608505	99.9	2	3.7	4268
SPAdes+	192	91893	221829	4593658	99.9	2	3.5	**4274**

The best assembler by each criteria is indicated in bold. EULER-SR 2.0.1, Velvet 0.7.60, Velvet-SC, and E+V-SC were run with vertex size 55. SOAPdenovo 1.0.4 was run with vertex size 27–31. SPAdes-single refers to SPAdes without repeat resolution, (without using read-pairs information) for comparison with E+V-SC, which does not use read-pairs information. SPAdes, SPAdes-single and SPAdes-rectangle iterated over edge sizes $k = 22, 34, 56$.

the multiple cell *E.coli* dataset with average insert size 215 bp, and denoted as ECOLI-MC. The second dataset is single cell *E.coli* dataset with average insert size 266 bp.

Benchmarking. We compare our SPAdes+ algorithm with EULER-SR [4], SOAPdenovo [9], Velvet [13], Velvet-SC [5], E+V-SC [5] and SPAdes [3]. See Table 1. Our rectangle graph algorithm outperforms other assemblers in most metrics. Improvement is more significant for single cell dataset. On ECOLI-SC dataset, SPAdes+ produces contigs with higher N50 (56,842 bp vs 49,623 by SPAdes), with higher largest contig (209,690 bp vs 177,944 by SPAdes) and no misassemblies, also captures 64 additional *E. coli* genes (3975 vs 3911 by SPAdes).

For both *E. coli* datasets, the rectangle graph module works for less than 10 seconds and using less than 100 MB RAM given (1) the de Bruijn graph has been already constructed and (2) mapping information of all paired-end reads to the de Bruijn graph has been calculated (using other modules in [3].

5   Conclusion

In this paper, we modeled the problem of genome assembly using read-pairs as a simple jigsaw puzzle and reintroduced the notion of rectangle graph [3] in a more intuitive way. We further addressed algorithmic challenges that arise in the application of rectangle graphs that have not been addressed in [3]. We demonstrated that by addressing these algorithmic challenges, the quality of the assembly significantly improves for both single cell and multicell bacterial datasets.

Acknowledgements. Authors would like to thank all members of SPAdes team for productive collaboration. We are especially indebted to Dmitry Antipov, Paul Medvedev, Glenn Tesler and Anton Korobeynikov for their insightful comments.

References

1. Aardenne-Ehrenfest, T., Bruijn, N.G.: Circuits and trees in oriented linear graphs. Classic papers in combinatorics, 149–163 (1987)
2. Abrham, J., Kotzig, A.: Transformations of euler tours. Annals of Discrete Mathematics 8, 65–69 (1980)
3. Bankevich, A., Nurk, S., Antipov, D., Gurevich, A.A., Dvorkin, M., Kulikov, A.S., Lesin, V.M., Nikolenko, S.I., Pham, S., Prjibelski, A.D., et al.: Spades: A new genome assembly algorithm and its applications to single-cell sequencing. Journal of Computational Biology 19(5), 455–477 (2012)
4. Chaisson, M.J., Pevzner, P.A.: Short read fragment assembly of bacterial genomes. Genome Research 18(2), 324 (2008)
5. Chitsaz, H., Yee-Greenbaum, J.L., Tesler, G., et al.: Efficient de novo assembly of single-cell bacterial genomes from short-read data sets. Nat. Biotechnol. 29(10), 915–921 (2011)

6. Demaine, E.D., Demaine, M.L.: Jigsaw puzzles, edge matching, and polyomino packing: Connections and complexity. Graphs and Combinatorics 23, 195–208 (2007)

7. Idury, R.M., Waterman, M.S.: A new algorithm for DNA sequence assembly. Journal of Computational Biology 2(2), 291–306 (1995)

8. Kampel, M., Sablatnig, R.: 3d puzzling of archeological fragments. In: Proc. of 9th Computer Vision Winter Workshop, vol. 2. Slovenian Pattern Recognition Society (2004)

9. Li, R., Zhu, H., Ruan, J., Qian, W., Fang, X., Shi, Z., Li, Y., Li, S., Shan, G., Kristiansen, K., et al.: De novo assembly of human genomes with massively parallel short read sequencing. Genome Research 20(2), 265 (2010)

10. Medvedev, P., Pham, S., Chaisson, M., Tesler, G., Pevzner, P.: Paired de bruijn graphs: A novel approach for incorporating mate pair information into genome assemblers. Journal of Computational Biology, 1625–1634 (2011)

11. Pevzner, P.A., Tang, H.: Fragment assembly with double-barreled data. Bioinformatics 17(suppl. 1), S225 (2001)

12. Pham, S.K., Antipov, D., Sirotkin, A., Tesler, G., Pevzner, P.A., Alekseyev, M.A.: Pathset Graphs: A Novel Approach for Comprehensive Utilization of Paired Reads in Genome Assembly. In: Chor, B. (ed.) RECOMB 2012. LNCS, vol. 7262, pp. 200–212. Springer, Heidelberg (2012)

13. Zerbino, D.R., Birney, E.: Velvet: algorithms for de novo short read assembly using de Bruijn graphs. Genome Research 18(5), 821 (2008)

MORPH-PRO: A Novel Algorithm and Web Server for Protein Morphing

Natalie E. Castellana[1], Andrey Lushnikov[4], Piotr Rotkiewicz[2],
Natasha Sefcovic[3], Pavel A. Pevzner[1], Adam Godzik[2], and Kira Vyatkina[4]

[1] Department of Computer Science, University of California-San Diego
[2] Burnham Institute for Medical Research, North Torrey Pines Road, La Jolla, CA
[3] Joint Center for Structural Genomics, Bioinformatics Core,
University of California-San Diego
[4] Algorithmic Biology Laboratory, Saint Petersburg Academic University,
Saint Petersburg, Russia

Abstract. Proteins are known to be dynamic in nature, changing from
one conformation to another while performing vital cellular tasks. It is
important to understand these movements in order to better understand
protein function. At the same time, experimental techniques provide us
with only single snapshots of the whole ensemble of available conformations. Computational protein morphing provides a visualization of
a protein structure transitioning from one conformation to another by
producing a series of intermediate conformations. We present a novel, efficient morphing algorithm, MORPH-PRO based on linear interpolation.
We also show that apart from visualization, morphing can be used to provide plausible intermediate structures. We test intermediate structures
constructed by our algorithm for a protein kinase and evaluate these
structures in a virtual docking experiment. The structures are shown to
dock with higher score to known ligands than structures solved using
X-Ray crystallography.

Keywords: protein morphing, virtual docking.

1 Introduction

The number of solved protein structures in PDB [1] has grown enormously in
recent years. However, the function of many proteins is highly correlated with
their movement. X-Ray crystallography, which contributes most of the structures
in PDB, gives us only a static view of protein structure. Recent developments
in computational protein morphing [2,3,4] provide visualization of a molecule
transitioning from one conformation to another by producing a series of intermediate conformations. In this paper we present a novel, computationally
efficient algorithm for generating intermediate structures between two solved
conformations of the same protein. In addition, we explore the possibility that
intermediate structures generated in the morphing procedure may also represent
realistic approximations of the actual protein conformational change, including
the structures of the intermediate conformations.

B. Raphael and J. Tang (Eds.): WABI 2012, LNBI 7534, pp. 262–273, 2012.
© Springer-Verlag Berlin Heidelberg 2012

Various attempts to predict the trajectory of proteins through conformational space have been made. Some success has been achieved through the use of elastic network models [5,6]. However, the accuracy of these methods depends on the chosen starting conformation (either apo- or holo-) and collectivity of the atoms in the motion [7]. Other attempts require numerous iterations of energy-minimization [8], which can be computationally expensive. Molecular dynamics simulations [9] may also be useful in determining the nature of conformational changes, but currently require significant computing power. Furthermore, motion planning techniques can be adapted to model molecular motions [10,11,12], providing an attractive alternative to the mentioned approaches due to their efficiency.

The most widely-used application to produce protein morphs is the Morph Server developed by Krebs and Gerstein [8]. The goal of the Morph Server is to provide visualization and classification of protein movements. Our emphasis is on the fast generation of intermediate structures that represent realistic conformations.

Given two aligned proteins as input, our MORPH-PRO algorithm produces a series of intermediate conformations. We use *linear interpolation*, so that at each step every residue will move along the straight line between its current position and its ending position. Unfortunately, this can lead to biologically infeasible intermediate structures with atoms occupying the same space, incorrect bond lengths, and incorrect bond angles. Therefore, we use the atom positions generated by linear interpolation as a first approximation to the correct solution, and use a dynamic programming algorithm to ensure that certain biological constraints are satisfied. This produces structures which better resemble real proteins. Because these techniques are very efficient, our algorithm can produce many intermediate structures very quickly.

The intermediate structures produced by morphing algorithms show great promise in molecular docking [13]. Molecular docking, which uses computer simulations to model and score protein-ligand binding, is a critical tool for drug discovery. Protein flexibility is believed to play a significant role in ligand binding [14]. One method for including flexibility in the docking experiment is to perform ensemble docking [15], which uses multiple conformations of the protein for evaluation. Performing docking against several conformations of a protein has been shown to provide better screening results, than against a single static structure [16]. The intermediate structures produced by morphing algorithms may improve our ability to detect these ligands, and therefore aide in the development of drug-like molecules [17].

2   Methods

In this section we analyze the simplest form of the morphing problem and present our MORPH-PRO algorithm. We designate P_{start} and P_{end} as the coordinates of the $C\alpha$ atoms for the starting and ending conformations. For simplicity, we assume that proteins P_{start} and P_{end} have an equal number of residues, and are aligned in 3-D. Later we will discuss the situation where P_{start} and P_{end} do not meet these conditions and will address various extensions to the simplest model of the protein morphing problem.

2.1   Morphing Algorithm

We represent a sequence of n points in 3-D (n-tuple) as a $3 \cdot n$ matrix (p_{ij}), where p_{ij} is the i-th coordinate of the j-th point. Let n be the number of residues in P_{start} and P_{end}. Given a parameter α, we define the α-intermediate of proteins P and P' as $(1 - \alpha) \cdot P + \alpha \cdot P'$. The simplest way to morph P_{start} into P_{end} is to generate intermediate reconstructions $(1 - \alpha) \cdot P_{start} + \alpha \cdot P_{end}$ for $0 < \alpha < 1$. However, some α-intermediates may not look like real proteins, for example they may consist of consecutive Cα atoms at biologically impossible distances. Below we show how to solve the protein morphing problem thereby transforming every intermediate reconstruction (being a sequence of n points) into a protein-like sequence of points. At each iteration, every point first moves by an appropriate distance towards its ending position, and then the obtained sequence of points is adjusted to become protein-like.

The pseudo code of the algorithm for generating K protein-like sequences $P_1 \ldots, P_K$ of points is as follows:

procedure $Morph(P_{start}, P_{end}, K)$
$\quad P_0 \leftarrow P_{start}$
$\quad$ **for** $m = 1$ **to** K **do**
$\quad\quad \alpha \leftarrow \frac{1}{K+2-m}$
$\quad\quad P \leftarrow \alpha$-intermediate of P_{m-1} and P_{end}
$\quad\quad P_m \leftarrow Proteinize(P)$
$\quad$ **end for**

Below we describe the algorithm for transforming a sequence of points P into a protein-like structure $Proteinize(P)$.

2.2   Optimal Equidistant Sequence Problem

Given a sequence P of n points, we define $d_j(P)$ as the distance between the (j)-th and the $(j+1)$-th points in P:
$d_j(P) = \sqrt{(p_{1,j+1} - p_{1,j})^2 + (p_{2,j+1} - p_{2,j})^2 + (p_{3,j+1} - p_{3,j})^2}$. A sequence P is (a, ϵ)-equidistant if $a - \epsilon \le d_j(P) \le a + \epsilon$ for $1 \le j \le n - 1$. Protein structures exhibit a strict distance constraint between consecutive Cα atoms that are 3.8 Å apart within an error margin of 0.1 Å. A sequence of points is protein-like if it is (3.8,0.1)-equidistant. We note that the consecutive Cα atoms in cis-proline do not adhere to this distance rule, and these cases are not handled by our algorithm.

We define the distance $d(P, P')$ between two sequences P and P', of n points each, as $\sum_{j=1}^{n} \sum_{i=1}^{3} (p_{i,j} - p'_{i,j})^2$. An (a, ϵ)-equidistant sequence P' is called an optimal (a, ϵ)-equidistant approximation of P if $d(P, P')$ is minimum among all possible (a, ϵ)-equidistant sequences P'. Below we describe an approximate solution to the following problem:

Optimal Equidistant Sequence Problem (OESP): Given a sequence of points, and parameters a and ϵ, find its optimal (a, ϵ)-equidistant approximation.

2.3   Solving OESP

Here we describe an approximate OESP algorithm that assumes the space of possible solutions is discretized. For each point from the sequence P, we construct a lattice of 3-D points centered around it, as shown in Figure 1. Thus, each lattice is local to its corresponding point from P, which distinguishes our approach from naive and out-dated attempts to understand protein folding which utilize a global lattice [18,19,20]. The selection of the number of points in the lattice and the edge length is discussed later. Let $v_{i,j}$ be the ith vertex in the lattice constructed around the jth point. Let $v_{0,j}$ be the vertex corresponding to the j-th point in P. Let Q be the number of vertices in each lattice.

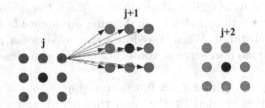

Fig. 1. Lattices constructed in 2-D. The black vertices ($v_{0,j}$, $v_{0,j+1}$, $v_{0,j+2}$) are the first approximations for the jth, $(j + 1)^{th}$, and $(j + 2)^{th}$ points. Each black vertex has a lattice constructed around it. Directed edges from a vertex $v_{i,j}$ to all vertices in lattice $(j + 1)$ are also shown.

We construct a directed edge from a vertex $v_{i,j}$ to a vertex $v_{g,j+1}$ for $1 \leq i, g \leq Q$ and $1 \leq j \leq n - 1$. The score of an edge is defined as:

$$EScore(v_{i,j}, v_{g,j+1}) = \begin{cases} 0, & \text{if } 3.7\text{Å} \leq d(v_{i,j}, v_{g,j+1}) \leq 3.9\text{Å} \\ \infty, & \text{otherwise} \end{cases}$$

We also assign a score to each vertex, $v_{i,j}$,

$$VScore(v_{i,j}) = (d(v_{i,j}, v_{0,j}))^2 \ for \ 1 \leq i \leq Q \ and \ 1 \leq j \leq n, \tag{1}$$

where $d(v_{i,j}, v_{0,j})$ gives the distance between $v_{i,j}$ and $v_{0,j}$. Finding a protein-like sequence P' of points which minimizes $d(P, P')$ translates into finding the path with the minimum score through the graph starting in the first lattice and ending in the nth lattice. The score of a path is defined as the sum of the scores of its edges and vertices. Let $PATH(v_{i,j})$ be the value of the minimum scoring path among those that start in the first lattice and end at vertex $v_{i,j}$. Variable $PATH(v_{i,j})$ can be computed using the following recurrence:

$$PATH(v_{i,1}) = VScore(v_{i,1}) \ for \ 1 \leq i \leq Q.$$

$$PATH(v_{i,j}) = VScore(v_{i,j}) + min_{1 \leq h \leq Q} \quad \quad (2)$$
$$\{PATH(v_{h,j-1}) + EScore(v_{h,j-1}, v_{i,j})\}$$

The score of the protein-like sequence of points which is closest to our original approximation is then

$$min_{1 \leq i \leq Q} PATH(v_{i,n}) . \quad \quad (3)$$

The solution of OESP can be determined by backtracking. The time complexity of generating a protein-like conformation of Cα atoms from a collection of n points, if one exists, is $O(nQ^2)$.

2.4   Angle and Proximity Constraints

The above approach solves OESP and produces a *(3.8,0.1)-equidistant sequence*. There is more, however, to consider when defining a *protein-like* structure than consecutive residue distance. We now redefine the notion of a *protein-like* sequence of points to take into account consecutive residue angles and proximity constraints.

Given 3-D points q_1, q_2, and q_3, a function $ang(q_1, q_2, q_3)$ is defined as the minor angle in degrees created by the lines through q_1 and q_2 and through q_2 and q_3, respectively. Given a sequence P of n points, we let $ang_j(P) = ang(p_{j-1}, p_j, p_{j+1})$ for $2 \leq j \leq n - 1$. A sequence P is *(a,b)-angle consistent* if $a° \leq ang_j(P) \leq b°$ for $2 \leq j \leq n - 1$. We observed that most Cα angles in real proteins fall in the range of 70° to 120°.

Furthermore, a sequence P of points is *z-distance consistent* if the distance between any two non-consecutive points in P is at least z Å. We determined that a distance of 2.0 Å was typical in real proteins.

Finally, a sequence P is *protein-like* if it is *(3.8,0.1)-equidistant, (70,120)-angle consistent*, and *2.0-distance consistent*.

We introduce a new score to evaluate the angle defined by three vertices, v_1, v_2, and v_3.

$$AScore(v_1, v_2, v_3) = \begin{cases} 0, & \text{if } 70° \leq ang(v_1, v_2, v_3) \leq 120° \\ \infty, & \text{otherwise} \end{cases}$$

In order to incorporate angles into our algorithm, we must use a more complex recurrence which relies on both the current vertex, $v_{i,j}$, and a preceding vertex, $v_{h,j-1}$. We define $PATH(v_{i,j}, v_{h,j-1})$ as the path with minimum score among all paths that start in the first lattice, end in $v_{i,j}$, and pass through $v_{h,j-1}$. We replace (2) with the following for $1 \leq i, h \leq Q$:

$$
\begin{aligned}
PATH(v_{i,2}, v_{h,1}) &= VScore(v_{i,2}) + EScore(v_{h,1}, v_{i,2}) \\
&+ VScore(v_{h,1})
\end{aligned}
$$

$$
\begin{aligned}
PATH(v_{i,j}, v_{h,j-1}) &= VScore(v_{i,j}) + EScore(v_{h,j-1}, v_{i,j}) \\
&+ min_{1 \leq g \leq Q}\{PATH(v_{h,j-1}, v_{g,j-2}) \\
&+ AScore(v_{i,j}, v_{h,j-1}, v_{g,j-2})\}
\end{aligned}
$$

To determine the score of the *protein-like* sequence of points which is closest to our original approximation, we find:

$$
min_{1 \leq i,h \leq Q} PATH(v_{i,n}, v_{h,n-1}) .
$$

This construction does not force the sequence of points to be *2.0-distance consistent*. For this, we apply a heuristic, which increases the *VScore* of vertices which are close to other lattices. We replace (1) with

$$
VScore(v_{i,j}) = \begin{cases} (d(v_{i,j}, v_{0,j}))^2, & \text{if } d(v_{i,j}, v_{0,j}) > 2.0\text{Å} \\ (d(v_{i,j}, v_{0,j}))^2 + 100 \sum_{m=1}^{j-2} (d(v_{i,j}, v_{0,m}))^{-2}, & \text{o.w.} \end{cases}
$$

We chose the multiplier 100 because it worked well to prevent Cα clashes in our morphs. The addition of the angle and distance constraints requires $O(n^3 Q^3)$.

However, the advanced strategy described above may be impractical if the proteins being examined are large or the conformational change is dramatic. Therefore, we also considered a simplified strategy which can significantly improve the running time. In the simplified strategy, (2) is replaced with

$$
\begin{aligned}
PATH(v_{i,j}) = VScore(v_{i,j}) &+ min_{1 \leq h \leq Q} \\
&\{PATH(v_{h,j-1}) + \\
&EScore(v_{h,j-1}, v_{i,j}) + \\
&AScore(prev_{PATH}(v_{h,j-1}), v_{h,j-1}, v_{i,j})\},
\end{aligned} \tag{4}
$$

where $prev_{PATH}(v_{h,j-1})$ is the vertex preceding $v_{h,j-1}$ in the best path ending at $v_{h,j-1}$, the score of which is determined by the value of $PATH(v_{h,j-1})$. Similar to the the basic method, the score of the optimal protein-like sequence of points is

$$
min_{1 \leq i \leq Q} PATH(v_{i,n}), \tag{5}
$$

and thus, the time complexity of the simplified strategy is also $O(nQ^2)$.

The simplified strategy may provide a sub-optimal intermediate structure, or fail to produce a structure at all when one exists. However, if a structure is produced, it obeys both the angle and proximity constraints. The heuristic employed in the simplified strategy works well in practice, and was used in the experiments described below. In addition, starting and ending conformations were preprocessed as described in the online Appendix at http://bioinf.spbau.ru/proteomics/morph-pro.

2.5   A Morphing Server

The MORPH-PRO server interface allows a user to upload two PDB files containing the starting and the ending conformations for a morph. After the intermediate conformations are computed, the morphing process can be visualized either as a movie or step-by-step. A transformation between two consecutive conformations is accomplished via linear interpolation. A publicly available archive of morphs is stored on the server. Further details of the MORPH-PRO server implementation are included in the online Appendix.

3   Results and Discussion

We evaluate our morphs by looking at both the biological feasibility of each individual structure, as well as the series of structures as a whole. We evaluate our morphs by comparing to proteins which have 3 or more solved structures in PDB, as proposed by [21]. In many instances, multiple conformations of the same protein are not available. Instead, we used proteins from the same family with nearly identical sequences as endpoints in our morph.

3.1   Pyrophosphokinases

We created a morph between two members of the pyrophosphokinase family (PDB codes: 1DY3, 1RAO). The alignment produced 158 residues with a maximum Cα displacement of 22 Å. The RMSD between the starting structure and the ending structure is 4.07 Å.

We examined each intermediate structure produced from this morph, and looked for clashing Cα atoms. None of the intermediate structures had atoms within 2 Å of another atom. We also looked at torsion angles created by Cα atoms. The Ramachandran plot of phi versus psi angles of the intermediate structure, which occurs halfway through the morph, is shown in Appendix Figure 1. The majority of the points in the plot fall within a region that is observed in real proteins. This indicates that our structure exhibits characteristics of real proteins.

It is also beneficial to look at the intermediate structures in the context of the entire morph. We have shown that our intermediates are protein-like, and we now demonstrate that the series of intermediate structures closely mimics the series of conformations a protein would visit. If multiple conformations of the same protein are known, then we can compare our predicted trajectory to the solved trajectory by calculating the RMSD between our intermediates and the experimentally solved intermediates. However, alternate conformations were not available for these proteins, so instead we used solved structures for proteins in the pyrophosphokinase family.

We chose two additional pyrophosphokinases to act as 'experimental' intermediates (PDB codes: 1RB0, 1HKA). We chose these proteins because they can be ordered by their RMSD between 1DY3 and 1RAO, and therefore are likely to be similar to the trajectory the morph should take. We plot the RMSD of our intermediate structures against each of these four proteins in Appendix Figure 2.

Intermediates which are produced early in the morph are closest to the starting protein, 1DY3, while those that are produced late in the morph are closest to the ending protein, 1RAO, as expected. Our intermediates from the middle of the morph become close to both 'experimental' intermediates, 1RB0 and 1RAO, suggesting that our movement closely follows the evolutionary changes which occurred between the two proteins. In addition, the intermediate structures generated by our algorithm come roughly as close, if not closer, to the known homologs as those produced by Morph Server, as demonstrated in Appendix Table 1. A direct speed test with the Morph Server was not possible because a fully functional standalone tool was not available.

3.2   F1-ATPase

The technique of looking at RMSD of the intermediate structures to known structures is most useful when X-Ray structures of actual intermediate conformations are available. There are three conformations solved for the F1-ATPase molecular motor (PDB code: 1E79) which exhibit a subtle change. The RMSD between the starting and ending conformations is 1.78 Å. The protein has 492 residues and the largest movement of a Cα is 11 Å. We produce a morph of 11 total structures from 1E79A to 1E79C.

The intermediate structures are very similar to all of the known structures, with RMSD consistently less than 2 Å. We do, however, see our intermediate structures become closer to the known intermediate 1E79B. One intermediate structure comes as close as 1.61 Å, while the starting structure (1E79A) is 1.85 Å and the ending structure (1E79C) is 1.73 Å. Appendix Figure 3 demonstrates how the predicted intermediates are similar to the starting structure early in the morph, become more similar to the known intermediate structure in the middle of the moprh, and then finally become similar to the ending structure. In Appendix Figure 3 we generated 30 intermediate structures to better illustrate this point.

3.3   GroEL

Our algorithm also performs well on large proteins. GroEL proteins chaperon the folding of other proteins. Two GroEL proteins (PDB codes: 1GRL and 1AON) exhibit a simple morph on 515 aligned residues, changing from a closed conformation to an open conformation. The RMSD between these two structures is 12.36 Å while the largest movement of a single Cα is 34.8 Å. Despite the large number of atoms and the significant movement, the morph took only a couple minutes to run. Figure 2 shows the initial conformation, the final conformation and 2 out of 34 intermediate structures produced in this morph.

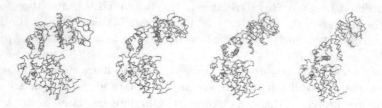

Fig. 2. The initial conformation, 2 intermediate structures, and the final conformation for GroEL

3.4   Virtual Screening

Virtual screening [22] is a technique which simulates the binding of a protein and a ligand, in order to determine the best ligand candidates from a large database. Most often, virtual screening is used as part of a drug development pipeline, guiding the selection of likely drug candidates. The predicted binding affinity of a ligand for a protein is determined by a docking algorithm, which finds the orientation and location of the ligand with respect to the protein. Modeling protein flexibility is very difficult due to the large degrees of freedom of a protein structure [13], [23]. One promising approach to implicitly incorporating protein flexibility is to dock against an ensemble of static protein structures [24].

If multiple conformations of the target protein are solved using NMR or X-Ray studies, these are good candidates for ensemble docking. However, in the more common case of unknown intermediate conformations a computational method can provide accurate models more quickly. Use of computationally-produced intermediates in virtual screening has shown promising results [25].

To test the potential for our intermediate structures in virtual screening we examined docking scores of our structures versus those solved experimentally against a small database of ligands. First, we produced a morph of the c-Jun N-terminal kinase 1 (JNK1). The starting conformation of this protein (1UKH) was solved complexed with a peptide (pepJIP1) derived from the binding portion of the scaffolding protein JIP1. The ending conformation (1UKI) was solved complexed with pepJIP1 and the ATP mimic SP600125. The binding of pepJIP1 to the JIP1 binding site on JNK1 causes a small conformational change at the ATP site. Though the movement is small, it produces a morph of 3 intermediates (P_2, P_3, P_4) in addition to the starting and ending conformations. The absent backbone atoms and side chains of each intermediate structure were reconstructed using Maxsprout [26], and energy minimization was performed using Swiss-PDB viewer [27]. As a basis for comparison, the X-Ray structures of 1UKH and 1UKI were also reduced to their $C\alpha$'s and then reconstructed in the same manner to produce P_1 and P_5, respectively.

Next, we performed docking with GOLD [28], a commonly used docking program and scoring scheme, on four ligands (extracted from PDB) known to bind to JNK1, as well as SP600125. Table 1 shows the rankings of the binding affinities from highest to lowest based on the GoldScore. The headings are the PDB codes for the solved structures of JNK1 complexed with each ligand.

Table 1. The rankings of binding affinities for 5 ligands against the predicted intermediate structures and the two solved structures for JNK1

SP600125	2G01	2N03	2H96	2GMX
$1UKI$	P_5	P_2	P_2	P_2
$1UKH$	P_2	P_5	P_5	P_5
P_2	P_4	$1UKI$	$1UKI$	$1UKH$
P_5	$1UKI$	P_3	P_3	$1UKI$
P_3	P_1	$1UKH$	$1UKH$	P_3
P_4	$1UKH$	P_4	P_4	P_1
P_1	P_3	P_1	P_1	P_4

The first column behaves as expected. The structure which has the highest binding affinity for SP600125 is 1UKI which is the structure of JNK1 complexed with SP600125. The X-Ray structures docked with SP600125 rank significantly higher than the reconstructed P_1 and P_5. This suggests that better side chain reconstruction could greatly improve the docking results.

For three of the other ligands, the second intermediate structure, P_2 scores higher than any other intermediate structure as well as any X-Ray structure. This demonstrates that our intermediate structures would be more likely to identify ligands which bind to JNK1 than either of the two X-Ray structures.

4 Conclusions

It is clear that there is much to learn about the nature of protein structure dynamics that is not addressed in the static information contained in PDB. The intermediate structures representing a protein as it moves from one conformation to another may yield much information about how a protein functions. Experimental techniques are inadequate for this task due to practical and technological limitations. For this reason, structural biology is in great need of algorithms which can accurately predict the intermediate structures as a protein undergoes a conformational change.

While other morphing algorithms require computationally expensive energy and elastic network modeling calculations, our morphing algorithm is based on a few simple observations of protein structure, and therefore produces multiple intermediate conformations very quickly. Our intermediate structures represent realistic protein structures, and demonstrate the motion of a protein as it changes between conformations.

The morphed structures also show promise in the area of virtual screening. Most techniques limit protein flexibility to the side chain atoms, and may allow limited flexibility of the substrate. Our morph produces intermediate structures which are hypotheses for possible backbone movements. For this reason, some ligands bound more favorably to our intermediate structures than the solved structures. These are strong implications for the potential of morphs in guiding drug development.

Acknowledgments. We would like to acknowledge the valuable input provided by Mallika Veeramalai, Piotr Cieplak, and Lukasz Jaroszewski (Joint Center for Structural Genomics). We thank Hongbin Yuan and Maurizio Pellechia at the Sanford-Burnham Medical Research Institute for input on protein docking.

References

1. Berman, H.M., Westbrook, J., Feng, Z., Gilliland, G., Bhat, T.N., Weissig, H., Shindyalov, I.N., Bourne, P.E.: The protein data bank. Nucleic Acids Research 28(1), 235–242 (2000)
2. Echols, N., Milburn, D., Gerstein, M.: Molmovdb: analysis and visualization of conformational change and structural flexibility. Nucleic Acids Research, 478–482 (2003)
3. Kim, M.K., Jernigan, R.L., Chirikjian, G.S.: Efficient generation of feasible pathways for protein conformational transitions. Biophys. J. 83, 1620–1630 (2002)
4. Kim, M.K., Chirikjian, G.S., Jernigan, R.L.: Elastic models of conformational transitions in macromolecules. J. Mol. Graph. Model. 21, 151–160 (2002)
5. Franklin, J., Koehl, P., Doniach, S., Delarue, M.: Minactionpath: maximum likelihood trajectory for large-scale structural transitions in a coarse-grained locally harmonic energy landscape. Nucleic Acids Research, 477–482 (2007)
6. Ahmed, A., Gohlke, H.: Multiscale modeling of macromolecular conformational changes combining concepts from rigidity and elastic network theory. Proteins 63, 1038–1051 (2006)
7. Yang, L., Song, G., Jernigan, R.L.: How well can we understand large-scale protein motions using normal modes of elastic network models? Biophys. J. 93, 920–929 (2007)
8. Krebs, W.G., Gerstein, M.: The morph server: a standardized system for analyzing and visualizing macromolecular motions in a database framework. Nucleic Acids Res. 28, 1665–1675 (2000)
9. Duan, Y., Kollman, P.A.: Pathways to a protein folding intermediate observed in a 1-microsecond simulation in aqueous solution. Science 282(5389), 740–744 (1998)
10. Amato, N.M., Song, G.: Using motion planning to study protein folding pathways. Journal of Computational Biology 9(2), 149–168 (2002)
11. Apaydin, M.S., Brutlag, D.L., Guestrin, C., Hsu, D., Latombe, J.C., Varma, C.: Stochastic roadmap simulation: an efficient representation and algorithm for analyzing molecular motion. Journal of Computational Biology 10(3-4), 257–281 (2003)
12. Raveh, B., Enosh, A., Schueler-Furman, O., Halperin, D.: Rapid sampling of molecular motions with prior information constraints. PLoS Comput. Biol. 5(2), e1000295 (2009)

13. Teodoro, M.L., Kavraki, L.E.: Conformational flexibility models for the receptor in structure based drug design. Curr. Pharm. Des. 9, 1635–1648 (2003)
14. Carlson, H.A.: Protein flexibility and drug design: how to hit a moving target. Curr. Opin. Chem. Biol. 6, 447–452 (2002)
15. Knegtel, R.M., Kuntz, I.D., Oshiro, C.M.: Molecular docking to ensembles of protein structures. J. Mol. Biol. 266, 424–440 (1997)
16. Craig, I.R., Essex, J.W., Spiegel, K.: Ensemble docking into multiple crystallographically derived protein structures: an evaluation based on the statistical analysis of enrichments. J. Chem. Inf. Model 50, 511–524 (2010)
17. Goh, C.S., Milburn, D., Gerstein, M.: Conformational changes associated with protein-protein interactions. Current Opinion in Structural Biology 14(1), 104–109 (2004)
18. Taketomi, H., Ueda, Y., Go, N.: Studies on protein folding, unfolding and fluctuations by computer simulation. International Journal of Peptide and Protein Research 7(6), 445–459 (1975)
19. Lau, K.F., Dill, K.A.: A lattice statistical mechanics model of the conformational and sequence spaces of proteins. Macromolecules 22(10), 3986–3997 (1989)
20. Sali, A., Shakhnovich, E., Karplus, M.: How does a protein fold? Nature 369, 248–251 (1994)
21. Weiss, D.R., Levitt, M.: Can morphing methods predict intermediate structures? J. Mol. Biol. 385, 665–674 (2009)
22. Walters, W.P., Stahl, M.T., Murco, M.A.: Cheminform abstract: Virtual screening an overview. ChemInform 29(38), 160–178 (1998)
23. Teague, S.J.: Implications of protein flexibility for drug discovery. Nat. Rev. Drug. Discov. 2, 527–541 (2003)
24. Wei, B.Q., Weaver, L.H., Ferrari, A.M., Matthews, B.W., Shoichet, B.K.: Testing a flexible-receptor docking algorithm in a model binding site. J. Mol. Biol. 337, 1161–1182 (2004)
25. Broughton, H.B.: A method for including protein flexibility in protein-ligand docking: improving tools for database mining and virtual screening. J. Mol. Graph. Model. 18, 247–257 (2000)
26. Holm, L., Sander, C.: Database algorithm for generating protein backbone and side-chain co-ordinates from a c alpha trace application to model building and detection of co-ordinate errors. Journal of Molecular Biology 218(1), 183–194 (1991)
27. Guex, N., Peitsch, M.C.: Swiss-model and the swiss-pdbviewer: an environment for comparative protein modeling. Electrophoresis 18(15), 2714–2723 (1997)
28. Jones, G., Willett, P., Glen, R.C., Leach, A.R., Taylor, R.: Development and validation of a genetic algorithm for flexible docking. Journal of Molecular Biology 267(3), 727–748 (1997)

How Accurately Can We Model Protein Structures with Dihedral Angles?

Xuefeng Cui[1], Shuai Cheng Li[2], Dongbo Bu[3],
and Babak Alipanahi Ramandi[1], and Ming Li[1],*

[1] University of Waterloo, Ontario, Canada
mli@cs.uwaterloo.ca
[2] City University of Hong Kong, Hong Kong, China
[3] Chinese Academy of Sciences, Beijing, China

Abstract. Previous study shows that the same type of bond lengths and angles fit Gaussian distributions well with small standard deviations on high resolution protein structure data. The mean values of these Gaussian distributions have been widely used as ideal bond lengths and angles in bioinformatics. However, we are not aware of any research work done to evaluate how accurately we can model protein structures with dihedral angles and ideal bond lengths and angles.

In this paper, we first introduce the protein structure idealization problem. Then, we develop a fast $O(nm/\epsilon)$ dynamic programming algorithm to find an approximately optimal idealized protein backbone structure according to our scoring function. Consequently, we demonstrate that idealized backbone structures always exist with small changes and significantly better free energy. We also apply our algorithm to refine protein pseudo-structures determined in NMR experiments.

1 Introduction

When studying the functions of a protein, it is crucial to know the three-dimensional structure that consists of the Cartesian coordinates of all the atoms of the protein. These atoms are bonded together by inter-atomic forces called chemical bonds. It has been observed that the bond lengths and angles of the same type assume a Gaussian distribution with small standard deviations (STDEVs) in high resolution protein structure data. Typically, the bond lengths on protein backbones have STDEVs between 0.019Å and 0.033Å, and the bond angles on protein backbones have STDEVs between 1.5° and 2.7° [1,2]. These results suggest the possibility of modeling protein structures with the mean values of bond lengths and angles, which are often referred to as *ideal values*.

Ideal bond lengths and angles have been widely used in Nuclear magnetic resonance (NMR) protein structure determination [3,4] and protein structure prediction [5,6,7,8,9]. Moreover, stereochemical restraints are also used in X-ray protein structure determination [10,11]. In protein structure prediction, the

---

* Corresponding author.

B. Raphael and J. Tang (Eds.): WABI 2012, LNBI 7534, pp. 274–287, 2012.
© Springer-Verlag Berlin Heidelberg 2012

main advantage of using ideal bond lengths and angles is a reduction in the search space for the target protein structure [12,13]. However, we are not aware of any research work done to evaluate how accurately we can model protein structures with dihedral angles. This motivates our problem, which we call the *protein structure idealization problem*: Given the coordinates of the target protein structure, find the coordinates of the optimal idealized protein structure. Here, an idealized protein structure is a protein structure with bond lengths and angles that are ideal with respect to a given scoring function; the function is to depend on the resultant structure's free energy, as well as its similarity with the target structure. Thus, the idealized protein structure is taken to be a protein-like structure that is close to the target protein structure.

We solve the protein structure idealization problem by first idealizing the backbone structure and then the side-chain structure. This approach is widely accepted because previous research suggests that the backbone conformation is archived before the side-chain conformations are archived [14]. In our work, Ω dihedral angles are rounded to be either $0°$ or $180°$, and some discussions on the properness of idealizing Ω dihedral angles can be found in [15,16].

In this paper, we introduce a novel dynamic programming algorithm with a run-time complexity of $O(n/\epsilon^8)$, where ϵ is a small constant, to find the optimal idealized protein backbone structure according to our scoring function. In practice, we observed that there is no need to remember the entire dynamic programming table. Thus, with a filtering technique, the run-time complexity is further reduced to $O(nm/\epsilon)$, where m is a constant integer.

As our initial study on the protein structure idealization problem, this paper focuses on the protein backbone structure idealization. Side chain structures are determined using an exhaustive search which assumes that side chain structures of different residues are independent from each other. The scoring function is similar to the one we used for backbone structure idealization. In practice, we observed that it is fast to regenerate idealized structures that are similar to a given idealized structure. We also refine the idealized backbone and side-chain structures according to our scoring functions iteratively.

Our algorithm is used to evaluate how accurately we can model protein structures with dihedral angles. We idealized all the X-ray protein structures from PDB [17] which satisfy the high resolution and the low sequence identity constraints downloaded on June 6, 2008 [18,19]. The results show that such idealized structures always exist, and they are very similar to the target structures in terms of the root mean square deviation (RMSD) of C_α or all atoms. Moreover, the idealized backbone structures tend to have dDFIRE free energy scores [20,21] significantly better than the target structures. The results well support our conclusion that one can model protein structures accurately with dihedral angles on all high resolution protein backbone structures.

One application of the protein structure idealization algorithm is to refine protein pseudo-structures either determined in experiments or predicted by computers. In this paper, we have demonstrated one such case to improve poor (Φ, Ψ) dihedral angles of protein structures determined by NMR. The experiment result is also consistent with the previous experiment that the idealized structure has a small RMSD and better backbone free energy. We also discussed several potential applications of our protein structure idealization algorithm in Section 4.

2 Protein Backbone Structure Idealization

Given the target protein backbone structure, we would like to find the optimal idealized backbone structure. For an idealized protein backbone structure, the coordinates of O, H and C_β backbone atoms can be calculated from the coordinates of N, C_α and C backbone atoms. Thus, we only describe how to generate coordinates of N, C_α and C atoms in this section. For sake of simple explanations, a structure is always referred to a protein backbone structure unless strictly specified in this section.

2.1 Idealized Backbone Structure Generation

Given the target structure, we would like to generate idealized structures fulfilling two generation goals. First, the idealized structures should be similar to the target structure. Second, each pair of idealized structures should be at least some distance away in order to avoid redundant computation. Furthermore, we are interested in generating as many of such idealized structures as possible.

Before describing how we fulfil the generation goals, we first need to describe a simple distance metric to measure the distance between two sets of coordinates representing the target protein. Let P_i be a set of coordinates representing the target protein, and $P_i^j \in P_i$ be the coordinate of the j-th atom of the target protein. Thus, there is $P_i = \{P_i^1, P_i^2, ..., P_i^{3n}\}$, where n is the number of amino acids of the target protein. For sake of simplicity, let P_0 always represent the target structure, and P_i represent a generated idealized structure for $i > 0$. Let $D(P_i^k, P_j^k)$ be the Euclidean distance between P_i^k and P_j^k. We describe the distance between P_i and P_j as the bottleneck distance as following:

$$D(P_i, P_j) = \max_k D(P_i^k, P_j^k). \tag{1}$$

Using the above distance metric, we fulfill both generation goals by satisfying the following generation constraints:

$$\begin{cases} D(P_0, P_i) \leq r \ \forall i > 0 \\ D(P_i, P_j) \geq \epsilon \ \forall i, j > 0 \end{cases}. \tag{2}$$

The first generation constraint brings the assumption that the accuracy of the coordinates of the target structure is reasonably good, and no-worse than r. If this constraint is satisfied, the distance between the target coordinate and any generated coordinate representing the same atom is upper bounded by r. Thus, it is reasonable for any generated idealized structure P_i to be considered similar to target structure P_0. If the second generation constraint is satisfied, for each pair of generated idealized structures, there exists a pair of coordinates, one from each structure representing the same atom, such that they are at least ϵ distance away from each other. Therefore, both generation goals are achieved.

The above generation constraints suggest to limit the search space inside a sphere with radius r, and to discrete the search space with grids of size ϵ. When $\epsilon = 0.001$Å, the accuracy of X-ray crystallography [22] and PDB (protein database) format [23] is reached. Thus, this method is capable to generate all possible idealized structures at the accuracy of X-ray crystallography and PDB format.

Given the limited and discrete search space of each atom, one can generate idealized structure coordinates from the first atom to the last atom. For the first atom, an idealized coordinate lies within a sphere. Thus, the number of generated coordinates is bounded by $O(1/\epsilon^3)$. For each generated coordinate P_i^1 of the first atom, an idealized coordinate of the second atom lies on a ball surface with constant distance to P_i^1. Thus, the number of generated coordinates is bounded by $O(1/\epsilon^2)$. For each generated coordinate pair (P_i^1, P_i^2) of the first two atoms, an idealized coordinate of the third atom lies on a circle with constant distances to P_i^1 and P_i^2. Thus, the number of generated coordinates is bounded by $O(1/\epsilon)$. Similarly, the number of generated coordinates for any of the following atoms is also bounded by $O(1/\epsilon)$. Moreover, since Ω dihedral angles are rounded to be either $0°$ or $180°$ in our work, the coordinate of any C_α atom is unique and can be calculated from the coordinates of the previous three atoms.

Therefore, the total number of coordinates generated for all atoms is bounded by $O(1/\epsilon^{2n+4})$ by induction. Here, it is fine to assume that r is a constant because it is only related to the first atom. For the following atoms, we did not limit the search space to be inside the sphere with radius r in the above calculation, and thus the actual number of generated coordinates should be much smaller in practice.

2.2 Idealized Backbone Structure Scoring Function

Given the generated idealized structures $\{P_i\}$, we need a scoring function $S_{BB}(P_i)$ to find the optimal idealized structure. The scoring function should evaluate not only the similarity between generated idealized structure P_i and target structure P_0, but also the free energy of P_i to ensure that P_i is protein-like. Thus, we define our scoring function as following:

$$S_{BB}(P_i) = S_f(P_i) - w_1 D_\alpha(P_i, P_0) - w_2 D_\beta(P_i, P_0) - w_3 D_H(P_i, P_0) - w_4 D_{\Phi,\Psi}(P_i, P_0), \tag{3}$$

where w_a are the weighting parameters, $S_f(P_i)$ is the free energy score, $D_\alpha(P_i, P_0)$ is the root mean square divergence (RMSD) of C_α atoms, $D_\beta(P_i, P_0)$ is the RMSD of C_β atoms, $D_H(P_i, P_0)$ is the RMSD of the hydrogen and oxygen atoms participated in hydrogen bonds, $D_{\Phi,\Psi}(P_i, P_0)$ is the RMSD of (Φ, Ψ) dihedral angles.

In our scoring function, the free energy is evaluated by a simple (Φ, Ψ) dihedral angle log-odd score as the free energy score $S_f(P_i)$. To be more specific, we discrete the Ramachandran plot into grids of 360 by 360, and draw one plot for each type of amino acid. Then, we calculated the log-odd score $S_f(P_i^{1,t})$ of idealized structure $P_i^{1,t}$ of the first t atoms as following:

$$S_f(P_i^{1,t}) = \sum_{5 \leq i \leq t, A_i = C_\alpha} \log \frac{P_{AA_{i-3}}(\Phi_{i-3}, \Psi_{i-3})}{P_{null}(\Phi_{i-3}, \Psi_{i-3})}, \tag{4}$$

where one log-odd score is calculated at each C_α atom (by checking that atom type A_i is C_α) for the previous amino acid (represented by the previous C_α atom at $i-3$), $P_{AA_{i-3}}(\Phi_{i-3}, \Psi_{i-3})$ is the probability of the grid containing (Φ_{i-3}, Ψ_{i-3}) on the Ramachandran plot of amino acid type AA_{i-3}, and $P_{null}(\Phi_{i-3}, \Psi_{i-3})$ is the probability of the null model with an uniform distribution such that $P_{null}(\Phi_{i-3}, \Psi_{i-3}) = \frac{1}{360}\frac{1}{360}$.

In addition to evaluating the free energy, the structure similarity is evaluated by other distance matrices in our scoring function. $D_\alpha(P_i, P_0)$ and $D_{\Phi,\Psi}(P_i, P_0)$ serves as distance metrics to conserve the backbone structures; $D_\beta(P_i, P_0)$ serves as a distance metric to conserve the side-chain structure compatibilities; and $D_H(P_i, P_0)$ serves as a distance metric to conserve the hydrogen bonds. Thus, some global dependencies are addressed implicitly by distance matrices $D_\beta(P_i^{1,t}, P_0^{1,t})$ and $D_H(P_i, P_0)$.

2.3   Dynamic Programming Algorithm

Theoretically, one can calculate scores for all generated idealized structures, and find the optimal one with the maximum score. This method works fine as long as similar structures always have similar scores. More formally, the method requires the assumption that $D(P_i, P_j) \leq \epsilon \implies |S_{BB}(P_i) - S_{BB}(P_j)| \leq \epsilon_s$, which is reasonable for small ϵ. Unfortunately, since the total number of generated idealized structures is bounded by $O(1/\epsilon^{2n+4})$, this method is computationally expensive. Thus, we introduce a dynamic programming algorithm with a filtering technique to find the optimal idealized structure efficiently in this section.

The dynamic programming algorithm has two assumptions. One assumption is that given two generated idealized structures $P_i^{1,t-1}$ and $P_j^{1,t-1}$ of the first $t-1$ atoms such that $D(P_i^{t-k,t-1}, P_j^{t-k,t-1}) \leq \epsilon$, for any generated coordinate P_i^t of the t'th atom, there always exists generated coordinate P_j^t such that $D(P_i^t, P_j^t) \leq \epsilon$. The other assumption is that the scoring function satisfies the additive property such that $S_{BB}(P_i^{1,t}) = S_{BB}(P_i^{1,t-k}) \oplus S_{BB}(P_i^{t-k+1,t})$ under some addition operator $\oplus$.

We observed that it is rare to see counter examples of the first assumption when $k \geq 5$, though counter examples exist theoretically. The second assumption holds for our scoring function described in Section 2.2. Distance matrices $D_\alpha(P_i^{1,t}, P_0^{1,t})$, $D_\beta(P_i^{1,t}, P_0^{1,t})$, $D_H(P_i^{1,t}, P_0^{1,t})$ and $D_{\Phi,\Psi}(P_i^{1,t}, P_0^{1,t})$ satisfies the additive property because RMSD $D_{RMS}(P_i^{1,t}, P_0^{1,t})$ satisfies the additive property as following:

$$\begin{aligned}
& D_{RMS}(P_i^{1,t}, P_0^{1,t}) \\
&= D_{RMS}(P_i^{1,t-k}, P_0^{1,t-k}) \oplus D_{RMS}(P_i^{t-k+1,t}, P_0^{t-k+1,t}) \\
&= \sqrt{\frac{D_{RMS}^2(P_i^{1,t-k}, P_0^{1,t-k})(t-k) + D_{RMS}^2(P_i^{t-k+1,t}, P_0^{t-k+1,t})k}{t}}.
\end{aligned} \quad (5)$$

Moreover, the free energy score $S_f(P_i^{1,t})$ satisfies the additive property as following:

$$\begin{aligned}
S_f(P_i^{1,t}) &= S_f(P_i^{1,t-k}) \oplus S_f(P_i^{t-k+1,t}) \\
&= S_f(P_i^{1,t-k}) + S_f(P_i^{t-k+1,t}).
\end{aligned} \quad (6)$$

The second assumption is the fundamental of our dynamic programming algorithm. By induction, the first assumption implies that if $D(P_i^{t-k,t-1}, P_j^{t-k,t-1}) \leq \epsilon$, for any generated idealized structure $P_i^{t,n}$, there always exists generated idealized structure $P_j^{t,n}$ such that $D(P_i^{t,n}, P_j^{t,n}) \leq \epsilon$. Recall that the scoring function assumes that $D(P_i^{t,n}, P_j^{t,n}) \leq \epsilon \implies |S_{BB}(P_i^{t,n}) - S_{BB}(P_j^{t,n})| \leq \epsilon_s$, and thus there is $S_{BB}(P_i^{t,n}) \approx S_{BB}(P_j^{t,n})$. If $S_{BB}(P_i^{1,t-1}) \geq S_{BB}(P_j^{1,t-1})$, there is approximately $S_{BB}(P_i) = S_{BB}(P_i^{1,t-1}) \oplus S_{BB}(P_i^{t,n}) \geq S_{BB}(P_j^{1,t-1}) \oplus S_{BB}(P_j^{t,n}) = S_{BB}(P_j)$. Therefore, if $D(P_i^{t-k,t-1}, P_j^{t-k,t-1}) \leq \epsilon$ and $S_{BB}(P_i^{1,t-1}) \geq S_{BB}(P_j^{1,t-1})$, these is no need to generate $P_i^{t,n}$ to find an approximately optimal solution.

Based on the above observation, we developed a novel dynamic programming algorithm. Idealized structures are still generated as described in Section 2.1, but the generation process is stopped for some idealized structures if we know it cannot lead us to the optimal one. First, the search space for each atom of the target protein is discretized to grids of size ϵ. When generating coordinates for atom t, if $P_i^{t-k+1,t}$ and $P_j^{t-k+1,t}$ are located in the same grid set $G_g^{t-k+1,t}$, we know that there is no need to continue the generation process on the lower scored one of $P_i^{1,t}$ and $P_j^{1,t}$. Thus, we defined the dynamic programming table $T_{BB}(t, G_g^{t-k+1,t})$ to be the optimal idealized structure for each observed tail grid set $G_g^{t-k+1,t}$ as followings:

$$\begin{cases}
T_{BB}(t, G_g^{t-k+1,t}) = \max\limits_{i,j} T_{BB}(t-1, G_i^{t-k,t-1}) \oplus S_{BB}(P_j^t) \\
T_{BB}(k, G_g^{1,k}) = \max\limits_{i} S_{BB}(P_i^{1,k})
\end{cases} \quad (7)$$

where $G_g^{t-k+1,t-1} = G_i^{t-k+1,t-1}$, $P_j^{t-k+1,t} \in G_g^{t-k+1,t}$ and $S_{BB}(P_j^{1,t-1}) \oplus S_{BB}(P_j^t) = S_{BB}(P_j^{1,t})$. Thus, the dynamic programming table can be calculated from the first atom to the last atom. Finally, the optimal idealized structure is the one with the highest score $\max_g G_g^{3n-k+1,3n}$.

The run-time complexity of our dynamic programming algorithm depends on the value of k. In order to keep all possible (Φ, Ψ) dihedral angles of the previous residue when generating C_α atoms, we have to choose $k \geq 5$. For sake of speed, we simply chose $k = 5$ in our implementation. In this case, the number of score calculations required to calculate $T_{BB}(t, G_g^{t-4,t})$ is no more than the maximum number of coordinates sampled for six consecutive backbone atoms using the method introduced in Section 2.1. Recall that there are exactly two C_α atoms in six consecutive backbone atoms, and the Ω dihedral angle is rounded. Thus, the coordinate of one C_α atom can be calculated from the coordinates of the other C_α atom and the two atoms between them. For this reason, the maximum number of sampled coordinates is bounded by $O(1/\epsilon^8)$. Moreover, the number of score calculations required to calculate $T_{BB}(k, G_g^{1,k})$ is no more than the maximum number of possible coordinates sampled for five consecutive backbone atoms, which is also $O(1/\epsilon^8)$. Therefore, the run-time complexity of our dynamic programming algorithm is $O(n/\epsilon^8)$.

To further speed up the dynamic programming algorithm, we applied an additional filtering technique to remember only the highly scored idealized structures. To be more specific, the algorithm only remembers the optimal idealized structure for the top m scored tail configurations instead of all possible ones. Thus, the run-time complexity is reduced to $O(nm/\epsilon)$. This approach works well in practice because it is rare to see an optimal idealized structure with a long poorly scored fragment. In other words, we assumed that the local quality of the idealized structure should be reasonably high (in the top m score list).

3   Result

In order to study the protein structure idealization problem and its applications, we implemented our protein structure idealization algorithm. In our implementation, we used the mean bond lengths and angles that had been reported in [2] as the ideal bond lengths and angles, respectively. When idealizing the protein backbone structure, we set the search space radius of an atom as $r = 1.6$Å and the discrete grid size as $\epsilon = r/5$. We found that $m = 50000$ had a reasonable balance between speed and accuracy. When idealizing the protein side-chain structure, we set the search space of a rotamer dihedral angle to be within 3σ distance from the mean value, where σ is the STDEV of the rotamer dihedral angle, and the discrete grid size to be $10°$. We also refine the idealized structure by iteratively reducing the search space and the discrete grid size by a constant factor of 0.5. Since finding the best scoring function for the protein structure idealization is out of the scope of this paper, we set all weights $w_a = 1.0$ for all a in our scoring function.

3.1   PDB Protein Structure Idealization

In this experiment, we addressed the question of how accurately we can model protein structures with dihedral angles. We idealized the high resolution protein structures with low sequence identities of the CULLPDB_PC30_RES1.6_R0.25 data set [18,19]. In fact, the CULLPDB_PC30_RES1.6_R0.25 data set is the complete set of X-ray protein structures in PDB [17] with a sequence identity cutoff of 30%, a resolution cutoff of 1.6Å and an R factor cutoff of 0.25. In summary, the data set contains 1898 proteins with an average length of 227 residues as downloaded on June 6, 2008.

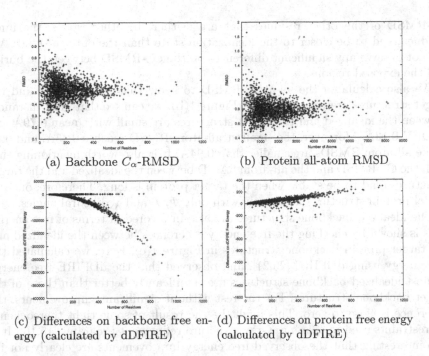

(a) Backbone C_α-RMSD   (b) Protein all-atom RMSD

(c) Differences on backbone free en- (d) Differences on protein free energy
ergy (calculated by dDFIRE)   (calculated by dDFIRE)

Fig. 1. Comparison of the idealized and the target protein structures of CULLPDB_PC30_RES1.6_R0.25

In order to show that the idealized and the target backbone structures are very similar, we calculated the C_α-RMSD as shown in Figure 1(a). The C_α-RMSD is a popular distance metric to evaluate the backbone distance between two protein backbone structures. The result shows that most distances between the idealized and the target backbone structures are small with mean 0.53Å and STDEV 0.08Å. To be more specific, the smallest C_α-RMSD reaches 0.16Å, and 90% of the C_α-RMSDs are smaller than 0.63Å. Moreover, the C_α-RMSD is upper bounded by 1.00Å though the search space radius for each atom is set to be 1.6Å. This result is consistent with the result of checking (Φ, Ψ) dihedral angles. The average difference between the idealized and the target (Φ, Ψ) dihedral angles is as small as 0.08°. Therefore, one can model protein backbone structures in CULLPDB_PC30_RES1.6_R0.25 accurately using only Φ and Ψ dihedral angles.

We further studied the C_α-RMSD in different regions of the target protein structures. We observed that the C_α-RMSD of the α-helix and the β-sheet regions are smaller than that of the complete protein by 0.28Å and 0.12Å, respectively. Indeed, these regions are more restricted because of using $D_H(P_i, P_0)$ to conserve hydrogen bonds of α-helices and β-sheets as described in Section 2.2. We also observed that the C_α-RMSD of residues that are closer to the geometric center of a target protein structure is 0.13Å smaller on average than the

C_α-RMSD of the other residues that are farther. In other words, the inner residues tend to be closer to the idealization state than the outer residues. We did not observe any significant differences on the C_α-RMSD between the buried and the exposed regions.

We also calculated the all-atom RMSD to show that the idealized and the target structures are very similar. In Figure 1(b), we can see that most distances between the idealized and the target structures are small with mean 0.79Å and STDEV 0.13Å. Moreover, the smallest all-atom RMSD reaches 0.45Å, and 90% of the all-atom RMSDs are smaller than 0.94Å. It is also worth mentioning that both the C_α-RMSD and the all-atom RMSD between the idealized and the target structures tend to be stable when the target protein is long. Therefore, one can model protein structures accurately with only Φ, Ψ and χ dihedral angles.

The idealized backbone structures are also in favored in terms of free energy. This is shown by checking the free energy differences between the idealized and the target protein backbone structures in Figure 1(c). Here, we calculated the free energy using dDFIRE [20,21], and observed that the dDFIRE free energy of most idealized backbone structures are significantly better than those of the target backbone structures. For the rest without significant improvements, the difference is close to zero. This might be the result of some tight thereochemical restraints used in existing X-ray structure refinement programs [15,16]. It is also interesting that the observed free energy improvements are clearly not independent from the protein length. The figure actually suggests the free energy difference to have a square dependence on the protein length.

After idealizing the side-chain structures, the free energy is either improved by a relatively bigger amount or worsen by a relatively smaller amount as shown in Figure 1(d). Unfortunately, in most cases, the free energy is worsen slightly but still in a stable state with negative values. Again, we used dDFIRE [20,21] to calculate the free energy here. We observed that the dDFIRE free energy is improved for 90 or 4.74% of the idealized protein structures, and is worsen slightly by 44 on average. Moreover, the dDFIRE free energy is improved by 1585 in the best case, and worsen by 293 in the worst case. The figure also suggests the free energy difference to have a linear dependence on the protein length.

The above results also demonstrate that the quality of side-chain structures modelled by χ dihedral angles is not as good as that of backbone structures modelled by Φ and Ψ dihedral angles. To further study this problem, we compared the predicted side-chain structures given the native backbone structures and the predicted side-chain structures given the predicted backbone structures in terms of free energy. Here, we treat the idealized backbone structures of the CULLPDB_PC30_RES1.6_R0.25 data set as the best possibly predicted ones. Moreover, we used SCWRL4 [9] to predict side-chain structures and dDFIRE [20,21] to calculate free energies. The result shows that the free energy is worsen slightly by 43 if the predicted backbone structures are used. This is consistent to the fact that the quality of side-chain structures tends to be unreliable than that of backbone structures determined by experiments.

3.2 NMR Protein Structure Refinement

In this experiment, we demonstrated an application of the protein structure idealization problem in NMR by idealizing 32 NMR protein structures. The NMR protein structures were randomly chosen from PDB [17] with a sequence identity cutoff of 30% and a gapless fragment length cutoff of 80 residues. In case of multiple chains or models of some NMR protein structures, only the first chain from the first model is used in this experiment. In addition to the conclusion of the previous experiment, the results show that poor (Φ, Ψ) dihedral angles of the NMR protein structures are improved by idealizing them.

Table 1. favored (Φ, Ψ) dihedral angle percentagies of 32 NMR protein structures before and after idealization

PDB	Native	Ideal	Diff	PDB	Native	Ideal	Diff
1SSK	44.6%	71.9%	27.3%	2LBN	59.7%	77.6%	17.9%
2KQP	62.9%	80.0%	17.1%	1WPI	64.4%	81.4%	17.0%
1EXE	60.5%	76.7%	16.2%	2LNV	58.6%	72.4%	13.8%
1X6F	64.1%	73.1%	9.0%	2L6B	72.2%	81.1%	8.9%
2GFU	72.3%	80.4%	8.1%	1PC2	79.3%	87.4%	8.1%
2LMR	70.7%	87.0%	7.3%	2KA0	72.6%	78.3%	5.7%
2L3O	71.3%	76.9%	5.6%	1O1W	67.2%	72.1%	4.9%
2CQ9	78.3%	82.6%	4.3%	2RQA	72.0%	75.4%	3.4%
2D86	89.0%	92.1%	3.1%	1NTC	80.5%	83.1%	2.6%
2JZT	76.6%	79.0%	2.4%	2CZN	76.5%	76.5%	0.0%
1RCH	75.4%	74.6%	-0.8%	2JU1	77.1%	75.9%	-1.2%
2KV7	85.5%	84.2%	1.3%	2JT2	83.6%	81.5%	-2.1%
2KYW	83.8%	81.1%	-2.7%	2OSR	82.7%	80.0%	-2.7%
2L6M	81.7%	78.5%	-3.2%	2CU1	81.1%	77.8%	-3.3%
1AJ3	93.3%	88.8%	-4.5%	1WI5	84.0%	78.0%	-6.0%
1NMW	85.0%	78.0%	-7.0%	2LBV	83.9%	74.8%	-9.1%

In order to demonstrate that (Φ, Ψ) dihedral angles of the NMR protein structures are improved after idealization, we simply compared the percentage of favored (Φ, Ψ) dihedral angles calculated by PROCHECK [24] in Table 1. After idealization, we can see that 19 out of 32 NMR protein structures have more favored (Φ, Ψ) dihedral angles. Overall, the percentage is increased by 4.34% on average and 27.30% in the best case towards 90%, which is the minimum percentage expected in a good quality model [24].

Unfortunately, for those NMR protein structures that already have more than approximately 75% of favored (Φ, Ψ) dihedral angles, idealization harms the percentage by -0.85% on average. There are at least two reasons behind this. First, our free energy score $S_f(P_i)$ defined in Section 2.2 is calculated from a data set that is different from the one used by PROCHECK. In fact, we used 1898 protein structures of the CULLPDB_PC30_RES1.6_R0.25 data set [18,19], while PROCHECK used 118 protein structures with a resolution cutoff of 2.0Å and an

R factor cutoff of 0.20 [24]. Although the percentages of favored (Φ, Ψ) dihedral angles are decreased in Table 1, our free energy scores of proteins 1WI5, 1NMW and 2LBV are increased by 0.22, 1.35 and 0.31, respectively, after idealization. Second, our implementation is trying to optimize our score function $S_{BB}(P_i)$ defined in Section 2.2 instead of the free energy score only. Thus, it is possible to see decreased free energy scores after idealization, especially when the target protein structure has a high percentage of favored (Φ, Ψ) dihedral angles.

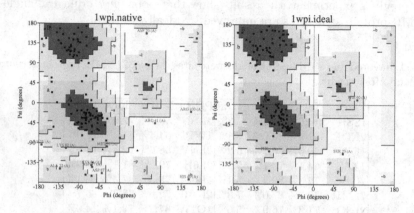

Fig. 2. Ramachandran plots of the native structure (left) and the idealized structure (right) of NMR protein structure 1WPI

Our conclusion is further supported by the Ramachandran plots of the NMR structure with PDB ID 1WPI [25] as a case study. In Figure 2, the Ramachandran plots were drawn by PROCHECK [24]. We can see that (Φ, Ψ) dihedral angles tend to move towards favored regions. To be more specific, the native structure contains only 64.4% (Φ, Ψ) dihedral angles in favored regions, while the idealized structure contains a significantly improved percentage of 81.4% of (Φ, Ψ) dihedral angles in favored regions. Moreover, the native structure contains three (Φ, Ψ) dihedral angles that are not in any feasible areas of the Ramachandran plot. However, there is only one such case found in the idealized structure. Thus, two infeasible (Φ, Ψ) dihedral angles are fixed by the (Φ, Ψ) dihedral angle log-odd score introduced in Section 2.2. Here, we did not, but certainly can, implement a hard constraint to disallow any infeasible (Φ, Ψ) dihedral angles.

4   Discussion

In this paper, we have introduced the protein structure idealization problem, and performed our first attempt on the problem. The experiment results show that idealized structures always exist with small changes on the coordinates. Furthermore, the idealized backbone structures have significantly better free energy and (Φ, Ψ) dihedral angle distributions. Therefore, protein structures can be modelled accurately with dihedral angles, ideal bond lengths and angles, and it is feasible to predict protein backbone and side-chain structures by searching the dihedral angle space.

Our protein structure idealization method can still be improved in several ways. Since our scoring functions are very simple with all weights $w_a = 1.0$ in the current implementation, it is clear that there is space for improvements here. We are also looking forward to add protein-ligand interaction energy to our score function and to study the effect of idealization on protein-ligand interactions. Moreover, since some atoms are more flexible than others, we can also set different search spaces for different atoms in our method. For example, when idealizing X-ray protein structures, the search space of each atom could be selected according to its B-factor. We can also adopt an divide-and-conquer algorithm in our method to find the global instead of local optimal idealized structure. To be more specific, we can divide the protein structure into small fragments, idealize each fragment separately, and merge idealized fragments. The keys here are to divide the protein structure by a tree decomposition of the interaction graph, and to remember the optimal idealized fragment for each possible configuration of atoms with interactions to external atoms. Actually, similar ideas have already been used to successfully improve the speed and the accuracy of backbone and side-chain structure predictions [8,9,26,27,28,29].

Our protein structure idealization method can also correct modelling errors of protein structures in PDB [17]. In fact, previous research indicate that many bond conformations and side-chain rotamers are likely incorrect in PDB, and it is useful to have an automated mechanism to fix these problems [30,31]. Thus, we can fix these problems by idealizing all protein structures in PDB with our protein structure idealization method and specially tuned scoring functions.

Acknowledgments. This work was supported by Startup Grant at City University of Hong Kong [7002731], National Basic Research Program of China [2012CB316500], NSERC Grant [OGP0046506], Canada Research Chair program, NSERC Collaborative Grant, OCRiT, Premier's Discovery Award, Killam Prize and SHARCNET.

References

1. Engh, R.A., Huber, R.: Accurate bond and angle parameters for x-ray protein structure refinement. Acta Crystallographica Section A 47, 392–400 (1991)
2. Engh, R.A., Huber, R.: Structure quality and target parameters. In: International Tables for Crystallography. International Union of Crystallograph, vol. F, 18.3, pp. 382–416 (2006)
3. Güntert, P., Wüthrich, K.: Improved efficiency of protein structure calculations from nmr data using the program diana with redundant dihedral angle constraints. J. Biomol. NMR 1(4), 447–456 (1991)
4. Güntert, P., Mumenthaler, C., Wüthrich, K.: Torsion angle dynamics for nmr structure calculation with the new program dyana. Journal of Molecular Biology 273(1), 283–298 (1997)

5. Simons, K.T., Kooperberg, C., Huang, E., Baker, D.: Assembly of protein tertiary structures from fragments with similar local sequences using simulated annealing and Bayesian scoring functions. J. Mol. Biol. 268(1), 209–225 (1997)
6. Simons, K.T., Strauss, C., Baker, D.: Prospects for ab initio protein structural genomics. J. Mol. Biol. 306, 1191–1199 (2001)
7. Li, S.C., Bu, D., Xu, J., Li, M.: Fragment-HMM: a new approach to protein structure prediction. Protein Science 17(11), 1925–1934 (2008)
8. Canutescu, A.A., Shelenkov, A.A., Dunbrack, R.L.: A graph-theory algorithm for rapid protein side-chain prediction. Protein Science 12(9), 2001–2014 (2003)
9. Krivov, G.G., Shapovalov, M.V., Dunbrack, R.L.: Improved prediction of protein side-chain conformations with SCWRL4. Proteins: Structure, Function, and Bioinformatics 77(4), 778–795 (2009)
10. Kuszewski, J., Gronenborn, A.M., Clore, G.M.: Improving the quality of NMR and crystallographic protein structures by means of a conformational database potential derived from structure databases. Protein Sci. 5(6), 1067–1080 (1996)
11. Kuszewski, J., Gronenborn, A.M., Clore, G.M.: Improvements and extensions in the conformational database potential for the refinement of NMR and x-ray structures of proteins and nucleic acids. Journal of Magnetic Resonance 125(1), 171–177 (1997)
12. Rice, L.M., Brünger, A.T.: Torsion angle dynamics: Reduced variable conformational sampling enhances crystallographic structure refinement. Proteins: Structure, Function, and Genetics 19(4), 277–290 (1994)
13. Stein, E.G., Rice, L.M., Brünger, A.T.: Torsion-angle molecular dynamics as a new efficient tool for NMR structure calculation. J. Magn. Reson. 124, 154–164 (1997)
14. Dunbrack, R.L., Cohen, F.E.: Bayesian statistical analysis of protein side-chain rotamer preferences. Protein Sci. 6(8), 1661–1681 (1997)
15. Evans, P.R.: An introduction to stereochemical restraints. Acta Crystallographica Section D 63(1), 58–61 (2007)
16. Jaskolski, M., Gilski, M., Dauter, Z., Wlodawer, A.: Stereochemical restraints revisited: how accurate are refinement targets and how much should protein structures be allowed to deviate from them? Acta Crystallographica Section D 63(5), 611–620 (2007)
17. Berman, H.M., Westbrook, J., Feng, Z., Gilliland, G., Bhat, T.N., Weissig, H., Shindyalov, I.N., Bourne, P.E.: The protein data bank. Nucleic Acids Res. 28, 235–242 (2000)
18. Wang, G., Dunbrack, R.L.: Pisces: a protein sequence culling server. Bioinformatics 19(12), 1589–1591 (2003)
19. Wang, G., Dunbrack, R.L.: Pisces: recent improvements to a pdb sequence culling server. Nucleic Acids Research 33(Web-Server-Issue), 94–98 (2005)
20. Yang, Y., Zhou, Y.: Ab initio folding of terminal segments with secondary structures reveals the fine difference between two closely related all-atom statistical energy functions. Protein Sci. 17(7), 1212–1219 (2008)
21. Yang, Y., Zhou, Y.: Specific interactions for ab initio folding of protein terminal regions with secondary structures. Proteins 72(2), 793–803 (2008)
22. Lattman, E., Loll, P.J., Loll, P.: Protein crystallography: a concise guide. In: Protein Crystallography. Johns Hopkins University Press (2008)
23. Rutgers, UCSD: Protein data bank contents guide (July 2011), http://www.wwpdb.org/documentation/format33/v3.3.html
24. Laskowski, R.A., MacArthur, M.W., Moss, D.S., Thornton, J.M.: PROCHECK: a program to check the stereochemical quality of protein structures. Journal of Applied Crystallography 26(2), 283–291 (1993)

25. Jung, J.W., Yee, A., Wu, B., Arrowsmith, C.H., Lee, W.: Solution structure of YKR049C, a putative redox protein from Saccharomyces cerevisiae. J. Biochem. Mol. Biol. 38(5), 500–504 (2005)
26. Xu, J.: Rapid Protein Side-Chain Packing via Tree Decomposition. In: Miyano, S., Mesirov, J., Kasif, S., Istrail, S., Pevzner, P.A., Waterman, M. (eds.) RECOMB 2005. LNCS (LNBI), vol. 3500, pp. 423–439. Springer, Heidelberg (2005)
27. Xu, J., Jiao, F., Berger, B.: A tree-decomposition approach to protein structure prediction. In: Proc. IEEE Comput. Syst. Bioinform. Conf., pp. 247–256 (2005)
28. Xu, Y., Xu, D.: Protein threading using PROSPECT: Design and evaluation. Proteins: Structure, Function, and Genetics 40(3), 343–354 (2000)
29. Kim, D., Xu, D., Guo, J.T., Ellrott, K., Xu, Y.: PROSPECT II: protein structure prediction program for genome-scale applications. Protein Engineering 16(9), 641–650 (2003)
30. Hooft, R.W.W., Vriend, G., Sander, C., Abola, E.E.: Errors in protein structures. Nature 381(6580), 272 (1996)
31. Joosten, R.P., Joosten, K., Cohen, S.X., Vriend, G., Perrakis, A.: Automatic rebuilding and optimization of crystallographic structures in the Protein Data Bank. Bioinformatics 27(24), 3392–3398 (2011)

Resolving Spatial Inconsistencies
in Chromosome Conformation Data

Geet Duggal[1], Rob Patro[1], Emre Sefer[1], Hao Wang[2], Darya Filippova[1],
Samir Khuller[1], and Carl Kingsford[1]

[1] Department of Computer Science
University of Maryland, College Park MD 20742, USA
{geet,rob,esefer,dfilippo,samir,carlk}@cs.umd.edu
[2] Department of Electrical and Computer Engineering
University of Maryland, College Park MD 20742, USA
hwang825@umd.edu

Abstract. We introduce a new method for filtering noisy 3C interactions that selects subsets of interactions that obey metric constraints of various strictness. We demonstrate that, although the problem is computationally hard, near-optimal results are often attainable in practice using well-designed heuristics and approximation algorithms. Further, we show that, compared with a standard technique, this metric filtering approach leads to (a) subgraphs with higher total statistical significance, (b) lower embedding error, (c) lower sensitivity to initial conditions of the embedding algorithm, and (d) structures with better agreement with light microscopy measurements.

Keywords: metric subgraph, chromosome conformation.

1 Introduction

Chromosome conformation capture (3C) is a recent experimental technique designed to observe how the genome folds in the nucleus of a cell [2]. Measurements from 3C experiments have been used to construct three-dimensional models of chromosomes at a higher resolution than what is possible with light microscopy [9], and these models are correlated with long-range regulation [1], chromatin accessibility [8], as well as cancer-related genome alterations [4]. Since its introduction, the 3C technique has become widely adopted and has been applied to bacterial, yeast, fruit fly, and human genomes [1,3,7,13,14,17].

Measured interactions between genomic locations are aggregated into a chromosome conformation or 3C graph. The frequency of an interaction between a particular pair of genomic locations in the assayed population of cells can be converted to a distance, and this mapping allows the graph to be embedded in three dimensions. Before embedding, interactions in 3C graphs are usually filtered so that only interactions with unusually high frequencies given their genomic distance are kept. For example, normalized contact matrices normalize the observed frequency of an interaction within a chromosome by the expected frequency within an entire genome (e.g. [8,18]) while others more explicitly model

B. Raphael and J. Tang (Eds.): WABI 2012, LNBI 7534, pp. 288–300, 2012.
© Springer-Verlag Berlin Heidelberg 2012

the distribution of interaction frequencies (e.g. [1,3]). In this sense, traditional statistical filtering methods retain high-confidence interactions.

However, because 3C measurements are aggregated over millions of cells, the distances associated with these high-confidence interactions are often metrically inconsistent. For example, among 2,257,241,015 triplets of measurements that form triangles in the yeast 3C data of Duan et al. [3], 679,480,886 (30%) do not satisfy the triangle inequality. These inconsistencies make it difficult to reason about conformational properties of the genome. Further, existing filtering procedures do not use relationships between the edges to, for example, discard high-confidence edges that are apparently inconsistent with many lower-confidence edges, or to include seemingly low-confidence edges that are nonetheless consistent with many others.

To address these shortcomings, we introduce the idea of metric filtering where we seek a high-confidence metrically consistent subset of the 3C graph. We frame the procedure as a family of optimization problems where we want to find a subgraph of high total weight (confidence) such that the set of chosen edges satisfies metric constraints of various stringency. We show that this family of problems is NP-hard, and provide four algorithms for the approximate solution of the least and most stringent versions.

We apply the metric filtering algorithms to 4C measurements for budding yeast [3] and show that the heuristics are able to find near-optimal solutions to the most useful variant of the problem where only triangles are consistent. Despite the additional metric constraints, the selected set of edges is often of higher total confidence than the data set considered in Duan et al. [3] that had an estimated 1% false discovery rate (FDR).

We show that embeddings based on these filterings have lower embedding error than those based on an existing filtering technique [3]. The structures also exhibit lower variation when different initial conditions are chosen for a previously proposed non-linear optimization embedding technique. Finally, we provide anecdotal evidence that the structure resulting from the metrically filtered interactions is in better agreement with known biology than the structure derived using standard filtering techniques. The improved agreement is a result of the metric filtering being able to include longer-distance, but lower-confidence, interactions.

2   Problem Definition

Problem 1 (Consistent-k-Paths). Given an integer $k \geq 2$, and a graph $G = (V, E)$, where each edge $e \in E$ is associated with a non-negative length $d(e)$ and a positive reward $r(e)$, find a subset $S \subseteq E$ of edges that maximizes $R(S) = \sum_{e \in S} r(e)$ and such that, for all $e \in S$ and for any path P_e^k of k or fewer edges in S joining the endpoints of e, the following condition holds:

$$\sum_{e' \in P_e^k} d(e') \geq d(e). \tag{1}$$

In other words, we seek the highest *total reward* subgraph where the length of every chosen edge is shorter than any path of length k joining the endpoints of that edge. If an edge satisfies condition (1) for a given k, we say it is k-*consistent*, or simply *consistent* if k is clear from the context. If the edge is not consistent, it is *violated*.

Consistent-k-Paths is a family of problems parameterized by k. The value of k allows the strictness of the metric condition to be varied. To obtain an idea as to how relatively stringent the filterings are, we focus on the two extreme case of $k = 2$ and $k = |V| - 1$. The strictest condition is $k = |V| - 1$, when every alternative path must be at least as long as the direct edge connecting the endpoints of the path, while the most lenient condition is $k = 2$, where consistency is only enforced for triangles. Because of their importance, we give names to these two special cases.

Definition 1. *ConsT is an instance of* Consistent-k-Paths *with* $k = 2$, *i.e. every triangle must satisfy inequality eq. (1). ConsP is an instance of* Consistent-k-Paths *with* $k = |V| - 1$ *implying that all paths are consistent.*

The ConsT formulation (and any formulation with $k < |V| - 1$) is motivated by the fact that each measured distance is associated with some uncertainty which propagates when summing distances over longer paths. The ConsP property is more stringent, requiring all paths to be consistent. Requiring the graph distances to be embeddable in $\mathbb{R}^3$ would be the most stringent property (but embeddability testing is NP-Hard [12]). While the more strict criteria of ConsP or embeddability in $\mathbb{R}^3$ are desireable, they both require consistency of longer paths and are thus associated with less certain distances.

3   NP-Hardness of Consistent-k-Paths

Theorem 1. Consistent-k-Paths *is NP-hard for* $k > 1$.

Proof. Reduction from **Independent Set**: given $\ell \in \mathbb{Z}_{\geq 0}$ and a graph $G = (V, E)$, construct a graph $H = (V \cup \{u\}, E \cup E')$ as follows. Let $d(e) = r(e) = 3 \; \forall e \in E$. Create a new vertex u, and a new set of edges $E' = \{\{u, v\} \mid v \in V\}$. Set $d(e) = r(e) = 1 \; \forall e \in E'$. Then, G has an independent set of size $\geq \ell \iff H$ has a solution of total reward $R \geq 3|E| + \ell$. Note that a violating path in H contains exactly 2 edges and, along with the violated edge, forms a triangle. This is because every edge in H has $d \geq 1$, so that every path containing 3 or more edges will have a total $d \geq 3$. Such a path is as long as any edge in H, and hence can violate no edge. It follows that this reduction applies for all $k \geq 2$.

$\implies$ Let $S \subseteq V$ be an independent set of size $\geq \ell$. Choose all edges of E and the edges $E_S = \{\{u, w\} \mid w \in S\}$. The total reward of this set is $R = 3|E| + |S| \geq 3|E| + \ell$. Since all of the edges in E have $d = 3$, all 2-hop paths formed by these edges have $d = 6$ and do not violate eq. (1). Further, since S is an independent set, no triangle involving u is selected. Therefore, the graph induced by the selected set of edges, $E \cup E_S$, is consistent.

$\impliedby$ Assume E^* is a solution to the `Consistent-k-Paths` problem with $R \geq 3|E| + \ell$. First, note that no triangle $\{\{u,w\}, \{w,v\}, \{v,u\}\}$ can be selected since this would violate $\{v,w\}$ because $d(\{v,u\}) + d(\{u,w\}) < d(\{v,w\})$. Due to the following argument, we assume, without loss of generality, that all edges of G are chosen: Suppose a pair of edges $\{u,v\}$ and $\{u,w\}$ was chosen and edge $\{v,w\}$ exists in E but was not chosen. Then we can remove $\{u,v\}$ and add $\{v,w\}$. This is still a solution with reward $\geq 3|E| + \ell$, since the swap only increases the value of the solution. Repeated application of this will produce a solution of cost $\geq 3|E| + \ell$ that includes all of E.

In the transformed solution, the edges of E contribute a reward of $3|E|$. Further, to avoid violating edges, the endpoints of selected edges adjacent to u must form an independent set, and to achieve a total reward $R \geq 3|E| + \ell$, there must have been an independent set size $\geq \ell$. $\square$

Corollary 1. `Consistent-k-Paths` *restricted to* $r(e) = 1$ *for all* $e \in E$ *is NP-hard for* $k > 1$.

Proof. Apply the same reduction as above, except that $r(e) = 1$ instead of 3 for all $e \in E$, and using a total reward threshold of $E^* \geq |E| + \ell$ instead of $3|E| + \ell$. In the new formulation, E^* might include two edges $\{u,w\}$ and $\{u,v\}$ without picking $\{v,w\} \in E$ if it exists. However, any such solution can be transformed into one of the form above with equal cost by removing $\{u,w\}$ and adding $\{v,w\}$. This neither changes the number of edges nor decreases the total reward of E^*, and the proof can proceed as in Theorem 1. $\square$

4 Algorithms

Since `ConsT` and `ConsP` are NP-hard, it is unlikely that there exist algorithms that solve these problems in polynomial time. Thus, we have developed several approximation algorithms and heuristics to tackle them in practice. We present four algorithms below; the first two apply to the `ConsT` problem while the latter two apply to the `ConsP` problem.

A Set-Cover-Based Algorithm. We formulate `ConsT` as a *minimum weight set cover problem* by removing the lowest weight set of edges that restores consistency, and therefore maximizes the weight of the remaining graph (the complement of the original problem). Let I be the set of violated triangles in G (where a triangle is *violated* if it does not obey the triangle inequality). For edge $\{u,v\}$ in E, let S_{uv} be the subset of triangles in I that contain $\{u,v\}$, and let $C = \{S_{uv} : \forall \{u,v\} \in E\}$. Define the cost $c(S_{uv}) = r(\{u,v\})$. We then seek the smallest weight collection of sets S_{uv} that cover all the violated triangles I, a direct application of minimum-weight set cover. Removing the edges corresponding to each chosen subset S_{uv} will resolve all of the violated triangles. This problem can be approximated using either an LP relaxation or a greedy algorithm [6]. Note that, since each violated triangle belongs to at most 3 sets of the collection C, there is an algorithm that finds a solution to this SET-COVER instance with a cost no more than 3 times OPT [6]. For the experiments described here, we use the greedy algorithm. There exist exact algorithms for the related hitting set problem [10], but these are only efficient when the number of edges that need to be removed is small, which is not what we observe in the 3C data we analyze.

A Hierarchical Maximum Cut Approach. Another approach to solving ConsT uses a solution to the MAX-CUT problem to find a maximum weight (i.e. maximum total reward) cut-set, E', of G. Because E' is bipartite, it will have no triangles, and thus, no violated triangles. The LOCAL-CUT algorithm [5] guarantees that E' has at least $1/2$ the total reward of G. Therefore, this algorithm is a $1/2$-approximation to ConsT. We add all edges in E' to the growing solution set E^*. Then, for every pair of edges $\{u, v\}, \{u, w\}$ in E', if there is an edge $\{v, w\} \in E \setminus E'$ that forms a violated triangle, we remove $\{v, w\}$ from G, and we recursively apply this procedure to the two partitions induced by the maximum cut. Because the subgraphs induced by V_1 and V_2 only contain the set of edges that form non-violating triangles with the edges in E', the constructed solution contains no triangle violations.

Taking the Union of Shortest Paths. Let $\mathcal{P}_{uv}$ be the set of edges in all shortest paths (according to d) going from node u to node v in $G = (V, E)$. A feasible solution to ConsP is to take the edges in $\bigcup_{\{u,v\} \in E} \mathcal{P}_{uv}$. We call this the SP-UNION heuristic. The intuition behind it is that, by definition, no edge that is part of some shortest path in G can be violated. Assume such an edge, $\{u, v\}$, was violated. Then, there must exist some path p between u and v with $d(p) < d(\{u, v\})$. However, this contradicts the fact that $\{u, v\}$ belongs to some shortest path, because we could replace $\{u, v\}$ with p and shorten this path.

Unfortunately, there may be an exponential number of shortest paths in G. However, by removing from E all edges that are not part of some shortest path, we can obtain the desired set of edges without explicitly enumerating all shortest paths. The SP-UNION heuristic first computes, for every edge $\{u, v\} \in E$, the length of the shortest path between its endpoints, $d(\mathrm{sp}_{uv})$. Then, the solution is simply given by $E^* = E \setminus \{\{u, v\} \in E \mid d(\mathrm{sp}_{uv}) < d(u, v)\}$.

Maximum Spanning Tree Heuristic. The final heuristic, MST-ADD, first constructs a maximum-reward spanning tree $T = (V, E_T)$ on $G = (V, E)$ and adds its edges to E^*. This can be computed using any standard maximum-weight spanning tree algorithm. By construction, T has a high total reward. Since it is a tree, it contains no cycles, and hence no violations. We sort the remaining edges $E \setminus E_T$ by their reward and, iterating through them in descending order, add them to E^* if they do not violate any shortest paths in the growing E^*.

5 Computational Results

5.1 Weights for 3C Interactions in Budding Yeast

We use the measurements from Duan et al. [3], who used a 3C variant called 4C to assay interactions for the entire *S. cerevisae* genome during interphase with two experiments based on the HindIII MseI and MspI restriction enzymes. In total, Duan et al. measured $4,097,539$ interactions across $4,193$ genome fragments

Table 1. Sizes of yeast chromosome conformation graphs. $|V|$ is the number of interaction fragments per chromosome, and $|E|$ is the number of intra-chromosomal edges.

Chr	1	2	3	4	5	6	7	8	9	10	11	12	13	14	15	16
$|V|$	54	311	100	521	184	92	404	192	149	257	253	361	331	283	368	333
$|E|$	1046	31600	3738	86126	11243	3050	52224	13226	7738	2234	20232	38052	36792	26531	43807	37054

(nodes). Each of these interactions e is associated with a frequency $f(e)$ — the number of times it was observed.

Duan et al. [3] process these raw frequency counts to derive several other measures for each interaction. A spatial distance $d(e)$, which we use as the edge length in eq. (1), is assigned to every interaction using a frequency-to-distance mapping based on the observed inverse relationship between genomic separation and frequency for intra-chromosomal interactions. Such a distance mapping is common to most approaches that seek embeddings of 3C data [1,3,11,17]. Because the distance mapping is based on intra-chromosomal interactions, we have more confidence in the spatial distances $d(e)$ for interactions within a chromosome. Therefore, in most of the experiments below, we consider each chromosome individually. Table 1 gives the sizes of the graphs for each chromosome of yeast.

Duan et al. [3] also compute a p-value — derived via a statistical null model — for every e. Using these p-values, they further derive a "q-value" that accounts for multiple hypothesis testing caused by the large number of edges sampled. See the "Computational methods" of the supplementary material of Duan et al. for a description of how the q-values are computed. We reproduced their p-value and q-value computations and use $r(e) = 1 - q(e)$ as the reward for including an edge in the solution in eq. (1). The value $1 - q(e)$ is a measure of confidence: high values indicate low p-values, which indicate that interactions occur with a frequency that one would not expect by chance.

The input to the filtering procedures is thus the graph $G = (V, E)$ where V is a set of restriction fragments and E is the set of interactions. The distance on an edge e is $d(e)$ and the benefit on an edge is the confidence $r(e)$. The goal of metric filtering is to find a subgraph with high confidence (i.e. generally low p-values) with no metric violations as defined by eq. (1).

5.2   Ability of the Algorithms to Find High-Weight Subgraphs

The heuristics of section 4 were tested on each of the 16 yeast chromosomes, and Fig. 1 summarizes the algorithms' ability to find high-confidence solutions. In 15 out of 16 cases, MAX-CUT finds a ConsT subgraph with the largest total confidence (Fig. 1, green triangles). However, in all 16 chromosomes, the SET-COVER method finds a graph of nearly the same quality, indicating that this method is competitive in terms of its ability to optimize the objective function.

The ConsT subgraphs have similar—and usually higher—total confidence than FDR 1% while eliminating all violated triangles (compare black circles with green triangles and red squares in Fig. 1). Both SET-COVER and MAX-CUT achieve

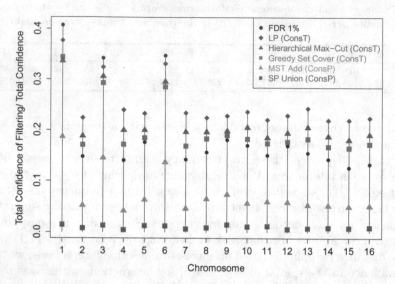

Fig. 1. Fraction of available confidence that each algorithm recovers. Higher values are better. The total confidence of each obtained interaction set is divided by the total confidence of the non-filtered interaction set. The objective value of the set cover linear program (blue diamonds) gives an upper bound for both the ConsT and ConsP solutions. We also compare our algorithms to the Duan et al. filtering method at FDR 1% (black circles) which is their largest and highest-confidence filtered interaction set.

total confidence that is higher than the Duan et al. FDR 1% filtering for all but the smallest chromosomes (1, 3, and 6). Even in those cases, SET-COVER and MAX-CUT solutions are no more than 25% away from the FDR 1% total confidence.

Due to the NP-hardness of the problems, optimal solutions for ConsT and ConsP are difficult to obtain. However, the ConsT problem can be expressed as an integer linear program (ILP) using the standard ILP for set cover. While this ILP is also difficult to solve, its linear relaxation is solvable in practice and provides a provable upper bound on the optimal solution, shown as blue diamonds in Fig. 1. This bound reveals that the SET-COVER and MAX-CUT approaches find solutions that are close to optimal. Experiments on all chromosomes achieve total confidence values that are at least 70% of the linear program upper bound and four cases achieve total confidence of around 90% of the upper bound. Since the LP overestimates the optimal value, it is likely that the heuristics provide solutions that are much closer than 70% of the true optimal solution.

The algorithms for ConsP (MST-ADD and SP-UNION) find graphs with far fewer edges and far lower total confidence than any of the solutions for ConsT (Fig. 1), and they sacrifice a significant proportion of the total confidence to obtain a completely metric subgraph. This is a strong indication of how much more strict the ConsP condition is compared with ConsT. In addition, the SP-UNION algorithm performed quite poorly compared with MST-ADD. The severe

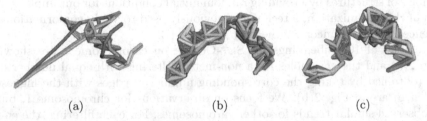

| (a) | (b) | (c) |

Fig. 2. Superposition of 10 embeddings for both ConsT and C-RANK filterings. (a) SET-COVER. (b) C-RANK of the same size as the SET-COVER. (c) SET-COVER after removing ≈ 20% of the lowest-confidence edges.

condition required by ConsP is likely too strict for the noisy 3C data, and ConsT provides a more reasonable trade-off between avoiding metric violations and keeping a useful fraction of the interactions.

5.3 Metric Filtering Produces Lower-Error Embeddings

The ConsT and ConsP filterings both result in lower-error embeddings than their associated confidence-ranked filterings when embedded using a nonlinear optimization technique. To control for the size of filterings we compare a metric filtering with m interactions to an associated set of the m highest confidence interactions (C-RANK). The optimization attempts to place nodes to minimize the sum-squared error of $\sum_{e \in E'}(o(e) - d(e))^2$ between the original $d(e)$ and the embedded $o(e)$ distance. The SET-COVER filtering of chromosome 1 resulted in a mean sum-squared error of 0.07 across 10 embeddings while C-RANK resulted in an error of 1.58. Similarly, MST-ADD had an average error of 0.067 while C-RANK produced an error of 0.39. Our improved performance may be due to the fact that metric violations result in distance contradictions that cannot be resolved by the optimization procedure.

To confirm the hypothesis that metric violations cause increased errors in the embeddings, we systematically re-introduced violated triangles using the following procedure. We choose a triangle $\{u, v, w\}$ at random. If $d(u, v) < d(u, w)$, then we set $d(u, v) = \alpha|d(v, w) - d(u, w)|$. Otherwise, we set $d(u, w) = \alpha|d(v, w) - d(u, v)|$, for some choice of $0 < \alpha < 1$. As the percentage of violated triangles increased, the embedding error increased as well with $1.2, 1.4, 1.6, 2.0, 2.3, 2.8$ average error for $10, 20, 30, 40, 50, 60\%$ violated triangles respectively with $\alpha = 0.9$. Metric filtering therefore has the desirable property of removing the embedding error that results from the existence of violated triangles.

5.4 Metric Filtering Produces Low-Variance, More Biologically Plausible Embeddings

We embedded the various sets of filtered constraints for the chromosomes using an established non-linear optimization technique [3,14] that incorporates chromatin packing constraints consistent with known biology in yeast. We obtained ensembles of structures by providing random initial conditions for our implementation of this optimization, a technique previously used to study conformational differences between cancer and healthy genomes [1].

Ensembles of 10 embeddings for SET-COVER on chromosome 1 are shown in Fig. 2(a) and the ensembles for a non-metric filtering with equal number of edges (obtained by taking the corresponding number of edges with the highest confidence) are in Fig. 2(b). We focus on observations for chromosome 1, but have observed similar trends for other chromosomes. For each filtering, the embeddings are aligned to each other using a maximum likelihood superpositioning technique [15].

Lower-variance embeddings. Both ConsT (Fig. 2) and ConsP filterings result in ensembles of embeddings that are more homogeneous than those from the associated C-RANK sets as indicated by the superposition of structures in Fig. 2. We can quantify the variance of an ensemble by computing the sum of the branch lengths of a minimum spanning tree of a complete graph where the nodes represent the embedded structures and the edge weights are the RMSD between the alignments of pairs of structures. The minimum spanning tree on this graph represents a parsimonious way to describe the variability among embedded structures. The MST-based variability between the SET-COVER and MST-ADD embeddings of chromosome 1 are 0.17 and 0.0093 respectively while the MST variability of the associated C-RANK embeddings are much larger, at 0.26 and 0.32 respectively.

Low variance among the embeddings of metric subgraphs indicates that selecting edges for their metric consistency allows fewer highly different solutions to be found. This is desirable, because we do not want embeddings to be sensitive to the initial conditions of the optimization procedure. Further, because they were taken from a population of cells, the 3C measurements are in fact taken from an ensemble of structures. The large variance among C-RANK edges may reflect this fact. In contrast, the metric filtering appears to be selecting subsets of constraints that could plausibly represent a single structure. Hence, metric filtering may be one way to partially deconvolve the population-averaged measurements.

Biologically plausible embeddings. The ConsT embeddings result in telomere distances that match known microscopy distances better than the associated C-RANK set. A recent experiment [16] establishes that the distance between the telomeres of chromosome 1 in budding yeast are often about $1\mu m$. The embeddings of C-RANK in Fig. 2(b) have an average distance of $0.45\mu m$ while the embeddings of SET-COVER (Fig. 2(a)) have an average distance of $0.96\mu m$, which is a much better match to the experimentally observed value.

Despite having edge sets of the same or larger total confidence, the metric filtering produces very different structures than the C-RANK filtering. However, removing the 71 lowest-confidence edges from the ConsT embedding does result in a structure similar to the C-RANK filterings 2(c). Thus, it seems these lower-confidence, but metrically-consistent, interactions are crucial to obtaining the more distended structures that are more consistent with microscopy experiments.

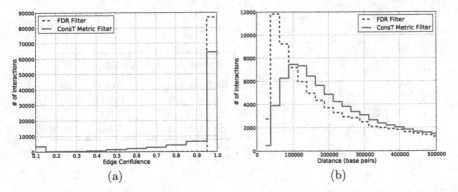

(a)   (b)

Fig. 3. Histogram of genomic distances and interaction confidences for the union of all intra-chromosomal interactions for both metric filterings and their associated C-RANK filterings

5.5   Analysis of the Types of Edges Kept by Metric Filtering

For all chromosomes, both ConsP and ConsT keep more low-confidence edges than the C-RANK filtering (ConsT shown in Fig. 3(a)). Although, in general, more low-confidence edges are kept, the ConsT filtering of chromosome 1 preserves overall higher distance interactions than the associated C-RANK filtering with a mean distance of 0.55 while the C-RANK filtering has a mean distance of 0.25 (all distances in μm). Of these interactions, the high-confidence ones in the ConsT filtering (i.e. those above 0.8) have a mean distance of 0.31 while the C-RANK filtering has a mean distance of 0.25. This is due to the fact that the interactions in the C-RANK filtering are concentrated in a small region of chromosome 1, while the SET-COVER filtering distributes the interactions across the entire chromosome: for the interactions in the C-RANK set but not in the ConsT filtering, 76 out of the 131 interactions lie inside the positions 75881 and 130646 of chromosome 1 while the densest region of similar size in the SET-COVER filtering has only 26 interactions. The preservation of larger-distance, higher-confidence interactions in the ConsT filtering is likely what results in the expanded structure where telomere distances are more in line with microscopy experiments. For the ConsP embedding, however, the large disparity in mean distance of interactions (C-RANK: 0.17, MST-ADD: 1.69) is due mostly to low-confidence edges. This creates an undesirable structure that contains very little useful information about long-range interactions. This is another indication that the strictness of the ConsP filtering may be too severe compared with the more relaxed and biologically plausible ConsT approach. The inclusion of long-range interactions resulting from lower-confidence edges represent some of the most interesting and desired information obtained from 3C experiments. C-RANK necessarily ignores many of these long-range constraints, while the metric filtering allows the inclusion of both metrically consistent and higher-confidence constraints.

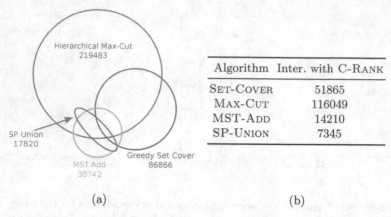

Algorithm	Inter. with C-RANK
SET-COVER	51865
MAX-CUT	116049
MST-ADD	14210
SP-UNION	7345

(a)   (b)

Fig. 4. (a) Intersection among metric edge sets and (b) intersection with C-RANK sets of equal size

In addition, the ConsT method generally keeps interactions with larger genomic distances than C-RANK. The average genomic distance of C-RANK is 243.5 kilobases while the average genomic distance of SET-COVER is 285.0 kilobases (Fig. 3(b)). Surprisingly, while ConsP keeps more low-frequency interactions, these tend to be at shorter genomic distances.

5.6   Various Heuristics Result in Very Different Sets of Edges

Although the MAX-CUT and SET-COVER algorithms aim to optimize the same objective and find subgraphs of approximately the same total weight, they result in very different edge sets (Fig. 4(a)). Further, their intersections with the most surprising edges are also different: of the 219, 483 edges returned by MAX-CUT, 53% are among the top 219, 483 most surprising edges, while 60% of the 86, 866 edges returned by SET-COVER are among the most surprising edges (Fig. 4(b)). The differing number of edges in solutions with similar total confidence also indicates that the edges in the MAX-CUT solution are of lower average confidence.

The structure of the graphs returned by MAX-CUT is also very different than that returned by SET-COVER. The MAX-CUT solution has very few triangles. For example, on chromosome 1 MAX-CUT retains only 27 out of the original 10091 triangles while SET-COVER keeps 495. This difference is somewhat intuitive since SET-COVER is explicitly trying to throw away few triangles while MAX-CUT is explicitly looking for triangle-free (i.e. bipartite) subgraphs. For fewer, higher-weight edges with many triangles the SET-COVER should be preferred. This is likely the scenario that is most applicable to 3C chromosome embedding. Because optimal solutions cannot be found for large instances, it is unclear at this point whether the large variation in the returned edge sets is due to the objective function admitting many solutions or whether, if optimal solutions could be found, they would all be similar.

The two algorithms designed for ConsP also result in very different graphs, but this is primarily because the MST-ADD algorithm is far more successful at finding a good solution than the SP-UNION approach. The two algorithms had similar intersections with the top-most surprising edges: for MST-ADD, 46% of the edges were among the top-most surprising edges, while for SP-UNION the fraction was 41%.

5.7 Practical Running Times of the Algorithms

When applied to the largest yeast chromosome (4), the SET-COVER and MAX-CUT implementations take 21 seconds and 21.5 minutes to run respectively on an Opteron 8431 processor. The SP-UNION and MST-ADD methods take 2.25 minutes and < 2 days respectively. The current implementation of MST-ADD re-computes the shortest paths after every edge addition, and this could be substantially sped up with a dynamic shortest-paths method. The SET-COVER implementation is fast enough to be run on the entire the entire yeast genome, including inter-chromosomal interactions, within 5 hours. In this case the ConsT filtering results in a significantly different edge set than the C-RANK embedding (the size of the intersection with C-RANK is only 350960 out of 657177 edges). The SET-COVER algorithm also yields a relatively high average confidence (0.88) when compared to the average confidence from C-RANK (0.97).

6   Conclusions

We have provided evidence that a filtering scheme for 3C data that uses both statistical confidence and metric consistency as criteria produces sets of interactions that are more embeddable, and create more consistent and more biologically plausible estimations for the 3D structures of the chromosomes. We show that such filtering in general is NP-hard, but by comparing to LP-based upper bounds, we empirically demonstrate that both a set cover approach and a hierarchical maximum cut algorithm produce nearly optimal solutions avoiding any violated triangles.

A natural extension adds a slack factor ρ to eq. (1): $\sum_{e' \in P_e^k} d(e') \geq \rho d(e)$ where $0 < \rho < 1$. This allows some amount of violation. Exploring the Consistent-k-Paths problem for various ρ may be especially helpful in the context of uncertain and noisy 3C distance estimates.

Acknowledgements. This work was partially supported by the the National Science Foundation [CCF-1053918, EF-0849899, and IIS-0812111], National Institutes of Health [1R21AI085376], and a University of Maryland Institute for Advanced Studies New Frontiers Award. C.K. received support as an Alfred P. Sloan Research Fellow.

References

1. Baù, D., et al.: The three-dimensional folding of the α-globin gene domain reveals formation of chromatin globules. Nat. Struct. & Mol. Biol. 18(1), 107–114 (2010)
2. Dekker, J., et al.: Capturing chromosome conformation. Science 295(5558), 1306–1311 (2002)
3. Duan, Z., et al.: A three-dimensional model of the yeast genome. Nature 465(7296), 363–367 (2010)
4. Fudenberg, G., et al.: High-order chromatin architecture determines the landscape of chromosomal alterations in cancer. Nat. Biotechnol. 29(12), 1109–1113 (2011), http://www.ncbi.nlm.nih.gov/pubmed/22101486, doi:10.1038/nbt.2049
5. Gomes, C., Williams, R.: Approximation algorithms. In: Burke, E.K., Kendall, G. (eds.) Search Methodologies: Introductory Tutorials in Optimization and Decision Support Techniques, ch. 18. Springer (2005)
6. Hochbaum, D.S. (ed.): Approximation algorithms for NP-hard problems. PWS Publishing Co, Boston (1997)
7. Kalhor, R., et al.: Genome architectures revealed by tethered chromosome conformation capture and population-based modeling. Nat. Biotechnol. 30(1), 90–98 (2012)
8. Lieberman-Aiden, E., et al.: Comprehensive mapping of long-range interactions reveals folding principles of the human genome. Science 326(5950), 289–293 (2009)
9. Marti-Renom, M.A., Mirny, L.A.: Bridging the resolution gap in structural modeling of 3D genome organization. PLoS Comput. Biol. 7(7), 1002125 (2011)
10. Niedermeier, R., Rossmanith, P.: An efficient fixed-parameter algorithm for 3-hitting set. Journal of Discrete Algorithms 1(1), 89–102 (2003)
11. Rousseau, M., et al.: Three-dimensional modeling of chromatin structure from interaction frequency data using Markov chain Monte Carlo sampling. BMC Bioinformatics 12(1), 414 (2011)
12. Saxe, J.: Embeddability of weighted graphs in k-space is strongly NP-hard. In: 17th Allerton Conference in Communications, Control and Computing, pp. 480–489 (1979)
13. Sexton, T., et al.: Three-dimensional folding and functional organization principles of the Drosophila genome. Cell 148(3), 458–472 (2012)
14. Tanizawa, H., et al.: Mapping of long-range associations throughout the fission yeast genome reveals global genome organization linked to transcriptional regulation. Nuc. Acids Res. 38(22), 8164–8177 (2010)
15. Theobald, D.L., Wuttke, D.S.: THESEUS: maximum likelihood superpositioning and analysis of macromolecular structures. Bioinformatics 22(17), 2171–2172 (2006)
16. Therizols, P., et al.: Chromosome arm length and nuclear constraints determine the dynamic relationship of yeast subtelomeres. Proc. Natl. Acad. Sci. USA 107(5), 2025–2030 (2010)
17. Umbarger, M.A., et al.: The three-dimensional architecture of a bacterial genome and its alteration by genetic perturbation. Mol. Cell 44(2), 252–264 (2011)
18. Yaffe, E., Tanay, A.: Probabilistic modeling of Hi-C contact maps eliminates systematic biases to characterize global chromosomal architecture. Nature Genetics 43(11), 1059–1065 (2011)

MS-DPR: An Algorithm for Computing Statistical Significance of Spectral Identifications of Non-linear Peptides

Hosein Mohimani[1], Sangtae Kim[2], and Pavel A. Pevzner[2,3,*]

[1] Department of Electrical and Computer Engineering, UC San Diego
[2] Department of Computer Science and Engineering, UC San Diego
ppevzner@cs.ucsd.edu

Abstract. While non-linear peptide natural products such as Vancomycin and Daptomycin are among the most effective antibiotics, the computational techniques for sequencing such peptides are still in infancy. Previous methods for sequencing peptide natural products are based on Nuclear Magnetic Resonance spectroscopy, and require large amounts (milligrams) of purified materials. Recently, development of mass spectrometry based methods has enabled accurate sequencing of non-linear peptidic natural products using picograms of materials, but the question of evaluating statistical significance of Peptide Spectrum Matches (PSM) for these peptides remains open. Moreover, it is unclear how to decide whether a given spectrum is produced by linear, cyclic, or branch-cyclic peptide. Surprisingly, all previous mass spectrometery studies overlooked the fact that a very similar problem has been succesfully addressed in particle physics In 1951. In this paper we develop a method for estimating statistical significance of PSMs defined by non-linear peptides, which makes it possible to identify whether a peptide is linear, cyclic or branch-cyclic, an important step toward identification of peptidic natural products.

1 Introduction

The dominant technique for sequencing cyclic peptides is nuclear magnetic resonance (NMR) spectroscopy, which requires large amount (milligrams) of highly purified materials that are often nearly impossible to obtain [1]. Tandem mass spectrometry (MS/MS) provides an attractive alternative to NMR because it allows one to sequence a peptide from picograms of non-purified material. Recently, new algorithms have been developed for interpreting mass spectra of cyclic peptides using de novo sequencing [2–4] and database search [5].

MS/MS coupled with database search is the most popular method for identification of (linear) peptides. A database search engine selects candidate peptides from a database of protein sequences that match the precursor mass from mass spectrum. Then for each candidate peptide, the software compares a theoretical MS/MS derived from the peptide to the experimental mass spectrum, and reports a peptide with best score.

---

* Corresponding author.

B. Raphael and J. Tang (Eds.): WABI 2012, LNBI 7534, pp. 301–313, 2012.
© Springer-Verlag Berlin Heidelberg 2012

In the last decade, many efforts have been invested in computing statistical significance of Peptide Spectrum Matches (PSMs). Many of these studies followed on the pioneering paper by Fenyo and Beavis [6] that proposed to approximate the statistical significance of PSMs by first modeling the distribution of PSM scores (e.g. by Gumbel distribution [6]) and further using this distribution to compute p-values [7–13]. Unfortunately, this approximation approach, while useful in many applications, often fails when one has to estimate extremely small p-values typical for mass spectrometry (e.g. PSM p-values of the order 10^{-10} are required to achieve 1% FDR [14]). The challenges of estimating the probability of extremely rare events was first addressed in particle physics in 1950s [15], and communication systems in 1980s [16]. Unfortunately, the mass spectrometry community has overlooked these fundamental studies (directly relevant to mass spectrometry) resulting in inaccurate p-value estimation in many mass spectrometry studies [17].

In late 1940s, many top mathematicians worked on *neutron shielding* problem that was crucial for designing nuclear facilities [18, 19]. In this problem, one has to compute the probability that a neutron, doing a random walk, would pass through slab, an extremely rare event. Two general methods emerged for evaluating extremely rare events by Monte-Carlo random sampling (using computers that became available in mid 1940s), *importance sampling* and *multilevel splitting*, both developed for nuclear-physics calculations by von Neumann, Ulam, Fermi, Kahn, Metropolis, and their colleagues, during the production of the first nuclear bomb [15, 18–21]. Importance sampling is based on the notion of modifying the underlying probability distribution in such a way that the rare events occur much more frequently. Multilevel splitting uses a selection mechanism to favor the trajectories deemed likely to lead to the rare events of interest. While importance sampling is the most popular rare event simulation method, the main advantage of multilevel splitting approach as compared to importance sampling is the fact that the former does not need to modify the probabilistic model governing the system, making it applicable to systems represented as a black box [19], and thus directly applicable to mass spectrometery studies. Kahn and Harris solved the neutron shielding problem using multilevel splitting in 1951 [15]. Later, similar rare event estimation techniques found applications in numerous fields, including communication systems [16], financial mathematics [22], air traffic management [23], and chemistry [24]. However, this powerful approach has never been applied to mass spectrometry. This is surprising because there is a clear analogy between statistical significance evaluation in mass spectrometery, and the neutron shielding problem, where a spectrum plays a role of a neutron, a peptide plays a role of a slab, and an event "spectrum gets a high score against a peptide" plays a role of an event "neutron passes through a slab".

Currently, the dominant technique for statistical evaluation of a set of PSMs is to compute the False Discovery Rate (FDR) using the Target Decoy Approach (TDA) [25]. TDA is attractive for proteomics studies because it is widely applicable to different instrument platforms and database search algorithms. However, TDA is not applicable to non-linear peptide studies, because in these studies researchers usually work on a few non-linear peptide at a time, whereas TDA is best suited for statistical analysis of large spectral datasets [25]. Even in the case of linear peptides, some popular database search tools are not TDA-compliant as discussed in [26].

An alternative technique is to compute a p-value of an *individual* PSM [17]. Given a PSM (*Peptide*, *Spectrum*) of score t, the p-value of (*Peptide*, *Spectrum*) is defined as the fraction of random peptides with a score equal to or exceeding t [17]. Unlike the FDR that is defined on a set of PSMs, the p-value is defined on a single PSM. Therefore computing the p-value is adequate for non-linear peptide studies, where a single or few non-linear peptides are considered at a time. Since our results can be applied to both cyclic peptides (e.g. surfactin) and branch-cyclic peptides (e.g. daptomycin), below we use the same term 'cyclic' to refer to both cyclic and branch-cyclic peptides.

For cyclic peptide studies, computing p-values offers additional advantages. In studies of peptide natural products, we are given a mixture of spectra of linear and cyclic peptides, from which a small number of spectra of cyclic peptides should be separated and investigated independently. Therefore we need a method that given a spectrum identifies whether it represents a linear or a cyclic peptide. This is difficult because different scoring functions are used for linear and cyclic peptides. Since scores from different scoring functions are not usually comparable, we need to convert them into p-values, because p-values are comparable across different scoring functions [27, 28].

In the case of linear peptides, Kim *et al.*, 2008 [17] presented a polynomial time MS-GF algorithm for computing p-values. To the best of our knowledge, MS-GF is the only existing method that calculates theoretical p-values under certain null hypothesis. However, MS-GF is only applicable to scoring functions that can be represented as a dot-product of vectors, i.e. *additive scoring functions*. Moreover, MS-GF is only applicable to linear peptides, and no one has generalized MS-GF to non-linear peptides yet.

Fenyo and Beavis constructed an empirical score distribution of low-scoring (erroneous) peptide identifications and extrapolated it to evaluate the p-value of high-scoring peptide identifications in the tail of the distribution [6]. Similar approaches are now used in many tools, that provide p-value or E-value of individual PSMs, e.g. OMSSA [30]. However, this approach was demonstrated to be inaccurate [17]. While the pitfalls of such approaches are well recognized in genomics, they remian under-appriciated in proteomics.[1] One may consider estimating p-values by a Monte-Carlo simulation generating a population of millions of peptides and estimating the probability distribution of scores on this population [31]. This approach becomes time-consuming for estimating extremely low

---

[1] Waterman and Vingron [29] argued that it is difficult to accurately estimate the extreme tails of a distribution in general, requiring accurate estimation of rare events probability.

p-values, since it requires calculating scores of billions of randomly generated peptides for accurate estimation of p-values as lows as 1 in a billion.[2]

In this paper we propose MS-DPR (MS-Direct Probability Redistribution), a new method for estimating p-values of PSMs based on rare event probability estimation by multilevel splitting. Our p-value are computed under the same null hypothesis used in MS-GF. We show that MS-DPR reports p-values similar to those reported by MS-GF in the case of linear peptides, confirming that it accurately estimates p-values. Furthermore, we show that unlike MS-GF, MS-DPR can compute p-values of PSMs when arbitrary (non-additive) scoring function is used or when the peptide is non-linear.

2   Method

In contrast to importance sampling, which changes the probability laws driving the model, multilevel splitting [15, 20] uses a selection mechanism to favor the trajectories deemed likely to lead to rare events. The multilevel splitting is composed of three steps. First decompose the trajectories to the rare events of interest into shorter subpaths whose probability is not so small. Second, encourage the realizations that take these subpaths (leading to the events of interest) by giving them a chance to reproduce. Third, discourage the realizations that go in the wrong direction by killing them with some positive probability. The sub-trajectories are usually delimited by levels. Starting from a given level, the realizations of the trajectories that do not reach the next level will not reach the rare event, but those that do will split into multiple copies when they reach the next level. Each copy pursues its evolution independently from then on. This creates an artificial drift toward the rare event by favoring the trajectories that go in the right direction. In the end, an unbiased estimator can be recovered by multiplying the contribution of each trajectory by the appropriate weight.[3]

---

[2] Estimating rare event probabilities using Monte Carlo simulation requires a prohibitively large number of trials to achieve an acceptable accuracy [31]. The book [31] discusses pitfalls of approaches similar to [6] for computing rare event probabilities.

[3] While the original paper by Kahn and Harris [15] presented multilevel splitting by trajectories as mentioned, many papers use the equivalent nested subspace notation. To compute the probability of a rare event A by multilevel splitting, one needs to construct a nested sequence of events $A \subset A_1 \subset \cdots \subset A_k$ satisfying $p\{A\} \ll p\{A_1\} \ll \cdots \ll p\{A_k\} \ll 1$. Then probability of the event A can be calculated by

$$p\{A\} = p\{A_k\}p\{A_{k-1}|A_k\} \cdots p\{A|A_1\}$$

Consider the simple case $k = 1$ and $p\{A\} = p\{A_1\}p\{A|A_1\}$. With the same number of trials, $p\{A_1\}$ can be estimated more accurate than $p\{A\}$, because $p\{A\} \ll p\{A_1\}$. If estimation of $p\{A|A_1\}$ is done by uniform sampling over the entire search space, there is no overall improvement in accuracy of $p\{A\}$ estimation. On the other hand, if there exists a way to efficiently sample from A_1, the conditional probability can be estimated more accurately, resulting in a more accurate estimation of the rare event probability.

While multilevel splitting has wide applicability across diverse fields, in mass spectrometry applications it is not clear how to select reproduction and killing probabilies, and the number of offsprings. Inspired by Kahn and Harris [15] and proposed by Haraszti and Townsend [32], *Direct Probability Redistribution* (DPR) is a realization of multilevel splitting for estimation of probability of rare states in a *Markov chain*. Given a Markov chain, DPR implicitly constructs a modified Markov chain where probabilities of states are increased by an arbitrary order of magnitude. For Markov chain with n states and (unknown) equilibrium probabilities $p_1, \cdots, p_n$, given *oversampling factors* $\mu_1, \cdots, \mu_n$, DPR constructs a Markov chain with equibrium probability $p'_1 = \mu_1 p_1 / \sum \mu_k p_k, \cdots, p'_n = \mu_n p_n / \sum \mu_k p_k$. For example, for a two-state Markov chain with equilibrium probabilities $p_1 = 0.999$ and $p_2 = 0.001$, if we choose $\mu_1 = 1$ and $\mu_2 = 999$, we end up with equilibrium probability $p'_1 = 0.5$ and $p'_2 = 0.5$ (Figure 1). Given the oversampling factors, DPR is capable of calculating reproduction and killing probabilities and the number of offsprings for each level, making it very suitable for mass spectrometry applications. Here we descibe how to apply DPR to the problem of estimating probability distribution of PSM scores. For simplicity, we

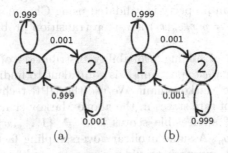

Fig. 1. a) Markov chain before DPR performed, with equilibrium probability distribution $(0.999, 0.001)$. b) Markov chain after performing DPR, with equilibrium probability distribution is $(0.5, 0.5)$.

define a spectrum as a set of integer masses. A peptide of length k is defined as a string of k integers $Peptide = m_1 m_2 \cdots m_k$. The mass of the peptide is defined as sum of all the integers in the string. A score between a peptide and a spectrum is denoted by $Score(Peptide, Spectrum)$.

Given $Peptide = (m_1, \ldots, m_i, m_{i+1}, \cdots, m_k)$ and integers $1 \le i \le k$ and $-m_i < \delta < m_{i+1}$ and, we define $Peptide(i, \delta)$ as a peptide $(m_1, \cdots, m_i + \delta, m_{i+1} - \delta, \cdots, m_k)$. The former peptide is called the *mother peptide*, and the latter is called the *daughter peptide*. Note that there are many alternative ways to define the notion of daughter peptides. $RandomTransition(Peptide)$ is a $Peptide(i, \delta)$ of $Peptide$, where i and δ are integer random variables, i chosen from a uniform distribution on $[1, k]$, and δ chosen from a uniform distribution on $[-m_i, m_{i+1}]$. Appendix 1 discusses the conditions on a $RandomTransition$ that are needed for proper work of MS-DPR. We define $PeptideSpace$ as the set of all peptides with length k and mass m. Consider the following Markov chain defined on *PeptideSpace*:

$$Peptide_0, Peptide_1, Peptide_2, \dots$$

where $Peptide_0$ is chosen from *PeptideSpace* with uniform distribution, and[4]

$$Peptide_{t+1} = RandomTransition(Peptide_t)$$

Now given a spectrum S, consider the following sequence of scores:

$$x_0, x_1, x_2, \dots$$

where $x_t = Score(Peptide_t, Spectrum)$.[5] To demonstrate that the sequence of scores is a stationary Markov chain, one needs to show that $p(x_t|x_{t-1} = s)$ does not depend on t (stationary property), and that $p(x_t|x_{t-1} = s, x_{t-2} = s')$ does not depend on s' (Markovian property). While the stationary property holds because *RandomTransition* is a time-homogeneous function, the Markovian property may not hold for the sequence of scores. However, in practice, we validated that $p(x_t|x_{t-1} = s, x_{t-2} = s')$ for the scoring functions we use (see below). The Markovian property is validated using Chapman Kolmogorov test, $\mathbf{P}_2 = \mathbf{P}_1{}^2$, where $\mathbf{P}_n = p(x_{t+n}|x_t)$ is n-step transition probability of the chain (see Table 1).

Then the problem of finding probability distribution of all peptide scores from *PeptideSpace* against spectrum S is equivalent to finding equilibrium distribution of the above Markov chain. We use the DPR technique to accurately estimate probability of rare states in the above Markov chain.

Assume the set of all feasible scores is $S = \{1, \dots, n\}$, with (unknown) probabilities $p_1, \cdots, p_n$. Assume arbitrary oversampling factors $\mu_1, \cdots, \mu_n$ are given. Then the DPR approach provides a way to modify the transition probabilities such that the equilibrium distribution of the resulting Markov chain is $p'_1 = \mu_1 p_1 / \sum \mu_k p_k, \cdots, \mu_n p'_n = p_n / \sum \mu_k p_k$. Figure 2 shows the algorithm adapted from [32]. A proof of convergence of this algorithm is given in [32].

Glasserman *et. al.*, 1998, [33] show the optimal choice of $\mu_1, \cdots, \mu_n$ (with respect to reducing the number of trials to achieve the required accuracy) is the one that makes all states equiprobable, *i.e.* $(\mu_1, \cdots, \mu_n) = (1/p_1, \cdots 1/p_n)$. However, no estimation of p_k is given beforehand. Our idea is to first run the algorithm with $\mu_1 = \cdots = \mu_n = 1$, and obtain a rough estimation of $p_1, \cdots, p_n$. Then we choose $\mu_k = 1/p_k$ in the next iteration.[6] This procedure is summarized in Figure 3.

---

[4] In this Markove chain, the set of states are all possible peptides in the *PeptideSpace*, and transition probability from *Peptide* to *Peptide'* is a constant p if *Peptide'* = *Peptide*(i, δ) for some i and δ, and zero otherwise. For cyclic peptides, every peptide has exactly $2M$ daughters, resulting in $p = 1/2M$, where M is parent mass.

[5] Note that both *Peptide*$_t$ and x_t are random variables.

[6] If $p_k = 0$ for some values of k, we replace them with p_{min}, where p_{min} is the minimum non-zero probability among them.

Table 1. Comparison of empirical 2-step transion probability matrix ($\mathbf{P_2}$), with square of empirical single-step transition matrix($\mathbf{P_1^2}$), for a million randomly generated linear peptides of length 7 and mass 787, and a spectrum of peptide KYIPGTK (the first spectrum from the standard ISB dataset [34]). The average deviation of $\mathbf{P_2}$ from $\mathbf{P_1^2}$ is 34% .

$$\mathbf{P_1^2} = \begin{bmatrix} 0.79 & 0.17 & 0.033 & 0.0026 & 0.00015 & 0 & 0 & 0 & 0 & 0 \\ 0.22 & 0.61 & 0.13 & 0.025 & 0.0017 & 0 & 0 & 0 & 0 & 0 \\ 0.097 & 0.29 & 0.49 & 0.10 & 0.016 & 0.0010 & 0 & 0 & 0 & 0 \\ 0.024 & 0.17 & 0.34 & 0.37 & 0.071 & 0.0010 & 0 & 0 & 0 & 0 \\ 0.0071 & 0.060 & 0.26 & 0.36 & 0.26 & 0.040 & 0.0068 & 0.0009 & 0 & 0 \\ 0 & 0.018 & 0.11 & 0.31 & 0.36 & 0.17 & 0.023 & 0.0045 & 0 & 0 \\ 0 & 0 & 0.087 & 0.22 & 0.41 & 0.19 & 0.027 & 0.067 & 0 & 0 \\ 0 & 0 & 0 & 0.097 & 0.19 & 0.44 & 0.020 & 0.25 & 0 & 0 \\ 0 & 0 & 0 & 0 & 0 & 0 & 0 & 0 & 0 & 0 \\ 0 & 0 & 0 & 0 & 0 & 0 & 0 & 0 & 0 & 0 \end{bmatrix}$$

$$\mathbf{P_2} = \begin{bmatrix} 0.79 & 0.17 & 0.032 & 0.0020 & 0.00017 & 0 & 0 & 0 & 0 & 0 \\ 0.22 & 0.62 & 0.13 & 0.022 & 0.0019 & 0 & 0 & 0 & 0 & 0 \\ 0.10 & 0.29 & 0.49 & 0.099 & 0.016 & 0.0011 & 0 & 0 & 0 & 0 \\ 0.028 & 0.17 & 0.33 & 0.39 & 0.070 & 0.011 & 0 & 0 & 0 & 0 \\ 0.0064 & 0.069 & 0.25 & 0.36 & 0.26 & 0.045 & 0.0085 & 0.0021 & 0 & 0 \\ 0 & 0.014 & 0.070 & 0.35 & 0.33 & 0.16 & 0.042 & 0.027 & 0 & 0 \\ 0 & 0 & 0.22 & 0.22 & 0.11 & 0.22 & 0.11 & 0.11 & 0 & 0 \\ 0 & 0 & 0 & 0 & 0 & 0.5 & 0 & 0.5 & 0 & 0 \\ 0 & 0 & 0 & 0 & 0 & 0 & 0 & 0 & 0 & 0 \\ 0 & 0 & 0 & 0 & 0 & 0 & 0 & 0 & 0 & 0 \end{bmatrix}$$

3   Results

Datasets. We used the Standard Protein Mix database consisting of 1.1 million spectra generated from 18 proteins using eight different mass spectrometers [34]. For this study, we considered the charge 2 spectra generated by Thermo Electron LTQ where 1388 linear peptides of length between 7 and 20 are identified with false discovery rate 2.5% using Sequest [35] and PeptideProphet [36] in the search against the *Haemophilus influenzae* database appended with sequences of the 18 proteins (567,460 residues). For testing MS-DPR on cyclic peptides, we use the cyclic peptide dataset from [5]. This dataset contains two identified cyclopeptides, SFTI-1 and SFTL-2, from *Helianthus annuus*, and a linear and a cyclic peptide, SDP and SKF, from *Bacillus subtilits*.

To apply MS-DPR, we first need to define scoring functions for linear and cyclic peptides. Linear theoretical spectrum of a peptide $Peptide = (m_1, \cdots, m_k)$, $LinearSpectrum(Peptide)$, is a set of $k - 1$ b-ions and $k - 1$ y-ions, where each b-ion is the mass of a prefix of the peptide plus rounded H^+ mass, $m_1 + \cdots + m_{j-1} + 1$, and each y-ion is the mass of a suffix of the peptide plus rounded H^+ and H_2O mass, $m_j + \cdots + m_k + 19$. Cyclic theoretical spectrum of the peptide, $CyclicSpectrum(Peptide)$, is defined as the set of masses of its $k(k - 1)$ substrings of the peptide, $m_i + \cdots + m_{j-1}$ ($m_i + \cdots + m_k + m_1 + \cdots + m_{j-1}$, if $i \geq j$). Score of linear peptide $Peptide$ and a spectrum $Spectrum$, $LinearScore(Peptide, Spectrum)$, is defined as the number of shared masses between $Spectrum$ and $LinearSpectrum(Peptide)$.

Similarly, $CyclicScore(Peptide, Spectrum)$ is defined as the number of shared masses between $Spectrum$ and $CyclicSpectrum(Peptide)$.

In addition to the p-value computed by MS-DPR (denoted by p_{DPR}), we also compute the empirical p-value (denoted by p_E), using a Monte Carlo

$$MS - DPR - Iteration(\mu_1, \cdots, \mu_n)$$

---
input: Spectrum $Spectrum$, score function $Score(Peptide) = Score(Peptide, Spectrum)$ with scores in $1, \cdots, n$ domain, random transition generator $RandomTransition(Peptide)$, number of output peptides N, and oversampling factors $\mu_1 \cdots \mu_n$.
output: An estimate of score probability distribution $p_1, \cdots, p_n$ on the peptide space.
initialization:
       select a random $Peptide_0$ from $PeptideSpace$
       $z \leftarrow 0$ and $\mu_{min} \leftarrow min_{k=1,n}\mu_k$
execution:
       run $SimulateDPR(Peptide_0, \mu_{min})$
procedure $SimulateDPR(Peptide, \Omega)$
       while $z < N$ **do**
              $Peptide' \leftarrow RandomTransition(Peptide)$
              if $\mu_{Score(Peptide')} < \Omega$
                   return
              if $\mu_{Score(Peptide')} > \mu_{Score(Peptide)}$
                   $Y \leftarrow \mu_{Score(Peptide')}/\mu_{Score(Peptide)}^{*}$
                   for $i = 1$ to $Y - 1$
                           choose Ω' from a uniform distribution on
$[\mu_{Score(Peptide)}, \mu_{Score(Peptide')}]$
                         run $SimulateDPR(Peptide', \Omega')$
                   end
              end
              $z \leftarrow z + 1$
              $Peptide_z \leftarrow Peptide'$
       end
       return
end
obtaining probability estimation:
       $n_k \leftarrow \#\{z|Score(Peptide_z) = k\}$.
       $p'_k \leftarrow n_k/N$.
       $p_k \leftarrow \frac{n_k/\mu_k}{\sum n_k/\mu_k}$.

Fig. 2. $MS - DPR - Iteration(\mu_1, \cdots, \mu_n)$ algorithm from [32] adapted for estimating statistical significance of PSMs. The algorithm produces peptide process $Peptide_0, Peptide_1, \cdots, Peptide_N$, and scores process $Score(Peptide_0), Score(Peptide_1), \cdots, Score(Peptide_N)$, with equilibrium probability distribution $p'_1, \cdots, p'_n$ satisfying $p'_k = c\mu_k p_k$ for a constant c. (*) Most of the times $\mu_{Score(Peptide')}/\mu_{Score(Peptide)}$ is not integer. In that case Y would be a random variable, taking $\lceil \mu_{Score(Peptide')}/\mu_{Score(Peptide)} \rceil$ with probability $p = \mu_{Score(Peptide')}/\mu_{Score(Peptide)} - \lfloor \mu_{Score(Peptide')}/\mu_{Score(Peptide)} \rfloor$ and $\lfloor \mu_{Score(Peptide')}/\mu_{Score(Peptide)} \rfloor$ with probability $1 - p$. Not that in case of $\mu_1 = \cdots = \mu_n = 1$, this reduces to simple Monte Carlo estimation of probability distribution from N peptides.

$$MS - DPR$$

---
goal : Estimating probabilty distribution of scores
input : Number of iterations K
output : an estimation of the probability distribution $p_1, \cdots p_n$
initialization : $(\mu_1 \cdots \mu_n) \leftarrow (1, \cdots, 1)$
for $iter = 1$ **to** K
       run $MS - DPR - Iteration(\mu_1, \cdots, \mu_n)$.
       $(\mu_1 \cdots \mu_n) \leftarrow (1/p_1, \cdots, 1/p_n)$
end

Fig. 3. MS-DPR algorithm for estimating the probability distribution of scores

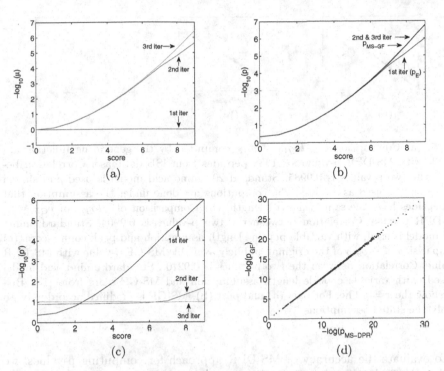

Fig. 4. Evolution of (a) μ_k (b) p_k (c) p'_k for three iterations of MS-DPR. The analysis is performed for $N = 1,000,000$ simulated peptides of length 7, and a spectrum of peptide KYIFGTK from standard ISB database with parent mass 787. Blue, red and green plot stands for first, second, and third iterations respectively. In part (b) MS-GF p-value is plotted by black. Note that the blue plot in part (b) corresponds to first iteration of MS-DPR, which simply gives the empirical p-value, p_E. From the second iteration on, the MS-DPR p-value is very similar to MSGF p-value. Part (c) shows the evolution of p' to uniform distribution (making all states of the Markov chain equiprobable). (d) Comparision of $-log_{10}$ of MS-GF p-value with MS-DPR p-value for 1388 peptides from ISB database. Red dot show the $x = y$ line. Correlation between the two p-values is 0.9998. Non-standard amino acid model is used, assuming each peptide has a fixed known length, and peak count score. MS-GF approach from [17] is modified accordingly, to satisfy these assumptions.

approach by generating millions (or even billions) of random peptides and estimating probability distribution. Moreover, we compute the p-value using the generating function approach (denoted by p_{GF}) [17].

Figure 4(a-c) shows the evolution of μ, **p** and **p**$'$ in three iterations of the algorithm. $\mathbf{p} = (p_1, \cdots p_n)$ is the original probability distribution, $\mathbf{p}' = (p'_1, \cdots p'_n)$ is the modified probability distribution, and $\mu = (\mu_1, \cdots \mu_n)$ is the vector of over-sampling factors. $\mathbf{p}'$ converges to uniform distribution, and **p** converges to the correct distribution given by MS-GF dynamic programming. According to our experiments, three iterations is usually enough for getting an accurate estimation of the probability distribution.

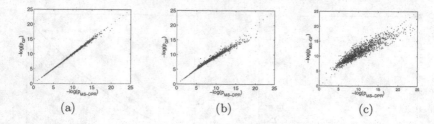

(a)   (b)   (c)

Fig. 5. (a) Comparision of $-log_{10}$ of p_{GF} computed by the generating function approach with MS-DPR p-value for 1388 peptides from ISB database. Correlation between the two p-values is 0.9985. Standard 20 amino acid model is used and shared peaks count is used as a score. The compuctions are done under the assumption that all peptides have the same (known) length. (b) Comparision of $-log_{10}$ of p_{GF} with MS-DPR p-value. Correlation between the two p-values is 0.9904. Standard amino acid model is used, with variable peptide length assumption and peak count score. (c) Comparision of $-log_{10}$ of the original, publicly available MS-GF p-value with MS-DPR p-value. Correlation between the two p-values is 0.9210 . Standard amino acid model is used, with variable peptide length assubmption and MS-GF score from [17]. Red dot show the $x = y$ line. For part (a) and part (b), MS-GF is modified accordingly, to satisfy the stated assumptions.

To evaluate the accuracy of MS-DPR approach for computing p-values, we used all 1388 identifications from ISB database, and compared p_{DPR} with p_{GF} (Figure 4(d)), under the following assumptions: (i) all integers are considered as possible masses of amino acids (non-standard amino acids assumption, typical for analyzing non-ribosomal peptides in extended amino acid alphabet [4]), (ii) p-values are computed under the assumption that peptides have fixed known length, and (iii) shared peak count is used as score. A correlation $R^2 = 0.9998$ between the two p-values shows that our method accurately estimates the probability distribution. Figure 5(a) shows the same plot while considering that only 20 standard amino acid masses (correlation of 0.9985), and Figure 5(b) shows the plot with the assumption that peptides have variable lengths[7] (correlation of 0.9904). Figure 5(c) shows the comparison with the actual MS-GF defined in [17] (correlation of 0.9210).

4   Conclusion

Most of the computational techniques developed in mass spectrometry are focusing on linear peptides, and a very small amount of computational research is done for non-linear peptides. Hence bioinformatics have not opened its way yet to the field of natural products, where majority of the interesting peptides are

---

[7] To be able to cover peptides with variable length, we need to modify transitions to make it possible to move from a peptide of length k to peptides of length $k - 1$ and $k + 1$.

cyclic or branch-cyclic. One of the important questions for natural product scientists is, how to determine the structure (linear/cyclic/branch-cyclic) and amino acid sequence of a peptide from its mass spectrum. While scoring schemes for linear, cyclic and branch-cyclic peptides are very different, one may be able to compare them after converting their scores to p-values. Thus, converting scores of non-linear peptides to p-values is the first step toward automated MS-based discovery of peptidic natural products.

We presented MS-DPR, a method for estimating statistical significance of PSMs in mass spectrometry. In contrast to existing methods for estimating p-values, MS-DPR can work with arbitrary scoring functions and non-linear peptides. To enable accurate estimation of p-values from a limited number of trials, it utitlizes DPR algorithm, a method for fast and accurate estimation of rare states in a Markov chain. Comparison of p-values given by this approach with correct p-values (given by the generating function approach from [17]) validates the method in the case of additive scoring function and linear peptides. Incorporating p-values in the recently develped Cycloquest algorithm [5] improved performance (e.g. identification of cyclic peptide SFTI-1 missed by Cycloquest in previous study). New version of Cycloquest web-server available at cyclo.ucsd.edu reports MS-DPR p-values for database search results. The source code for MS-DPR is available at proteomics.ucsd.edu .

Acknowledgement. We would like to thank Pieter Dorrestein for making the spectral dataset for SKF and SDP available, and Michelle Colgrave, Aaron Poth and Joshua Mylne for making the spectral dataset for SFTI 1 and SFT-L1 available. This work was supported by the U.S. National Institutes of Health 1-P41-RR024851-01. Appexndix is available at cyclo.ucsd.edu/publications.

References

1. Li, J.W., Vederas, J.C.: Drug discovery and natural products: end of an era or an endless frontier? Science 325, 161–165 (2009)
2. Ng, J., Bandeira, N., Liu, W.T., Ghassemian, M., Simmons, T.L., Gerwick, W.H., Linington, R., Dorrestein, P.C., Pevzner, P.A.: Dereplication and de novo sequencing of nonribosomal peptides. Nature Methods 6, 596–599 (2009)
3. Mohimani, H., Liu, W.T., Liang, Y., Gaudenico, S., Fenical, W., Dorrestein, P.C., Pevzner, P.: Multiplex de novo sequencing of peptide antibiotics. J. Comp. Biol. 18(11), 1371–1381 (2011)
4. Mohimani, H., Liang, Y., Liu, W.T., Hsieh, P.W., Dorrestein, P.C., Pevzner, P.: Sequencing cyclic peptides by multistage mass spectrometry. J. Proteomics 11(18), 3642–3650 (2011)
5. Mohimani, H., Liu, W.T., Mylne, J.S., Poth, A.G., Tran, D., Selsted, M.E., Dorrestein, P.C., Pevzner, P.A.: Cycloquest: Identification of cyclopeptides via database search of their mass spectra against genome databases. J. Prot. Res. 10(10), 4505–4512 (2011)
6. Fenyo, D., Beavis, R.: A method for assessing the statistical significance of mass spectrometry-based protein identifications using general scoring schemes. Anal. Chem. 75, 768–774 (2003)

 7. Sadygov, R.G., Liu, H., Yates, J.R.: Statistical Models for Protein Validation Using Tandem Mass Spectral Data and Protein Amino Acid Sequence Databases. Anal. Chem. 76(6), 1664–1671 (2004)
 8. Matthiesen, R., Trelle, M.B., Højrup, P., Bunkenborg, J., Jensen, O.N.: VEMS 3. 0: Algorithms and Computational Tools for Tandem Mass Spectrometry Based Identification of Post-translational Modifications in Proteins. J. Proteome Res. 4(6), 2338–2347 (2005)
 9. Chamrad, D.C., Koerting, G., Gobom, J., Thiele, H., Klose, J., Meyer, H.E., Blueggel, M.: Interpretation of mass spectrometry data for high-throughput proteomics. Analytical and Bioanalytical Chemistry 376(7), 1014–1022 (2007)
10. Nesvizhskii, A., Vitek, O., Aebersold, R.: Analysis and validation of proteomic data generated by tandem mass spectrometry. Nature Methods 4, 787–797 (2007)
11. Nesvizhskii, A., Aebersold, R.: Analysis, statistical validation and dissemination of large-scale proteomics datasets generated by tandem MS. Drug Discovery Today 9(4), 173–181 (2004)
12. Spirin, V., Shpunt, A., Seebacher, J., Gentzel, M., Shevchenko, A., Gygi, S., Sunyaev, S.: Assigning spectrum-specific P-values to protein Identifications by mass spectrometry. Bioinformatics 27(8), 1128–1134 (2011)
13. Weatherly, B., Atwood, J.A., Minning, T.A., Cavola, C., Tarleton, R.L., Orlando, R.: A Heuristic Method for Assigning a False-discovery Rate for Protein Identifications from Mascot Database Search Results. Mol. Cell. Proteomics 4, 762–772 (2005)
14. Kim, S., Mischerikow, N., Bandeira, N., Navarro, J.D., Wich, L., Mohammed, S., Heck, A.J.R., Pevzner, P.A.: The generating function of CID, ETD and CID/ETD pairs of tandem mass spectra: Applications to database search. Molecular and Cellular Proteomics 9, 2840–2852 (2010)
15. Kahn, H., Harris, T.E.: Estimation of Particle Transmission by Random Sampling. National Bureau of Standards Applied Mathematics Series (1951)
16. Villen-Altamirano, M., Villen-Altamirano, J.: RESTART: A method for accelerating rare events simulations. Queueing Performance and Control in ATM. In: Proceedings of ITC, vol. 13, pp. 71–76 (1991)
17. Kim, S., Gupta, N., Pevzner, P.: Spectral Probabilities and Generating Functions of Tandem Mass Spectra: A Strike against Decoy Databases. J. Prot. Res. 7(8), 3354–3363 (2008)
18. Hammersley, J.M., Handscomb, D.C.: Monte carlo methods. Methuen, London (1964)
19. Rubino, G., Tuffin, B.: Rare event simulation using Monte Carlo methods. Wiley (2009)
20. Kahn, H., Marshall, A.W.: Methods for reducing sample size in Monte Carlo computations. Oper. Res. Soc. Amer, 263–278 (1953)
21. Kahn, H.: Use of different Monte Carlo sampling techniques. RAND corporation (1956)
22. Glasserman, P., Heidelberger, P., Shahabuddin, P.: Asymptotically optimal importance sampling and stratification for pricing path dependent options. Mathematical Finance 9(2), 117–152 (1999)
23. Blom, H.A.P., Krystul, J., Bakker, G.J., Klompstra, M.B., Obbink, B.K.: Free flight collision risk estimation by sequential MC simulation. In: Cassandras, C.G., Lygeros, J. (eds.) Stochastic Hybrid Systems. CRC Press, Boca Raton (2007)
24. Sandmann, W.: Applicability of importance sampling to coupled molecular reactions. In: Proceedings of the 12th International Conference on Applied Stochastic Models and Data Analysis (2007)

25. Elias, J.E., Gygi, S.P.: Target-decoy search strategy for increased confidence in large-scale protein identifications by mass spectrometry. Nature Methods 4(3), 207–214 (2007)

26. Gupta, N., Bandeira, N., Keich, U., Pevzner, P.A.: Target-decoy search strategy for increased confidence in large-scale protein identifications by mass spectrometry. J. Am. Soc. Mass Spectrom. 22, 1111–1120 (2011)

27. Nesvizhskii, A.: Survey of computational methods and error rate estimation procedures for peptide and protein identification in shotgun proteomics. J. Prot. Res. 73(11), 2092–2123 (2010)

28. Kwon, T., Choi, H., Vogel, C., Nesvizhskii, A.I., Marcotte, E.M.: MSblender: A Probabilistic Approach for Integrating Peptide Identifications from Multiple Database Search Engines. J. Prot. Res. 10(7), 2949–2958 (2011)

29. Waterman, M., Vingron, M.: Rapid and accurate estimates of statistical significance for sequence data base searches. Proc. Natl. Acad. Sci. U.S.A. 91, 4625–4628 (1994)

30. Geer, L.Y., Markey, S.P., Kowalak, J.A., Wagner, L., Xu, M., Maynard, D.M., Yang, X., Shi, W., Bryant, S.H.: Open mass spectrometry search algorithm. J. Proteome Res. 3(5), 958–964 (2004)

31. Asmussen, S., Glynn, P.W.: Stochastic simulation: algorithms and analysis. Springer (2007)

32. Haraszti, Z., Townsend, J.K.: The theory of direct probability redistribution and its application to rare even simulation. ACM Trans. Modeling and Computer Simulation 9(2), 105–140 (1999)

33. Glasserman, P., Heidelberger, P., Shahabuddin, P.: A large deviations perspective on the efficiency of multilevel splitting. IEEE Trans. Automat. Contr. 43(12), 1666–1679 (1998)

34 Klimek, J., Eddes, J.S., Hohmann, L., Jackson, J., Peterson, A., Letarte, S., Caflren, P.R., Katz, J.E., Mallick, P., Lee, H., Schmidt, A., Ossola, R., Eng, J.K., Aebersold, R., Martin, D.B.: The standard protein mix database: a diverse data set to assist in the production of improved peptide and protein identification software tools. J. Proteome Res. 7, 96–103 (2008)

35. Eng, J., McCormack, A., Yates, J.: An approach to correlate tandem mass spectral data of peptides with amino acid sequences in a protein database. J. Am. Soc. Mass Spectrom. 5, 976–989 (1994)

36. Keller, A., Nesvizhskii, A., Kolker, E., Aebersold, R.: Empirical statistical model to estimate the accuracy of peptide identifications made by MS/MS and database search. Anal. Chem. 74, 5383–5392 (2002)

37. Tanner, S., Shu, H., Frank, A., Wang, L., Zandi, E., Mumby, M., Pevzner, P., Bafna, V.: InsPecT: identification of posttranslationally modified peptides from tandem mass spectra. Anal. Chem. 77, 4626–4639 (2005)

FinIS: Improved *in silico* Finishing Using an Exact Quadratic Programming Formulation

Song Gao[1], Denis Bertrand[2], and Niranjan Nagarajan[2]

[1] NUS Graduate School for Integrative Sciences and Engineering, National
University of Singapore, 21 Lower Kent Ridge Road, Singapore
[2] Computational and Systems Biology, Genome Institute of Singapore,
60 Biopolis Street, Singapore
`nagarajann@gis.a-star.edu.sg`

Abstract. With the increased democratization of sequencing, the reliance of sequence assembly programs on heuristics is at odds with the need for black-box assembly solutions that can be used reliably by nonspecialists. In this work, we present a formal definition for *in silico* assembly validation and finishing and explore the feasibility of an exact solution for this problem using quadratic programming (FinIS). Based on results for several real and simulated datasets, we demonstrate that FinIS validates the correctness of a larger fraction of the assembly than existing *ad hoc* tools. Using a test for unique optimal solutions, we show that FinIS can improve on both precision and recall values for the correctness of assembled sequences, when compared to competing programs. Source code and executables for FinIS are freely available at http://sourceforge.net/projects/finis/.

Keywords: Genome Assembly, Finishing, Quadratic Programming, Graph Algorithms.

1 Introduction

The task of assembling a genome is no longer an area of specialized interest, limited to genome centers. Decreased sequencing costs and the advent of "desktop sequencers" (e.g. Ion Torrent PGM, Illumina MiSeq and GS Junior) has dramatically increased the set of people who need to use sequence assembly tools. In addition, assembly tools are now applied to a range of tasks from the analysis of structural variants [1] to transcriptome [2] and metagenomic assembly [3]. On the other hand, sequence assembly continues to remain more of an art rather than a science, with assembly programs relying on a diverse set of heuristics in response to the complexity of the analysis and often requiring fine-tuning of parameters to "optimize" the assembly [4,5]. This dichotomy between ease of data generation (microbial genomes can now be sequenced for < \$100) and assembly challenges has renewed interest in improved algorithms and in particular in the feasibility of exact algorithms for assembly [6,7].

In a typical genome assembly project reads from shotgun-sequencing form the basis for a first-pass assembly into *contigs* (ungapped sequences representing

B. Raphael and J. Tang (Eds.): WABI 2012, LNBI 7534, pp. 314–325, 2012.
© Springer-Verlag Berlin Heidelberg 2012

parts of the genome). These are then often combined with mate-pair sequences to construct *scaffolds* i.e. ordered sets of contigs with unknown sequences (*gaps*) between them [8,7]. Where available, information from restriction maps [9] and similar genomes [10] can also help to generate better and more complete scaffolds. Despite these efforts, the resulting assembly is still fragmented with numerous scaffolds and gaps affecting the utility of the draft genome.

Not so long ago, the draft genome was only a starting point for significant directed-sequencing and validation efforts in a process known as *finishing*. This is still the "gold-standard" as it confirms the correctness of the contigging and scaffolding process, and the completeness of the genome is a valuable resource for downstream analysis of genes and synteny, repeat sequences and genomic rearrangements. The daunting cost of finishing efforts in terms of time and resources, though, has emphasized the need for computational means to speed up the process [11]. As many scaffolding algorithms ignore repeat sequences [8,12] and typical short-read WGS datasets afford high coverage of the genome, significant computational improvement of the assembly is still feasible.

In this work, we propose a formulation for finishing *in-silico* (called "FinIS") by carefully exploiting the shotgun-sequencing information that was left unused by the assembly. We show that this framework can be used to simultaneously verify and close gaps in the genome, taking into account repeat-based conflicts and coverage imposed constraints. While several existing assemblers and tools aim to achieve a similar gap-closure goal through *ad hoc* heuristics [13,14], we demonstrate in this work that, by formulating and optimizing a clear objective function, we can simultaneously improve the contiguity and reliability of the assembly. In combination with other work on exact algorithms for assembly, this brings us one more step closer to the idea of obtaining finished genomes *in silico* and realizing a vision of sequence assembly as a "black box", based on exact algorithms that need no tuning.

2   Methods

2.1   Definitions

The draft assembly from a shotgun-sequencing project (genome, transcriptome or metagenome) is typically in the form of a set of scaffolds, where each *scaffold* is an ordered and oriented (positive or negative strand) set of *contigs* (assembled ungapped sequences). Let $S_C = \{c_1, \ldots, c_n\}$ be the ordered set of contigs for a scaffold S. For every $c_i \in S_C$, we denote the two possible orientations as c_i and $-c_i$. For the gap between contigs c_i and c_{i+1} (also referred to as s_i and t_i), the size and standard-deviation estimates from the scaffold are assumed to parameterize a normal distribution (i.e. $N(\mu_i, \sigma_i)$) and the unknown sequence is denoted as g_i.

Given a draft assembly, the set of all gaps G in the assembly can be viewed as potential endpoints for missing sequenced reads in the genome (Fig. 1). The missing reads can be encoded in a graph structure representing the overlap relation between the reads, analogous to what is done in overlap graph [15] and

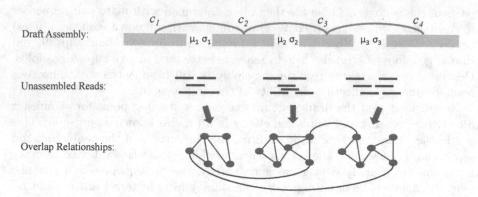

Fig. 1. Unassembled reads and gaps in a draft assembly. Nodes in the graph represent reads and edges represent overlap relationships.

de Bruijn graph assemblers [16]. Furthermore, the graph can be simplified to encode essential information as a *contig graph* (or string graph [17]), where reads that assemble unambiguously are compressed into contig nodes (these contigs are in addition to those in the original scaffold) and edges represent contigs that overlap in their ends (Fig. 2a,b). Note that as each contig has two orientations, there are four ways in which they can overlap, as shown in Figure 2a and we encode this with bidirected edges in the graph (see Section 2.5 for details on graph construction).

Given a draft assembly and the contig graph for the rest of the sequences the problem of *in silico* finishing can be viewed as that of simultaneously finding "good" paths P_i for each gap i in the genome that validate the scaffold and also provide gap sequences (g_i's). The goodness of paths can be measured in a probabilistic setting given the scaffold estimates for μ_i and σ_i as:

$$\Pi_{i \in G} \mathrm{Pr}_{N(\mu_i, \sigma_i)}(|P_i|) = \Pi_{i \in G} \frac{1}{\sqrt{2\pi\sigma_i^2}} e^{-\frac{(|P_i| - \mu_i)^2}{\sigma_i^2}}$$

Transforming this into log-space, we can maximize this probability by minimizing the following natural quadratic function for a set of paths P:

$$f(P) = \sum_{i \in G} \frac{(|P_i| - \mu_i)^2}{\sigma_i^2}$$

In addition, information about the *copy-number* of a contig ($\mathrm{cn}(c_i)$) i.e. the number of times it is expected to be represented in the genome, can be further used to constrain the paths. In this work, we use a straightforward coverage-based estimate[1],

---

[1] A more accurate approach would be to model coverage biases using the negative-binomial distribution.

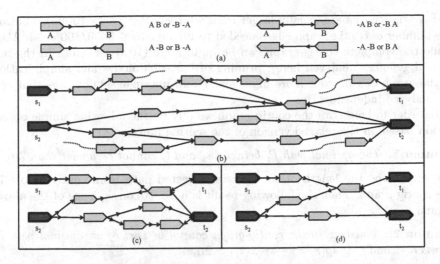

Fig. 2. Contig-graph and its simplification. (a) Encoding of orientations for overlapping contigs in the contig-graph (b) A contig-graph representing sequences connecting two gaps in the genome. Dotted lines indicate edges to other parts of the contig-graph. (c) A "trimmed" version of the graph in (b). Note that the solutions for the two gaps need to be considered together to identify the unique optimal solution. (d) The final graph after further simplifications.

$$cn(c_i) = \left\lceil \frac{\text{average-base-coverage}(c_i)}{\text{average-base-coverage-of-genome}} \right\rceil,$$

but information from other sources can also be incorporated in this framework, giving us the following definition for the *Finishing Problem* (FP):

Definition 1 (Finishing Problem). *Given a set of gaps G, a contig-graph CG and a threshold T, is there a set of paths P in CG, where P_i connects contigs s_i and t_i, and each c_i is visited $\leq cn(c_i)$ times in P, such that $f(P) \leq T$.*

In practice, requiring that every gap have a path in the contig-graph may be too stringent and we relax this constraint by allowing for gaps to be left unfilled, with the length of the resulting path $|P_i| = 0$. Note that our formulation here can be viewed as a constrained form of graph-based assembly [15,6] and a straightforward reduction from the Longest-Path Problem can be used to show that it is also NP-complete.

2.2 Graph Trimming and Simplification

While FP is not expected to have an algorithmic solution with runtime polynomial in the size of the contig-graph in the worst-case, in practice, several aspects of the problem can make exact algorithms feasible. In particular, gap sizes are often restricted by the size of mate-pair libraries (typically < 3Kbp) and contig lengths are usually larger than read lengths (i.e. $> 100bp$), limiting the number of contig-graph edges needed to fill gaps to $\leq 3000/100 = 30$. Also, while the complete contig-graph can be quite large (1000s of nodes in the microbial genomes analyzed here), pruning of infeasible nodes and simplification of the graph can make it more manageable and untangle sets of gaps that can be analyzed independently.

In order to help trim the contig-graph we rely on the following simple observation based on the relaxed version of the scoring function f:

Lemma 1. *The optimal path P_i between s_i and t_i cannot be longer than $2\mu_i$.*

Let $sp(a, b)$ be the length of the shortest directed path from contig a to b in the contig graph. Then the following result is a direct consequence of the above lemma.

Lemma 2. *A node x in the contig-graph cannot be part of an optimal path P_i between s_i and t_i if $sp(s_i, x) + sp(x, t_i) > 2\mu_i$.*

This lemma can then be used to identify nodes that are irrelevant for a particular gap to construct a subgraph containing all it's feasible paths. In order to precompute the shortest paths, we rely on two searches using a modified Dijkstra's algorithm (search is limited to paths $\leq 2\mu_i$), one starting from s_i and another from t_i where edge directions have been reversed in the graph. After construction of subgraphs for each gap independently, we can then merge subgraphs that share nodes and therefore should be jointly analyzed (Fig. 2c). As shown in Section 3 this allows for substantial reduction in graph size (by 2 orders of magnitude) and in decoupling gaps that do not have any dependencies.

After the trimming step, we employed two graph simplification steps to compress the graph further: "path-compression" and "bubble-popping" [18]. For path-compression, we identified simple paths where the internal nodes had only one incoming and one outgoing edge. The internal nodes and edges were then replaced with a single edge from the start node to the end node in the path (it is trivial to show that path-compression does not change the optimal solution for FP). For bubble-popping, we used graph-traversal to identify contig-pairs a, b such that all outgoing edges from a lead to contigs c_i that have a unique incoming edge and a unique outgoing edge to b. If all such paths are of the same length[2], we can replace them with a single edge from a to b without altering the optimal solution to FP. By iteratively applying the simplification steps (till no changes are possible), complex structures in the graph can be resolved to give a much simpler contig-graph for analysis (see Section 3 and Fig. 2d).

2.3 Quadratic Programming Formulation

In this work, we focussed on exploring the feasibility of an exact solution to FP. As the contig-graph is typically sparse and of modest size (after trimming and

---

[2] In sequencing datasets where indel errors are likely (e.g. 454 sequencing), more aggressive forms of bubble-popping may be needed in the pre-processing of the contig graph provided to FinIS.

simplification) an exact solution may be possible in many sequence assembly projects. In addition, complexity results, approximation algorithms and greedy algorithms for problems related to FP such as the Longest-Path problem and the Disjoint-Paths problem have been explored extensively [19,20] and these results can be extrapolated to FP, where needed. Finally, as each connected component of the contig-graph can be analyzed independently, we can rely on exact solutions for graphs where it is feasible and turn to approximations or heuristics only when it is absolutely necessary.

Analogous to the "flow" formulations for the disjoint path problem [20], we construct a Mixed-Integer Quadratic Program (MIQP) for FP such that any solution to FP is also a solution for the MIQP[3]. Specifically, let $V(CG)$ be the set of vertices, $E_j^{in}(CG), E_j^{out}(CG)$ be the sets of incoming and outgoing edges for vertex j in the contig-graph CG and l_j be the length of contig c_j. Also, let x_{ik} be the number of times edge k will be used in the path P_i (the "flow" on this edge). Then, we can formulate and solve (for x) the following MIQP for each connected component CG:

$$\text{Aim: } \min_x \sum_{i \in G} ((\sum_{c_j \in V(CG)} (\sum_{k \in E_j^{in}(CG)} x_{ik}) \times l_j) - \mu_i)^2$$

$$s.t.$$
$$\sum_{i \in G} \sum_{k \in E_j^{in}(CG)} x_{ik} \leq \text{cn}(c_j), \qquad \forall c_j \in V(CG)$$
$$\sum_{k \in E_j^{in}(CG)} x_{ik} = \sum_{k \in E_j^{out}(CG)} x_{ik}, \forall i \forall c_j \in V(CG) - (\{s_i\} \cup \{t_i\})$$
$$x_{ik} \in \mathbb{N}^0$$

Note that the objective function is a translation of $f(P)$ in this setting and while the first set of constraints ensure that the paths do not violate contig copy-numbers, the second set enforces a balanced-flow in the graph. To reconstruct paths from the flow, we first construct a graph for gap i with x_{ik} copies of edge k. We then used Hierholzer's algorithm [21] for finding an Eulerian path in this graph from s_i to t_i. Briefly, the algorithm works by first finding a path from s_i to t_i with a depth-first-search. As long as unused edges remain, the algorithm continues by finding cycles from nodes with unused edges adjacent to them, and replacing the node with the cycle in the final solution.

2.4   Uniqueness and Correctness of Paths

While our formulation of FP requires *a* solution, in practice, the solution can be non-unique and this ambiguity can affect the correctness of the result. Another measure of correctness is given by the Z-score $= |\frac{|P_i| - \mu_i}{\sigma_i}|$ for a path. Large Z-scores may indicate an incorrectly filled gap or errors in the draft assembly. Also, a non-unique path can have a low Z-score and these results are still useful as they confirm the correctness of the assembly and narrow down the options for the gap sequence. As gap estimates are normally distributed, an appropriately

---

[3] The opposite is unfortunately not true and the rare instances where circular flows are obtained have to be analyzed with alternate approaches.

chosen threshold on the Z-score (say 3) can be used to flag potentially erroneous gaps.

As MIQP solvers can be made to report multiple optimal solutions, non-uniqueness of the MIQP solution can be tested in a straightforward way. A given MIQP solution can however correspond to several paths, corresponding to the Eulerian tours in Section 2.3. In addition, different Eulerian tours can still produce the same gap sequence [22]. To account for this, we use equation (1) in [22] to count the number of unique sequences that arise from the Eulerian tours and are therefore potential solutions for FP.

2.5   Graph Construction

The unused and partially-used reads in a draft assembly can be re-analyzed using a de Bruijn graph or an overlap graph based assembler to construct the contig-graphs that are needed in this study. However, this process can be computationally intensive even if done in the context of just a local re-assembly [13,14]. As an alternative, we exploited the fact that many assemblers output some form of "assembly graph" in addition to the final assembly [18,13]. This graph can be extracted for two popular assemblers, Velvet [18] and SOAPdenovo [13], and parsed into a consistent contig-graph representation for use in FinIS. Where needed, FinIS can also take in a user-supplied graph for this analysis.

3   Results

3.1   Datasets and Comparisons

To evaluate FinIS, we compared it to the two general and freely-available gap-fillers that we are aware of – GapCloser [13] (version 1.12) and IMAGE [14] (version 2). Both methods are *ad hoc* and rely on heuristic local assembly, where each gap is analyzed independently. All methods were tested on several simulated (for *E. coli*, *B. subtilis* and *C. crescentus*) and real sequencing (*S. aureus*: SRA accession number SRX007712, and *P. stipitis*) [27]) datasets. Simulated datasets were generated using Metasim [23] based on NCBI reference genomes. For each dataset, we simulated a high-coverage paired-read library as well as a long-insert mate-pair library. Detailed statistics for all datasets can be found in Table 1.

Draft assemblies for each dataset were constructed as follows: paired-reads were assembled using Velvet [18] (k=49), with reads being error-corrected for the real datasets using Quake [24] (coverage-cutoff of 5). Contigs for simulated datasets were scaffolded using mate-pair data and Opera [7], while a reference-based scaffold was created for the real datasets (using MUMmer [25]). Draft assemblies were aligned to the reference genome using MUMmer to identify *valid* gaps (i.e. those where adjacent contigs have correct orientations and size estimates are within 1kbp of the correct answer). IMAGE and GapCloser (k=31) were run with default parameters[4] and quadratic programs were constructed in

---

[4] IMAGE ran for more than 3 days on the *P. stipitis* assembly and correspondingly we terminated it after 4 iterations, instead of the default 10.

FinIS based on the Velvet contig graph and solved using the MOSEK C++ API (http://www.mosek.com).

Table 1. Datasets and Sequencing Statistics

	S. aureus	*B. subtilis*	*C. crescentus*	*E. coli*	*P. stipitis*
Genome Size (Mbp)	2.9	4.0	4.0	4.6	15.4
# of Chromosomes	1	1	1	1	8
Paired-reads (length, insert-size, coverage)	76bp, 300bp, 668X	80bp, 300bp, 100X	80bp, 300bp, 100X	80bp, 300bp, 100X	75bp, 279bp, 304X
Mate-pairs (length, insert-size, coverage)	n.a.	50bp, 10kbp, 2X	50bp, 10kbp, 2X	50bp, 10kbp, 2X	n.a.

3.2   Finishing Completeness

The ability to finish more gaps *in silico* is, of course, an important metric for finishing programs and ideally gap sizes should have limited effect on this ability. As the *ad hoc* methods used more sensitive parameters to detect overlaps, we expected them to be more successful in this aspect. Despite this, in our evaluation, FinIS was consistently more effective than IMAGE and GapCloser in reconstructing gap sequences, in particular, closing 100% of the gaps in the *B. subtilis* genome (Fig. 3). On real datasets, GapCloser and FinIS were able to close fewer gaps and on closer inspection this seems to be due to the lack of read data for these gaps. This discrepancy could be due to several reasons, including biological variation in strains and GC-composition-associated sequencing biases.

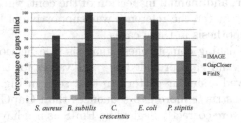

Fig. 3. Percentage of gaps finished *in silico*. Note that the species are sorted by genome size.

A breakdown of these results in terms of gap sizes suggests that the performance of IMAGE and GapCloser is influenced significantly by this factor (Fig. 4). While IMAGE has a limit of ~2Kbp, GapCloser's performance degrades quickly for gap sizes longer than ~3Kbp. In contrast, FinIS does well for short as well as long gaps. For example, in the *E. coli* dataset, FinIS constructed sequences for 12 out of 14 gaps longer than $2kbp$ while GapCloser could only reconstruct 3.

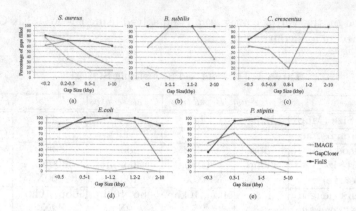

Fig. 4. Size distribution of *in silico* finished gaps. Values plotted show the percentage of gaps in each gap-size bin that were filled by FinIS. Note that the species are sorted by genome size.

3.3 Finishing Correctness

To evaluate the correctness of a reconstructed sequence for a gap, we aligned it with the reference sequence (using MAFFT [26]) and required > 95% similarity and a length difference of < 5%. Based on this criteria, in our evaluation of precision, GapCloser and FinIS performed consistently well, while IMAGE results were more variable (Fig. 5a). As expected, sequences reconstructed by FinIS had very high precision in the simulated datasets (> 90%) and the few incorrect sequences that we saw were likely due to gap size estimates (from Opera) that were slightly off. On real datasets, FinIS still had higher precision (> 85%) than IMAGE and GapCloser, and manual inspection of the contig-graph for gaps where the sequence did not match the reference was supportive of biological variation as an alternative hypothesis in these cases.

Restricting our finishing completeness analysis (see Section 3.2) to uniquely and correctly filled, valid gaps provided us *recall* values for the various methods (Fig. 5b) that followed a similar pattern to that seen in Fig. 3. In all datasets, FinIS correctly reconstructs more gap sequences than GapCloser and IMAGE, and all invalid gaps were correctly flagged by FinIS as not having a gap sequence that met the MIQP criteria.

3.4 Program Sizes and Runtime

To evaluate the utility of the graph trimming and simplification steps, we computed the number of variables and constraints in the MIQP for each contig-graph and at various stages of the process. As can be seen from Table 2, the number of variables and constraints for the largest MIQP typically went down by about 2 orders of magnitude. Overall the largest program solved was for the *P. stipitis* genome and had more than a 1000 constraints and variables.

In terms of overall runtime, FinIS has modest requirements that are comparable to what is needed for GapCloser, on the genomes that we tested (Table 3).

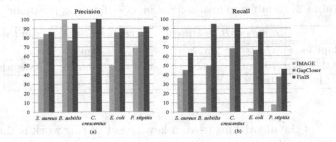

Fig. 5. Finishing performance measured as precision and recall. Note that we use the following definitions: Precision $= \dfrac{\text{\# of correctly finished gaps}}{\text{\# of finished gaps}}$; Recall $= \dfrac{\text{\# of correctly finished gaps}}{\text{\# of valid gaps}}$.

In effect, this demonstrates that the assembly graph information provided by assemblers is sufficient for *in silico* finishing and that computational time can be better spent in analyzing this graph.

Table 2. Size of the MIQP for various datasets. We report the number of variables and constraints in the largest program at each stage of analysis in FinIS.

| | Complete Graph | | Trimmed Graph | | Final Graph | |
Genomes	Variables	Constraints	Variables	Constraints	Variables	Constraints
S. aureus	35090	27199	481	474	478	470
B. subtilis	2300	1806	267	253	230	221
C. crescentus	5976	4847	67	98	52	78
E. coli	20856	15946	393	362	281	266
P. stipitis	529348	396389	1467	2104	1160	1654

4 Discussion

The analysis and results in this work serve to establish the utility of a formal definition for *in silico* finishing and the feasibility of an exact solution for FP in the assembly of microbial genomes. Our experiments confirm the intuition that an exact analysis can simultaneously improve precision and recall of the reconstructed sequences. In addition, the ability to count and, if needed, enumerate equally good solutions can help to confirm the correctness of an assembly and reduce wet-lab finishing efforts.

Note that while for large eukaryotic genomes we are likely to encounter contiggraphs that cannot be solved using existing MIQP solvers, a majority of the gaps are still likely to benefit from this analysis. Where needed, we can also reformulate the MIQP as a semi-definite program (SDP) and exploit the fact that ϵ-approximate solutions can be found for SDPs in time polynomial in the size of the program and $\log(1/\epsilon)$ [28].

Table 3. Runtime comparison for *in silico* finishing programs

	S. aureus	*B. subtilis*	*C. crescentus*	*E. coli*	*P. stipitis*
GapCloser	7m7s	2m57s	1m42s	1m37s	21m57s
IMAGE	10h2m	2h42m	4h35m	4h50m	> 3 days
FinIS	1m29s	7s	6s	8s	26m

Independent of the algorithms used, a key aspect of this work is that it highlights the utility of the contig-graph and a graph-based analysis for assembly validation and finishing. While we have explored one particular approach for analyzing this graph, alternate formulations (such as one where copy-numbers are optimized but gap sizes serve as path constraints) and solutions (such as greedy algorithms) could potentially lead to a more robust approach in practice.

Acknowledgments. This work was supported by the Biomedical Research Council of A*STAR (Agency for Science, Technology and Research) Singapore. S.G. is supported by a NUS graduate scholarship.

References

1. Li, Y., Zheng, H., Luo, R., et al.: Structural variation in two human genomes mapped at single-nucleotide resolution by whole genome *de novo* assembly. Nature Biotechnology 29, 6723–6730 (2011)
2. Birol, I., Jackman, S.D., Nielsen, C.B., et al.: *De novo* transcriptome assembly with ABySS. Bioinformatics 25(21), 2872–2877 (2009)
3. Woyke, T., Teeling, H., Ivanova, N.N., et al.: Symbiosis insights through metagenomic analysis of a microbial consortium. Nature 443, 950–955 (2006)
4. Nagarajan, N., Pop, M.: Sequencing and genome assembly using next-generation technologies. Methods in Molecular Biology 673, 1–17 (2010)
5. Baker, M.: *De novo* genome assembly: what every biologist should know. Nature Methods 9, 333–337 (2012)
6. Nagarajan, N., Pop, M.: Parametric complexity of sequence assembly: theory and applications to next generation sequencing. Journal of Computational Biology 16(7), 897–908 (2009)
7. Gao, S., Sung, W.K., Nagarajan, N.: Opera: reconstructing optimal genomic scaffolds with high-throughput paired-end sequences. Journal of Computational Biology 18(11), 1681–1691 (2011)
8. Pop, M., Kosack, S.D., Salzberg, S.L.: Hierarchical scaffolding with bambus. Genome Research 14, 149–159 (2004)
9. Nagarajan, N., Read, T.D., Pop, M.: Scaffolding and validation of bacterial genome assemblies using optical restriction maps. Bioinformatics 24(10), 1229–1235 (2008)
10. Pop, M., Phillipy, A., Delcher, A.L., Salzberg, S.L.: Comparative genome assembly. Briefings in Bioinformatics 5(3), 237–248 (2004)
11. Nagarajan, N., Cook, C., Bonaventura, M.D., et al.: Finishing genomes with limited resources: lessons from an ensemble of microbial genomes. BMC Genomics 11(242) (2010)

12. Zerbino, D.R., McEwen, G.K., Marguiles, E.H., Birney, E.: Pebble and rock band: heuristic resolution of repeats and scaffolding in the velvet short-read de novo assembler. PLoS ONE 4(12) (2009)
13. Li, R.H., Zhu, J., Ruan, W., et al.: *De novo* assembly of human genomes with massively parallel short read sequencing. Genome Research 20, 265–272 (2010)
14. Tsai, I.J., Otto, T.D., Berriman, M.: Improving draft assemblies by iterative mapping and assembly of short reads to eliminate gaps. Genome Biology 11, R41 (2010)
15. Kececioglu, J.D., Myers, E.W.: Combinatorial algorithms for DNA sequence assembly. Algorithmica 13, 7–51 (1993)
16. Pevzner, P.A., Tang, H., Waterman, M.S.: A Eularian path approach to DNA fragment assembly. Proceedings of the National Academy of Sciences 98(17), 9748–9753 (2001)
17. Myers, E.W.: The fragment assembly string graph. Bioinformatics 21(2), 79–85 (2005)
18. Zerbino, D., Birney, E.: Velvet: algorithms for de novo short read assembly using de Bruijn graphs. Genome Research (2008), doi:10.1101/gr.074492.107
19. Karger, D., Motwani, R., Ramkumar, G.D.S.: On approximating the longest path in a graph. Algorithmica 18, 421–432 (1993)
20. Kleinberg, J.M.: Approximation algorithms for disjoint path problems. Ph.D Thesis, Dept. of EECS. MIT (1996)
21. Fleischner, H.: Algorithms for Eulerian Trails, Eulerian Graphs and Related Topics. Annals of Discrete Mathematics, Part 1 2(50), X.1C13 (1991)
22. Kingsford, C., Schatz, M.C., Pop, M.: Assembly complexity of prokaryotic genomes using short reads. BMC Bioinformatics 11(21) (2010)
23. Richter, D.C., Ott, F., Schmid, R., Huson, D.H.: Metasim: a sequencing simulator for genomics and metagenomics. PloS One 3(10) (2008)
24. Kelley, D.R., Schatz, M.C., Salzberg, S.L.: Quake: quality-aware detection and correction of sequencing errors. Genome Biology 11, R116 (2010)
25. Kurtz, S.A., Phillippy, A., Delcher, A.L., et al.: Versatile and open software for comparing large genomes. Genome Biology 5, R12 (2004)
26. Katoh, K., Misawa, K., Kuma, K., Miyata, T.: MAFFT: a novel method for rapid multiple sequence alignment based on fast Fourier transform. Nucleic Acids Research 30(14) (2002)
27. Jarrod, A.C., Isaac, H., Sirisha, S., Shujun, L., Gary, P.S., Daniel, S.R.: Meraculous: *De Novo* Genome Assembly with Short Paired-End Reads. PLoS ONE 6(8), e23501 (2011), doi:10.1371/journal.pone.0023501
28. Vandenberghe, L., Boyd, S.: Semidefinite Programming. SIAM Review 38, 49–95 (1996)

Lightweight LCP Construction
for Next-Generation Sequencing Datasets

Markus J. Bauer[1], Anthony J. Cox[1],
Giovanna Rosone[2], and Marinella Sciortino[2]

[1] Illumina Cambridge Ltd., United Kingdom
{mbauer,acox}@illumina.com
[2] University of Palermo, Dipartimento di Matematica e Informatica, Italy
{giovanna,mari}@math.unipa.it

Abstract. The advent of "next-generation" DNA sequencing (NGS) technologies has meant that collections of hundreds of millions of DNA sequences are now commonplace in bioinformatics. Knowing the longest common prefix array (LCP) of such a collection would facilitate the rapid computation of maximal exact matches, shortest unique substrings and shortest absent words. CPU-efficient algorithms for computing the LCP of a string have been described in the literature, but require the presence in RAM of large data structures. This prevents such methods from being feasible for NGS datasets.

In this paper we propose the first lightweight method that simultaneously computes, via sequential scans, the LCP and BWT of very large collections of sequences. Computational results on collections as large as 800 million 100-mers demonstrate that our algorithm scales to the vast sequence collections encountered in human whole genome sequencing experiments.

Keywords: BWT, LCP, text indexes, next-generation sequencing datasets, massive datasets.

1 Introduction

The longest common prefix array (LCP) of a string contains the lengths of the longest common prefixes of the suffixes pointed to by adjacent elements of the suffix array (SA) of the string [15]. The most immediate utility of the LCP is to speed up suffix array algorithms and to simulate the more powerful, but more resource consuming, suffix tree. When combined with the suffix array or the Burrows-Wheeler transform (BWT) of a string the LCP facilitates, among other things, the rapid search for maximal exact matches, shortest unique substrings and shortest absent words [14,5,10,1]. Existing algorithms for computing the LCP require data structures of size proportional to the input data to be held in RAM, which has made it impractical to compute the LCP of massive datasets such as the collections of hundreds of millions of reads produced by so-called Next-Generation Sequencing (NGS) technologies.

B. Raphael and J. Tang (Eds.): WABI 2012, LNBI 7534, pp. 326–337, 2012.

In this context, the aim of our paper is designing an algorithm for the computation of the LCP of large collections of strings which works on an external memory system, by performing disk data accesses only via sequential scans, and is lightweight in the sense that its working space requirement is very low.

Computing the LCP of a collection of strings has been considered in the literature [17]. Defining N and K as respectively the sum of the lengths of all strings and the length of the longest string in the collection, the described approach requires $O(N \log K)$ time, but the $O(N \log N)$ bits of internal memory needed to store the collection and its SA in internal memory prevents the method from scaling to massive data.

One can note that several algorithms to compute the LCP of a single string in semi-external memory (see for instance [11]) or directly via BWT (see [5]) could be adapted to solve the problem of computing the LCP of a collection of strings. It could be sufficient to concatenate all the members of the collection into a single string and use distinct end marker symbols as separators. However, assigning a different end marker to each string is not feasible when the collection is very large, but the alternative of terminating each member with the same symbol could lead to LCP values that exceed the lengths of the strings and that depend on the order in which the strings are concatenated. In our approach, we compute the LCP of the collection directly from the strings, without needing to concatenate them and without requiring precomputed auxiliary information such as the SA or BWT of the collection.

In fact, building upon the method of BWT computation introduced in [2], our algorithm adds some lightweight data structures and allows the LCP and BWT of a collection of m strings to be computed simultaneously in $O((m + \sigma^2) \log(N))$ bits of memory, with a worst-case time complexity of $O(K(m + \mathrm{sort}(m)))$, where $\mathrm{sort}(m)$ is the time taken to sort m integers, σ is the size of the alphabet, N is the sum of the lengths of all strings and K is the length of the longest string.

The low memory requirement enables our algorithm to scale to the size of dataset encountered in human whole genome sequencing datasets: in our experiments, we compute the BWT and LCP of collections as large as 800 millions 100-mers.

Section 2 gives preliminaries that we will use throughout the paper, whereas Section 3 describes the sequential computation of the LCP. We present details on the efficient implementation of the algorithm and computational results on real data in Sections 4 and 5, respectively.

2   Preliminaries

Let $\Sigma = \{c_1, c_2, \ldots, c_\sigma\}$ be a finite ordered alphabet with $c_1 < c_2 < \ldots < c_\sigma$, where $<$ denotes the standard lexicographic order. We append to a finite string $w \in \Sigma^*$ an end marker symbol \$ that satisfies \$ $< c_1$. We denote its characters by $w[0], w[1], \ldots, w[k]$, where $k + 1$ is the *length* of w, denoted by $|w|$. Note that, for $i < k$, $w[i] \in \Sigma$ and $w[k] = \$$. A *substring* of a string w is written as $w[i,j] = w[i] \cdots w[j]$, with a substring $w[0,j]$ being called a *prefix*, while a substring $w[i,k]$ is referred to as a *suffix*.

The *suffix array* of a string w is an array SA containing the permutation of the integers $0 \ldots |w| - 1$ that arranges the starting positions of the suffixes of w into lexicographical order. There exist some natural extensions of the suffix array to a collection of sequences (see [17]). We denote by S the collection of m strings $\{w_0, w_1, \ldots, w_{m-1}\}$. We append to each sequence w_i an end marker symbol $\$_i$ smaller than c_1, and $\$_i < \$_j$ if $i < j$. Let us denote by N the sum of the lengths of all strings in S.

Let us denote by $S_{(Pos,Seq)}$ the suffix starting at the position Pos of the string w_{Seq}. We define the *generalized suffix array GSA* of the collection S as the array of N pairs (Pos, Seq), sorted by the lexicographic order of their corresponding suffixes $S_{(Pos,Seq)}$. In particular, $GSA[i] = (t, j)$ is the pair corresponding to the i-th smallest suffix of the strings in S.

The *longest common prefix array* (denoted by LCP) of a collection S of strings is an array storing the length of the longest common prefixes between two consecutive suffixes of S in the lexicographic order. For every $j = 1, \ldots, N - 1$, if $GSA[j-1] = (p_1, p_2)$ and $GSA[j] = (q_1, q_2)$, $LCP[j]$ is the length of the longest common prefix of suffixes starting at positions p_1 and q_1 of the words w_{p_2} and w_{q_2}, respectively. We set $LCP[0] = 0$.

Note that the generalization of the suffix array to a collection S of strings is related to an extension of the notion of the *Burrows-Wheeler transform* to a collection of strings that is a reversible transformation introduced in [12] (see also [13]). Actually, in its original definition, such a transformation produces a string that is a permutation of the characters of all strings in S but it does not make use of any end marker.

In this paper we suppose that a different end marker is appended to each string of S. Let us denote by $BWT(S)$ the Burrows-Wheeler transform of the collection S and its output is produced according to the generalized suffix array of S. In particular, if $GSA[i] = (t, j)$ then $BWT[i] = w_j[(t-1) \bmod |w_j|]$.

Note that the output of $BWT(S)$ differs, for at least m symbols, from BWT applied to the string obtained by concatenating all strings in S. External memory methods for computing $BWT(S)$ are given in [2].

3   LCP Computation of a Collection of Strings via BWT

The main goal of this section is to describe the strategy to compute the LCP of a massive collection of strings via sequential scans of the disk data. In particular, the main theorem of the section enables the simultaneous computation of both LCP and BWT of a collection S of m strings $\{w_0, w_1, \ldots, w_{m-1}\}$. We suppose that the last symbol of each sequence w_i is the end marker $\$_i$. Our method scans all the strings from right to left and both LCP and BWT are incrementally built by simulating, step by step, the insertion of all suffixes having the same length in the generalized suffix array.

We refer to the suffix starting at the position $|w_i| - j - 1$ of a string w_i as its j-*suffix*. With the end marker $\$_i$ included, the j-suffix is of length $j + 1$; the 0-*suffix* contains $\$_i$ alone. Let us denote by $\mathcal{S}_j$ the collection of the j-suffixes of all the strings of $\mathcal{S}$.

Let us denote by K the maximal length of the strings in $\mathcal{S}$ and by $\mathsf{lcp}_j(\mathcal{S})$ the longest common prefix array of the collection $\mathcal{S}_j$. It is easy to see that when $j = K$, $\mathsf{lcp}_j(\mathcal{S})$ coincides with the LCP of $\mathcal{S}$. Since all m end-markers are distinct, the longest common prefix of any pair of the 0-suffixes is 0, so the first m positions into $\mathsf{lcp}_j(\mathcal{S})$ are 0 for any $j \geq 0$.

Note that $\mathsf{lcp}_j(\mathcal{S})$ can be considered to be the concatenation of $\sigma + 1$ arrays $L_j(0), L_j(1), \ldots, L_j(\sigma)$ where, for $h = 1, \ldots, \sigma$, the array $L_j(h)$ is the LCP of the suffixes of $\mathcal{S}_j$ that start with $c_h \in \Sigma$, while $L_j(0)$ (corresponding to the 0-suffixes) is an array of m zeroes. It is easy to see that $\mathsf{lcp}_0(\mathcal{S}) = L_0(0)$ and that $L_0(h)$ is empty for $h > 0$. For sake of simplicity, each segment is indexed starting from 1. We note that, for each $0 < h \leq \sigma$, $L_j(h)[1] = 0$ and $L_j(h)[i] \geq 1$ for $i > 1$.

Similarly, we define the string $\mathsf{bwt}_j(\mathcal{S})$ as the Burrows-Wheeler transform of the collection of the j-suffixes of $\mathcal{S}$. This can be partitioned in an analogous way into segments $B_j(0), B_j(1), \ldots, B_j(\sigma)$, where the symbols in $B_j(0)$ are the characters preceding the lexicographically sorted 0-suffixes of $\mathcal{S}_j$ and the symbols in $B_j(h)$, with $h \geq 1$, are the characters preceding the lexicographically sorted suffixes of $\mathcal{S}_j$ starting with $c_h \in \Sigma$. Moreover, $\mathsf{bwt}_0(\mathcal{S}) = B_0(0)$ and the segments $B_0(h)$ are empty for $h > 0$.

In this section we show that, for each $j > 0$, $\mathsf{lcp}_j(\mathcal{S})$ can be sequentially constructed by using $\mathsf{bwt}_{j-1}(\mathcal{S})$ and $\mathsf{lcp}_{j-1}(\mathcal{S})$ (in previous work [2], three of the present authors showed how $\mathsf{bwt}_j(\mathcal{S})$ may be computed from $\mathsf{bwt}_{j-1}(\mathcal{S})$). Note that $\mathsf{bwt}_0(\mathcal{S})$ and $\mathsf{lcp}_0(\mathcal{S})$ are defined above.

Given the segments $B_j(h)$ and $L_j(h)$, $h = 0, \ldots, \sigma$, for the symbol x occurring at position r of $B_j(h)$ we define the (j, h)-*LCP Current Interval* of x in r (denoted by $LCI_j^h(x, r)$) as the range $(d_1, r]$ in $L_j(h)$, where d_1 is the greatest position smaller than r of the symbol x in $B_j(h)$, if such a position exists. If such a position does not exist, we define $LCI_j^h(x, r) = L_j(h)[r]$. Analogously, we define for the symbol x the (j, h)-*LCP Successive Interval* of x in r (denoted by $LSI_j^h(x, r)$) as the range $(r, d_2]$ in $L_j(h)$, where d_2 is the smallest position greater than r of the symbol x in $B_j(h)$, if it exists. If such a position does not exist we define $LSI_j^h(x, r) = L_j(h)[r]$. In our notation, a range is delimited by a square bracket if the correspondent endpoint is included, whereas the parenthesis means that the endpoint of the range is excluded.

Actually, it is easy to verify that $d_1 = \mathsf{select}(\mathsf{rank}(x, r) - 1, x)$ and $d_2 = \mathsf{select}(\mathsf{rank}(x, r) + 1, x)$, where $\mathsf{rank}(x, r)$ counts the number of x's until position r and $\mathsf{select}(p, x)$ finds the position of the p-th occurrence of x in a segment B_j.

The following theorem shows how to compute the segments $L_j(h)$, with $j > 0$, by using $L_{j-1}(h)$ and $B_{j-1}(h)$ for any $h > 0$. We denote by $\mathrm{Suf}_j(0)$ the lexicographically sorted 0-suffixes and by $\mathrm{Suf}_j(h)$, for $h > 0$, the lexicographically sorted t-suffixes of $\mathcal{S}_j$, with $t \leq j$ starting with $c_h \in \Sigma$.

Theorem 1. *Let* $\mathcal{I} = \{r_0 < r_1 < \ldots < r_{q-1}\}$ *be the set of the positions in* $Suf_j(z)$ *of the j-suffixes starting with the letter c_z. For each position $r_p \in \mathcal{I}$ $(0 \le p < q)$,*

$$L_j(z)[r_p] = \begin{cases} 0 & \text{if } r_p = 1 \\ 1 & \text{if } r_p > 1 \text{ and } LCI^v_{j-1}(c_z, t) = L_{j-1}(v)[t] \\ \min LCI^v_{j-1}(c_z, t) + 1 & \text{otherwise} \end{cases}$$

where c_v is the first character of the $(j-1)$-suffix of w_i, and t is the position in $B_{j-1}(v)$ *of symbol c_z preceding the $(j-1)$-suffix of w_i.*

For each position $(r_p + 1) \notin \mathcal{I}$ (where $r_p \in \mathcal{I}$ and $0 \le p < q$), then

$$L_j(z)[r_p + 1] = \begin{cases} 1 & \text{if } LSI^v_{j-1}(c_z, t) = L_{j-1}(v)[t] \\ \min LSI^v_{j-1}(c_z, t) + 1 & \text{otherwise} \end{cases}$$

For each position s, where $1 \le s < r_p$ (for $p = 0$), $r_{p-1} < s < r_p$ (for $0 < p < q - 1$), $s > r_p$ (for $p = q - 1$) then

$$L_j(z)[s] = L_j(z)[s - p]$$

For lack of space, the proof of the theorem is omitted and we defer it in the full paper.

A consequence of the theorem is that the segments B_j and L_j can be constructed sequentially and stored in external files. This fact will be used in the next section.

4   Lightweight Implementation via Sequential Scans

Based on the strategy described in the previous section, here we propose an algorithm (named extLCP) that simultaneously computes the BWT and LCP of a set of strings $\mathcal{S}$. Memory use is minimized by reading data sequentially from files held in external memory: only a small proportion of the symbols of $\mathcal{S}$ need to be held in RAM and we do not need to keep the generalized suffix array of $\mathcal{S}$. Obviously, the generalized suffix array can be a side output of our implementation.

Our method extends previous work [2,3] on computing the BWT of a collection of strings and we follow the notation therein.

Although our algorithm is not restricted to collections of strings of uniform length, for sake of simplicity our description supposes that $\mathcal{S}$ comprises m strings of length k and we assume that $j = 0, 1, \ldots, k$ and $i = 0, 1, \ldots, m - 1$. We simulate m distinct end-markers by using a single end-marker $\$ = c_0$ and setting $w_s[k] < w_t[k]$ if and only if $s < t$, so that if two strings w_s and w_t share the j-suffix, then $w_s[k - j, k] < w_t[k - j, k]$ if and only if $s < t$. Moreover, we assume that the values of $\mathsf{lcp}_j(\mathcal{S})$ do not exceed j and the first m positions into $\mathsf{lcp}_j(\mathcal{S})$ are 0 for any $j \ge 0$.

The main part of the algorithm consists of k consecutive iterations. At iteration j, we consider all the j-suffixes of S and simulate their insertion in the GSA. In other words, for each i, we have to find the position of the suffix $w_i[k-j,k]$ according to the lexicographic order of all the suffixes of S of length at most j, then insert the new symbol circularly preceding the j-suffix of w_i into $B_j(z)$, where $c_z = w_i[k-j]$, for some $z = 1, \ldots, \sigma$, and update the values in $L_j(z)$. Consequently, both $\mathsf{bwt}_j(S)$ and $\mathsf{lcp}_j(S)$ are updated accordingly. Note that, at each iteration j, both the segments B_j and L_j, initially empty, are stored in different external files that replace the files used in the previous iteration.

In order to compute $\mathsf{bwt}_j(S)$ and $\mathsf{lcp}_j(S)$, the algorithm needs to hold six arrays of m integers in internal memory. Four of these (P_j, Q_j, N_j and U_j) are useful to compute the BWT (see [2]), a further two (C_j and S_j) are needed to compute and update the values of the longest common prefixes. we will give a description of all these arrays but, for brevity, we will focus on the computation of C_j and S_j.

Each of the arrays P_j, Q_j, N_j and U_j contains m elements, as detailed in the following. At the end of iteration j, if $w_i[k-j,k]$ is the q-th j-suffix then:

- $N_j[q]$ contains the index i. It uses $O(m \log m)$ bits of workspace.
- $Q_j[q]$ stores the index z where $c_z = w_i[k-j]$, i.e. the first symbol of the j-suffix. It uses $O(m \log \sigma)$ bits of workspace.
- $P_j[q]$ contains the position in $B_j(z)$ of the symbol circularly preceding the j-suffix $w_i[k-j,k]$, such a symbol is $w_i[k-j-1]$ and it is stored at the position $N_j[q]$ of U_j. So, it needs $O(m \log(mk))$ bits of workspace.
- U_j stores the new characters to be inserted, one for each sequence in S, so it uses $O(m \log \sigma)$ bits of workspace.

The arrays C_j and S_j each contain m integers useful to compute $\mathsf{lcp}_j(S)$ for $j > 0$. In particular, $C_j[q]$ stores the value in LCP between the j-suffix $w_i[k-j,k]$ and the previous suffix in the GSA with respect to the lexicographic order of all the suffixes of S of length at most j, whereas $S_j[q]$ contains the value in LCP between the j-suffix and the next suffix in GSA (if it exists). Such values will be computed at the iteration $j-1$ according to Theorem 1. We observe that C_j and S_j contain exactly one integer for each sequence in the collection and they use $O(m \log k)$ bits of workspace.

At the iteration $j = 0$, the algorithm initializes the segments B_0 and L_0 as described in previous section, i.e. $B_0(0) = w_0[k-1]w_1[k-1] \cdots w_{m-1}[k-1]$ and for each $q = 0, \ldots, m-1$, we set $L_0(0)[q] = 0$. Consequently, for $q = 0, \ldots, m-1$, the arrays are initialized by setting $N_0[q] = q$, $P_0[q] = q+1$, $Q_0[q] = 0$, $C_1[q] = 1$ and $S_1[q] = 1$.

For I/O efficiency, each iteration $j > 0$ can be divided into two consecutive phases: during the first one we read only the segments B_{j-1} in order to find the arrays P_j, Q_j N_j and U_j. In phase 2, the segments B_{j-1} and L_{j-1} are read once sequentially both for the construction of new segments B_j and L_j and for the computation of the arrays C_{j+1} and S_{j+1}, as they will be used in the next iteration. In the following we describe both the phases of the generic iteration

$j > 0$. Figure 1 illustrates the execution of the algorithm for a simple collection at the iterations 12 and 13.

In the first phase the arrays P_j, Q_j and N_j are computed. In particular, if $w_i[k - j - 1]$ (or the end marker $ for the last step) is the new symbol to be inserted, its position r is obtained by computing the number of occurrences of $c_z = w_i[k - j]$ in $B_{j-1}(0), \ldots, B_{j-1}(v - 1)$ and in $B_{j-1}(v)[1, t]$, where $c_v = w_i[k - (j - 1)]$ and t is the position of c_z in $B_{j-1}(v)$. Hence, the index z is stored into Q_j at some position q, the computed position r (where storing the new symbol) is added to the array $P_j[q]$ and i is added to $N_j[q]$. Note that in order to find the positions, a table of $O(\sigma^2 \log(mk))$ bits of memory is used. Finally we sort Q_j, P_j, N_j, C_j, S_j where the first and the second keys of the sorting are the values in Q_j and P_j respectively.

Here we focus on the second phase in which the computation of the segments L_j is performed by using the arrays C_j and S_j constructed during the previous step. Note that the sorting of the arrays allows us to open and sequentially read the pair files ($B_{j-1}(h)$ and $L_{j-1}(h)$ for $h = 0, \ldots, \sigma$) at most once.

For all symbols in U_j that we have to insert in the segment $B_j(h)$, the crucial point is to compute C_{j+1} by using LCI_j^h and S_{j+1} by using LSI_j^h while the new files are being constructed, instead of using auxiliary data structures to compute rank and select. For each index z, we consider all the elements in Q_j equal to z. Because of the sorting, such elements are consecutive. Let $0 \leq l, l' \leq m - 1$ be their first and the last positions, respectively. Hence for each $l \leq p \leq l'$, we have $Q_j[p] = z$ and $P_j[l] < \ldots < P_j[l']$. In order to apply Theorem 1, we need to compute LCI_j^h and LSI_j^h of each new symbol in its new position. Since each $B_j(h)$ and $L_j(h)$ are constructed sequentially, we do not know a priori the opening positions LCI_j^h and the closing positions LSI_j^h that are used to compute C_{j+1} and S_{j+1}. However we can observe that when we write a symbol x into $B_j(h)$, its occurrence could be the opening or closing positions of some LCI_j^h and LSI_j^h of x, if x is a new symbol. Such considerations are outlined in detail in the following.

For each symbol α that we insert at position s in $B_j(z)$, with $1 \leq s < P_j[l]$, it is easy to see that $B_j(z)[s] = B_{j-1}(z)[s]$ and $L_j(z)[s] = L_{j-1}(z)[s]$. Moreover, the position of α could be the opening position of $LCI_j^z(\alpha, y)$, if α is the new symbol that will be inserted at some next position y.

For each new symbol β that we insert at position $P_j[q]$ in $B_j(z)$ ($l \leq q \leq l'$), we have $\beta = U_j[N_j[q]]$ and, by Theorem 1, it follows that $L_j(z)[P_j[q]] = 0$ if $P_j[q] = 1$ or $L_j(z)[P_j[q]] = C_j[q]$ otherwise. Moreover:

- The position $P_j[q]$ surely is the closing position of $LCI_j^z(\beta, P_j[q])$. If the position $P_j[q]$ is the first occurrence of β in $B_j(z)$, then $LCI_j^z(\beta, P_j[q]) = L_j(z)[P_j[q]]$ and we set $C_{j+1}[q] = L_j(z)[P_j[q]] + 1$ according to Theorem 1. Otherwise, we set $C_{j+1}[q] = \min(LCI_j^z(\beta, P_j[q])) + 1$, whose computation has been started when the interval was opened.
- The position $P_j[q]$ could be the opening position of $LCI_j^z(\beta, y)$, if β will be inserted, as new symbol, at some next position y.

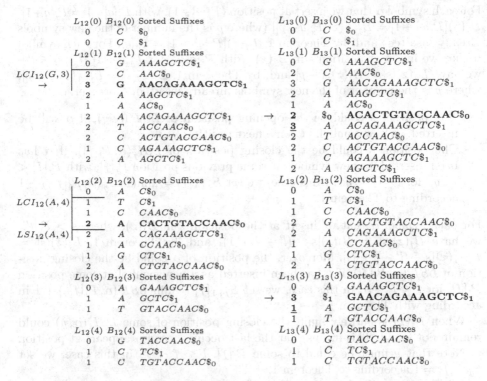

Fig. 1. Iteration 12 (on the left) and iteration 13 (on the right) on the collection $S = \{ACACTGTACCAAC, GAACAGAAAGCTC\}$. We append different end-marker to each string ($\$_0$ and $\$_1$, respectively) to make the explanation more immediate but the same situation would occur using the same symbol. The first two columns represent the partial LCP and the partial BWT after the iterations. The positions of the new symbols corresponding to the 13-suffixes (shown in bold on the right) are computed from the positions of the 12-suffixes (in bold on the left), which were retained in the array P after the iteration 12. The new values in LCP (shown in bold on the right) are computed during the iteration 12 and are contained in C_{12}. The updated values in LCP (shown in bold and underlined on the right) are computed during the iteration 12 and are contained in S_{12}.

- The position $P_j[q]$ could be the closing position of $LSI_j^z(\beta, y)$, where y represents, eventually, the largest position $P_j[f]$, with $P_j[f] < P_j[q]$, for $l \leq f < q$, where β has been inserted. In this case, we set $S_{j+1}[f] = \min(LSI_j^z(\beta, P_j[q])) + 1$ in according with Theorem 1.

- The position $P_j[q]$ surely is the opening position of $LSI_j^z(\beta, P_j[q])$. We observe that if the position $P_j[q]$ is the last occurrence of β in $B_j(z)$ (we will discover this at the end of the file), it means that $LSI_j^z(\beta, P_j[q]) = L_j(z)[P_j[q]]$, i.e. $S_{j+1}[q] = 1$.

For each symbol α that we insert at position $(P_j[q]+1)$, with $P_j[q]+1 \neq P_j[q+1]$, $B_j(z)[P_j[q]+1] = B_{j-1}(z)[P_j[q-p]]$ (where p is the number of the new symbols already inserted) and, by Theorem 1, $L_j(z)[P_j[q]+1] = S_j[q]$. For each symbol α that we insert at position s in $B_j(z)$, with $P_j[q] < s < P_j[q+1]$ $(l < q \leq l')$, we have $B_j(z)[s] = B_j(z)[s-p]$ and, by Theorem 1, $L_j(z)[s] = L_{j-1}(z)[s-p]$, where p is the number of the new symbols already inserted. Moreover:

- The position s could be the opening position of $LCI_j^z(\alpha, y)$, if α will be inserted, as new symbol, at some next position y.
- The position s could be the closing position of $LSI_j^z(\alpha, P_j[f])$, if α has been inserted, as new symbol, at some previous position $P_j[f]$, with $P_j[f] < P_j[q]$, for $l \leq f < q$. In this case, we set $S_{j+1}[f] = \min(LSI_j^z(\alpha, P_j[f])) + 1$ according to Theorem 1.

For each symbol α that we insert at the position s in $B_j(z)$, where $s > P_j[l']$, we have $B_j(z)[s] = B_j(z)[s - (l'-l+1)]$ and, by Theorem 1, $L_j(z)[s] = L_{j-1}(z)[s - (l'-l+1)]$. Moreover, the position of α could be the closing position of $LSI_j^z(\alpha, P_j[f])$, if α has been inserted as a new symbol at some position $P_j[f]$, for $l \leq f \leq l'$. In this case, we set $S_{j+1}[f] = \min(LSI_j^z(\alpha, P_j[f])) + 1$ in according with Theorem 1.

When $B_j(z)$ is entirely built, the closing position of some $LSI_j^z(\alpha, y)$ could remain not found. This means that the last occurrence of α appears at position y. Note that y must be equal to some $P_j[f]$, $l \leq f \leq l'$. In this case, we set $S_{j+1}[f] = 1$ according to Theorem 1.

It is easy to verify that we can run these steps in a sequential way. Moreover, one can deduce that, while the same segment is considered, for each symbol $\alpha \in \Sigma$ at most one $LCI_j^h(\alpha, t)$ for some t, and at most one $LSI_j^h(\alpha, r)$ for some $r \leq t$, will have not their closing position. For this reason we use two arrays $minLCI$ and $minLSI$ of σ integers that store, for each symbol α in Σ, the minimum among the values of LCP in the possible corresponding LCI or LSI without closing position, respectively.

From the size of the data structures and from the above description of the phases of the extLCP algorithm, we can state the following theorem.

Theorem 2. *Given a collection S of m strings of length k over an alphabet of size σ, the* extLCP *algorithm computes BWT and LCP of S by using $O(mk^2 \log \sigma)$ disk I/O and $O((m + \sigma^2) \log(mk))$ bits of memory in $O(k(m + sort(m)))$ CPU time, where $sort(m)$ is the time taken to sort m integers.*

The following corollary describes the performance of the method when the collection contains strings of different length.

Corollary 1. *Given a collection S of m strings over an alphabet of size σ, the LCP and BWT of S are computed simultaneously in $O((m + \sigma^2) \log(N))$ bits of memory, with a worst-case time complexity of $O(K(m + sort(m)))$, where $sort(m)$ is the time taken to sort m integers, N is the sum of the lengths of all strings and K is the length of the longest string.*

5   Computational Experiments and Discussion

To assess the performance of our algorithm on real data, we used a publicly available collection of human genome sequences from the Sequence Read Archive [8] at ftp://ftp.sra.ebi.ac.uk/vol1/ERA015/ERA015743/srf/ and created subsets containing 43, 85, 100, 200 and 800 million reads, each read being 100 bases in length. We developed extLCP, an implementation of the algorithm described in Section 4, which is available upon request from the authors. Our primary goal was to analyze the additional overhead in runtime and memory consumption of simultaneously computing both BWT and LCP via extLCP compared with the cost of using BCR ([2]) to compute the BWT alone.

Table 1 shows the results for the instances that we created. We do see increase in runtime since extLCP writes the values of LCP after that the symbols in BWT are written, so it effectively increases the I/O operations. So, a time optimization could be obtained if we read/write at the same time both the elements in BWT and LCP by using two different disks. All tests except the 800 million read instance were done on the same machine, having 16Gb of memory and two quad-core Intel Xeon E5450 3.0GHz processors. Although the LCP of a collection of 700 million 100-mers was successfully computed on the same machine using 15Gb of RAM, the collection of 800 million reads needed slightly more than 16Gb so was processed on a machine with 64Gb of RAM and four quad-core Intel Xeon E7330 2.4GHz processors. On both machines, only a single core was used for the computation. Moreover, to examine the behavior of our algorithm on reads longer than 100bp, we created a set of 50 million 200bp long reads based on the 100 million 100bp instance. It turns out that, although the sheer data volume is the same, extLCP uses 1.2Gb and takes 10.3 microseconds per input base.

Our algorithm represents the first lightweight method that simultaneously computes, via sequential scans, the LCP and BWT of a vast collection of sequences. Recall that the problem of the LCP computation of a collection of strings has been faced in [17], but such a strategy works in internal memory. Recently, however, some lightweight approaches for the LCP computation of a single string were described in the literature. Some of them use of the suffix array of the string [11,7,15,9], but the space needed to hold this in RAM is prohibitive for NGS datasets. However, in [5], the authors give an algorithm for the construction of the LCP of a string that acts directly on the BWT of the string and does not need its suffix array. A memory-optimized version of this algorithm [4] (called bwt_based_laca2) needs to hold the BWT of the string in internal memory plus a further $1.5n$ bytes, where n is the length of the input string.

Notice that an entirely like-for-like comparison between our implementation and the above existing implementation for BWT and LCP computation of a string would imply the concatenation of the strings of the collection by different end markers. However, for our knowledge, the existing implementations do not support the many millions of distinct end markers our test collections would require.

An alternative is to concatenate each of strings with the same end marker. This leads to values in the LCP that may possibly exhibit the undesirable properties

Table 1. The input string collections were generated on an Illumina GAIIx sequencer, all reads are 100 bases long. Size is the input size in gigabytes, wall clock time—the amount of time that elapsed from the start to the completion of the instance—is given as microseconds per input base, and memory denotes the maximal amount of memory (in gigabytes) used during execution. The efficiency column states the CPU efficiency values, i.e. the proportion of time for which the CPU was occupied and not waiting for I/O operations to finish, as taken from the output of the /usr/bin/time command.

instance	size	program	wall clock	efficiency	memory
0043M	4.00	BCR	0.99	0.84	0.57
	4.00	extLCP	3.29	0.98	1.00
0085M	8.00	BCR	1.01	0.83	1.10
	8.00	extLCP	3.81	0.87	2.00
0100M	9.31	BCR	1.05	0.81	1.35
	9.31	extLCP	4.03	0.83	2.30
0200M	18.62	BCR	1.63	0.58	4.00
	18.62	extLCP	4.28	0.79	4.70
0800M	74.51	BCR	3.23	0.43	10.40
	74.51	extLCP	6.68	0.67	18.00

of exceeding the lengths of the strings and depending on the order in which the strings are concatenated, but does allow the BWT of the resulting string to be computed in external memory by using the algorithm bwte proposed in [6].

The combined BWT/LCP computation provided by extLCP has a faster runtime than bwte. In particular, for the $0085M$ instance, bwte uses 14Gb of memory and needs 3.84 microseconds per input base vs extLCP that uses 2Gb of memory and 3.81 microseconds per input base.

We have also used BCR by suitable preprocessing steps, to simulate the computation of the BWT of the concatenated strings. We compared BCR, extLCP and bwt_based_laca2 on the $0200M$ instance. Since the memory consumption of bwt_based_laca2 exceeded 16Gb on this dataset, we ran the tests on a machine of identical CPU to the 16Gb machine, but with 64Gb RAM.

With BCR, the BWT was created in under 5 hours of wallclock time taking only 4Gb of RAM, while bwt_based_laca2 required 18Gb of RAM to create the LCP in about 1 hour 45 minutes. Our new method extLCP needed 4.7Gb of RAM to create both BWT and LCP in just under 18 hours. Attempting to use bwt_based_laca2 to compute the LCP of the $0800M$ instance exceeded the available RAM on the 64Gb RAM machine.

The experimental results show that our algorithm is a competitive tool for the lightweight simultaneous computation of LCP and BWT on the string collections produced by NGS technologies. Actually, the LCP and BWT are two of the three data structures needed to build a compressed suffix tree (CST) [16] of a string. The strategy proposed in this paper could enable the lightweight construction of CSTs of strings collections for comparing, indexing and assembling vast datasets of sequences when memory is the main bottleneck. Our current prototype can be further optimized in terms of memory by performing the sorting step in external memory. Further saving of the working space could be obtained if we embody our strategy in BCRext or BCRext++ (see [3]). These methods, although slower than BCR, need to store only a constant and (for the DNA alphabet) negligibly small number of integers in RAM regardless of the size of the input data.

References

1. Abouelhoda, M.I., Kurtz, S., Ohlebusch, E.: Replacing suffix trees with enhanced suffix arrays. Journal of Discrete Algorithms 2(1), 53–86 (2004)
2. Bauer, M.J., Cox, A.J., Rosone, G.: Lightweight BWT Construction for Very Large String Collections. In: Giancarlo, R., Manzini, G. (eds.) CPM 2011. LNCS, vol. 6661, pp. 219–231. Springer, Heidelberg (2011)
3. Bauer, M.J., Cox, A.J., Rosone, G.: Lightweight algorithms for constructing and inverting the bwt of string collections. Theor. Comput. Sci. (in press, 2012)
4. Beller, T., Gog, S., Ohlebusch, E., Schnattinger, T.: Computing the longest common prefix array based on the Burrows-Wheeler transform. Journal of Discrete Algorithms (to appear)
5. Beller, T., Gog, S., Ohlebusch, E., Schnattinger, T.: Computing the Longest Common Prefix Array Based on the Burrows-Wheeler Transform. In: Grossi, R., Sebastiani, F., Silvestri, F. (eds.) SPIRE 2011. LNCS, vol. 7024, pp. 197–208. Springer, Heidelberg (2011)
6. Ferragina, P., Gagie, T., Manzini, G.: Lightweight Data Indexing and Compression in External Memory. In: López-Ortiz, A. (ed.) LATIN 2010. LNCS, vol. 6034, pp. 697–710. Springer, Heidelberg (2010)
7. Fischer, J.: Inducing the LCP-Array. In: Dehne, F., Iacono, J., Sack, J.-R. (eds.) WADS 2011. LNCS, vol. 6844, pp. 374–385. Springer, Heidelberg (2011)
8. National Center for Biotechnology Information. Sequence Read Archive, http://trace.ncbi.nlm.nih.gov/Traces/sra/sra.cgi?
9. Gog, S., Ohlebusch, E.: Fast and Lightweight LCP-Array Construction Algorithms. In: ALENEX, pp. 25–34. SIAM (2011)
10. Herold, J., Kurtz, S., Giegerich, R.: Efficient computation of absent words in genomic sequences. BMC Bioinformatics 9(1), 167 (2008)
11. Kärkkäinen, J., Manzini, G., Puglisi, S.J.: Permuted Longest-Common-Prefix Array. In: Kucherov, G., Ukkonen, E. (eds.) CPM 2009 Lille. LNCS, vol. 5577, pp. 181–192. Springer, Heidelberg (2009)
12. Mantaci, S., Restivo, A., Rosone, G., Sciortino, M.: An extension of the Burrows-Wheeler Transform. Theor. Comput. Sci. 387(3), 298–312 (2007)
13. Mantaci, S., Restivo, A., Rosone, G., Sciortino, M.: A new combinatorial approach to sequence comparison. Theory Comput. Syst. 42(3), 411–429 (2008)
14. Ohlebusch, E., Gog, S., Kügel, A.: Computing Matching Statistics and Maximal Exact Matches on Compressed Full-Text Indexes. In: Chavez, E., Lonardi, S. (eds.) SPIRE 2010. LNCS, vol. 6393, pp. 347–358. Springer, Heidelberg (2010)
15. Puglisi, S., Turpin, A.: Space-Time Tradeoffs for Longest-Common-Prefix Array Computation. In: Hong, S.-H., Nagamochi, H., Fukunaga, T. (eds.) ISAAC 2008. LNCS, vol. 5369, pp. 124–135. Springer, Heidelberg (2008)
16. Sadakane, K.: Compressed suffix trees with full functionality. Theor. Comp. Sys. 41(4), 589–607 (2007)
17. Shi, F.: Suffix Arrays for Multiple Strings: A Method for On-line Multiple String Searches. In: Jaffar, J., Yap, R.H.C. (eds.) ASIAN 1996. LNCS, vol. 1179, pp. 11–22. Springer, Heidelberg (1996)

Sign Assignment Problems on Protein Networks

Shay Houri and Roded Sharan

Blavatnik School of Computer Science, Tel Aviv University, Tel Aviv 69978, Israel
{shayhour,roded}@post.tau.ac.il.

Abstract. In a maximum sign assignment problem one is given an undirected graph and a set of signed source-target vertex pairs. The goal is to assign signs to the graph's edges so that a maximum number of pairs admit a source-to-target path whose aggregate sign (product of its edge signs) equals the pair's sign. This problem arises in the annotation of physical interaction networks with activation/repression signs. It is known to be NP-complete and most previous approaches to tackle it were limited to considering very short paths in the network. Here we provide a sign assignment algorithm that solves the problem to optimality by reformulating it as an integer program. We apply our algorithm to sign physical interactions in yeast and measure our performance using edges whose activation/repression signs are known. We find that our algorithm achieves high accuracy (89%), outperforming a state-of-the-art method by a significant margin.

Keywords: network annotation, protein-protein interaction, activation, repression, integer linear program.

1 Introduction

A holy grail of biological research is obtaining a working model of the cell. Protein-protein interactions (PPIs) form the skeleton of signal processing circuitry. Despite their importance, current models of PPI networks are mostly topological, lacking the underlying logic of those circuits. In this paper we tackle combinatorial problems arising in the annotation of PPI networks with signs of activation/repression. Constructing such an annotation is a key step in deciphering the logic of those networks.

Current technologies for PPI mapping do not provide information on the direction of signal flow or the activation/repression effect of the measured interactions. Such information can be gained from additional, indirect data such as gene expression knockout experiments. The latter pinpoint pairs of genes such that the deletion of a gene (cause) leads to a change in the expression level of the other gene (effect) with a certain sign (down- or up-regulation). Yeang et al. [7] were the first to suggest a computational framework for annotating networks with directions and signs given cause-effect data. Later work by Ourfali et al. [4] formulated the sign assignment problem as an integer programming problem, aiming to maximize the expected number of cause-effect pairs that can be explained. The main caveat of these works is the need to consider an

B. Raphael and J. Tang (Eds.): WABI 2012, LNBI 7534, pp. 338–345, 2012.

exponential space of paths in the network during the optimization process, necessitating heuristic approaches that focus on very short paths (with at most 3 edges) and miss information contained in longer ones. Finally, Peleg et al. [5] gave an algorithm that assigns signs to nodes in a manner that is independent of a PPI network; these signs are then used in order to predict edge signs in the network. Their algorithm was shown to compare favorably to the previous sign assignment algorithms of [7,4].

Here we revisit the problem of assigning signs to the edges of the PPI network so as to best explain a given cause-effect data set. Following previous work, we consider a cause-effect pair of a certain sign to be *explained* by an assignment if the network contains a path from the cause to the effect whose aggregate sign (product of the signs along its edges) is opposite (due to the knockout) to the sign of the pair. Our goal is to assign signs to the edges of the network so that a maximum number of cause-effect pairs can be explained. We study the resulting Maximum Sign Assignment (MSA) problem, which was shown to be NP-hard in [5]. We provide a polynomial 0.878-approximation algorithm for it and for a constrained variant where some of the edges are pre-assigned with signs. We further provide an integer programming formulation of the problem that allows us to solve it to optimality on current networks. We apply our algorithm to annotate a network of physical interactions in yeast, obtaining high success rates against known sign data. In comparison to our previous state-of-the-art method by Peleg et al. [5] the current approach attains significantly higher accuracy levels (89% vs. 68-73%).

2   Preliminaries

Let $G = (V, E)$ be an undirected graph, representing a protein network, with a set V of vertices and a set E of edges. We assume that each edge $e \in E$ is associated with a sign $s(e) \in \{-1, +1\}$ (which may be unknown) describing its activation/repression effect. For a path P, we define its sign as the product of the signs of its edges (assuming their signs are known), i.e., $s(P) = \prod_{e \in P} s(e)$. Our goal is to infer the edge signs from knockout data which can be summarized as triples of a knockout gene u, an affected gene v and the effect sign s. Our assumption, following [7], is that each such triple can be explained by a path in G from u to v whose sign is $-s$ (and due to the knockout the observed effect is s). The problem is formally stated as follows:

Definition 1 (Maximum Sign Assignment (MSA)). *Given an undirected graph $G = (V, E)$ and a collection of signed vertex pairs $\{(u, v, s)\}_{u,v \in V, s \in \{+1,-1\}}$, assign signs to the edges of G so that a maximum number of signed pairs (u, v, s) admit a path whose sign is s.*

In [5] it is shown that MSA is NP-complete and hard to approximate to within a factor of 11/12. In the next section we shall develop approximation algorithms for the problem and a constrained variant of it.

We assume that the input graph $G = (V, E)$ is connected, otherwise one can operate independently on each of its connected components. We note that any cycle in G can be contracted without affecting the optimum solution. This follows from the observation that by assigning -1 to one of its edges and +1 to the remaining edges, every pair of vertices that admit a path that visits this cycle can be satisfied. Thus, we may assume, w.l.o.g., that G is a tree. Given a sign assignment to the edges of G, we say that two vertices admit a *positive path* (resp., negative path) between them if there exists a path connecting the vertices whose sign is +1 (resp., -1).

3   An Approximation Algorithm for MSA

We approximate the problem by reducing it into a MAX-E2-LIN2 problem. In MAX-E2-LIN2 the input consists of a set of linear equations over Z_2 with at most two variables per equation; the goal is to find a solution that maximizes the number of satisfied equations. MAX-E2-LIN2 is known to be hard to approximate to within 11/12. On the positive side, it can be 0.878-approximated using the semi-definite programming approach of Geomans and Williamson [2].

Given an instance of MSA, we reduce it by representing every signed pair (u, v, s) as a linear equation over Z_2: $x_u \oplus x_v = f(s)$, where $f(s) = (1 - s)/2$.

Theorem 1. *The reduction is approximation preserving.*

Proof. We prove that there exists a solution of the MAX-E2-LIN2 instance that satisfies k equations iff there is a solution to the MSA instance that satisfies k of the pairs. Let X be a solution to the MAX-E2-LIN2 instance which satisfies k of the input equations. X induces a partition of V into two parts: $\{v \in V : X_v = 1\}$ and $\{v \in V : X_v = 0\}$. Vertices that did not participate in the equations are arbitrarily assigned to one of the parts. Now, we can sign the edges as follows: edges that cross between the two parts are assigned a minus sign; the rest are assigned a plus sign. It is easy to check that under this assignment (and since G is connected) every pair (u, v, s) whose corresponding equation was satisfied admits a path of sign s.

Conversely, let S be a sign assignment which satisfies k pairs. Take an arbitrary vertex v and set $X_v = 0$. For any other vertex $u \in V$ that has a positive path to v, set $X_u = 0$; otherwise, u must have a negative path to v and we set $X_u = 1$. Since G is a tree, every two vertices have a unique path between then, so its sign is well defined. It is again easy to check that the proposed assignment satisfies at least k equations (corresponding to the k satisfied pairs).

Corollary 1. *MSA can be 0.878-approximated.*

4   Dealing with Assignment Constraints

The discussion thus far assumed that any assignment is legal. In practice, some edge signs are known in advance. Under such constraints the problem cannot be reduced to a tree anymore, since some cycles cannot be contracted without affecting the optimum solution. For a given sign assignment, a pair of vertices are called *ambiguous* if they admit both a positive and a negative path between them. In the following we call a graph *strongly signed* if there exists an assignment to its yet unsigned edges such that every pair of vertices is ambiguous.

Lemma 1. *A cycle is strongly signed iff it admits an assignment whose aggregate sign is -1.*

Proof. If a cycle is strongly signed then by definition the product of signs along its edges is -1, as every two vertices on the cycle have exactly two paths connecting them with one being negatively-signed and the other being positively-signed.

Conversely, if the product of edge signs along the cycle is -1 then every two vertices on the cycle are ambiguous.

Our algorithm for the constrained case relies on decomposing the input graph to its blocks. Recall that a *block* is a 2-vertex connected component. A block may contain a single edge, or else it contains a cycle. Any block of size at least 3 admits an *open ear decomposition* of its edges $(P_0, \ldots, P_k)$. P_0 may be any cycle of the block. Each $P_i, i > 0$ is a path whose two end-points are distinct and included in previous ears $P_j, j < i$.

Lemma 2. *A block G of size at least 3 is strongly signed iff it admits an assignment such that the aggregate sign of some cycle in G is -1.*

Proof. If G is strongly signed then pick any two ambiguous vertices and their connecting positive and negative paths span a cycle as desired.

We prove the opposite direction by induction on the number of ears in the decomposition of G. The base case trivially holds as the first ear is a cycle. Suppose we constructed a partial decomposition $P_0, \ldots, P_{k-1}$, where P_0 is the strongly signed cycle, and let P_k be a new ear to be added to the decomposition. By the induction hypothesis, every pair of vertices in $P_0 \cup \ldots \cup P_{k-1}$ is ambiguous. By this property, it is also trivial to see that every pair of vertices in P_k are ambiguous – simply use as one path the path connecting them in P_k and as the other path a path of opposite sign that visits the previous ears. Such a path exists by the ambiguity of the end-points of P_k. Finally, consider two vertices $u \in P_0 \cup \ldots \cup P_{k-1}$ and $v \in P_k$. Let w be one of the end-nodes in P_k such that $w \neq u$. Since u and w are ambiguous the claim follows.

Since every unsigned edge in a block (of size at least 3) is on some cycle and can be used to force the aggregate sign of this cycle, we get the following corollary:

Corollary 2. *A block of size at least 3 that is not strongly signed must be completely pre-assigned with signs to its edges. Moreover, every pair of vertices in the block is unambiguous.*

Given an input graph G, we can build a tree-like decomposition of G into its blocks and cut-vertices. In this decomposition, the tree vertices are the blocks and cut vertices of G; the tree edges connect cut vertices to the blocks that contain them. Every path in this tree can be translated to a path in G. If a block in G is strongly signed, all input pairs whose connecting path visits this block can be satisfied and the block can be contracted. Every other block must be a single edge or pre-assigned with signs. In particular, if we consider any input pair whose connecting path visits a pre-assigned block, then any sub-path used by the connecting path through the block has a pre-defined sign (since the corresponding pair of cut vertices used are unambiguous). By multiplying the sign of every input pair by the pre-defined signs of the sub-paths through such pre-assigned blocks along its connecting path, we can contract all pre-assigned blocks. This contraction reduces G into a tree and thus we can apply the approximation algorithm of the previous section to it.

Corollary 3. *Constrained-MSA can be 0.878-approximated.*

4.1   An ILP Formulation

As we have seen, both the constrained and unconstrained versions of MSA can be reduced to solving a system of linear equations over Z_2. The latter problem can be easily translated to an integer linear program (ILP) formulation and solved to optimality using an industrial solver such as CPLEX. Let $x_{i_1} \oplus x_{i_2} = r_i$ denote the i-th equation in the reduced instance. The translation is done by defining auxiliary variables y_i denoting whether equation i is satisfied. The exact formulation is:

$$\max \quad \sum_i y_i$$
$$\text{s.t. } y_i = (x_{i_1} + x_{i_2} + r_i + 1 - 2\gamma_i) \ \forall i$$
$$y_i, x_{i_1}, x_{i_2}, \gamma_i \in \{0, 1\} \quad \forall i$$

where the γ_i-s are auxiliary variables that allow computing the parity of x_{i_1}, x_{i_2} and r_i.

5   Experimental Results

5.1   Data Acquisition and Integration

We gathered yeast physical interactions, including PPIs, protein-DNA interactions (PDIs) and kinase-substrate and phosphatase-substrate interactions (KPIs), from different sources. We used the PPI data set "Y2H-union" from Yu et al. [8], which contains 2,930 highly reliable undirected interactions among 2,018 proteins. The PDI data were taken from MacIsaac et al. [3]. We used the collection of PDIs with $p < 0.001$ conserved over at least two other yeast species, which consists of 4,095 unique PDIs spanning 2,079 proteins. The KPIs were collected from Breitkreutz et al. [1] and included 1,361 KPIs among 802 proteins. We complemented the interaction data by information on cause-effect pairs. To this end, a set of 110,487 knockout pairs among 6,228 proteins was taken from Reimand et al. [6]. We integrated the data to obtain a physical network of 3,659 proteins, 2,649 PPIs, 4,095 PDIs and 1,361 KPIs, which spans 52,650 of the cause-effect pairs.

For validation purposes we collected sign information on the PDIs and KPIs in our data set. For a given PDI, we assumed that it is positive (resp., negative) if the transcription-factor involved is an activator (resp., repressor). We retrieved activator-repressor information from the gene ontology (GO:0045893 for activators and GO:0045892 for repressors), obtaining signs for 1,938 PDIs. For a given KPI, we assumed that it is positive if one of the two interacting proteins was a kinase and the other was not a phosphatase; we assumed that it is negative if one of the proteins was a phosphatase and the other was not a kinase. Overall, we estimated signs for 1,148 KPIs. Note that the signs of kinase-phosphatase edges remain undecided.

5.2   Evaluation Procedure

The algorithm can assign signs to two types of edges: (i) an edge within a strongly signed block when it is the *last* unsigned edge of the block; (ii) and an edge "participating" in the ILP. The former assignments are very scarce in our setting as typically most of the edges reside in a single huge block. The latter assignments are sensitive to the exact solution chosen by the ILP. To ensure that we focus on non-arbitrary assignments, we test our confidence in each edge assignment in the following way: for an edge e that is assigned sign s by the ILP, we rerun the ILP while forcing the sign of e to be $-s$. If the resulting objective (number of satisfied pairs) is equal to the original one, we view the assignment as arbitrary; otherwise, the new objective is smaller than the optimal one and we say that the assignment of e is *confident*. Finally, we focus the evaluation on those confidently assigned edges.

5.3   Results

We applied our algorithm in three main settings and evaluated its results (see Table 1). First, we used the entire unsigned network. In this case there are no "last" edges whose signs are determined. Out of 167 edges participating in the ILP, the assignment to 127 of them was confident. 56 of the 127 edges were PDIs with known signs; none was a KPI with a known sign (with only 3 KPIs among the 167 participating edges and the rest in strongly signed blocks). The algorithm predicted correctly the signs of 50 of the 56 edges, yielding a success rate of 89.3%.

Next, we reran the algorithm while using the known signs of the KPIs and focusing the evaluation on the PDIs. While the additional information slightly increased the number of confidently assigned edges (to 135), the intersection with the known PDIs remained the same (56) and again 50 of these were predicted correctly.

Finally, we reran the algorithm on two smaller network instances: one containing PPIs and PDIs but no KPIs, and the other containing PDIs only. As the network gets smaller, the strongly signed components cover less edges and, thus, a larger fraction of the edges participate in the ILP and are assigned with signs. Indeed, in the first application (PPIs and PDIs) more confident sign assignments were made (191) compared to the original application. Out of 64 PDIs with known signs that were confidently assigned, 57 were correctly predicted, yielding a success rate of 89%. When operating on the PDI network only, the largest number of confident sign assignments were made (263). Out of 79 PDIs with known signs that were confidently signed, 70 were correctly predicted, yielding a success rate of 89%.

Notably, in all cases reported above the percent of positive PDIs among the confidently signed ones ranged from 68-72%, significantly lower than the success rate of the algorithm. We further compared our results to the state-of-the-art approach of Peleg et al. [5]. The approach of Peleg et al. works in two stages. In the first stage, the cause-effect pair information is used to split the vertices into groups so that signs of pairs within the same group tend to be positive while signs of pairs crossing groups tend to be negative. Notably, this step is independent of a physical network. The second stage uses the group information to annotate the network's edges with signs. Precisely, the sign of an edge is determined based on the majority sign of pairs that come from the same groups as the edge's endpoints. This implies that in the setting we have used it suffices to apply the method of Peleg et al. once to the entire knockout data, use it to annotate the edges of the integrated network, and then evaluate its performance with respect to each one of the test networks. The results are summarized in Table 1 and show a marked advantage to our approach with success rates higher by 16-22% across the different test cases.

Table 1. Assessment of results and comparison to the network-free approach of Peleg et al. [5]. The best result in each row appears in bold.

Network	Size	#Assignments	#Confident	#Known	%Success	%Success of [5]
KPI+PDI+PPI	8,105	167	127	56	**89.3**	67.9
KPI (signed)+ PDI+PPI	8,105	164	135	56	**89.3**	67.9
PDI+PPI	6,798	228	191	64	**89.1**	70.3
PDI	4,095	290	263	79	**88.6**	73.4

6   Conclusions

We provided an ILP based algorithm to predict edge signs in signaling-regulatory networks. Our algorithm can account for known signs and evaluate the confidence of the assignment. In application to real data it exhibits high success rates of 89%, outperforming the state-of-the-art method of Peleg et al. [5] by a significant margin.

While the annotation results are promising, a current limitation of the algorithm is the low percent of edges that are confidently assigned. One way to tackle this problem is by combining sign prediction with direction prediction which could potentially eliminate many degrees of freedom in the solution and yield a higher percent of annotated edges.

Acknowledgments. We thank Richard Karp for fruitful discussions about this work. RS was supported by a research grant from the Israel Science Foundation (grant no. 241/11).

References

1. Breitkreutz, A., Choi, H., Sharom, J.R., Boucher, L., Neduva, V., Larsen, B., Lin, Z., Breitkreutz, B., Stark, C., Liu, G., Ahn, J., Dewar-Darch, D., Reguly, T., Tang, X., Almeida, R., Qin, Z.S., Pawson, T., Gingras, A., Nesvizhskii, A.I., Tyers, M.: A global protein kinase and phosphatase interaction network in yeast. Science 328(5981), 1043–1046 (2010)
2. Goemans, M., Williamson, D.: Improved approximation algorithms for maximum cut and satisfiability problems using semidefinite programming. J. ACM 42, 1115–1145 (1995)
3. MacIsaac, K., Wang, T., Gordon, D.B., Gifford, D., Stormo, G., Fraenkel, E.: An improved map of conserved regulatory sites for saccharomyces cerevisiae. BMC Bioinformatics 7(1), 113 (2006)
4. Ourfali, O., Shlomi, T., Ideker, T., Ruppin, E., Sharan, R.: SPINE: a framework for signaling-regulatory pathway inference from cause-effect experiments. Bioinformatics 23(13), i359–i366 (2007)
5. Peleg, T., Yosef, N., Ruppin, E., Sharan, R.: Network-free inference of knockout effects in yeast. PLoS Computational Biology 6(1), e1000635 (2010), PMID: 20000032
6. Reimand, J., Vaquerizas, J.M., Todd, A.E., Vilo, J., Luscombe, N.M.: Comprehensive reanalysis of transcription factor knockout expression data in saccharomyces cerevisiae reveals many new targets. Nucleic Acids Research 38(14), 4768–4777 (2010)
7. Yeang, C., Ideker, T., Jaakkola, T.: Physical network models. Journal of Computational Biology 11(2-3), 243–262 (2004)
8. Yu, H., Braun, P., Yildirim, M.A., Lemmens, I., Venkatesan, K., Sahalie, J., Hirozane-Kishikawa, T., Gebreab, F., Li, N., Simonis, N., Hao, T., Rual, J., Dricot, A., Vazquez, A., Murray, R.R., Simon, C., Tardivo, L., Tam, S., Svrzikapa, N., Fan, C., de Smet, A., Motyl, A., Hudson, M.E., Park, J., Xin, X., Cusick, M.E., Moore, T., Boone, C., Snyder, M., Roth, F.P., Barabasi, A., Tavernier, J., Hill, D.E., Vidal, M.: High-Quality binary protein interaction map of the yeast interactome network. Science 322(5898), 104–110 (2008)

Sparse Learning Based Linear Coherent Bi-clustering

Yi Shi[1,*], Xiaoping Liao[2], Xinhua Zhang[1], Guohui Lin[1], and Dale Schuurmans[1]

[1] Department of Computing Science,
University of Alberta, Edmonton, Alberta, Canada
[2] Department of Agricultural, Food and Nutritional Science,
University of Alberta, Edmonton, Alberta, Canada
`{ys3,xliao2,xinhua2,guohui,daes}@ualberta.ca`

Abstract. Clustering algorithms are often limited by an assumption that each data point belongs to a single class, and furthermore that all features of a data point are relevant to class determination. Such assumptions are inappropriate in applications such as gene clustering, where, given expression profile data, genes may exhibit similar behaviors only under some, but not all conditions, and genes may participate in more than one functional process and hence belong to multiple groups. Identifying genes that have similar expression patterns in a common subset of conditions is a central problem in gene expression microarray analysis. To overcome the limitations of standard clustering methods for this purpose, Bi-clustering has often been proposed as an alternative approach, where one seeks groups of observations that exhibit similar patterns over a subset of the features. In this paper, we propose a new bi-clustering algorithm for identifying linear-coherent bi-clusters in gene expression data, strictly generalizing the type of bi-cluster structure considered by other methods. Our algorithm is based on recent sparse learning techniques that have gained significant attention in the machine learning research community. In this work, we propose a novel sparse learning based model, SLLB, for solving the *linear coherent* bi-clustering problem. Experiments on both synthetic data and real gene expression data demonstrate the model is significantly more effective than current bi-clustering algorithms for these problems. The parameter selection problem and the model's usefulness in other machine learning clustering applications are also discussed. The on-line appendix for this paper can be found at `http://www.cs.ualberta.ca/~ys3/SLLB`.

Keywords: Bi-clustering, Microarray, Sparse Learning, Gene Expression, Linear Coherent.

1 Introduction

Gene expression microarrays measure the expression levels of thousands of genes across multiple conditions (conditions are also often referred to as *samples*).

---

* Corresponding author.

B. Raphael and J. Tang (Eds.): WABI 2012, LNBI 7534, pp. 346–364, 2012.

Identifying groups of genes that have similar expression patterns in a common subset of conditions is a central problem in gene expression microarray data analysis. Unfortunately, traditional clustering methods, such as those deployed in [5,24,25], are ill-suited to this purpose for two reasons: that genes may exhibit similar behaviors only under some, but not all conditions, and that genes may participate in more than one functional process and hence belong to multiple groups.

To overcome the limitations of standard clustering methods, *bi-clustering* [10,18] has been proposed to identify groups of data points that exhibit similar patterns on a subset of features. The first work to apply bi-clustering to gene expression analysis is [4], which has motivated many other bi-clustering based approaches. Although the general bi-clustering problem is NP-hard [4], many papers have proposed heuristic methods for finding bi-clusters of different types. In particular, as illustrated in Figure 4 in Appendix, there are six different types of bi-clusters that have been sought in previous work, including: (a) the constant value model, (b) the constant row model, (c) the constant column model, (d) the additive coherent model, where each row (or column) is obtained by adding a constant to another row (or column, respectively), (e) the multiplicative coherent model, where each row (or column) is obtained by multiplying another row (or column, respectively) by a constant value, and (f) the linear coherent model [7], in which each row (or column) is obtained by multiplying another row (or column) by a constant value and then adding a constant [23]. Mathematically, the linear coherent model (f) is strictly more general than the other five models, considered either row-wise or column-wise. In this paper, we design an algorithm that discovers linear coherent bi-clusters that are arbitrarily positioned and possibly even overlapping [16]. Note that, although bi-clusters cannot be simultaneously row-wise and column-wise linear coherent, one is usually more interested in clustering one dimension than the other [9,2]. For example, in the case of gene expression analysis, the main purpose is to identify groups of *genes* that co-participate in certain genetic regulatory process, hence grouping conditions (samples) is only a secondary consideration. Most bi-clustering algorithms implicitly address non time series microarray data and only few address time series microarray data [17]. For time series data, the time lag between mRNA transcription and transcription factor translation needs to be considered. In this paper, we address non time series data.

The motivation for considering linear coherent bi-clusters for gene expression analysis specifically is illustrated in Figure 5 in Appendix [23]. The participation of a pair of genes in a linear coherent bi-cluster must be evidenced by a non-trivial subset of samples in which these two genes are co-up-regulated (or co-down-regulated). Due to data noise, the linear coherence exhibit beams rather than lines in a gene pairwise 2D plot [23,9].

We compare our Sparse Learning based Linear Coherent Bi-clustering (SLLB) algorithm to seven representative bi-clustering algorithms that have been predominant in the field. The first method is a recent bi-clustering algorithm, QUBIC, which finds bi-clusters by a combination of (semi-) qualitative measures of gene expression data and a combinatorial optimization technique [14]. The second method is "Linear Coherent Bi-cluster Discovery via Beam Detection and Sample Set Clustering" (LinCoh) [23], which detects linear bi-clusters by first evaluating the correlation of gene pairs, and then clustering the sample sets that evidence the correlation. The third method is "Linear Coherent Bi-cluster Discovery via Line Detection and Sample Majority Voting" (LCBD) [22], which is the line detection version of LinCoh. Then, we compare to the maximum similarity bi-clustering algorithm (MSBE) [15], which is the first polynomial time bi-clustering algorithm that finds optimal solutions under certain constraints. Then, we compare to the iterative signature algorithm (ISA) [11], which is based on a bi-cluster quality evaluation scheme that uses gene and condition signatures. One advantage of this method is that it can handle incomplete data by imputing a randomized ISA in locations where the expression value is not available. Then, we compare to the order preserving sub-matrix algorithm (OPSM) [3], which attempts to find bi-clusters within a gene expression matrix that contains genes having the same linear ordering of expression levels. Finally, we compare to the method of Cheng and Church (CC) [4], which evaluates the quality of a bi-cluster by a proposed merit score called *mean squared residue*, and then applies a greedy algorithm to find bi-clusters with a score greater than some given threshold. the last three methods have all been highlighted and implemented in a recent survey [19].

The remainder of this paper is organized as follows. Section 2 first introduces the details of our SLLB method we propose. Then, Section 3 introduces the quality measurements we will use to assess the bi-clustering results, provides an experimental evaluation on synthetic data sets, and finally presents bi-clustering results on two real datasets, namely yeast and e.coli. Section 4 then concludes this work with some remarks on the advantages and disadvantages of the proposed SLLB algorithm.

2   Methods

The goal of this work is that given a matrix M (n observations $\times$ p features), find row-wise linear coherent bi-clusters so that each cluster exhibits row-wise linear coherence under a subset of common features.

Let us first consider a pair of $1 \times p$ observation vectors $\mathbf{m}_{i:}$ and $\mathbf{m}_{j:}$. Here $\mathbf{m}_{i:}$ is defined as the ith row vector of matrix M. Other row/column vectors appearing later in this section will be written in the same way. For a given subset of features we can always find the linear regression of this pair of observations in a 2D space that gives us least sum of residuals. We denote the linear regression by slope a_{ij} and intercept b_{ij}. Now the problem is to select a subset of feature so that the sum of residuals from the best regression is minimized. For the bi-cluster that is generated based on the ith observation, we introduce a $1 \times p$ feature selection vector $\mathbf{s}_i \in \{0, 1\}$, where $s_{ik} = 1$ if the kth feature is selected and 0 otherwise. Without any constraint, this problem will always give a trivial solution $\mathbf{s}_i = \mathbf{0}$, yielding a zero sum of residuals. Therefore, we add a regularizer $\beta_1 \|\mathbf{1} - \mathbf{s}_i\|_1$ to penalize any solution with too few $s_{ik} = 1$ values, where $\mathbf{1}$ denotes a vector of all 1s and β_1 is the coefficient of the regularizer. In the subsequent formulations we choose the L_1 norm because it gives us a sparse solution in $\mathbf{1} - \mathbf{s_i}$ once $\mathbf{s_i}$ has been relaxed to $[0, 1]$. For a single row i, the problem can then be formulated as an optimization as follows:

$$\min_{\mathbf{s_i}, a_{ij}, b_{ij}} \sum_k s_{ik}(m_{ik} - a_{ij}m_{jk} - b_{ij})^2 + \beta_1 \|\mathbf{1} - \mathbf{s_i}\|_1$$
$$s.t. \quad s_{ik} \in \{0, 1\} \tag{1}$$

Now, consider the whole matrix M from which we want to detect a set of row-wise linear coherent bi-clusters. We introduce a $n \times n$ binary matrix W, where $w_{ij} = 1$ indicates there is strong linear coherence between the observation pair (i, j) and $w_{ij} = 0$ otherwise. By extending 1 in terms of the whole matrices M, S, A, B and introducing W, we obtain the complete formulation:

$$\min_{W,S,A,B} \sum_{i,j} w_{ij} \sum_k s_{ik}(m_{ik} - a_{ij}m_{jk} - b_{ij})^2$$
$$+\beta_1 \|\mathbf{1} \cdot \mathbf{1}^T - S\|_{1,1} + \beta_2 \|\mathbf{1} \cdot \mathbf{1}^T - W\|_{1,1} \tag{2}$$
$$s.t. \quad w_{ij} \in \{0, 1\}, s_{ik} \in \{0, 1\}$$

where W can be interpreted as observation (data) selection matrix, and S can be interpreted as the feature (sample) selection matrix. β_2 is the coefficient of the W-wise regularizer. Here S is a $n \times p$ binary matrix with the ith row corresponding to the feature selection vector for the ith observation. Note that the sparse regularizer $\beta_1 \|\mathbf{1} - \mathbf{s_i}\|_1$ becomes $\beta_1 \|\mathbf{1} \cdot \mathbf{1}^T - S\|_{1,1}$. Similarly, we add another sparse regularizer $\beta_1 \|\mathbf{1} \cdot \mathbf{1}^T - W\|_{1,1}$ to penalize trivial solutions where W is set too close to the zero matrix.

We want to favor the case that the scatter points (feature points) of a pairwise 2D plot do not stick together so as to exhibit better linear coherence. Towards this end, we introduce a $n \times n \times p$ matrix D, where $d_{ijk} \in [0, 1]$ indicates the importance of the kth feature under the observation pair (i, j). In the gene expression matrix case, because it is desired to favor co-up-regulated and co-down-regulated gene expression samples, we assign $d_{ijk} = e^{d'_{ijk}}$, where d'_{ijk} is the Euclidean distance of the kth data point to the central point $(\bar{m}_{i:}, \bar{m}_{j:})$. Different prior knowledge can be introduced to form D from other data sources. Therefore, after relaxing $W \in \{0, 1\}$ to $W \in [0, 1]$ and $S \in \{0, 1\}$ to $S \in [0, 1]$, we get:

$$\min_{W,S,A,B} \sum_{i,j} w_{ij} \sum_k s_{ik} \frac{1}{d_{ijk}}(m_{ik} - a_{ij}m_{jk} - b_{ij})^2$$
$$+\beta_1 \|\mathbf{1} \cdot \mathbf{1}^T - S\|_{1,1} + \beta_2 \|\mathbf{1} \cdot \mathbf{1}^T - W\|_{1,1} \tag{3}$$
$$s.t. \quad w_{ij} \in [0, 1], s_{ik} \in [0, 1]$$

By introducing some new notation, we can re-express this problem in an equivalent form that proves to be more convenient for formulating an efficient iterative procedure below. Let $\otimes$ denote Kronecker product, let $\triangle(\mathbf{m})$ denote putting a vector $\mathbf{m}$ on the main diagonal of a square matrix, and let $\div$ denote component-wise division. Then 3 can be equivalently re-written in terms of $\mathbf{s_{i:}}$ and $\mathbf{w_{i:}}$ as:

$$\min_{W,S,A,B} \sum_i \left\| \triangle(\mathbf{w_{i:}})^{1/2}(\mathbf{1} \otimes \mathbf{m_{i:}} - \triangle(\mathbf{a_{i:}})M - \triangle(\mathbf{b_{j:}})\mathbf{1} \otimes \mathbf{1}^T) \div D_i^* \triangle(\mathbf{s_{i:}})^{1/2} \right\|_F^2$$
$$+\beta_1\|\mathbf{1}\cdot\mathbf{1}^T - S\|_{1,1} + \beta_2\|\mathbf{1}\cdot\mathbf{1}^T - W\|_{1,1}$$

$$s.t. \quad w_{ij} \in [0,1], s_{ik} \in [0,1]$$

$$(4)$$

where D_i^* has the same dimension as D_i with each element equal to the square root of the corresponding element in D_i.

Unfortunately, 4 is not jointly convex in W, S, A and B, so we are currently unable to formulate an efficient global optimization procedure. Nevertheless, an efficient iterative procedure can be devised that finds a reasonable local solution.

2.1   Initialization

Because of the potential difficulty of local minima, initialization of W, S, A, and B becomes very important for solving 4 iteratively. To simplify the initialization, and allow a generally effective approach, we first normalize the data matrix M so that each row $\mathbf{m_{i:}} \in [0,1]$. In the case of gene expression analysis, A is initialized to $\mathbf{1}\cdot\mathbf{1}^T$ since a gene pair that has strong correlation will have a sufficient number of samples (features) under which the gene pair has a co-up-regulated and co-down-regulated pattern, which implies that on normalized data, the slope is near 1. The intercept b_{ij} is normalized in a way that the linear regression line for each observation pair passes through the central point $(\bar{m}_{i:}, \bar{m}_{j:})$ with slope a_{ij}. After A and B are initialized, $\mathbf{s_{i:}}$ is initialized such that $s_{ik} = 1$ if the distance d'_{ijk} of kth data point of the (i,j) pair to the line (a_{ij}, b_{ij}) is within some threshold. Since the data is normalized, an appropriate threshold can be set for data of the same type and will not affect the results to a large extent. In the case of gene expression data, since we want to favor sample points that are far away from the central point $(\bar{m}_{i:}, \bar{m}_{j:})$, we set the threshold as a monotonically increasing function of the distance d''_{ijk} between $(\bar{m}_{i:}, \bar{m}_{j:})$ and the projection of the (m_{ik}, m_{jk}) on the regression line. We do not initialize W as it will be immediately determined from the initial S, A, and B.

2.2   Iterative Update of W, S, A and B

Updating W. Denote the objective function in 4 by $f(W, S, A, B)$. Assume that S, A, and B are fixed (initialized as mentioned above for the first iteration). Then the objective function is a convex (linear) function of W and we can optimize W element by element in a closed form. In particular, for each w_{ij}, by ignoring constant terms, the problem is equivalent to minimizing $f(w_{ij})$:

$$\min_{w_{ij}} w_{ij} \left(\sum_k \frac{s_{ik}}{d_{ijk}}(m_{ik} - a_{ij}m_{jk} - b_{ij})^2 - \beta_2 \right)$$
$$s.t. \quad w_{ij} \in [0,1]$$

$$(5)$$

Because $f(w_{ij})$ is a linear function of w_{ij}, we obtain $w_{ij} = 1$ if $\sum_k \frac{s_{ik}}{d_{ijk}}(m_{ik} - a_{ij}m_{jk} - b_{ij})^2 < \beta_2$ and $w_{ij} = 0$ otherwise.

Updating S. When W, A, and B are fixed, $f(W, S, A, B)$ becomes a convex (linear) function of S, so similar to updating W, we can update S element by element in a closed form. In this case, s_{ik} can be calculated by minimizing $f(s_{ik})$ as follows:

$$\min_{s_{ik}} \left(\sum_j \frac{w_{ij}}{d_{ijk}}(m_{ik} - a_{ij}m_{jk} - b_{ij})^2 - \beta_1 \right) \tag{6}$$
$$s.t. \quad s_{ik} \in [0, 1]$$

Hence, $s_{ik} = 1$ if $\sum_j \frac{w_{ij}}{d_{ijk}}(m_{ik} - a_{ij}m_{jk} - b_{ij})^2 < \beta_1$ and $s_{ik} = 0$ otherwise.

Updating A and B. When W and S are fixed the minimization over A and B becomes a standard least squares linear regression problem for each observation pair. In particular, we have:

$$(a_{ij}, b_{ij})^T = (X_{ij}^T \triangle(\mathbf{s_{i:}} \bullet \mathbf{d_{ij:}})X_{ij})^{-1} X_{ij}^T \triangle(\mathbf{s_{i:}} \bullet \mathbf{d_{ij:}})\mathbf{y_{ij}} \tag{7}$$

where $\bullet$ denotes inner product, $X_{ij} = (1, \mathbf{m_{j:}}^T)$, and $\mathbf{y} = \mathbf{m_{i,:}}^T$

Finally, each of W, S, A, and B are iteratively updated until the objective function converges. Algorithm 1 in Appendix gives the details of the SLLB algorithm. Note that the time complexity of the SLLB is $O(n^2)$ per iteration. Later experiments on synthetic datasets show that SLLB converges after 6-8 iterations, which takes less than 10 seconds in total. On real datasets, good results can be obtained after 10-20 iterations, which take tens of hours.

3   Results and Discussion

We compare the SLLB with seven existing representative bi-clustering algorithms, QUBIC, LinCoh, LCBD, CC, OPSM, ISA, and MSBE on synthetic datasets and two real gene expression microarray datasets on *Saccharomyces cerevisiae* (yeast) and *Escherichia coli* (e.coli) respectively. The parameter settings for the compared algorithms mostly follow the previous works [19,15,23].

3.1   Synthetic Datasets

On synthetic datasets, Prelić's observation (gene) match score and overall match score [19] are adopted to evaluate the ability of bi-clustering algorithms in discovering the implanted (true) bi-clusters. Let $\mathcal{C}$ and $\mathcal{C}^*$ denote the set of output bi-clusters from an algorithm and the set of true bi-clusters for a dataset respectively. The observation match score of $\mathcal{C}$ with respect to the target $\mathcal{C}^*$ is defined as $\text{score}_G(\mathcal{C}, \mathcal{C}^*) = \frac{1}{|\mathcal{C}|} \sum_{(G_1, S_1) \in \mathcal{C}} \max_{(G_1^*, S_1^*) \in \mathcal{C}^*} \frac{|G_1 \cap G_1^*|}{|G_1 \cup G_1^*|}$, which is the average of the maximum observation match scores of bi-clusters in $\mathcal{C}$ with respect to the target bi-clusters. The feature match score $\text{score}_S(\mathcal{C}, \mathcal{C}^*)$ can be similarly defined by replacing observation sets with the corresponding feature sets in the above. The overall match score is then defined as their geometric mean, *i.e.*

$$\text{score}(\mathcal{C}, \mathcal{C}^*) = \sqrt{\text{score}_G(\mathcal{C}, \mathcal{C}^*) \times \text{score}_S(\mathcal{C}, \mathcal{C}^*)}.$$

As for the parameter setting of SLLB, we set $\beta_1 = 0.1$, $\beta_2 = 0.3$ on all the overlapping experiments, and we set $\beta_1 = 0.1$, $\beta_2 = 0.05$ on all the noise resistance experiments.

Overlapping Test: Bi-clusters may overlap in terms of either observations or features. Take gene expression as an example, some genes can participate in multiple biological processes which result in bi-clusters that overlap with common genes in an expression matrix. It is also the case in sample overlapping. This experiment intends to examine the ability of bi-clustering algorithms in recovering overlapping bi-clusters. We again consider type-(f) linear coherent bi-clusters and type-(d) additive bi-clusters, at a fixed noise level of $\ell = 0.1$. We generate ten 100×50 matrices based on a standard normal distribution. In each matrix, two 10×10 bi-clusters are embedded, with overlapping size: 0×0, 1×1, 2×2, 3×3, 4×4, and 5×5. In the case of gene expression, we assume that these overlapping genes obey a reasonable logic such as the AND gate and the OR gate which leads to a behavior of *union* and an *additive* respectively. So the overlapped entries in the union overlapping area preserve linear coherency in both bi-clusters and in the additive overlap model, these entries are assigned by the sum of the gene expression levels from both bi-clusters. The observation match scores of the

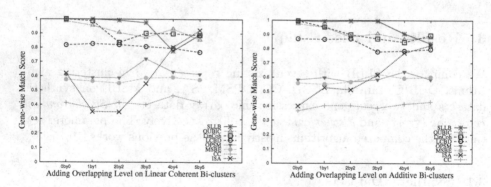

Fig. 1. The observation match scores of the eight algorithms for recovering the overlapping linear coherent and additive bi-clusters, under the adding overlap model

eight bi-clustering algorithms in this adding overlapping experiment are shown in Figure 1. Figures 9 and 10 in the Appendix plot the overall match scores and observation discovery rates under the adding overlap model. The results of the additive overlap model are shown in Figures 11, 12, and 13 in the Appendix. From all these results, we can conclude that SLLB outperformes the other seven algorithms. LinCoh's performance is slightly worse than SLLB; QUBIC, OPSM and MSBE perform worse, but similarly to each other; LCBD and CC performed the worst; and ISA demonstrates varying performance.

Noise Resistance Test: This experiment investigates the ability of different bi-clustering algorithms in recovering implanted bi-clusters with different noise level. Following Prelić's testing strategy, we first generate a 100×50 background matrix based on a standard normal distribution and then embed ten 10×5 non-overlapping linear coherent bi-clusters along the diagonal. Then, for each vector of the five expression values, we set the first two to be down-regulated, the last two to be up-regulated, and the middle one to be non-regulated. Lastly, we add noise of six different levels ($\ell = 0.00, 0.05, 0.10, 0.15, 0.20, 0.25$) to the embedded bi-clusters by perturbing the entry values so that the resultant values are ℓ away from the original values. The generation is repeated ten times. Based on the same simulation process, we generate additive bi-clusters on synthetic datasets when we compare the bi-clustering algorithms on their performance in discovering additive bi-clusters only (which is a special case of linear coherent bi-clusters).

Figure 6 in Appendix shows the observation match scores of the bi-clusters discovered by the eight algorithms at six different noise levels. Figures 7 and 8 in the Appendix demonstrate their overall match scores and observation discovery rates (defined as the percentage of observations in the output bi-clusters over all the observations in the true bi-clusters). From these figures, it is clearly shown that SLLB outperformes all the other seven algorithms; QUBIC, LinCoh and ISA rank the second, third, and forth, and the other three performed quite poorly. Note that by simply outputting more bi-clusters, observation discovery rate can be trivially lifted up. Therefore it is only a useful measurement in conjunction with match scores.

3.2   Real Datasets

On real datasets, the quality of bi-clusters is evaluated by known biological pathways, defined in the GO functional classification scheme [1], the KEGG pathways [12], the MIPS yeast functional categories [20] (for yeast dataset), and the EcoCyc database [13] (for e.coli dataset), in order to obtain their *gene functional enrichment score* as implemented in [14]. The average correlation coefficient is also used for evaluating the generated bi-clusters on real datasets.

We obtain the yeast dataset from [8]. It contains 2993 genes on 173 samples; the e.coli dataset is obtained from [6], (version 4 built 3). It contains initially 4217 genes on 264 samples. For the e.coli dataset, after removing genes with too small expression deviations, we get 3016 genes. This pre-process ensures that all eight bi-clustering algorithms can be run on the dataset. We use the gene functional enrichment score [14] to measure the performance of different algorithms. First, the P-value of each output bi-cluster is defined using its most enriched functional class (biological process). The probability of having r genes of the same functional class in a bi-cluster of size n from a genome with a total of N genes can be computed using the hypergeometric function, where p is the percentage of that functional class of genes over all functional classes of genes encoded in the whole genome. Numerically [14],

$$Pr(r|N, p, n) = \binom{pN}{r} \cdot \binom{(1-p)N}{n-r} / \binom{N}{n}.$$

Such a probability is taken as the P-value of the output bi-cluster enriched with genes from that functional class [14]. The P-value of the output bi-cluster is defined as the smallest P-value over all functional classes. The smaller the P-value of a bi-cluster the more likely do its genes come from the same biological process. We calculate for each algorithm the fraction of its output bi-clusters whose P-values are smaller than a significance cutoff α. As for the parameter setting of SLLB, we set $\beta_1 = 0.3$, $\beta_2 = 1.5$ for the yeast dataset, and $\beta_1 = 0.1$, $\beta_2 = 0.5$ for the e.coli dataset.

In Figure 2 the eight algorithms are compared using six different P-value cutoffs, evaluated on the GO database. Results on the KEGG, MIPS, and Regulon databases are in Figures 14 and 15 in Appendix. These results indicates that SLLB performs consistently well; QUBIC and LinCoh performs stable but worse than SLLB, OPSM and ISA does not perform consistently on the two datasets across databases; and that LCBD, MSBE and CC does not perform as well as the other three algorithms.

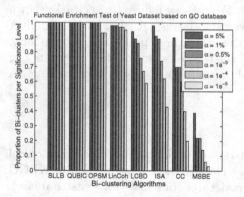 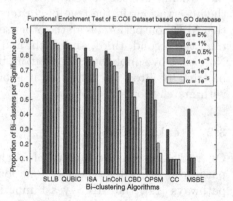

Fig. 2. Portions of discovered bi-clusters by the eight algorithms on the two real datasets that are significantly enriched in the GO biological process, using six different P-value cutoffs

One potential issue with the P-value based performance measurement is that P-values are sensitive to the bi-cluster size [14]; in general, this measurement favors bi-clusters with a larger size. For example, in Table 1, it is shown that OPSM finds bi-clusters that contain extremely large number of genes and very few samples. Bi-clusters of this kind are close to trivial bi-clusters (gene or sample set size close to 0) but with large number of gene, its enrichment P value can be easily lifted up. On the contrary, although our SLLB algorithm generates bi-clusters with large number of genes, the number of samples it generates is also large which indicates more confident linear coherence. In the last column of the Table 1, the numbers of unique functional terms enriched by the produced bi-clusters are listed. When measured by the gene enrichment significance score, OPSM performed very well on yeast dataset (Figure 2, left), but its bi-clusters only cover one functional term on the GO and KEGG databases and two terms on MIPS database. This suggests that the bi-clustering result can be biased to a group of correlated genes, which are missed by the P-value based significance test.

Considering these two potential issues, we can see that the P-value based evaluation is meaningful but has limitations. So we propose using the average absolute correlation coefficient over all gene pairs in a bi-cluster as an alternative assessment of the quality of a linear coherent bi-cluster. However, note that the numbers of samples in the bi-clusters generated by some algorithms are much smaller others, Table 1. Therefore, to compare algorithms in a less sample-size biased way, we replaced for each bi-cluster its average absolute correlation coefficient by the 99% confidence threshold using the number of samples in the bi-cluster [21,23]. These values are plotted in Figure 3.

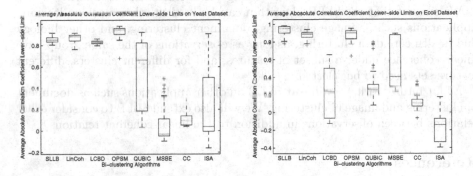

Fig. 3. Box plots of the average absolute correlation coefficients obtained by the eight bi-clustering algorithms on yeast and e.coli datasets, respectively

Figure 3 shows that SLLB, LinCoh, and OPSM have similarly good performance while QUBIC, LCBD, MSBE, CC, and ISA performs worse than these three. Note that due to noise effect when profiling genes, itt is hard to reach a very large value of the correlation coefficient.

4   Conclusion

In this article, we proposed a novel bi-clustering algorithm, SLLB, that can discover linear coherent bi-clusters based on a sparse learning optimization model. The experimental results on both synthetic and real datasets indicate that SLLB is not only able to discover linear coherent bi-clusters effectively, but able to discover meaningful linear coherent bi-clusters that can be verified by biological ground truth. Actually, for many bi-clusters discovered by SLLB, all their corresponding gene groups (with size 30-100) belong amazingly to the same gene ontology term. The time complexity of the SLLB algorithm is $O(n^2k)$ where n is the number of observations and k is the number of iterations that SLLB takes to converge, which is very fast compared to algorithms like LinCoh. Note that while discovering linear coherent bi-clusters, SLLB favors data points corresponding to features that are far away from each other in the observation pair 2D space. This nice property can be used for downstream data analysis such as feature clustering, observation-feature relation studies and observation/feature selection.

To set appropriate values for β_1 and β_2, We binary searched $\beta_1 \in [0, 1000]$ and $\beta_2 \in [0, 1000]$ and found the value ranges that produce non-trivial bi-clusters are $\beta_1 \in [0, 0.5]$ and $\beta_2 \in [0, 1.5]$. We then tested different combinations of $\beta_1 = [0.1, 0.5, 1]$ and $\beta_2 = [0.1, 0.5, 1, 1.5]$ and found the results are quite robust to different settings. The final β_1 and β_2 are chosen so that SLLB performs best. When come to practice, considering that β_1 actually controls the size of observation and β_2 controls the size of features in the result bi-clusters, β_1 and β_2 can be determined when prior knowledge of bi-cluster size is known.

We suggest that the SLLB algorithm can be used in other machine learning applications such as image clustering, document clustering, and other biology and health care data clustering, as long as observations of the same group have linear coherence under a subset of features, and for different clusters, different feature sets need to be selected.

As for future work, we will test SLLB on other applications such as document bi-clustering and image bi-clustering. We will also extend SLLB to consider other relations between observations in addition to the linear coherent relations.

References

1. Ashburner, M., Ball, C.A., Blake, J.A., et al.: Gene ontology: tool for the unification of biology. Nature Genetics 25, 25–29 (2000)
2. Ayadi, W., Elloumi, M., Hao, J.K.: Pattern-driven neighborhood search for biclustering of microarray data. BMC Bioinformatics 13(suppl. 7), S11 (2012)
3. Ben-Dor, A., Chor, B., Karp, R., Yakhini, Z.: Discovering local structure in gene expression data: The order-preserving sub-matrix problem. In: RECOMB 2002, pp. 49–57 (2002)
4. Cheng, Y., Church, G.M.: Biclustering of expression data. In: ISMB 2000, pp. 93–103 (2000)
5. Eisen, M.B., Spellman, P.T., Brown, P.O., Botstein, D.: Cluster analysis and display of genome-wide expression patterns. PNAS 95, 14863–14868 (1998)
6. Faith, J.J., Driscoll, M.E., Fusaro, V.A., et al.: Many microbe microarrays database: uniformly normalized Affymetrix compendia with structured experimental metadata. Nucleic Acids Research 36, D866–D870 (2008)
7. Gan, X., Liew, A.W.-C., Yan, H.: Discovering biclusters in gene expression data based on high-dimensional linear geometries. BMC Bioinformatics 9, 209 (2008)

8. Gasch, A.P., Spellman, P.T., Kao, C.M., et al.: Genomic expression programs in the response of yeast cells to environmental changes. Nucleic Acids Research 11, 4241–4257 (2000)

9. Gupta, R., Kumar, V., Rao, N.: Discovery of error-tolerant biclusters from noisy gene expression data. Bioinformatics 12(suppl. 12), S1 (2011)

10. Hartigan, J.A.: Direct clustering of a data matrix. Journal of the American Statistical Association 67, 123–129 (1972)

11. Ihmels, J., Bergmann, S., Barkai, N.: Defining transcription modules using large scale gene expression data. Bioinformatics 20, 1993–2003 (2004)

12. Kanehisa, M.: The KEGG database. In: Novartis Foundation Symposium, vol. 247, pp. 91–101 (2002)

13. Keseler, I.M., Collado-Vides, J., Gama-Castro, S., et al.: EcoCyc: a comprehensive database resource for escherichia coli. Nucleic Acids Research 33, D334–D337 (2005)

14. Li, G., Ma, Q., Tang, H., Paterson, A.H., Xu, Y.: QUBIC: A qualitative biclustering algorithm for analyses of gene expression data. Nucleic Acids Research 37, e101 (2009)

15. Liu, X., Wang, L.: Computing the maximum similarity bi-clusters of gene expression data. Bioinformatics 23, 50–56 (2006)

16. Madeira, S.C., Oliveira, A.L.: Biclustering algorithms for biological data analysis: A survey. Journal of Computational Biology and Bioinformatics 1, 24–45 (2004)

17. Meng, J., Huang, Y.: Biclustering of time series microarray data. Methods Mol. Biol. 802, 87–100 (2012)

18. Mirkin, B.: Mathematical classification and clustering. Kluwer Academic Publishers (1996)

19. Prelić, A., Bleuler, S., Zimmermann, P., Wille, A.: A systematic comparison and evaluation of biclustering methods for gene expression data. Bioinformatics 22, 1122–1129 (2006)

20. Ruepp, A., Zollner, A., Maier, D., et al.: The FunCat, a functional annotation scheme for systematic classification of proteins from whole genomes. Nucleic Acids Research 32, 5539–5545 (2004)

21. Shen, D., Lu, Z.: Computation of correlation coefficient and its confidence interval in SAS, http://www2.sas.com/proceedings/sugi31/170-31.pdf

22. Shi, Y., Cai, Z., Lin, G., Schuurmans, D.: Linear Coherent Bi-cluster Discovery via Line Detection and Sample Majority Voting. In: Du, D.-Z., Hu, X., Pardalos, P.M. (eds.) COCOA 2009. LNCS, vol. 5573, pp. 73–84. Springer, Heidelberg (2009)

23. Shi, Y., Hasan, M., Cai, Z., Lin, G., Schuurmans, D.: Linear coherent bi-cluster discovery via beam detection and sample set clustering. In: International Conference on Combinatorial Optimization and Applications, vol. 1, pp. 85–103 (2010)

24. Tamayo, P., Slonim, D., Mesirov, J., et al.: Interpreting patterns of gene expression with self-organizing maps: Methods and application to hematopoietic differentiation. PNAS 96, 2907–2912 (1999)

25. Tavazoie, S., Hughes, J.D., Campbell, M.J., Cho, R.J., Church, G.M.: Systematic determination of genetic network architecture. Nature Genetics 22, 281–285 (1999)

Appendix

x	y	z	w
1.0	1.0	1.0	1.0
1.0	1.0	1.0	1.0
1.0	1.0	1.0	1.0
1.0	1.0	1.0	1.0

(a)

x	y	z	w
1.2	1.2	1.2	1.2
0.8	0.8	0.8	0.8
1.5	1.5	1.5	1.5
0.6	0.6	0.6	0.6

(b)

x	y	z	w
1.2	0.8	1.5	0.6
1.2	0.8	1.5	0.6
1.2	0.8	1.5	0.6
1.2	0.8	1.5	0.6

(c)

x	y	z	w
1.2	0.8	1.5	0.6
1.0	0.6	1.3	0.4
2.0	1.6	2.3	1.4
0.7	0.3	1.2	0.3

(d)

x	y	z	w
2.0	4.0	8.0	1.0
1.0	2.0	4.0	0.5
4.0	8.0	16.0	2.0
1.0	2.0	4.0	0.5

(e)

x	y	z	w
2.0	4.0	3.0	5.0
1.5	2.5	2.0	3.0
2.3	4.3	3.3	5.3
4.5	8.5	6.5	10.5

(f)

Fig. 4. The six different types of bi-clusters: (a) constant block bi-cluster; (b) constant row bi-cluster; (c) constant column bi-cluster; (d) additive coherent bi-cluster; (e) multiplicative coherent bi-cluster; (f) row-wise linear coherent bi-cluster

---

Algorithm 1. The SLLB Algorithm

Input: M, β_1, β_2, ϵ.
Output: a set of linear coherent bi-clusters.

$A = \mathbf{1} \cdot \mathbf{1}^T$
$b_{ij} = \bar{m}_{i:} - a_{ij}\bar{m}_{j:}$ for $i, j \in [1, 2, ...n]$.
$s_{ik} = 1$ if $d'_{ijk} <= e^{d''_{ijk}} - 1$, $s_{ik} = 0$ otherwise, for $i \in [1, 2, ...n]$, $k \in [1, 2, ...p]$.
while $\Delta L > \epsilon$ **do**
    for each i, j **do**
        if $\sum_k \frac{s_{ik}}{d_{ijk}}(m_{ik} - a_{ij}m_{jk} - b_{ij})^2 < \beta_2$ **then**
          $w_{ij} = 1$
        else
          $w_{ij} = 0$
        end if
    end for
    for each i, j **do**
        if $\sum_j \frac{w_{ij}}{d_{ijk}}(m_{ik} - a_{ij}m_{jk} - b_{ij})^2 < \beta_1$ **then**
          $s_{ik} = 1$
        else
          $s_{ik} = 0$
        end if
    end for
    for each i, j **do**
        $(a_{ij}, b_{ij})^T = (X_{ij}^T \triangle(\mathbf{s}_{i:} \bullet \mathbf{d}_{ij:}) X_{ij})^{-1} X_{ij}^T \triangle(\mathbf{s}_{i:} \bullet \mathbf{d}_{ij:}) \mathbf{y}_{ij}$
    end for
    Calculate loss change ΔL.
end while
Construct bi-clusters from W, S and remove redundant bi-clusters.

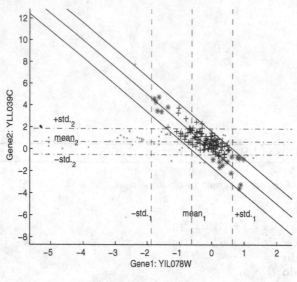

(a) Negative correlation.

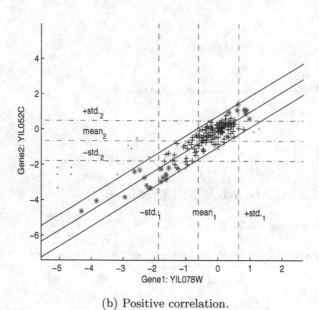

(b) Positive correlation.

Fig. 5. Scatter plots for the expression level of two different gene pairs across samples. Here each point (x, y) represents a sample in which the two genes have expression levels x and y respectively. (a) illustrates two yeast genes *YIL078W* and *YLL039C* that have negative expression correlation under a subset of conditions; the red conditions provide a stronger evidence than the blue conditions, whereas the green conditions do not suggest any correlation. Similarly in (b), genes *YIL078W* and *YIL052C* show a positive expression correlation.

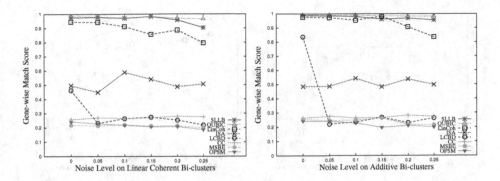

Fig. 6. The observation match scores of the eight algorithms on recovering linear coherent bi-clusters and additive bi-clusters at six different noise levels.

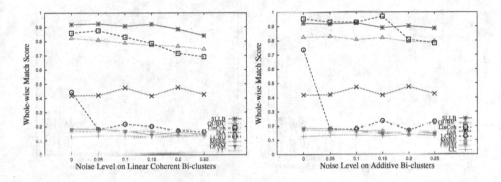

Fig. 7. The overall match scores of the eight algorithms for recovering linear coherent and additive bi-clusters, at six different noise levels.

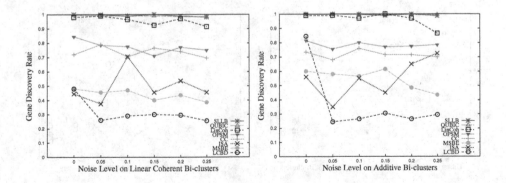

Fig. 8. The gene discovery rates of the eight algorithms for recovering linear coherent and additive bi-clusters, at six different noise levels.

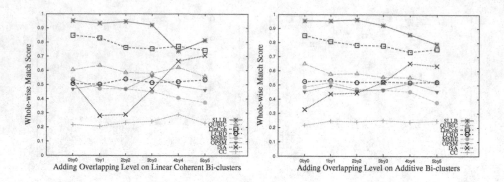

Fig. 9. The whole match scores of the eight algorithms for recovering the overlapping linear coherent and additive bi-clusters, under the adding overlap model.

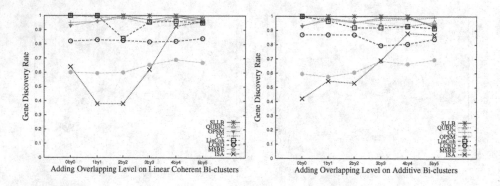

Fig. 10. The observation discovery rate of the eight algorithms for recovering the over-lapping linear coherent and additive bi-clusters, under the adding overlap model.

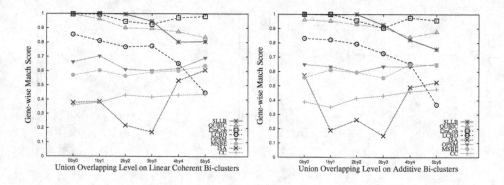

Fig. 11. The observation match scores of the eight algorithms for recovering the over-lapping linear coherent and additive bi-clusters, under the union overlap model.

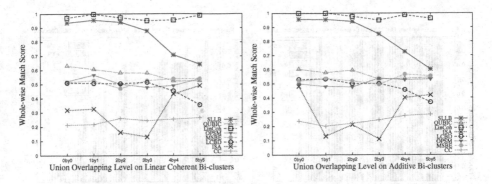

Fig. 12. The whole match scores of the eight algorithms for recovering the overlapping linear coherent and additive bi-clusters, under the union overlap model

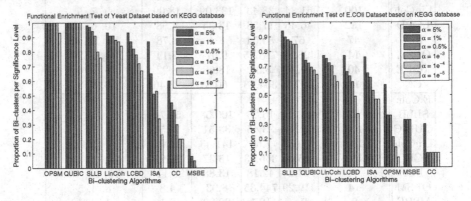

Fig. 13. The observation discovery rate of the eight algorithms for recovering the overlapping linear coherent and additive bi-clusters, under the union overlap model

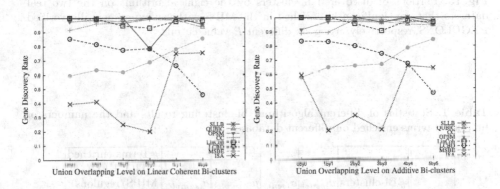

Fig. 14. Portions of discovered bi-clusters by the eight algorithms on the two real datasets that are significantly enriched in the KEGG pathway, using six different P-value cutoffs

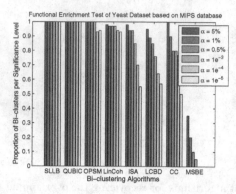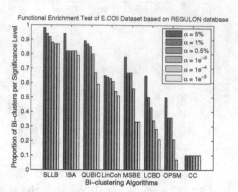

Fig. 15. Portions of discovered bi-clusters by the eight algorithms on the two real datasets that are significantly enriched in the MIPS pathway experimentally verified REGULONS, respectively, using six different P-value cutoffs.

Table 1. Statistics of different algorithms' bi-clustering results and the numbers of functional terms enriched on different databases

	#Bi-clusters	$\mu_{\lvert gene \rvert}$	$\sigma_{\lvert gene \rvert}$	$\mu_{\lvert sample \rvert}$	$\sigma_{\lvert sample \rvert}$	#Terms enriched (GO, KEGG, MIPS/regulons)
Yeast:						
SLLB	101	83.40	36.40	85.56	25.98	3, 7, 5
QUBIC	100	119.08	59.69	23.21	6.21	3, 2, 3
LinCoh	100	61.84	38.43	133.09	18.09	5, 7, 5
LCBD	132	46.46	17.53	13.35	4.58	10, 6, 11
ISA	47	67	34.54	8.4	1.78	15, 13, 18
OPSM	14	423.29	728.95	9.07	5.14	1, 1, 2
MSBE	40	19.25	8.32	18.68	8.22	8, 4, 6
CC	10	297.7	304.18	60.8	23.46	6, 4, 8
E.Coli:						
SLLB	52	43.79	16.23	106.58	51.70	8, 11, 13
QUBIC	100	73.91	33.45	33.51	14.51	14, 8, 15
LinCoh	100	9.63	7.66	141.43	34.04	24, 24, 22
LCBD	155	485.05	366.63	15.37	22.95	23, 22, 33
ISA	34	124.21	42.18	13.88	6.11	11, 10, 13
OPSM	14	419.29	744.35	8.93	4.8	8, 4, 5
MSBE	9	82.67	18.1	80.22	19.18	1, 3, 4
CC	10	309.9	950.15	31.4	81.74	2, 2, 2

A Simplified View of DCJ-Indel Distance

Phillip E.C. Compeau

University of California, San Diego Department of Mathematics

Abstract. The introduction of the double cut and join (DCJ) operation
and the derivation of its associated distance caused a flurry of research
into the study of multichromosomal rearrangements. However, little of
this work has incorporated indels (i.e., insertions and deletions) into the
calculation of genomic distance functions, with a particular exception of
Braga et al., who provided a linear time algorithm ([1]) for computing
the DCJ-indel distance. Although this algorithm only takes linear time,
its derivation is lengthy and depends on a large number of possible cases.
In this paper, we provide a simplified indel model that solves the problem
of DCJ-indel sorting in linear time directly from the classical breakpoint
graph, an approach that allows us to describe the solution space of DCJ-
indel sorting, thus resolving an existing open problem.

1 Introduction

Chromosomal rearrangements were first directly observed by Dobzhansky and
Sturtevant in 1938 ([2]), but extensive studies into quantifying their study did
not take off until the early 1990s. In the last two decades, a number of discrete
genomic models have been proposed and studied (see [3] for an overview of the
combinatorics of genome rearrangements).

Having selected a genomic model and a collection of genome operations to
consider, the standard algorithmic problem is the computation of the *distance*
between two genomes Π and Γ, or the minimum number of allowable operations required to transform Π into Γ; the related problem of *sorting* demands a
minimum set of such operations. The first historical example of such a discrete
genomic distance is the *prefix reversal distance* for permutations, introduced in
[4] and bounded in [5], [6], and [7]. The computation of prefix reversal distance
has been proposed to be NP-Hard ([8]).

Permutations effectively model single linear chromosomes, but recent research
has moved toward multichromosomal genomic models that incorporate both linear and circular chromosomes. The double cut and join operation (DCJ) was
introduced in [9] and incorporates segment reversals with a number of other
operations. Interestingly, a linear time algorithm exists for computing the DCJ
distance between two genomes with equal gene content via a straightforward
greedy sorting algorithm presented in [10].

The incorporation of insertions and deletions of chromosomes and chromosomal intervals (collectively called *indels*) into DCJ distance was discussed in
[11] and quantified rigorously in [1]. The latter authors provided a linear time

B. Raphael and J. Tang (Eds.): WABI 2012, LNBI 7534, pp. 365–377, 2012.
© Springer-Verlag Berlin Heidelberg 2012

algorithm for computing the associated *DCJ-indel distance*, which counts the minimum total number of DCJ and indel operations required to transform one genome into another. Yet their argument is case-ridden, and so we wish to provide a much simpler presentation of DCJ-indel sorting that still yields a linear-time solution to the problem.

2 Preliminaries

Say that we are given a perfect matching on $2N$ labeled vertices $\mathscr{V}$, forming a set $\mathcal{G}$ of N edges called *genes*; the vertices of each gene form its *head* and *tail*. We define a *genome* Π as the edge-disjoint union of two matchings. The (black) *genes* of Π, denoted $g(\Pi)$, form a matching on $\mathscr{V}$ such that $g(\Pi) \subseteq \mathcal{G}$; the *adjacencies* of Π, denoted $a(\Pi)$, form a matching on $V(g(\Pi))$ (see Fig. 1(a)).

A consequence of these definitions is that Π comprises a disjoint collection of paths and cycles, where each component alternates between genes and adjacencies. Each component of Π is called a *chromosome*; paths (cycles) of Π correspond to *linear* (*circular*) chromosomes of Π. The endpoint v of a path in Π is called a *telomere* of Π; v is not incident to an adjacency, and so for clerical purposes, we say that v has the *null adjacency* $\{v, \emptyset\}$. A genome consisting of only circular (linear) chromosomes is called a *circular* (*linear*) *genome*. Note that Π is circular if and only if the edges of $a(\Pi)$ form a perfect matching on $V(\Pi)$.

Henceforth, we only consider pairs of genomes such that $g(\Pi) \cup g(\Gamma) = \mathcal{G}$. A workhorse data structure encoding the relationship between two genomes is the *breakpoint graph* ([12]), denoted by $B(\Pi, \Gamma)$ and defined as the edge-disjoint union of $a(\Pi)$ and $a(\Gamma)$, where adjacencies of Γ are colored red (Fig. 1(b)). Observe that $B(\Pi, \Gamma)$ is also a disjoint union of paths and cycles, which alternate between red and blue edges. The *length* of a component of $B(\Pi, \Gamma)$ is its number of edges; we consider an isolated vertex in $B(\Pi, \Gamma)$ to be a path of length 0.

A *double cut and join* operation (DCJ) on Π ([9]) *uses* one or two adjacencies of Π via one of the following four operations to produce a new genome Π':

1. $\{v, w\}, \{x, y\} \longrightarrow \{v, x\}, \{w, y\}$
2. $\{v, w\}, \{x, \emptyset\} \longrightarrow \{v, x\}, \{w, \emptyset\}$
3. $\{v, \emptyset\}, \{w, \emptyset\} \longrightarrow \{v, w\}$
4. $\{v, w\} \qquad \longrightarrow \{v, \emptyset\}, \{w, \emptyset\}$

The DCJ incorporates an array of genome rearrangements, as shown in Fig. 2.

For the particular case that Π and Γ have the same genes (i.e., $g(\Pi) = g(\Gamma) = \mathcal{G}$), the *DCJ distance* between Π and Γ, written $d_{\mathrm{DCJ}}(\Pi, \Gamma)$, is the minimum number of DCJs required to transform Π into Γ. A closed formula for DCJ distance was derived in [10] and translated into breakpoint graph notation in [13]:

$$d_{\mathrm{DCJ}}(\Pi, \Gamma) = N - c(\Pi, \Gamma) - \frac{p_{\mathrm{even}}(\Pi, \Gamma)}{2} \tag{1}$$

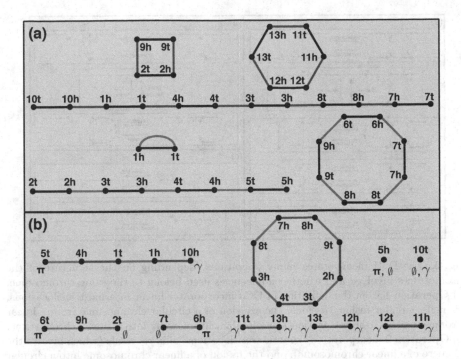

Fig. 1. (a) Genomes Π and Γ on a collection of 12 genes. We use "h" and "t" to denote the head and tail of a gene. (b) The breakpoint graph of Π and Γ. We have labeled the endpoint v of a path with π if v is π-open, with γ if v is γ-open, and with $\emptyset$ if v is a telomere of at least one genome.

Here, $c(\Pi, \Gamma)$ and $p_{\text{even}}(\Pi, \Gamma)$ denote the number of cycles and even paths in $B(\Pi, \Gamma)$, respectively (where an even path is a path of even length).

For the more general case that Π and Γ do not share the same genes, a *deletion* of a chromosomal interval of Π replaces adjacencies $\{v, w\}$ and $\{x, y\}$ (contained in the order (v, w, x, y) along a chromosome of Π) with the adjacency $\{v, y\}$ and removes the path connecting w to x. We also allow deletions of entire chromosomes; our only stipulation (introduced in [1]) is that every vertex removed from Π must belong to $\mathcal{V} - V(\Gamma)$.[1] The *insertion* of a chromosome or chromosomal interval into Π to obtain Π' is defined as the inverse of a corresponding deletion from Π' to obtain Π. Note that a consequence of this definition is that we may not insert a gene unless it is contained in $\mathcal{G}$. Insertions and deletions are collectively called *indels*; thus, we define the *DCJ-indel distance* between Π and Γ, written $d_{\text{DCJ}}^{\text{ind}}(\Pi, \Gamma)$, as the minimum number of DCJs and indels required to transform Π into Γ.

---

[1] This requirement bars the trivial transformation of Π into Γ in which every chromosome from Π is deleted, and then all the chromosomes of Γ inserted.

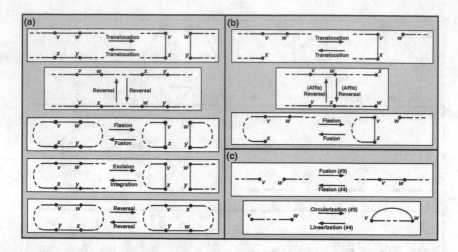

Fig. 2. The DCJ incorporates many operations, depending on the structure of the chromosomes involved and whether adjacencies used belong to the same chromosome. (a) Operation 1 from the definition of DCJ incorporates linear internal translocations, reversals, circular fusions/fissions, the excision of a circular chromosome from a linear chromosome, and the integration of a circular chromosome into a linear chromosome. (b) Operation 2 incorporates telomeric translocations, affix reversals (which involve the telomere of a linear chromosome), and the fission of a linear chromosome into a circular and linear chromosome (together with its inverse). (c) Operations 3 and 4 include linear fusions/fissions as well as the linearization/circularization of a chromosome.

3   DCJ-Indel Sorting

3.1   Handling Circular Singletons

We begin our discussion of DCJ-indel sorting by defining a *circular singleton* of Π ([1]) as a chromosome C such that $V(C) \cap V(\Gamma) = \emptyset$. Ideally, we could delete (insert) all circular singletons of Π (Γ) immediately when transforming Π into Γ; fortunately, this is indeed the case.

Proposition 1. *If Π' is formed by removing a circular singleton C from Π, then $d_{DCJ}^{ind}(\Pi', \Gamma) = d_{DCJ}^{ind}(\Pi, \Gamma) - 1$. Furthermore, when transforming Π into Γ via a minimum collection of DCJs and indels, no gene belonging to a circular singleton of Π can ever appear in the same chromosome as a gene of Γ.*

Proof. Any collection of k DCJs and indels transforming Π' into Γ can be supplemented by the deletion of C to yield $k + 1$ DCJs and indels transforming Π into Γ; thus, $d_{DCJ}^{ind}(\Pi', \Gamma) \geq d_{DCJ}^{ind}(\Pi, \Gamma) - 1$.

To obtain the reverse bound, let us view a transformation $\mathbb{T}$ of Π into Γ as a sequence $(\Pi_0, \Pi_1, \ldots, \Pi_n)$ $(n \geq 1)$, where $\Pi_0 = \Pi$, $\Pi_n = \Gamma$, and Π_{i+1} is obtained from Π_i as the result of a single DCJ or indel. Consider a sequence $(\Pi_0', \Pi_1', \ldots, \Pi_n')$, where Π_i' is constructed from Π_i by removing the subgraph of

Π_i induced by the vertices of C under the stipulation that whenever we remove a path P connecting v to w, we replace adjacencies $\{v, x\}$ and $\{w, y\}$ in Π with $\{x, y\}$ in Π'_i. It is easy to see that $\Pi'_0 = \Pi'$, $\Pi'_n = \Gamma$, and for every i in range, either Π'_{i+1} is the result of a DCJ or indel applied to Π'_i or $\Pi'_{i+1} = \Pi'_i$; thus, $(\Pi'_0, \Pi'_1, \ldots, \Pi'_n)$ encodes a transformation of Π' into Γ using at most n DCJs and indels. Furthermore, one can verify that $\Pi'_{i+1} = \Pi'_i$ only when an adjacency of C is used by the corresponding DCJ in $\mathbb{T}$ or when the vertices removed in a deletion all belong to C. At least one such operation must always occur in $\mathbb{T}$; hence, $d_{\text{DCJ}}^{\text{ind}}(\Pi', \Gamma) \le d_{\text{DCJ}}^{\text{ind}}(\Pi, \Gamma) - 1$.

The proposition's second conclusion follows from the fact that if for some j ($1 \le j \le n - 1$), a chromosome of Π_j contains a gene g_1 of Π and a gene g_2 of Γ, then one DCJ was required to combine g_1 and g_2 into the same chromosome, and another will be required to separate them, yielding two distinct values of i for which $\Pi'_{i+1} = \Pi'_i$. From the first part of the proof, we may conclude that $d_{\text{DCJ}}^{\text{ind}}(\Pi, \Gamma) < n$. □

Letting $\text{sing}(\Pi, \Gamma)$ denote the total number of circular singletons of Π and Γ, we have an immediate corollary.

Corollary 2. *The DCJ-indel distance is given by the following:*

$$d_{\text{DCJ}}^{\text{ind}}(\Pi, \Gamma) = \text{sing}(\Pi, \Gamma) + d_{\text{DCJ}}^{\text{ind}}(\Pi^0, \Gamma^0) \tag{2}$$

where Π^0 (Γ^0) is formed by removing all circular singletons from Π (Γ). □

With respect to DCJ-indel sorting, Corollary 2 allows us to assume without loss of generality that Π and Γ do not contain any circular singletons.

We next make an observation noted in [14], which is that the deletion of a chromosomal interval of Π connecting v to w may be viewed as a DCJ: $\{u, v\}, \{w, x\} \rightarrow \{u, x\}, \{v, w\}$; this operation produces a circular chromosome containing v and w that is scheduled for removal, including the case that u or x equals $\emptyset$ (the deletion of an entire linear chromosome is handled by $u = x = \emptyset$). Because insertions are the inverses of deletions, we would like to conclude that indels may be placed in a one-to-one correspondence with the removal of circular chromosomes. Ironically, the apparent exception to this proposed rule is the deletion of an entire circular chromosome.

Yet if a deleted circular chromosome C is not produced as the result of a DCJ, then C must be a circular singleton of Π in order to be deleted. Otherwise, C is produced as the result of a DCJ, which we can represent as a deletion by the method just described, unless the DCJ also creates another circular chromosome C' to be deleted. However, this sequence of operations cannot arise in a minimum collection of DCJs and indels transforming Π into Γ, as we could simply delete the chromosome(s) from which C and C' were produced by the DCJ in question, thus using strictly fewer operations.

3.2   Toward a New Model of Indels

Combining our correspondence between indels and circular chromosomes with the observation (made in [14]) that the actual removal of deleted chromosomes can occur as a final step in the transformation of Π into Γ, we may introduce the following framework.

Define a *completion* of Π as a genome Π' having $g(\Pi') = \mathcal{G}$ and for which $a(\Pi')$ is composed of $a(\Pi)$ together with a perfect matching on $V(\Pi') - V(\Pi)$. We call the adjacencies of $a(\Pi') - a(\Pi)$ *new*. Note that the chromosomes of Π embed as chromosomes of Π' and that the components of $\Pi' - \Pi$ form cycles because new adjacencies form a perfect matching on $V(\Pi') - V(\Pi)$; we may now without ambiguity call these circular chromosomes of Π' the *indels* of Π'. A *completion* of a pair of genomes (Π, Γ) is simply a pair (Π', Γ') for which Π' and Γ' are completions of Π and Γ, respectively. Our correspondence yields the following equation for DCJ-indel distance:

$$d_{\mathrm{DCJ}}^{\mathrm{ind}}(\Pi, \Gamma) = \min_{(\Pi', \Gamma')} \{d_{\mathrm{DCJ}}(\Pi', \Gamma')\} \tag{3}$$

where the minimum is taken over all completions of (Π, Γ). A completion (Π^*, Γ^*) is *optimal* if it attains the minimum in (3). Applying the closed form for the DCJ distance in (1) to (3) immediately produces the following result.

Theorem 3. *The DCJ-indel distance is given by the following equation:*

$$d_{\mathrm{DCJ}}^{\mathrm{ind}}(\Pi, \Gamma) = N - \max_{(\Pi', \Gamma')} \left\{ c(\Pi', \Gamma') + \frac{p_{\mathrm{even}}(\Pi', \Gamma')}{2} \right\} \tag{4}$$

where the maximum is taken over all completions of (Π, Γ).   □

3.3   Constructing an Optimal Completion

In light of Theorem 3, we have reduced DCJ-indel sorting to the problem of constructing indels intelligently to maximize a weighted sum of breakpoint graph components. Once we have produced an optimal completion (Π^*, Γ^*), we can simply invoke the $O(N)$-time sorting algorithm described in [10] to transform Π^* into Γ^* via a minimum collection of DCJs.

Our goal is to construct (Π^*, Γ^*) by direct analysis of $\mathrm{B}(\Pi, \Gamma)$. Because Π and Γ do not necessarily share the same genes, $\mathrm{B}(\Pi, \Gamma)$ may contain path endpoints that are not telomeres. Accordingly, we define a vertex v to be π-*open* (γ-*open*) if $v \notin \Pi$ ($v \notin \Gamma$). In other words, v must be matched to some other π-open vertex when constructing the indels of Π^*.[2] The paths of $\mathrm{B}(\Pi, \Gamma)$ are therefore classified by their endpoints: a π-*path* (γ-*path*) ends in a telomere and a π-open (γ-open) vertex; a $\{\pi, \gamma\}$-*path* ends in a π-open vertex and a γ-open vertex (such a path must be even of length at least 2); a $\{\pi, \pi\}$-*path* ($\{\gamma, \gamma\}$-*path*) ends in two π-open (γ-open) vertices and must therefore have odd length. We should also provide statistics for counting these different components. Define $p^{\pi, \gamma}$ as the number of $\{\pi, \gamma\}$-paths in $\mathrm{B}(\Pi, \Gamma)$; p_{even}^{π} as the number of even π-paths

---

[2] Note that v cannot be simultaneously π- and γ-open, although it may be a telomere of both Π and Γ or be π-open and a telomere of Γ (in both cases, v is an isolated vertex of $\mathrm{B}(\Pi, \Gamma)$, i.e., a path of length 0).

in B(Π, Γ); and p^0_{even} as the number of even paths in B(Π, Γ) containing no open vertices (i.e., containing one telomere from each genome). Similar statistics counting odd-length paths can be defined analogously. We have dropped the genomes from these statistics for the sake of simplicity; all component statistics will be taken with respect to B(Π, Γ) unless otherwise noted.

We first present a proposition regarding the parity of the paths of B(Π, Γ).

Proposition 4. *The component statistics of* B(Π, Γ) *satisfy the following condition:*

$$p^{\pi,\gamma} \equiv |p^{\pi}_{\text{odd}} - p^{\pi}_{\text{even}}| \equiv |p^{\gamma}_{\text{odd}} - p^{\gamma}_{\text{even}}| \mod 2 \tag{5}$$

Proof. The total number of π-open vertices is equal to $V(\Pi') - V(\Pi)$ and must therefore be even. Counting π-open and γ-open vertices of B(Π, Γ) over its components produces the following equivalences:

$$p^{\pi}_{\text{odd}} + p^{\pi}_{\text{even}} + p^{\pi,\gamma} \equiv 0 \mod 2 \tag{6}$$
$$p^{\gamma}_{\text{odd}} + p^{\gamma}_{\text{even}} + p^{\pi,\gamma} \equiv 0 \mod 2 \tag{7}$$

Adding $p^{\pi,\gamma}$ to both sides of (6) and (7) gives the following:

$$p^{\pi,\gamma} \equiv (p^{\pi}_{\text{odd}} + p^{\pi}_{\text{even}}) \equiv (p^{\gamma}_{\text{odd}} + p^{\gamma}_{\text{even}}) \mod 2 \tag{8}$$

The equivalence of (8) and (5) is an arithmetical fact. $\quad\square$

We next establish two necessary conditions on optimal completions by culling the set of possible adjacencies of any such completion. Our general strategy is to consider the addition of a new adjacency $\{v, w\}$ to a completion Π' as *linking* the component(s) of B(Π, Γ) ending in (π-open) vertices v and w. Our first result is that we must always link the endpoints of any $\{\pi, \pi\}$-path to each other.

Lemma 5. *If* (Π^*, Γ^*) *is an optimal completion of* (Π, Γ), *then every* $\{\pi, \pi\}$- *path* ($\{\gamma, \gamma\}$-path) *of length* $2k - 1$ *in* B(Π, Γ) *($k \geq 1$) embeds into a cycle of length $2k$ in* B(Π^*, Γ^*).

Proof. Let P be a path of length $2k - 1$ connecting π-open vertices v and w in B(Π, Γ). Suppose for the sake of contradiction that we have a completion (Π', Γ') such that P does not embed into a cycle of length $2k$ in B(Π', Γ'); in this case, we must have adjacencies $\{v, x\}$ and $\{w, y\}$ in $a(\Pi')$, where all four vertices are distinct.

Consider the completion Π'' that is identical to Π' except that $\{v, x\}$ and $\{w, y\}$ are replaced by new adjacencies $\{v, w\}$ and $\{x, y\}$. In B(Π'', Γ'), we have closed P into a cycle of length $2k$, and at the same time, we have changed neither the parity nor the linearity (respectively, circularity) of the component containing x and y. Thus, it follows from (1) that $d_{\text{DCJ}}(\Pi'', \Gamma') = d_{\text{DCJ}}(\Pi', \Gamma') - 1$, and so (Π', Γ') cannot be optimal. $\quad\square$

Having dealt with $\{\pi, \pi\}$- and $\{\gamma, \gamma\}$-paths of $B(\Pi, \Gamma)$, any remaining component of $B(\Pi^*, \Gamma^*)$ must be either a *j-bracelet*, which is a cycle linking j $\{\pi, \gamma\}$-paths (where $j \geq 2$ and j is even), or a *k-chain*, in which two π-paths or two γ-paths are linked via an intermediate number of $\{\pi, \gamma\}$-paths to form a path containing k components from $B(\Pi, \Gamma)$ $(k \geq 2)$. When k is even, a k-chain C must contain either two π-paths or two γ-paths, and when k is odd, C must contain one π-path and one γ-path.

For the sake of simplicity, we represent a j-bracelet by $(P_1 : P_2 : \cdots : P_j)$ and a k-chain by $[P_1 : P_2 : \cdots : P_k]$, where in both cases, every P_i is linked to P_{i+1}, and in the case of a j-bracelet, P_1 is linked to P_j. The length of a bracelet or chain in the breakpoint graph of an optimal completion is heavily restricted by the following lemma.

Lemma 6. *If (Π^*, Γ^*) is an optimal completion, then a component C^* of $B(\Pi^*, \Gamma^*)$ can only contain two or more $\{\pi, \gamma\}$-paths when C^* is a 2-bracelet.*

Proof. Again, say for the sake of contradiction that we have an optimal completion (Π', Γ') for which a component C' of $B(\Pi', \Gamma')$ contains two or more $\{\pi, \gamma\}$-paths. If C' is not a 2-bracelet, then it must contain two $\{\pi, \gamma\}$-paths P_1 and P_2 that are linked by precisely one adjacency. Say that P_1 joins π-open vertex v to γ-open vertex w and that P_2 joins π-open vertex x to γ-open vertex y. Furthermore, suppose that $\{v, x\} \in a(\Pi')$ but $\{w, y\} \notin a(\Gamma')$, where instead $\{w, w'\}$ and $\{y, y'\}$ are in $a(\Gamma')$. Replacing these two adjacencies with $\{w, y\}$ and $\{w', y'\}$ defines a new completion Γ'' for which $B(\Pi', \Gamma'')$ contains $(P_1 : P_2)$. Viewed as an operation on $B(\Pi', \Gamma')$ to yield $B(\Pi', \Gamma'')$, we have two cases.

First, if C' was a bracelet, then we have formed two new bracelets from C', one of which is $(P_1 : P_2)$. Otherwise, C' was a chain, in which case we have formed a chain in addition to $(P_1 : P_2)$. In either case, we may check that $d_{\mathrm{DCJ}}(\Pi', \Gamma'') < d_{\mathrm{DCJ}}(\Pi', \Gamma')$, and so (Π', Γ') cannot be optimal.   □

As a result of Lemma 6, we only allow 2-bracelets, 2-chains, and 3-chains in $B(\Pi^*, \Gamma^*)$. After a simple result about the parity of 2-chain components, we will be ready to state our main sorting result.

Proposition 7. *The breakpoint graph of an optimal completion cannot have one 2-chain joining two odd π-paths and another 2-chain joining two even π-paths. The same holds for γ-paths.*

We are now ready to state our main result on DCJ-indel sorting.

Proof. Say that (Π', Γ') is a completion with such 2-chains $[P_1 : P_2]$ and $[P_3 : P_4]$. Replacing these 2-chains with $[P_1 : P_3]$ and $[P_2 : P_4]$ forms two odd paths from two even paths; hence, (Π', Γ') cannot be optimal.   □

Theorem 8. *Algorithm 9, given below, describes an $O(N)$ time algorithm for DCJ-indel sorting. For pairs $\{\Pi, \Gamma\}$ having $\operatorname{sing}(\Pi, \Gamma) = 0$, the DCJ-indel distance is given by the following equation:*

$$d_{\mathrm{DCJ}}(\Pi, \Gamma) = N - \left[\left(c + p^{\pi,\pi} + p^{\gamma,\gamma} + \left\lfloor \frac{p^{\pi,\gamma}}{2} \right\rfloor \right) + \frac{1}{2} \left(p_{\mathrm{even}}^0 + \min \{ p_{\mathrm{odd}}^\pi, p_{\mathrm{even}}^\pi \} \right. \right.$$
$$\left. \left. + \min \{ p_{\mathrm{odd}}^\gamma, p_{\mathrm{even}}^\gamma \} + \delta \right) \right] \quad (9)$$

where $\delta = 1$ only if $p^{\pi,\gamma}$ is odd and either $p_{\mathrm{odd}}^\pi > p_{\mathrm{even}}^\pi, p_{\mathrm{odd}}^\gamma > p_{\mathrm{even}}^\gamma$ or $p_{\mathrm{odd}}^\pi < p_{\mathrm{even}}^\pi, p_{\mathrm{odd}}^\gamma < p_{\mathrm{even}}^\gamma$; otherwise, $\delta = 0$.

Proof. We aim to construct an optimal completion (Π^*, Γ^*) having

$$c(\Pi^*, \Gamma^*) = c + p^{\pi,\pi} + p^{\gamma,\gamma} + \left\lfloor \frac{p^{\pi,\gamma}}{2} \right\rfloor \quad (10)$$

$$p_{\mathrm{even}}(\Pi^*, \Gamma^*) = p_{\mathrm{even}}^0 + \min \{ p_{\mathrm{odd}}^\pi, p_{\mathrm{even}}^\pi \} + \min \{ p_{\mathrm{odd}}^\gamma, p_{\mathrm{even}}^\gamma \} + \delta \quad (11)$$

First, we count the cycles of $\mathrm{B}(\Pi^*, \Gamma^*)$. By Lemma 5, every $\{\pi, \pi\}$-path or $\{\gamma, \gamma\}$-path of $\mathrm{B}(\Pi, \Gamma)$ must be closed into a cycle by adding a single new adjacency (Step 1 of Algorithm 9). We now claim that there exists an optimal completion containing $\left\lfloor \frac{p^{\pi,\gamma}}{2} \right\rfloor$ 2-bracelets. Note that we may always replace 3-chains $[P_1 : P_2 : P_3]$ and $[P_4 : P_5 : P_6]$ (where P_1 and P_4 are π-paths) with $[P_1 : P_4], (P_2 : P_5),$ and $[P_3 : P_6]$, without increasing the DCJ-indel distance of the associated completion. This argument implies Step 2 of Algorithm 9 and produces the value of $c(\Pi^*, \Gamma^*)$ stated above.

As for the even paths of $\mathrm{B}(\Pi^*, \Gamma^*)$, if $p^{\pi,\gamma}$ is odd, then after forming a maximal collection of 2-bracelets, we will be left with one additional $\{\pi, \gamma\}$-path P. We claim that (Π^*, Γ^*) is optimal if we link as many π-paths (γ-paths) of opposite parity as possible. On the one hand, Proposition 7 states that we cannot have 2-chains $[P_1 : P_2]$ and $[P_3 : P_4]$, where P_1 and P_2 are even π-paths and P_3 and P_4 are odd π-paths. On the other hand, say that we have a 2-chain $[P_1 : P_2]$ and a 3-chain $[P_3 : P : P_4]$, where without loss of generality P_1 and P_2 are odd π-paths, P_3 is an even π-path, and P_4 is a γ-path. Replacing these chains with $[P_1 : P_3]$ and $[P_2 : P_3 : P_4]$ does not change the number of even paths in $\mathrm{B}(\Pi^*, \Gamma^*)$, implying Step 3 of Algorithm 9.

By Proposition 4, after linking as many π-paths (γ-paths) of opposite parity as possible, we are left with P, in addition to an odd number of π-paths and an odd number of γ-paths. All the π-paths must have the same parity, as must all the γ-paths; thus, we may choose any π-path P_1 and γ-path P_2 to link to P (Step 4 of Algorithm 9). If the parity of each remaining π-path equals the parity of each remaining γ-path, then the resulting 3-chain $[P_1 : P : P_2]$ is even (giving $\delta = 1$); otherwise, $[P_1 : P : P_2]$ is odd (giving $\delta = 0$). All remaining paths must be 2-chains linking pairs of π-paths or pairs of γ-paths (Step 5 of Algorithm 9).

If instead $p^{\pi,\gamma}$ is even, then $\delta = 0$, and the argument for constructing an optimal completion proceeds similarly, except that no $\{\pi, \gamma\}$-paths remain after forming a maximal collection of 2-bracelets, eliminating the need for Step 4. □

Algorithm 9. *Given genomes* (Π, Γ), *the following algorithm constructs an optimal completion* (Π^*, Γ^*) *in* $O(N)$ *time.*

0. *Remove all circular singletons from* Π *and* Γ.
1. *Close every* $\{\pi, \pi\}$-*path* $(\{\gamma, \gamma\}$-*path*) *into a cycle by adding a single new adjacency to* Π^* (Γ^*).
2. *Form a maximum set of 2-bracelets.*
3. *Form a maximum set of even 2-chains by linking pairs of* π-*paths* $(\gamma$-*paths*) *having opposite parity.*
4. *If* $p^{\pi, \gamma}$ *is odd, then link the remaining* $\{\pi, \gamma\}$-*path with any remaining* π-*path and* γ-*path.*
5. *Arbitrarily link pairs of remaining* π-*paths, all of which have the same parity. Do the same for remaining* γ-*paths.*

4   The Solution Space of DCJ-Indel Sorting

The problem of DCJ sorting is well understood, its solution space having been described in [15]. Thus, by Theorem 3, to identify the solution space of DCJ-indel sorting (an open problem), we simply need to count the number of ways to construct the indels of an optimal completion.

4.1   Handling Circular Singletons

By Proposition 1, we may consider the circular singletons of Π and Γ independently of other chromosomes; for that matter, because insertions and deletions are defined symmetrically, we may assume that Π contains k chromosomes and that Γ is the empty genome. Then by Corollary 2 and the fact that any DCJ applied to Π changes the total number of chromosomes of Π by at most 1 (see [9]), we may obtain Γ from Π in k steps if and only if we perform j successive DCJs ($0 \le j < k$), each of which fuses two circular chromosomes into one, followed by applying $k - j$ chromosome deletions.

Assuming that k is relatively small, the enumeration of all such transformations of Π into Γ poses a tedious but straightforward task, as a fusion of two circular chromosomes corresponds to a DCJ using two adjacencies from different chromosomes.

4.2   Genomes Lacking Circular Singletons

Having handled circular singletons, we may assume that $\mathrm{sing}(\Pi, \Gamma) = 0$. Fortunately, the lemmas presented before Theorem 8 have greatly reduced the collection of possible optimal completions, which we now continue to pare down.

Proposition 10. *Every* π-*path* $(\gamma$-*path*) *embedding into a 3-chain of an optimal completion must have the same parity.*

Proof. Say for the sake of contradiction that we have a completion (Π', Γ') such that $\mathrm{B}(\Pi', \Gamma')$ contains 3-chains $[P_1 : P_2 : P_3]$ and $[P_4 : P_5 : P_6]$, where P_1 and P_4 are π-paths of opposite parity. Then consider (Π'', Γ''), which is formed by rejoining adjacencies to form $[P_1 : P_4]$, $(P_2 : P_5)$, and $[P_3 : P_6]$ in $\mathrm{B}(\Pi'', \Gamma'')$. The 2-chain $[P_1 : P_4]$ must be even, and $(P_2 : P_5)$ is a cycle; thus, $d_{\mathrm{DCJ}}(\Pi'', \Gamma'') < d_{\mathrm{DCJ}}(\Pi', \Gamma')$, and so (Π', Γ') cannot be optimal. □

Proposition 11. *If $p^{\pi,\gamma}$ is even, then the breakpoint graph of an optimal completion must contain a maximum set of even 2-chains.*

Proof. We proceed by contradiction. Say that (Π', Γ') is an optimal completion for which an odd π-path P_1 and an even π-path P_2 are contained in different components of $\mathrm{B}(\Pi', \Gamma')$, neither of which is an even 2-chain. By Propositions 7 and 10, we may assume that P_1 and P_2 embed into an odd 2-chain $[P_1 : P_5]$ and a 3-chain $[P_2 : P_3 : P_4]$. Because $p^{\pi,\gamma}$ is even, we must have at least one additional 3-chain $[P_6 : P_7 : P_8]$, where (again by Proposition 10) P_6 is an even π-path and the γ-paths P_5 and P_8 have the same parity. With these assumptions in hand, we may rejoin adjacencies to form the four components $[P_1 : P_2]$, $[P_5 : P_6]$, $(P_3 : P_7)$, and $[P_4 : P_8]$, producing a cycle and at least two even 2-chains from our original three paths. Hence, (Π', Γ') cannot be optimal. □

We are now ready to fully describe the collection of optimal completions when $p^{\pi,\gamma}$ is even. To construct an optimal completion, after closing each $\{\pi,\pi\}$-path and $\{\gamma,\gamma\}$-path, we must form a maximum collection of even 2-chains by Proposition 11. Consider the following two subcases.

First, we could have that $p^{\pi}_{\mathrm{odd}} \leq p^{\pi}_{\mathrm{even}}$ and $p^{\gamma}_{\mathrm{odd}} \geq p^{\gamma}_{\mathrm{even}}$. After forming $p^{\pi}_{\mathrm{odd}} + p^{\gamma}_{\mathrm{even}}$ even 2-chains, we will be left with $p^{\pi}_{\mathrm{even}} - p^{\pi}_{\mathrm{odd}}$ even π-paths and $p^{\gamma}_{\mathrm{odd}} - p^{\gamma}_{\mathrm{even}}$ odd γ-paths. It is impossible to create any more even chains, and so we must form a maximum collection of $p^{\pi,\gamma}/2$ 2-bracelets from the $\{\pi,\gamma\}$-paths, then link arbitrary π-paths to each other and arbitrary γ-paths to each other.

Second, if $p^{\pi}_{\mathrm{odd}} > p^{\pi}_{\mathrm{even}}$ and $p^{\gamma}_{\mathrm{odd}} > p^{\gamma}_{\mathrm{even}}$, then we will have $p^{\pi}_{\mathrm{odd}} - p^{\pi}_{\mathrm{even}}$ odd π-paths and $p^{\gamma}_{\mathrm{odd}} - p^{\gamma}_{\mathrm{even}}$ odd π-paths after forming a maximum set of 2-chains. Setting $m = \min\{p^{\pi,\gamma}/2, p^{\pi}_{\mathrm{odd}} - p^{\pi}_{\mathrm{even}}, p^{\gamma}_{\mathrm{odd}} - p^{\gamma}_{\mathrm{even}}\}$, we may attain the formula in (9) by forming j (even) 3-chains for any even j satisfying $0 \leq j \leq m$; we then form $p^{\pi,\gamma}/2 - j$ total 2-bracelets from the remaining $\{\pi, \gamma\}$-paths. Any remaining odd π-paths (γ-paths) must then be linked to each other to form odd 2-chains. Reversing the inequalities in these two cases will yield analogous arguments.

If instead $p^{\pi,\gamma}$ is odd, then select one $\{\pi, \gamma\}$-path P to set aside that must belong to a 3-chain, after which we proceed as when $p^{\pi,\gamma}$ was even. There are four possibilities for the parity of the paths to which P may be linked. If $p^{\pi}_{\mathrm{odd}} < p^{\pi}_{\mathrm{even}}$ and $p^{\gamma}_{\mathrm{odd}} > p^{\gamma}_{\mathrm{even}}$, then one can verify that the only way we cannot attain (9) is if we link P to an odd π-path and an even γ-path. Furthermore, if $p^{\pi}_{\mathrm{odd}} > p^{\pi}_{\mathrm{even}}$ and $p^{\gamma}_{\mathrm{odd}} > p^{\gamma}_{\mathrm{even}}$, then the only way we can obtain (9) is if we link P to an odd π-path and an odd γ-path. Reversing these inequalities will result in analogous arguments for the remaining two cases.

Enumerating the number of optimal completions poses a tedious task, one that would take up too much space here yet can be conducted precisely for each subcase, heavily using the independence of selecting components to be linked and relying on very simple counting problems, such as the number of ways to form a matching on a collection of $2n$ vertices.

5   Conclusion

In this paper, we have demonstrated how the problem of DCJ-indel sorting, first computed in [1], can be solved via direct inspection of the breakpoint graph. Unfortunately, we do not yet see a natural correspondence between the two approaches to DCJ-indel sorting, which appear to be very different.

Furthermore, modeling an indel as a circular chromosome resulting from a DCJ has uncovered the solution space of DCJ-indel sorting, thus resolving an open problem. We wonder if other operations could be adapted to a similar model to yield a straightforward calculation of other genomic distances involving indels.

References

1. Braga, M.D.V., Willing, E., Stoye, J.: Genomic Distance with DCJ and Indels. In: Moulton, V., Singh, M. (eds.) WABI 2010. LNCS, vol. 6293, pp. 90–101. Springer, Heidelberg (2010)
2. Dobzhansky, T., Sturtevant, A.H.: Inversions in the chromosomes of drosophila pseudoobscura. Genetics 23(1), 28–64 (1938)
3. Fertin, G., Labarre, A., Rusu, I., Tannier, E., Vialette, S.: Combinatorics of Genome Rearrangements, pp. 205–206. MIT Press (2009)
4. Dweighter, H. (pseudonym of Goodman, J.): Problem E2569. American Mathematical Monthly 82, 1010 (1975)
5. Gates, W.H., Papadimitriou, C.H.: Bounds for sorting by prefix reversal. Discrete Mathematics 27(1), 47–57 (1979)
6. Heydari, M.H., Sudborough, I.H.: On the diameter of the pancake network. Journal of Algorithms 25(1), 67–94 (1997)
7. Chitturi, B., Fahle, W., Meng, Z., Morales, L., Shields, C.O., Sudborough, I.H., Voit, W.: An upper bound for sorting by prefix reversals. Theoretical Computer Science 410(36), 3372 (2009); Graphs, Games and Computation: Dedicated to Professor Burkhard Monien on the Occasion of his 65th Birthday
8. Bulteau, L., Fertin, G., Rusu, I.: Pancake flipping is hard. CoRRabs/1111.0434 (preprint)
9. Yancopoulos, S., Attie, O., Friedberg, R.: Efficient sorting of genomic permutations by translocation, inversion and block interchange. Bioinformatics 21(16), 3340–3346 (2005)
10. Bergeron, A., Mixtacki, J., Stoye, J.: A Unifying View of Genome Rearrangements. In: Bücher, P., Moret, B.M.E. (eds.) WABI 2006. LNCS (LNBI), vol. 4175, pp. 163–173. Springer, Heidelberg (2006)
11. Yancopoulos, S., Friedberg, R.: DCJ path formulation for genome transformations which include insertions, deletions, and duplications. Journal of Computational Biology 16(10), 1311–1338 (2009)
12. Bafna, V., Pevzner, P.A.: Genome rearrangements and sorting by reversals. SIAM J. Comput. 25(2), 272–289 (1996)

13. Tannier, E., Zheng, C., Sankoff, D.: Multichromosomal median and halving problems under different genomic distances. BMC Bioinformatics 10(1), 120 (2009)
14. Ma, J., Ratan, A., Raney, B.J., Suh, B.B., Miller, W., Haussler, D.: The infinite sites model of genome evolution. Proceedings of the National Academy of Sciences of the United States of America 105(38), 14254–14261 (2008)
15. Braga, M.D.V., Stoye, J.: The solution space of sorting by DCJ. Journal of Computational Biology 17(9), 1145–1165 (2010)

DCJ-indel Distance with Distinct Operation Costs*

Poly H. da Silva[1,2], Marília D.V. Braga[2],
Raphael Machado[2], and Simone Dantas[1]

[1] IME, Universidade Federal Fluminense, Brazil
[2] Inmetro - Instituto Nacional de Metrologia, Qualidade e Tecnologia, Brazil

Abstract. The double-cut and join (DCJ) is a genomic operation that generalizes the typical mutations to which genomes are subject. The distance between two genomes, in terms of number of DCJ operations, can be computed in linear time. More powerful is the DCJ-indel model, which handles genomes with unequal contents, allowing, besides the DCJ operations, the insertion and/or deletion of pieces of DNA – named indel operations. It has been shown that the DCJ-indel distance can also be computed in linear time, assuming that the same cost is assigned to any DCJ or indel operation. In the present work we consider a new DCJ-indel distance in which the indel cost is distinct from and upper bounded by the DCJ cost. Considering that the DCJ cost is equal to 1, we set the indel cost equal to a positive constant $w \leq 1$ and show that the distance can still be computed in linear time. This new distance generalizes the previous DCJ-indel distance considered in the literature (which uses the same cost for both types of operations).

1 Introduction

The distance between two genomes is often computed using only the common markers, that occur in both genomes. Such distance takes into consideration only *organizational* rearrangement operations, that change the organization of the genome, that is, the positions and orientations of markers, number and types of chromosomes. Inversions, translocations, fusions and fissions are some of these operations [1]. All these rearrangements can be generically represented as a *double-cut-and-join* (DCJ) operation [2]. The DCJ distance, which takes into consideration only DCJ operations, can be computed in linear time [3].

Nevertheless, genomes with the same content are rare, and differences in gene content may reflect important evolutionary aspects. In order to handle genomes with unequal contents, one has to take into consideration operations that change the contents of the genomes. These operations can be an *insertion* or a *deletion* of a block of contiguous markers. Insertions and deletions are also called *indels*.

In 2001, El Mabrouk [4] extended the classical sorting by inversions approach [5] and developed a method to compare unichromosomal genomes with

---

* This research was partially supported by the Brazilian agencies CNPq and FAPERJ.

B. Raphael and J. Tang (Eds.): WABI 2012, LNBI 7534, pp. 378–390, 2012.

unequal contents, considering only inversions and indels. She provided an exact algorithm that deals with insertions and deletions asymmetrically, and a heuristic that handles the operations symmetrically. More recently a model to sort multi-chromosomal genomes with unequal contents, but without duplicated markers, using both DCJ and indel operations was introduced by Yancopoulos and Friedberg [6]. Later, Braga et al. [7] presented an exact formula for the DCJ-indel distance, that can be computed in linear time handling indels symmetrically.

The approaches of El Mabrouk [4] and Braga et al. [7] assign the same cost to an organizational and to an indel operation. However, during the evolution of many organisms, indel operations are said to occur more often than organizational operations and, consequently, should be assigned to a lower cost. Examples are bacteria that are obligate intracellular parasites, such as *Rickettsia* [8]. The genomes of such intracellular parasites are observed to have a reductive evolution, that is, the process by which genomes shrink and undergo extreme levels of gene degradation and loss.

In the present work, we refine the DCJ-indel model, by adopting a distinct indel cost that is upper bounded by the DCJ cost. For simplicity, we assign a cost of 1 to DCJ and a positive cost of $w \leq 1$ to indel operations. We are then able to give an exact formula for the DCJ-indel distance for any positive $w \leq 1$ and, for $\frac{2}{3} < w \leq 1$, our formula is equivalent to the formula given by Braga et al. [7]. Indels are applied to pieces of DNA of any size, and a side effect of this fact is that the triangular inequality often does not hold for distances that consider indels [4,6,7,9]. In the case of our model, it is possible to do an *a posteriori* correction, using an approach similar to the one described in [9]. This paper is organized as follows. In Sections 2 and 3, we give definitions and previous results used in this work. In Section 4 we develop the formula for the distance with distinct DCJ and indel costs and in Section 5 we show how to establish the triangular inequality. Finally, in Section 6, we summarize our results.

2   The DCJ Model

A genome is composed of chromosomes and can be represented by a set of strings as follows. For each chromosome $\mathcal{C}$ of each genome, we build a string obtained by the concatenation of all markers in $\mathcal{C}$. Each marker g is a DNA fragment and is represented by the symbol g, if it is read in direct orientation, or by the symbol $\overline{g}$, if it is read in reverse orientation. Each one of the two extremities of a linear chromosome is called a *telomere*, represented by the symbol $\circ$. In our model, each marker occur once in a genome (duplicated markers are not allowed).

Given two genomes A and B, possibly with unequal content, let $\mathcal{G}$, $\mathcal{A}$ and $\mathcal{B}$ be three disjoint sets, such that the set $\mathcal{G}$ contains the markers that occur once in A and once in B, the set $\mathcal{A}$ contains the markers that occur only in A and the set $\mathcal{B}$ contains the markers that occur only in B. As an example, consider the genomes $A = \{obsu\overline{c}av\overline{d}e\circ\}$ and $B = \{oawb\overline{x}\overline{c}o, oydze\circ\}$, represented in Figure 1. Here we have $\mathcal{G} = \{a, b, c, d, e\}$, $\mathcal{A} = \{s, u, v\}$ and $\mathcal{B} = \{w, x, y, z\}$.

Given two genomes A and B, we denote the two extremities of each $g \in \mathcal{G}$ by g^t (tail) and g^h (head). Then, a $\mathcal{G}$-adjacency or simply adjacency [7] in genome A

$$A \quad \xrightarrow{b} \xrightarrow{s} \xrightarrow{u} \xleftarrow{\bar{c}} \xrightarrow{a} \xrightarrow{v} \xrightarrow{\bar{d}} \xrightarrow{e}$$

$$B \quad \xrightarrow{a} \xrightarrow{w} \xrightarrow{b} \xleftarrow{\bar{x}} \xrightarrow{c} \qquad \xrightarrow{y} \xrightarrow{d} \xrightarrow{z} \xrightarrow{e}$$

Fig. 1. Genomes $A = \{obsu\bar{c}av\bar{d}e\circ\}$, composed of one single linear chromosome, and $B = \{\circ awb\bar{x}c\circ, \circ ydze\circ\}$, composed of two linear chromosomes. The markers in $\mathcal{G}$ are represented in black, while the ones in $\mathcal{A}$ and in $\mathcal{B}$ are represented in red.

(respectively in genome B) is a string $v = \gamma_1 \ell \gamma_2 \equiv \gamma_2 \bar{\ell} \gamma_1$, such that each γ_i can be a telomere or an extremity of a marker from $\mathcal{G}$ and ℓ is a substring composed of the markers that are between γ_1 and γ_2 in A (respectively in B) and contains no marker that also belongs to $\mathcal{G}$. The substring ℓ is the *label* of v. If ℓ is empty, the adjacency is said to be *clean*, otherwise it is said to be *labeled*. If a linear chromosome is composed only of markers that are not in $\mathcal{G}$, it is represented by an adjacency $\circ \ell \circ$. Similarly, a circular chromosome composed only of markers that are not in $\mathcal{G}$ is represented by a (circular) adjacency ℓ. In general, a genome is either composed of circular or of linear chromosomes. For the linear genomes in Figure 1, the set of adjacencies in A is $\{\circ b^t, b^h suc^h, c^t a^t, a^h v d^h, d^t e^t, e^h \circ\}$ and the set of adjacencies in B is $\{\circ a^t, a^h wb^t, b^h \bar{x} c^t, c^h \circ, \circ y d^t, d^h z e^t, e^h \circ\}$.

2.1   DCJ Operations

A *cut* performed on a genome A separates two adjacent markers of A. A cut affects a single adjacency v in A: it is done between two symbols of v, creating two open ends[1]. A *double-cut and join* or *DCJ* applied on a genome A is the operation that performs cuts in two different adjacencies in A, creating four open ends, and joins these open ends in a different way. In other words, a DCJ rearranges two adjacencies in A, transforming them into two new adjacencies. As an example consider a DCJ applied to genome A (from Figure 1), that rearranges the adjacencies $a^h v d^h$ and $d^t e^t$ into the new adjacencies $a^h v d^t$ and $d^h e^t$. Observe that this operation corresponds to the inversion of marker d in genome A. Indeed, a DCJ operation can correspond to several rearrangement events, such as an inversion, a translocation, a fusion or a fission [2].

2.2   Adjacency Graph and the DCJ Distance

Given two genomes A and B, the *adjacency graph* $AG(A,B)$ [3] is the bipartite graph that has a vertex for each adjacency in A and a vertex for each adjacency in B. For each $g \in \mathcal{G}$, we have one edge connecting the vertex in A and the vertex in B that contain g^h and one edge connecting the vertex in A and the vertex in B that contain g^t. The connected components of $AG(A,B)$ alternate vertices in genome A and in genome B. Each component can be either a cycle, or an *AB-path* (that has one endpoint in genome A and the other in B), or an *AA-path* (that has both endpoints in genome A), or a *BB-path* (that has both

---

[1] In general a cut can be performed between two markers of a label, but the DCJ-indel distance can be computed considering only cuts that do not "break" labels.

endpoints in B). A special case of an AA or a BB-path is a *linear singleton*, that is a linear chromosome represented by an adjacency of type $\circ\ell\circ$. Paths occur when the genomes are linear. For circular genomes, the graph $AG(A,B)$ is composed of cycles only, and may also have a special type of component composed of a single vertex, that corresponds to a circular chromosome composed only of markers that are not in $\mathcal{G}$, called *circular singleton*. In Figure 2 we show the adjacency graph built over the linear genomes represented in Figure 1.

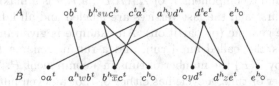

Fig. 2. For genomes A and B (Figure 1), the graph has one BB and two AB-paths

Components with 3 or more vertices need to be reduced, by applying DCJs, to components with only 2 vertices, that can be cycles or AB-paths [10]. This procedure is called *DCJ-sorting* of A into B. The number of AB-paths in $AG(A,B)$ is always even and a DCJ can be of three types [7]: it can either increase the number of cycles by one, or the number of AB-paths by two (*optimal*); or it does not affect the number of cycles and AB-paths (*neutral*); or it can either decrease the number of cycles by one, or the number of AB-paths by two (*counter-optimal*). The DCJ distance of A and B, denoted by $d_{DCJ}(A,B)$, is the minimum number of steps required to do a DCJ-sorting of A into B, given by the following theorem.

Theorem 1 ([3]). *Given two genomes A and B without duplicated markers, we have $d_{DCJ}(A,B) = |\mathcal{G}| - c - \frac{b}{2}$, where $\mathcal{G}$ is the set of common markers and c and b are, respectively, the number of cycles and of AB-paths in $AG(A,B)$.*

3 Handling Insertions and Deletions

3.1 Indel Operations

In order to deal with genomes that have unequal contents, we need operations that change the content of a genome. These operations can be an *insertion* or a *deletion* of a block of markers. We refer to insertions and deletions as *indel* operations. An indel only affects the label of one single adjacency, by deleting or inserting contiguous markers in this label, with the restriction that an insertion cannot produce duplicated markers [7]. Thus, while sorting A into B, the indels are the steps in which the markers in $\mathcal{A}$ are deleted and the markers in $\mathcal{B}$ are inserted. At most one chromosome can be entirely deleted or inserted at once.

In a different model, proposed in [11], a deletion and a subsequent insertion performed at the same position of the genome count for a single *substitution*. In the present work, however, substitutions are not allowed. Here a deletion and a subsequent insertion always count for two sorting steps, even if they are performed at the same position. We illustrate an indel with the following example: the deletion of markers su from adjacency $b^h suc^h$ of genome A (Figure 2), which results in the clean adjacency $b^h c^h$. The opposite operation would be an insertion.

3.2   Runs and Indel-Potential

Let us recall the concept of *run*, introduced by Braga *et al.* [7]. Given two genomes A and B and a component P of $AG(A,B)$, a *run* is a maximal subpath of P, in which the first and the last vertices are labeled and all labeled vertices belong to the same genome (or partition). An example is given in Figure 3. A run in genome A is also called an $\mathcal{A}$-run, and a run in genome B is called a $\mathcal{B}$-run. We denote by $\Lambda(P)$ the number of runs in a component P. While a path can have any number or runs, a cycle has either 0, 1, or an even number of runs.

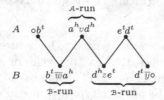

Fig. 3. An AB-path with 3 runs (extracted from Figure 2)

A set of labels of one genome can be accumulated with DCJs. For example, take the adjacencies $d^h z e^t$ and $d^t \bar{y} \circ$ from genome B (Figure 3). A DCJ applied to these two adjacencies could result into $d^t e^t$ and $d^h z \bar{y} \circ$, in which the label $z\bar{y}$ resulted from the accumulation of the labels of the two original adjacencies. In particular, when we apply optimal DCJs on only one component of the adjacency graph, we can accumulate an entire run into a single adjacency [7].

It is possible to do a separate DCJ-sorting using only optimal DCJs in any component P of $AG(A,B)$ [10]. We denote by $d_{DCJ}(P)$ the number of optimal DCJ operations used for DCJ-sorting P separately ($d_{DCJ}(P)$ depends only on the number of vertices or, equivalently, the number of edges of P [10]). The DCJ distance can also be re-written as $d_{DCJ}(A,B) = \sum_{P \in AG(A,B)} d_{DCJ}(P)$.

Runs can be merged by DCJ operations. Consequently, during the optimal DCJ-sorting of a component P, we can reduce its number of runs. The *indel-potential* of P, denoted by $\lambda(P)$, is defined by Braga *et al.* [7] as the minimum number of runs that we can obtain doing a separate DCJ-sorting in P with optimal DCJ operations. An example is given in Figure 4.

The indel-potential of a component depends only on its number of runs:

Proposition 1 ([7]). *Given two genomes A and B and a component P of $AG(A,B)$, the indel-potential of P is given by $\lambda(P) = \lceil \frac{\Lambda(P)+1}{2} \rceil$, if $\Lambda(P) \geq 1$. Otherwise, if $\Lambda(P) = 0$, then $\lambda(P) = 0$.*

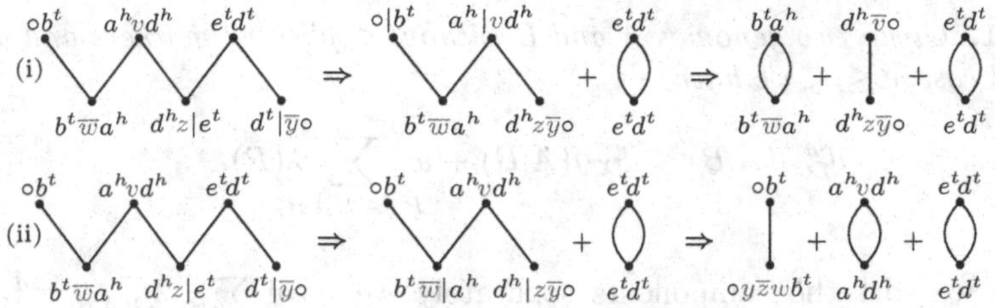

Fig. 4. Two optimal sequences for DCJ-sorting an AB-path with $\Lambda = 3$ (the cuts of each DCJ in each sequence are represented by "$|$"). In (i) the overall number of runs in the resulting components is three, while in (ii) the resulting components have only two runs. Indeed, in this case, the best we can have is the indel-potential $\lambda = 2$.

Let λ_0 and λ_1 be, respectively, the sum of the indel-potentials for the components of the adjacency graph before and after a DCJ operation ρ, and let $\Delta\lambda(\rho) = \lambda_1 - \lambda_0$. If ρ is an optimal DCJ acting on two adjacencies of a single component of the graph, the definition of indel-potential implies $\Delta\lambda(\rho) \geq 0$. It remains to examine $\Delta\lambda(\rho)$ if ρ is neutral or counter-optimal.

Proposition 2 ([7]). *Given a DCJ operation ρ acting on a single component, we have $\Delta\lambda(\rho) \geq 0$, if ρ is counter-optimal, and $\Delta\lambda(\rho) \geq -1$, if ρ is neutral.*

4 Computing the DCJ-indel Distance

We assign the cost of 1 to each DCJ and a positive cost $w \leq 1$ to each indel operation. Given two genomes A and B, we define the *DCJ-indel distance* of A and B, denoted by $d_{DCJ}^{id}(A,B)$, as the minimum cost of a DCJ-indel sequence of operations that sorts A into B. If $w = 1$, the DCJ-indel distance corresponds exactly to the minimum number of steps required to sort A into B. To compute the distance in this case, a linear algorithm was given by Braga *et al.* [7]. In this section, we present a more general method to compute the DCJ-indel distance for any positive $w \leq 1$.

We already know that each component P of $AG(A,B)$ can be sorted separately with DCJ and indel operations. Let $d_{DCJ}^{id}(P)$ be the DCJ-indel distance of P, which can be computed according to the following proposition.

Proposition 3. *For each $P \in AG(A,B)$, $d_{DCJ}^{id}(P) = d_{DCJ}(P) + w\lambda(P)$.*

Proof. By the definition of λ, the best we can do with optimal DCJs is $d_{DCJ}(P) + w\lambda(P)$. From Proposition 2, we have $\Delta\lambda(\rho) \geq 0$ if ρ is counter-optimal, thus we can only get more expensive sorting scenarios if we use such operation. We also know that $\Delta\lambda(\rho) \geq -1$ if ρ is neutral, and this kind of operation increases the sorting scenario by one DCJ with respect to the scenario with only optimal DCJs. If we use n neutral DCJs, this gives at least $d_{DCJ}(P) + n + w(\lambda(P) - n) \geq d_{DCJ}(P) + w\lambda(P)$, for any positive $w \leq 1$. ∎

With the previous observations, we have a tight upper bound for the DCJ-indel distance with distinct operation costs:

Lemma 1. *Given two genomes A and B without duplicated markers and a positive indel cost $w \leq 1$, we have*

$$d_{DCJ}^{id}(A,B) \leq d_{DCJ}(A,B) + w \sum_{P \in AG(A,B)} \lambda(P).$$

Proof. If we sort the components separately we have $\sum_{P \in AG(A,B)} d_{DCJ}^{id}(P)$, which corresponds exactly to $d_{DCJ}(A,B) + w \sum_{P \in AG(A,B)} \lambda(P)$.   ∎

4.1   Recombinations

Until this point, we have explored the possible effects of any DCJ that is applied to two adjacencies belonging to a single component of the graph. However, another type of DCJ must be considered. A DCJ ρ applied to adjacencies belonging to two different components is called a *recombination* and has $\Delta\lambda(\rho) \geq -2$ [7]. The components on which the cuts are applied are called *sources* and the components obtained after the joinings are called *resultants* of the recombination.

Since optimal DCJs are already counted in the upper bound given by Lemma 1 and the cost of a DCJ is 1, let $\Delta_{DCJ}(\rho)$ be respectively 0, 1 and 2 depending on whether the recombination ρ is optimal, neutral or counter-optimal. Any recombination applied to a vertex of an AA-path and a vertex of a BB-path is optimal [10]. A recombination applied to vertices of two different AB-paths can be either neutral, when the resultants are also AB-paths, or counter-optimal, when the resultants are an AA-path and a BB-path. All other path recombinations are neutral and all recombinations involving at least one cycle are counter-optimal. We define $\Delta d(\rho) = \Delta_{DCJ}(\rho) + w\Delta\lambda(\rho)$. Any counter-optimal recombination has $\Delta d \geq 2 - 2w \geq 0$, thus only path recombinations can have $\Delta d < 0$.

Let $\mathcal{A}$ (respectively $\mathcal{B}$) be a sequence with an odd (≥ 1) number of runs, starting and ending with an $\mathcal{A}$-run (respectively $\mathcal{B}$-run). We can then make any combination of $\mathcal{A}$ and $\mathcal{B}$, such as $\mathcal{AB}$, that is a sequence with an even (≥ 2) number of runs, starting with an $\mathcal{A}$-run and ending with a $\mathcal{B}$-run. An empty sequence (with no run) is represented by ε. Then each one of the notations AA_ε, $AA_\mathcal{A}$, $AA_\mathcal{B}$, $AA_{\mathcal{AB}} \equiv AA_{\mathcal{BA}}$, BB_ε, $BB_\mathcal{A}$, $BB_\mathcal{B}$, $BB_{\mathcal{AB}} \equiv BB_{\mathcal{BA}}$, AB_ε, $AB_\mathcal{A}$, $AB_\mathcal{B}$, $AB_{\mathcal{AB}}$ and $AB_{\mathcal{BA}}$ represents a particular type of path (AA, BB or AB) with a particular structure of runs (ε, $\mathcal{A}$, $\mathcal{B}$, $\mathcal{AB}$ or $\mathcal{BA}$). By convention, an AB-path is always read from A to B. We adopt these notations due to the observation that, besides the DCJ type of the recombination (optimal, neutral or counter-optimal), the only properties that matter are whether the components have an odd or an even number of runs and whether the first run is in A or in B. A neutral recombination, possibly with $\Delta d < 0$, is represented in Figure 5.

Each type of recombination can lead to different resultants, depending on where the cuts are applied. However, it is always possible to choose the "best" resultants in each case: we take the recombination with the smallest Δd, whose resultants can be better reused in further recombinations. The main observations to guide this task are: only recombinations of paths whose runs are $\mathcal{AB}$ or $\mathcal{BA}$

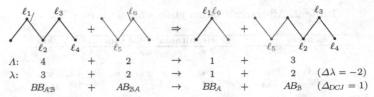

$$\Lambda: \quad 4 \quad + \quad 2 \quad \to \quad 1 \quad + \quad 3$$
$$\lambda: \quad 3 \quad + \quad 2 \quad \to \quad 1 \quad + \quad 2 \quad (\Delta\lambda = -2)$$
$$BB_{AB} \quad + \quad AB_{BA} \quad \to \quad BB_A \quad + \quad AB_B \quad (\Delta_{DCJ} = 1)$$

Fig. 5. Neutral recombination that has $\Delta d = 1 - 2w$ (we represent only the labels of the adjacencies, the cuts of the recombination are represented by "/" and "\")

have $\Delta\lambda = -2$ and only recombinations of type $AA + BB$ are optimal and have $\Delta_{DCJ} = 0$. In Table 1, we list all path recombinations that can have $\Delta d < 0$, together with neutral recombinations that have $\Delta d = 1 - w \geq 0$, but produce an AA_{AB} or a BB_{AB} path. We denote by $\bullet$ an AB-path that never appears as a source of a recombination in this table (these paths are AB_ε, AB_A and AB_B).

Table 1. Path recombinations of type o_{-2} (optimal with $\Delta\lambda = -2$), o_{-1} (optimal with $\Delta\lambda = -1$) and n_{-2} (neutral with $\Delta\lambda = -2$) can have $\Delta d < 0$. Recombinations of type n_{-1} (neutral with $\Delta\lambda = -1$) have $\Delta d = 1 - w \geq 0$, but produce an AA_{AB} or a BB_{AB} path.

Sources	Resultants	$\Delta\lambda$	Δ_{DCJ}	Δd	Sources	Resultants	$\Delta\lambda$	Δ_{DCJ}	Δd
o_{-2} $AA_{AB} + BB_{AB}$	$\bullet + \bullet$	-2	0	$-2w$					
					n_{-2} $AA_{AB} + AA_{AB}$	$AA_A + AA_B$	-2	1	$1-2w$
o_{-1} $AA_A + BB_{AB}$	$\bullet + AB_{AB}$	-1	0	$-w$	n_{-2} $BB_{AB} + BB_{AB}$	$BB_A + BB_B$	-2	1	$1-2w$
o_{-1} $BB_A + AA_{AB}$	$\bullet + AB_{BA}$	-1	0	$-w$	n_{-2} $AA_{AB} + AB_{BA}$	$\bullet + AA_A$	-2	1	$1-2w$
o_{-1} $AA_B + BB_{AB}$	$\bullet + AB_{BA}$	-1	0	$-w$	n_{-2} $AA_{AB} + AB_{BA}$	$\bullet + AA_B$	-2	1	$1-2w$
o_{-1} $BB_B + AA_{AB}$	$\bullet + AB_{AB}$	-1	0	$-w$	n_{-2} $BB_{AB} + AB_{AB}$	$\bullet + BB_B$	-2	1	$1-2w$
o_{-1} $AA_A + BB_A$	$\bullet + \bullet$	-1	0	$-w$	n_{-2} $BB_{AB} + AB_{BA}$	$\bullet + BB_A$	-2	1	$1-2w$
o_{-1} $AA_B + BB_B$	$\bullet + \bullet$	-1	0	$-w$	n_{-2} $AB_{AB} + AB_{BA}$	$\bullet + \bullet$	-2	1	$1-2w$
n_{-1} $AA_A + AB_{BA}$	$\bullet + AA_{AB}$	-1	1	$1-w$	n_{-1} $BB_A + AB_{AB}$	$\bullet + BB_{AB}$	-1	1	$1-w$
n_{-1} $AA_B + AB_{AB}$	$\bullet + AA_{AB}$	-1	1	$1-w$	n_{-1} $BB_B + AB_{BA}$	$\bullet + BB_{AB}$	-1	1	$1-w$

4.2   The Distance Formula

By analyzing the whole universe of operations, we could identify groups of recombinations, as listed in Table 2. Since some resultants of recombinations can be used in other recombinations, the groups can have more than one recombination. Groups $\mathcal{P}$, $\mathcal{S}_1$ and $\mathcal{S}_2$ are composed of a single recombination, while groups $\mathcal{T}$, $\mathcal{N}_1$ and $\mathcal{N}_2$ are composed of two recombinations and groups $\mathcal{Q}$ and $\mathcal{M}$ are composed of three recombinations. Remark that the recombination is not an associative operation, thus, in column 'DCJ seq.' of Table 2, we indicate how the sequence of DCJs must be applied in each group (the symbol $\prec$ separates preceeding and succeeding recombinations).

While in groups $\mathcal{Q}$ and $\mathcal{T}$ the preceeding recombinations have lower Δd, in groups $\mathcal{M}$, $\mathcal{N}_1$ and $\mathcal{N}_2$ we need to use operations of type n_{-1} in order to prepare better recombinations. Another important observation concerning groups $\mathcal{Q}$ and $\mathcal{T}$ is that, although their Δd indicate that $\mathcal{Q}$ could be applied for $w > 1/4$ and $\mathcal{T}$ could be applied for $w > 1/3$, the last operation of these groups is of type n_{-2} and actually increases Δd for $w \leq 1/2$. For this reason, we skip groups $\mathcal{Q}$ and $\mathcal{T}$

Table 2. All recombination groups obtained from Table 1

	Sources	Resultants	DCJ seq.	Δd	skip if
$\mathcal{P}$	$AA_{AB} + BB_{AB}$	2 •	o_{-2}	$-2w$	
$\mathcal{Q}$	$2AA_{AB} + BB_A + BB_B$	4 •	$2o_{-1} \prec n_{-2}$	$1-4w$	$w \leq \frac{1}{2}$
	$2BB_{AB} + AA_A + AA_B$	4 •		$1-4w$	
$\mathcal{T}$	$AA_{AB} + BB_A + AB_{AB}$	3 •	$o_{-1} \prec n_{-2}$	$1-3w$	$w \leq \frac{1}{2}$
	$AA_{AB} + BB_B + AB_{BA}$	3 •		$1-3w$	
	$BB_{AB} + AA_A + AB_{AB}$	3 •		$1-3w$	
	$BB_{AB} + AA_B + AB_{AB}$	3 •		$1-3w$	
	$2BB_{AB} + AA_A$	2 • $+BB_B$		$1-3w$	
	$2BB_{AB} + AA_B$	2 • $+BB_A$		$1-3w$	
	$2AA_{AB} + BB_A$	2 • $+AA_B$		$1-3w$	
	$2AA_{AB} + BB_B$	2 • $+AA_A$		$1-3w$	
$\mathcal{S}_1$	$AA_A + BB_A$	2 •	o_{-1}	$-w$	
	$AA_B + BB_B$	2 •		$-w$	
	$AA_{AB} + BB_A$	• $+ AB_{BA}$		$-w$	
	$AA_{AB} + BB_B$	• $+ AB_{AB}$		$-w$	
	$BB_{AB} + AA_A$	• $+ AB_{AB}$		$-w$	
	$BB_{AB} + AA_B$	• $+ AB_{BA}$		$-w$	
$\mathcal{S}_2$	$AB_{AB} + AB_{BA}$	2 •	n_{-2}	$1-2w$	$w \leq \frac{1}{2}$
	$AA_{AB} + AB_{AB}$	• $+ AA_A$		$1-2w$	
	$AA_{AB} + AB_{BA}$	• $+ AA_B$		$1-2w$	
	$BB_{AB} + AB_{AB}$	• $+ BB_B$		$1-2w$	
	$BB_{AB} + AB_{BA}$	• $+ BB_A$		$1-2w$	
	$AA_{AB} + AA_{AB}$	$AA_A + AA_B$		$1-2w$	
	$BB_{AB} + BB_{AB}$	$BB_A + BB_B$		$1-2w$	
$\mathcal{M}$	$2AB_{AB} + AA_B + BB_A$	4 •	$2n_{-1} \prec o_{-2}$	$2-4w$	$w \leq \frac{1}{2}$
	$2AB_{BA} + AA_A + BB_B$	4 •		$2-4w$	
$\mathcal{N}_1$	$AB_{AB} + AA_B + BB_A$	2 • $+AB_{BA}$	$n_{-1} \prec o_{-1}$	$1-2w$	$w \leq \frac{1}{2}$
	$AB_{BA} + AA_A + BB_B$	2 • $+AB_{AB}$		$1-2w$	
$\mathcal{N}_2$	$2AB_{AB} + AA_B$	2 • $+AA_A$	$n_{-1} \prec n_{-2}$	$2-3w$	$w \leq \frac{2}{3}$
	$2AB_{AB} + BB_A$	2 • $+BB_B$		$2-3w$	
	$2AB_{BA} + AA_A$	2 • $+AA_B$		$2-3w$	
	$2AB_{BA} + BB_B$	2 • $+BB_A$		$2-3w$	

for $w \leq 1/2$ (there is no loss with this approach, since their optimal operations are then counted in $\mathcal{S}_1$).

The deductions shown in Table 2 can be computed with an approach that greedily maximizes the number of occurrences in $\mathcal{P}$, $\mathcal{Q}$, $\mathcal{T}$, $\mathcal{S}_1$, $\mathcal{S}_2$, $\mathcal{M}$, $\mathcal{N}_1$ and $\mathcal{N}_2$ in this order. The two groups in $\mathcal{Q}$ are mutually exclusive after maximizing $\mathcal{P}$. The lines in $\mathcal{T}$ are subgroups of the lines in $\mathcal{Q}$, that is, they are only computed when there are enough remaining components after maximizing $\mathcal{Q}$. Similarly, each one of the remaining groups are computed when there are enough remaining components after maximizing the upper groups. With the results presented in this section we have an exact formula to compute the DCJ-indel distance:

Theorem 2. *Given two genomes A and B without duplicated markers and a positive indel cost $w \leq 1$, we have*

$$d_{DCJ}^{id}(A,B) = d_{DCJ}(A,B) + w \sum_{P \in AG(A,B)} \lambda(P) - 2w\mathcal{P} - (4w-1)\mathcal{Q} - (3w-1)\mathcal{T}$$

$$- w\mathcal{S}_1 - (2w-1)(\mathcal{S}_2 + 2\mathcal{M} + \mathcal{N}_1)$$

$$- (3w-2)\mathcal{N}_2$$

where $\mathcal{P}$, $\mathcal{Q}$, $\mathcal{T}$, $\mathcal{S}_1$, $\mathcal{S}_2$, $\mathcal{M}$, $\mathcal{N}_1$ and $\mathcal{N}_2$ are computed as described above.

As we mentioned before, the groups $\mathcal{Q}$ and $\mathcal{T}$ are skipped ($\mathcal{Q} = \mathcal{T} = 0$) for $w \leq 1/2$. Furthermore, we also have $\mathcal{S}_2 = \mathcal{M} = \mathcal{N}_1 = 0$ if $w \leq 1/2$ and $\mathcal{N}_2 = 0$ if $w \leq 2/3$. Although some groups have reusable resultants, those are actually never reused (if groups that are lower in the table use as sources resultants from higher groups, the sources of all referred groups would be previously consumed in groups that occupy even higher positions in the table). Due to this fact, the number of occurrences in each group depends only on w and the initial number of each type of component. Both $AG(A,B)$ and $d_{DCJ}(A,B)$ can be computed in linear time [3]. The types of paths can be obtained by a single walk through each path of $AG(A,B)$, thus the whole procedure takes linear time.

Observe that, for $w = 1$, our formula is identical to the one proposed by Braga *et al.* [7]. Actually, for any $2/3 < w \leq 1$, the two formulas are equivalent, since the same occurrences of groups of recombinations and an equivalent upper bound are taken into account. We illustrate the result of our formula with an example. Let $AG(A,B)$ have only the following labeled paths: two AA_{AB}, one BB_A and one BB_B. In this case, there are no occurrences of $\mathcal{P}$, thus we have $\mathcal{P} = 0$. If we take $w > \frac{1}{2}$, all labeled paths are consumed in one occurrence of $\mathcal{Q}$. We have $\mathcal{Q} = 1$, while all other values are zero, resulting in $\Delta d - 1 - 4w$. On the other hand, if $w \leq \frac{1}{2}$, we automatically set $\mathcal{Q} = \mathcal{T} = \mathcal{S}_2 = \mathcal{M} = \mathcal{N}_1 = \mathcal{N}_2 = 0$. The labeled paths are consumed in two occurrences of $\mathcal{S}_1$, that is, $\mathcal{S}_1 = 2$, resulting in $\Delta d = -2w$. For sure, $-2w \leq 1 - 4w$ only if $w \leq \frac{1}{2}$.

5   Establishing the Triangular Inequality

Let A, B and C be three genomes and denote by $\mathcal{A}$, $\mathcal{B}$, $\mathcal{C}$, $\mathcal{D}$, $\mathcal{E}$, $\mathcal{F}$ and $\mathcal{G}$ the disjoint sets of markers such that: $\mathcal{A}$, $\mathcal{B}$ and $\mathcal{C}$ are the sets of markers that occur respectively only in A, B and C, the markers in $\mathcal{D}$ are common only to A and B, the markers in $\mathcal{E}$ are common only to B and C, the markers in $\mathcal{F}$ are common only to A and C, and, $\mathcal{G}$ is the set of markers that are common to A, B and C. Without loss of generality, let $d_{DCJ}^{id}(A,B) \geq d_{DCJ}^{id}(A,C)$ and $d_{DCJ}^{id}(A,B) \geq d_{DCJ}^{id}(B,C)$. The triangular inequality is then the property which guarantees that the inequality $d_{DCJ}^{id}(A,B) \leq d_{DCJ}^{id}(A,C) + d_{DCJ}^{id}(B,C)$ also holds. This is indeed the case when $\mathcal{D} = \emptyset$, meaning that genomes A and B have no common marker that does not occur in C [9]. However, if $\mathcal{D} \neq \emptyset$, the triangular inequality can be disrupted for d_{DCJ}^{id}, and this may be an obstacle if one intends to use the DCJ-indel distance to compute the median of three or more genomes and in phylogenetic reconstructions. Take for example the genomes $A = \{oabcdeo\}$, $B = \{oac\overline{d}beo\}$ and $C = \{oaeo\}$ [6]. While the cost of sorting A (or B) into C is w (one indel), the minimum number of DCJs (that are inversions in this case) required to sort A into B is three. We have $d_{DCJ}^{id}(A,B) = 3$, $d_{DCJ}^{id}(A,C) = w$, $d_{DCJ}^{id}(B,C) = w$ and the triangular inequality is disrupted.

It is possible to establish the triangular inequality in our model *a posteriori*, by adapting an approach proposed in [9]: we simply sum to the distance a surcharge that depends on the number of unique markers. For genomes A and B and a positive constant k, let $m(A,B) = d_{DCJ}^{id}(A,B) + k \cdot u(A,B)$, where $u(A,B)$ is the number of unique markers between A and B [7,9]. We then have $m(A,B) = d_{DCJ}^{id}(A,B) + k(|\mathcal{A}| + |\mathcal{F}| + |\mathcal{B}| + |\mathcal{E}|)$, $m(A,C) = d_{DCJ}^{id}(A,C) + k(|\mathcal{A}| + |\mathcal{D}| + |\mathcal{C}| + |\mathcal{E}|)$ and $m(B,C) = d_{DCJ}^{id}(B,C) + k(|\mathcal{B}| + |\mathcal{D}| + |\mathcal{C}| + |\mathcal{F}|)$. From this definition, we can derive a simpler inequality that can be used to determine the value of the constant k:

Proposition 4 ([9]). *Given three genomes A, B and C without duplicated markers, the inequality $m(A,B) \leq m(A,C) + m(B,C)$ holds if, and only if, $d_{DCJ}^{id}(A,B) \leq d_{DCJ}^{id}(A,C) + d_{DCJ}^{id}(B,C) + 2k|\mathcal{D}|$, where $\mathcal{D}$ is the set of markers common only to A and B.*

The problem now is to find the minimum value of k for which the inequality of Proposition 4 holds. In order to accomplish this task, the first step is to determine the diameter of the DCJ-indel distance.

Lemma 2. *Given a positive indel cost $w \leq 1$ and two genomes A and B with n common markers, then*

$$d_{DCJ}^{id}(A,B) \leq (w+1)n + w(L_A + S_A + L_B + S_B),$$

where L_A, S_A and L_B, S_B are, respectively, the number of linear chromosomes and circular singletons in genomes A and B.

Proof. Let $|P|$ be the number of vertices in component P, that is DCJ-sorted with $\lfloor \frac{|P|-1}{2} \rfloor$ DCJs [10]. If $|P|$ is even, P is sorted with $\frac{|P|}{2} - 1$ DCJs and $\lambda(P) \leq \frac{|P|}{2} + 1$ indels, then $d_{DCJ}^{id}(P) \leq \frac{|P|}{2} - 1 + w(\frac{|P|}{2} + 1) = \frac{(w+1)|P|}{2} + w - 1$. If $|P|$ is odd, P is sorted with $\frac{|P|-1}{2}$ DCJs and $\lambda(P) \leq \frac{|P|+1}{2}$ indels, then $d_{DCJ}^{id}(P) \leq \frac{|P|-1}{2} + w\frac{|P|+1}{2} = \frac{(w+1)|P|+w-1}{2}$. As $w \leq 1$ implies $w - 1 \leq \frac{w-1}{2} \leq 0$, for any component P we have $d_{DCJ}^{id}(P) \leq \frac{(w+1)|P|+w-1}{2}$. Then, $d_{DCJ}^{id}(A,B) \leq \sum_{P \in AG(A,B)} d_{DCJ}^{id}(P) \leq \sum_{P \in AG(A,B)} \frac{(w+1)|P|+w-1}{2} = \frac{w+1}{2} \sum_{P \in AG(A,B)} |P| + \sum_{P \in AG(A,B)} \frac{w-1}{2}$. Each linear chromosome corresponds to one path in $AG(A,B)$, thus the number of components is at least $(L_A + S_A + L_B + S_B)$ and $\sum_{P \in AG(A,B)} \frac{w-1}{2} \leq \frac{(L_A + S_A + L_B + S_B)(w-1)}{2} \leq 0$. Furthermore, from [9] we know that $\sum_{P \in AG(A,B)} |P| = 2n + L_A + S_A + L_B + S_B$. ∎

We are ready to generalize the result of [9], and determine the minimum possible value of k.

Theorem 3. *For any positive indel cost $w \leq 1$, the function m satisfies the triangular inequality if and only if $k \geq \frac{w+1}{2}$.*

Proof. Recall that, to prove the triangular inequality for m, we only need to find a k such that $d_{DCJ}^{id}(A,B) \leq d_{DCJ}^{id}(A,C) + d_{DCJ}^{id}(B,C) + 2k|\mathcal{D}|$ holds (Proposition 4).

We know that the inequality holds when $\mathcal{D} = \emptyset$ [9]. It remains to examine the case in which $\mathcal{D} \neq \emptyset$. The worst case would be to have an empty genome C [9]. Let X_A and X_B be the number of chromosomes in A and B. Since C is empty, we know that $d_{DCJ}^{id}(A,C) = wX_A$ and $d_{DCJ}^{id}(B,C) = wX_B$. From Lemma 2, we have $d_{DCJ}^{id}(A,B) \leq (w+1)|\mathcal{D}|+w(L_A+S_A+L_B+S_B)$. This gives $(w+1)|\mathcal{D}|+w(L_A+S_A+L_B+S_B) \leq w(X_A+X_B)+2k|\mathcal{D}|$. Since $L_A+S_A+L_B+S_B \leq X_A+X_B$, we have $(w+1)|\mathcal{D}| \leq 2k|\mathcal{D}|$, which holds for any $k \geq \frac{w+1}{2}$.

For the necessity, take A and B with n common markers and let each genome be composed of one circular chromosome, meaning that we have one adjacency per common marker in each genome (or n adjacencies per genome). Then let $AG(A,B)$ have one single cycle with $2n$ vertices and let each vertex be labeled, so that the number of runs in the cycle is $2n$ and the number of unique markers in each genome is n. Thus, we have $d_{DCJ}^{id}(A,B) = (n-1)+w(n+1) = (w+1)n+(w-1)$ and the corrected distance is $m(A,B) = (w+1)n+(w-1)+2kn$. Take C as an empty genome, so that $d_{DCJ}^{id}(A,C) = d_{DCJ}^{id}(B,C) = w$ and $m(A,C) = m(B,C) = w+2kn$. The inequality $m(A,B) \leq m(A,C)+m(B,C)$ corresponds to $(w+1)n+(w-1)+2kn \leq 2w+4kn$ or, equivalently, $2kn \geq (w+1)n-w-1$, that is $k \geq \frac{w+1}{2}\left(1-\frac{1}{n}\right)$, which holds for all n only if $k \geq \frac{w+1}{2}$. ∎

6 Conclusions

In this work we have presented a method to compute in linear time the DCJ-indel distance between two genomes without duplicated markers, when the indel cost is distinct from and upper bounded by the DCJ cost. Indels can be applied to pieces of DNA of any size, and a side effect of this property is that the triangular inequality does not hold for our distance formula. However we have shown that a correction can be applied to establish the triangular inequality *a posteriori*.

References

1. Hannenhalli, S., Pevzner, P.: Transforming men into mice (polynomial algorithm for genomic distance problem). In: Proc. of FOCS, pp. 581–592 (1995)
2. Yancopoulos, S., Attie, O., Friedberg, R.: Efficient sorting of genomic permutations by translocation, inversion and block interchange. Bioinformatics 21, 3340–3346 (2005)
3. Bergeron, A., Mixtacki, J., Stoye, J.: A Unifying View of Genome Rearrangements. In: Bücher, P., Moret, B.M.E. (eds.) WABI 2006. LNCS (LNBI), vol. 4175, pp. 163–173. Springer, Heidelberg (2006)
4. El-Mabrouk, N.: Sorting signed permutations by reversals and insertions/deletions of contiguous segments. Journal of Discrete Algorithms 1(1), 105–122 (2001)
5. Hannenhalli, S., Pevzner, P.: Transforming cabbage into turnip: polynomial algorithm for sorting signed permutations by reversals. Journal of the ACM 46, 1–27 (1999)
6. Yancopoulos, S., Friedberg, R.: DCJ path formulation for genome transformations which include insertions, deletions, and duplications. Journal of Computational Biology 16(10), 1311–1338 (2009)

7. Braga, M.D.V., Willing, E., Stoye, J.: Double cut and join with insertions and deletions. Journal of Computational Biology 18(9), 1167–1184 (2011); A preliminary version appeared in: Braga, M.D.V., Willing, E., Stoye, J.: Genomic Distance with DCJ and Indels. In: Moulton, V., Singh, M. (eds.) WABI 2010. LNCS (LNBI), vol. 6293, pp. 90–101. Springer, Heidelberg (2010)
8. Blanc, G., Ogata, H., Robert, C., et al.: Reductive genome evolution from the mother of rickettsia. PLoS Genetics 3(1), e14 (2007)
9. Braga, M.D.V., Machado, R., Ribeiro, L.C., Stoye, J.: On the weight of indels in genomic distances. BMC Bioinformatics 12(suppl. 9), S13 (2011)
10. Braga, M.D.V., Stoye, J.: The solution space of sorting by DCJ. Journal of Computational Biology 17(9), 1145–1165 (2010)
11. Braga, M.D.V., Machado, R., Ribeiro, L.C., Stoye, J.: Genomic distance under gene substitutions. BMC Bioinformatics 12(suppl. 9), S8 (2011)

Hidden Breakpoints in Genome Alignments

Birte Kehr[1,2], Knut Reinert[1], and Aaron E. Darling[3]

[1] Department of Computer Science, Freie Universität Berlin,
Takustr. 9, 14195 Berlin, Germany
[2] Max Planck Institute for Molecular Genetics, Ihnestr. 63-73, 14195 Berlin, Germany
[3] Genome Center, University of California-Davis, Davis, CA 95616
birte.kehr@fu-berlin.de

Abstract. During the course of evolution, an organism's genome can
undergo changes that affect the large-scale structure of the genome.
These changes include gene gain, loss, duplication, chromosome fusion,
fission, and rearrangement. When gene gain and loss occurs in addition
to other types of rearrangement, breakpoints of rearrangement can exist
that are only detectable by comparison of three or more genomes. An
arbitrarily large number of these "hidden" breakpoints can exist among
genomes that exhibit no rearrangements in pairwise comparisons.

We present an extension of the multichromosomal breakpoint median
problem to genomes that have undergone gene gain and loss. We then
demonstrate that the median distance among three genomes can be used
to calculate a lower bound on the number of hidden breakpoints present.
We provide an implementation of this calculation including the median
distance, along with some practical improvements on the time complexity
of the underlying algorithm.

We apply our approach to measure the abundance of hidden breakpoints in simulated data sets under a wide range of evolutionary scenarios. We demonstrate that in simulations the hidden breakpoint counts
depend strongly on relative rates of inversion and gene gain/loss. Finally
we apply current multiple genome aligners to the simulated genomes,
and show that all aligners introduce a high degree of error in hidden
breakpoint counts, and that this error grows with evolutionary distance
in the simulation. Our results suggest that hidden breakpoint error may
be pervasive in genome alignments.

1 Introduction

Genome rearrangement plays a fundamental role in biological processes including cancer [8], gene regulation, and development [11] and a better understanding
of genome rearrangement is expected to lend insight into these biological processes. The primary evidence for genome rearrangement in modern genomes
comes from identifying rearrangement breakpoints in alignments of two or more
genomes. Genome alignments and the rearrangement breakpoints they encode
provide a basis for reconstructing genome rearrangement histories. The accurate
reconstruction of rearrangement history depends intimately on whether genome
alignments can accurately identify breakpoints.

B. Raphael and J. Tang (Eds.): WABI 2012, LNBI 7534, pp. 391–403, 2012.
© Springer-Verlag Berlin Heidelberg 2012

When genomes have undergone gene loss in addition to rearrangement some breakpoints may only be detectable by comparison of three or more genomes. This class of breakpoints has received limited attention to date. We refer to such breakpoints as "hidden" breakpoints, since usage of these rearrangement breakpoints in an organism's evolutionary history is not apparent from pairwise comparison of available genomes. Hidden breakpoints occur in regions of the genome conserved in some, but not all of the genomes under comparison. A simplest possible example of hidden breakpoints involves the three genomes A, B, C, with blocks 1, 2, 3 (see Fig. 1).

By extending a recent solution to the breakpoint median problem [14], we demonstrate that it is possible to calculate a hidden breakpoint count in three genomes when there is gain and loss. Our method is founded on the premise that a parsimonious reconstruction of the median genome would contain any sequence content present in at least two of the three genomes, while sequence content unique to a single genome would not be in the median. By comparing pairwise distances to the median distance, our method reveals some of the hidden breakpoints. Having implemented the method to calculate hidden breakpoint counts, we then conduct a simulation study to measure the abundance of hidden breakpoints under a range of different genome evolution parameters. We apply current multiple genome alignment systems to calculate alignments of the simulated sequences. Finally, we count hidden breakpoints in the calculated alignments and compare with the true number of hidden breakpoints in the simulated sequences. We show that all tested genome alignment algorithms introduce error in hidden breakpoint estimation and characterize their error rates under a range of evolutionary parameters in simulation.

2   Pairwise and Hidden Breakpoints

Genome alignments consist of sequence regions that are aligned in two or more genomes. Each aligned sequence region, often called a synteny block or locally collinear block, is internally free from any rearrangement but may contain gaps. Here we simply call these regions *blocks*. We assign an integer-valued identifier to each block of an alignment, using the sign of the integer to represent whether the block occurs in a genome in the forward or in reverse orientation. In this way, the arrangement of a single genome g in a genome alignment can be represented as a sequence of signed integers: $g = b_1\, b_2\, \ldots\, b_n$, $b_i \in \mathbb{Z}$. Note that this sequence of integers does not necessarily represent all parts of the genome, i.e. there may be unaligned, unique segments in-between blocks. Genomes themselves may consist of one or multiple chromosomes with either linear or circular topology. When a genome contains multiple chromosomes, it cannot be represented as a single sequence of integers, but rather as a set of sequences: $g = \{b_1 \ldots b_i, b_{i+1} \ldots b_{i+j}, \ldots\}$. Following previous literature we define linear chromosomes to be bounded by zero length *telomeres*, acknowledging the discrepancy between this definition and its use in the biological literature to describe a string of nucleotides of length > 0 near the ends of a linear chromosome. We use • to denote telomeres in a linear chromosome: • $b_1\, \ldots\, b_n$ •.

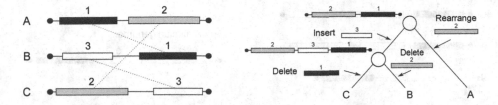

Fig. 1. Multiple alignment (left) and possible evolutionary scenario (right) of genomes A, B, and C containing a rearrangement not detectable in pairwise comparison. In all three pairwise projections two of the blocks become unique leading to a breakpoint distance of 0.

In the present work we investigate rearrangement breakpoints in groups of two or three genomes, although an alignment might contain many more genomes $g_1, \ldots, g_m$. This requires us to define a *projection* of an alignment to those blocks present in the current set of genomes. For three genomes g_x, g_y, and g_z with $x, y, z \in \{1, \ldots, m\}$ we denote the projection of the alignment as π_{xyz}, and a genome in the projected alignment by $\pi_{xyz}(g)$. We define the projected genome $\pi_{xyz}(g_x)$ as the subsequence of g_x containing those blocks present in at least two of the genomes used for projection. We thus remove blocks that are unique to g_x, g_y, or g_z from consideration in the three-way comparison since they can not reveal direct evidence of rearrangement breakpoints.

Two blocks b_i, b_j that are consecutive within a genome (possibly with unique segments in between) define an *adjacency*. In a pairwise projection of a genome alignment, a *pairwise breakpoint* is an adjacency from one genome that is missing in the other. By counting breakpoints in one genome g_x relative to another g_y we can calculate the *breakpoint distance* $d(g_x, g_y)$. Breakpoints at telomeres count only $\frac{1}{2}$ in $d(g_x, g_y)$, as only breakage and no fusion to another loose end has happened. When g_x and g_y have equal content $d(g_x, g_y)$ is equivalent to the classic breakpoint distance. If different sets of blocks are present in g_x and g_y, the breakpoint distance may be asymmetric, i.e. $d(g_x, g_y) \neq d(g_y, g_x)$. For brevity we define $d_{xyz}(g_x, g_y) := d(\pi_{xyz}(g_x), \pi_{xyz}(g_y))$.

Previous work has applied pairwise breakpoint distance in a sum-of-pairs score for multiple genome alignment [4]. However, multiple alignments may reveal rearrangements that are not visible in pairwise projections (see example in Fig. 1), information that sum-of-pairs scores do not capture. The breakage events of these rearrangements are hidden by gain or loss of rearranged blocks or by breakpoint re-use. We thus use the term *hidden breakpoint* to refer to this class of breakpoints.

3   The Median Approach for Counting Breakpoints

We now describe a method to compute a hidden breakpoint count $\mathcal{H}$ for three genomes. For alignments with $m > 3$ genomes, $\mathcal{H}$ can be calculated for all projections to three genomes.

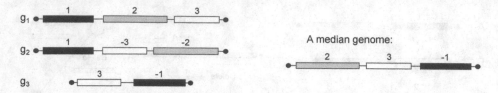

Fig. 2. Alignment of three genomes and a corresponding median genome. The pairwise distances to the median M are $d(g_1, M) = 1.5$, $d(g_2, M) = 0$, and $d(g_3, M) = 0.5$.

A median M for three genomes g_x, g_y, g_z with $x, y, z \in \{1, \ldots, m\}$ is a genome formed on all blocks in π_{xyz}. The blocks in M are arranged such that the sum

$$d_M := \sum_{k \in \{x,y,z\}} d_{xyz}(g_k, M)$$

is minimized (see Fig. 2 for an example). This represents a generalization of previous definitions of a median genome, e. g. [14], to include all blocks in at least two out of three genomes g_x, g_y, g_z. Previous breakpoint median methods required blocks to be conserved among *all* genomes. As we describe in Section 3.3, removing the portion of d_M attributed to pairwise breakpoints yields a hidden breakpoint count. We note that our method only requires computation of d_M and does not depend on accurate reconstruction of the actual median genome, for which many possibilities with minimal d_M may exist. We also note that duplications are forbidden; each block may occur at most once per genome.

We compute the median distance d_M by applying a recent polynomial time solution [14] to our generalization of the median definition. Following the proof of Theorem 1 in [14], we construct a graph from the genomes (see Sect. 3.1 for details), and compute a maximum weight perfect matching which corresponds to a median genome [14]. Because counting hidden breakpoints requires only the median distance d_M and not the actual median, we are able to improve the time complexity of the solution from [14] by reducing the number of graph edges (see Sect. 3.2) input to the perfect matching. Note that this reduction eliminates some valid medians, but the median distance d_M is unchanged.

3.1   Graph Construction

We first describe the graph as used in [14] to compute a median genome. Figure 3 (left) shows an example graph for the alignment from Fig. 2. Given a genome alignment, the graph $G = (V \cup T, E_V \cup E_T \cup E_U)$ contains four vertices from two different vertex sets V and T for each block b_i of the alignment: A *head* vertex $v_i^h \in V$ representing the start of the block, a *tail* vertex $v_i^t \in V$ representing the end of the block, a *head telomere* vertex $t_i^h \in T$, and a *tail telomere* vertex $t_i^t \in T$. The telomere vertices t_i^h and t_i^t represent the possibilities that chromosomes start or end with block b_i. It is necessary to have separate telomere vertices for each block to be able to find all (possibly multichromosomal) medians by the perfect

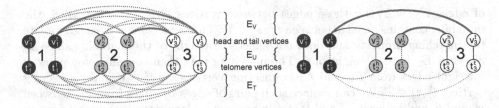

Fig. 3. Full graph (left) and a simplified graph (right) for the example alignment from Fig. 2. The simplified graph is sufficient to calculate the median distance d_M. Line width indicates edge weights. Dotted edges have a weight of 0. Red edges form a maximum weight perfect matching that corresponds to the median genome shown in Fig. 2.

matching computation. All vertices from V are connected among each other by edges $e_V \in E_V$, and all vertices from T are connected among each other by edges $e_T \in E_T$, i. e. the two subgraphs are complete. In addition, each head vertex v_i^h is connected to its corresponding head telomere vertex t_i^h, and each tail vertex v_i^t to its corresponding tail telomere vertex t_i^t by edges $e_U \in E_U$. Initially, all edges have a weight of 0. We increase edge weights according to adjacencies in the genomes. For telomere adjacencies, we increase weights of edges in E_U by $\frac{1}{2}$ following [14]. For all other adjacencies we increase weights of edges from E_V by 1. Depending on the orientation (sign) of the adjacent blocks, we increase the weight of the edge between a tail and head vertex, tail and tail, head and tail, or head and head vertex by 1. All edges in E_T keep a weight of 0.

3.2 Reducing the Number of Edges

Having constructed the graph, we compute a maximum weight perfect matching. The running time of the most efficient maximum weight perfect matching algorithms (e. g. [9]) depends on the number of edges in the graph, which in our case is quadratic in the number of blocks. In the following we show how it is possible to reduce the number of edges by a linear factor.

A perfect matching is a subset of the graph's edges $E_M \subseteq E_V \cup E_T \cup E_U$ such that each vertex of the graph is incident to exactly one edge in the matching. A maximum weight perfect matching is the perfect matching that maximizes the sum of weights of all edges $e \in E_M$. Figure 3 (left) demonstrates that many of the edges in a graph constructed as described in 3.1 have a weight of 0, and hence, will not contribute to the weight of a maximum weight perfect matching. Still, some of those edges are necessary to allow a perfect matching of maximum weight, but not all of them. In the following paragraph we explain why edges with a weight of 0 in E_V between vertices v_i^p and v_j^q, with $p, q \in \{h, t\}$ and with b_i, b_j being blocks, can be removed from the graph as well as corresponding edges in E_T between vertices t_i^p and t_j^q without affecting the weight of the resulting maximum weight perfect matching.

A graph that was constructed as described above can have multiple maximum weight perfect matchings. Given any of these matchings, we can replace the set of edges $E_M \cap E_T$ by those edges between vertices $t_i^p, t_j^q \in T$, for which the matching contains an edge between $v_i^p, v_j^q \in V$. This replacement does not violate the matching property and does not affect the weight of the matching since all edges in E_T have a weight of 0. The matching may then contain some pairs of 0-weight edges from E_V and E_T connecting vertices $v_i^p, v_j^q \in V$ and connecting vertices $t_i^p, t_j^q \in T$. We can replace such pairs of edges by edges from E_U between v_i^p and t_i^p and between v_j^q and t_j^q, again without violating the matching property and affecting the weight of the matching. After this replacement none of the 0-weight edges from E_V and the corresponding edges from E_T are part of the matching. Therefore, a maximum weight perfect matching exists that does not use these pairs of edges, and we can remove them from the initial graph without affecting the weight of the resulting maximum weight perfect matching.

3.3   Hidden Breakpoint Counts

A maximum weight perfect matching on the above described graph assigns each head or tail vertex to its telomere vertex or to another head or tail vertex by edges. These edges correspond to the adjacencies of a median genome. The edges that are not part of the matching but have a weight greater than 0, correspond to breakpoints among the median and one or more other genomes. Thus, by calculating the weight difference of the original graph and the matching, we obtain the total number of breakpoints d_M between the three original genomes g_x, g_y, and g_z and the median genome M. However, d_M overestimates the actual number of breakpoints among the three genomes. The median genome may contain blocks that are not present in one of the genomes, which disrupt adjacencies of blocks that occur in the same order without rearrangement. These disrupted adjacencies are neither pairwise nor hidden rearrangement breakpoints. Therefore, we calculate a separate hidden breakpoint count

$$\mathcal{H} = d_M - f \ .$$

The second term f removes the fraction of d_M which is due to breakpoints observable in pairs of genomes including those disrupted adjacencies caused by unequal block content. f is defined as:

$$f = \frac{1}{2} \sum_{a<b \in \{x,y,z\}} d_{xyz}(g_a, M_{ab}) + d_{xyz}(g_b, M_{ab}) \ .$$

For the computation of f we use a median M_{xy} of two genomes g_x, g_y, which is formed on all blocks of the triplet projection π_{xyz}. We can compute M_{xy} using the same approach as for a median of three genomes, i. e. construction of the graph and computation of a maximum weight perfect matching. The multiplication by $\frac{1}{2}$ is required because the distance from each genome to a median is counted exactly twice (to different medians).

In the results section we discuss hidden breakpoints in alignments containing arbitrarily many genomes. In alignments with more than three genomes, we calculate $\mathcal{H}$ for all projections to three genomes and report the sum over all projections. Likewise, for pairwise breakpoints we report the sum over all breakpoints in pairwise projections.

4   Simulating Evolution

We simulated genome evolution to generate sets of evolved sequences and genome alignments relating them using the previously described method sgEvolver [3]. Briefly, sgEvolver applies the standard Markov process model of evolution (specifically a marked Poisson process), with mutation events including substitutions, insertions and deletions of three different size distributions, and inversions. In our simulation each of the mutation types has the following properties:

Nucleotide Substitutions follow an HKY process with a transition/transversion ratio of 4 and background nucleotide frequencies A = 0.265, C = 0.235, T = 0.265, G = 0.235, as implemented in the program seq-gen [12]. Gamma-distributed heterogeneity in substitution rates across sites was simulated with the shape parameter $k = 1$.

Indels comprise the first class of insertions and deletions. Indel lengths follow a Poisson distribution with $\lambda = 3$.

Small Gain/Loss events comprise the second insertion/deletion class and have lengths geometrically distributed with $p = 200$.

Large Gain/Loss events comprise the final size class of insertions and deletions and have lengths uniformly distributed between 10000 and 50000.

Inversions follow a geometric length distribution with $p = 50000$.

These events were simulated on a phylogeny containing nine extant taxa. All insertion/deletion events are simulated to have uniformly random distributed positions in the genome. The rate of each event type is specified by a relative rate parameter; the expected event count per site on a particular branch of a phylogeny is the product of branch length and relative rate. The ancestral genome in our simulations is 1 Mbp in size, and since insertions and deletions occur with equal probability, the expected genome size remains constant throughout the simulation process (though the variance grows).

The approach used by sgEvolver to simulate insertions can also introduce duplications even though they are not directly modeled. In order to simulate insertions, sgEvolver maintains a finite pool of "donor" genomic material. When an insertion occurs, the sequence to insert is sampled uniformly at random from the donor material pool. This approach is intended to model the total pool of gene content available in a population, sometimes called a pan-genome [10]. When the simulation includes a large enough number of insertion events, especially "large gain/loss", it becomes likely that the same material from the donor pool will be inserted more than once, effectively creating a duplication in the simulated genome. sgEvolver does not keep track of the donor pool positions for insertion, and therefore the simulated genome alignments ignore possible duplications.

When simulating 1 Mbp genomes, the simulation and subsequent genome alignment process is fast enough that we can analyze a large range of parameters in parallel on a compute cluster. We report results on various parameter ranges below.

5   Hidden Breakpoints in Simulated Data

In this section we examine the effect of inversion, gain/loss, and nucleotide substitution rate on pairwise and hidden breakpoints. We expect hidden breakpoints to be driven by inversion and gain/loss, while nucleotide substitution may affect breakpoint counts in calculated alignments by increasing the difficulty of accurate alignment.

We describe two simulated data sets below, *InvNt* and *InvGL*, consisting of 400 and 200 genome alignments respectively. On each alignment of the datasets we computed the true number of pairwise breakpoints and hidden triplet breakpoints (see Sect. 5.1). Then in Sect. 5.3, we present calculated alignments of the simulated sequences done with three different genome alignment systems. We discuss how calculated alignments generally overestimate breakpoint counts and relate this error to evolutionary rates and other measures of alignment accuracy.

5.1   True Breakpoint Counts in Simulated Genome Alignments

The *InvNt* data set enables examination of the influence of the inversion and nucleotide substitution rate on the breakpoint counts. We expect nucleotide substitutions to have no effect on breakpoint counts in simulated alignments, although as we discuss later in calculated alignments a high substitution rate affects estimated breakpoint counts by making alignment difficult. We simulated 400 alignments with inversion rates between 0 and 2×10^{-4} in steps of 1×10^{-5} and nucleotide substitution rates between 0 and 1 in steps of 0.05. All three insertion/deletion rates were fixed: indels at 1×10^{-3}, small gain/loss at 5×10^{-4}, and large gain/loss at 2×10^{-5}. Previous work has demonstrated that alignment algorithms can reconstruct highly accurate genome alignments at these relatively low insertion/deletion rates [4]. In the alignments we observe 0 to 3100 pairwise breakpoints and 0 to 1150 hidden triplet breakpoints (see Fig. 4A). Despite these large ranges, the ratio between pairwise and hidden breakpoints stays constant. The coloring in Fig. 4A (top) shows that with growing inversion rates, we obtain more breakpoints, both pairwise and hidden triplet breakpoints. As one would expect, the nucleotide substitution rate has no direct effect on the number of breakpoints (Fig. 4A, bottom).

In the *InvGL* data set we examine breakpoint counts for different large gain/loss and different inversion rates. *InvGL* consists of 200 alignments with large gain/loss rates between 0 and 1×10^{-4} step size 1×10^{-5} and again inversion rates between 0 and 2×10^{-4} in steps of 1×10^{-5}. The other two insertion/deletion rates were fixed at the same values as above, and the nucleotide substitution rate at 0.01. As with *InvNt*, we observe more pairwise and hidden triplet breakpoints for growing inversion rates (Fig. 4D, top), but in contrast to *InvNt*, the ratio between pairwise and hidden breakpoint counts varies. The coloring in Fig. 4D (bottom) demonstrates that this variation is due to different gain/loss rates: With growing gain/loss rate, we observe more hidden triplet breakpoints whereas the number of pairwise breakpoints goes down slightly. This result is consistent with our definition of hidden breakpoints. The slight decrease in pairwise breakpoints at higher gain/loss rates is due to removal of breakpoints by gene loss. The hidden breakpoint count is also affected by gene loss in the *InvGL* simulation, reaching a maximum at a gain/loss rate of about 0.8.

5.2   Tested Alignment Programs and Evaluation Metrics

Having established the true counts of hidden triplet breakpoints under a range of evolutionary scenarios, we continue by evaluating whether some current algorithms can infer genome alignments with the correct number of hidden breakpoints. The algorithm to calculate hidden triplet breakpoints requires positional homology alignments, the method cannot be applied to genome alignments containing paralog alignments. For this reason we selected and applied three current programs that generate positional homology genome alignments: TBA, progressiveMauve, and Mugsy.

The **Threaded Blockset Aligner (TBA)** [2] constructs multiple genome alignments by a process of merging and filtering pairwise alignments generated by BLASTZ [13]. TBA requires specification of a phylogenetic guide tree, and for this we gave it the same topology used for simulation. We used TBA version 2009-Jan-21, the most recent version publicly available at the time of this work. The **progressiveMauve** genome alignment algorithm [4] begins by identifying approximate multi-matches among input genomes, then progressively grouping these into locally collinear blocks and aligning nucleotides within these blocks. For grouping matches progressiveMauve applies a sum-of-pairs breakpoint scoring scheme, which penalizes pairwise breakpoints. In the present work we used progressiveMauve version 2011-02-02 with default options. **Mugsy** is a newer aligner [1] that also uses three steps as progressiveMauve, but differs in the details of each step. For example, it uses a min-cut max-flow algorithm to partition the multi-matches into blocks under a synteny score. We used Mugsy v1r2.2 with default options.

For each alignment from the three aligners we computed the sum of hidden triplet breakpoint counts $\mathcal{H}$. To assess the error that aligners introduce to alignments in terms of breakpoints, we calculated the difference between the number of breakpoints in calculated alignments and the number of breakpoints in the true alignments. We also calculated a score that measures nucleotide alignment accuracy independently of genome arrangement: We compared each pair of aligned nucleotides in the calculated alignments with the true alignment to obtain precision and recall of nucleotide positional homology prediction, and combined the two values to an F_1 score.

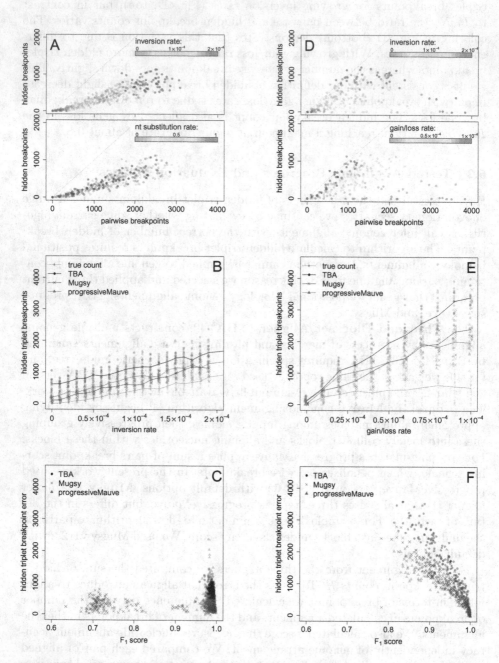

Fig. 4. Hidden triplet breakpoint counts in simulated alignments with different inversion and nucleotide substitution rates (*InvNt*, A–C), and 200 simulated alignments with different inversion and large gain/loss rates (*InvGL*, D–F)

5.3   Breakpoint Counts in Calculated Genome Alignments

We computed alignments with TBA, progressiveMauve, and Mugsy on all 600 sets of nine evolved genomes in the two data sets *InvNt* and *InvGL*. Figures 4B and 4C show hidden triplet breakpoint counts of the alignments from *InvNt*, and Fig. 4E and 4F for *InvGL*. In Fig. 4B and 4E, we display the median of the hidden triplet breakpoint counts per program for alignments simulated with equal inversion (4B) or gain/loss rate (4E) as lines.

In alignments calculated on *InvNt* we found that all three programs over-estimate the hidden breakpoint counts by a constant value independent of the inversion rate, suggesting a relationship with the constant gain/loss rate in this dataset. The two aligners that use an algorithm designed to minimize pairwise breakpoints introduce fewer additional hidden triplet breakpoints than TBA. The F_1 score of most alignments is very good (> 0.95) except for some Mugsy alignments. We recognize a trend that alignments with F_1 scores closer to 1 have lower hidden triplet breakpoint errors. An exception are Mugsy alignments on sequences simulated with nucleotide substitution rates greater than 0.4. These Mugsy alignments have a very low recall suggesting the simulated genomes were too divergent to be aligned by Mugsy with its default settings.

Calculated alignments of *InvGL* data show that the amount of gene gain/loss has a strong effect on the hidden triplet breakpoint error. The error is small at low rates of gain/loss, but grows quickly with the gain/loss rate. In Figure 4F we see that accurate alignments as measured by F_1 have lower hidden triplet breakpoint error. We find that the F_1 score drops with increasing gain/loss rates (data not shown), suggesting that genomes with high rates of simulated gain/loss may be generally difficult to align.

6   Conclusions

We have introduced the concept of hidden breakpoints in genome alignments and presented a method to calculate hidden triplet breakpoint counts. We have examined hidden breakpoint counts in simulated alignments evolved over a wide range of evolutionary parameters, and in alignments calculated by three different genome alignment algorithms on the same simulated datasets. We find that all tested genome aligners introduce a high degree of error in hidden breakpoint counts compared to the true count, and that this error grows with evolutionary distance in the simulation. Our results suggest that hidden breakpoint error may be pervasive in genome alignments. The error may in turn lead to erroneous inference of ancestral genomes and rearrangement history. Therefore, studies of the relationship between genome rearrangement history and biological processes could be improved by considering this error during the computation of genome alignments.

Overall, the error in triplet breakpoint counts is of the same order of magnitude as the pairwise breakpoint error. Methods such as the sum-of-pairs breakpoint score in progressiveMauve reduce the pairwise breakpoint error, but may not effectively reduce hidden breakpoint error. The max-flow min-cut algorithm employed by Mugsy appears to yield the most precise block and breakpoint estimates, but only when substitution rates are low enough. It is possible that the use of a more sensitive matching method such as promer [6] or another approximate matching approach would enable Mugsy to produce high accuracy results also at higher levels of sequence divergence. Future alignment systems would benefit from approaches that either explicitly or implicitly consider hidden breakpoints.

Limitations. The presented concept of hidden breakpoints has some limitations. First, the pairwise breakpoint distance is a lower bound for the number of breakage events that happened during evolution because several rearrangement events may use the same breakpoint. The hidden breakpoint count described here improves the estimate of breakage events, but it remains a lower bound to the actual number of breakage events that happened during evolution. A further limitation is that the current approach does not directly give rise to a method for phylogenetic inference on genome arrangement. Others have demonstrated practical algorithms for phylogenetic inference on genome arrangement using DCJ medians with gene gain and loss [16]. Although our median algorithm could be used in such a context, there is no constraint on the number or topology of chromosomes in median genomes and this might yield inferred ancestral genomes that are extremely biologically unlikely. Another limitation is the perfect-matching based median approaches cannot handle duplicated blocks in the alignments. Finally, we have only derived a formula to calculate hidden breakpoints among three genomes but some hidden breakpoints may be identifiable only among 4 or more genomes. We leave identification of hidden breakpoints in 4 or more genomes, and analysis of genomes with duplications as future work.

Acknowledgements. This work was enabled by funding from the DFG through the Dahlem Research School. We have implemented the calculation of breakpoint counts using the SeqAn C++ library (http://www.seqan.de), and the maximum weight perfect matching implementation from the Lemon graph library (http://lemon.cs.elte.hu).

References

1. Angiuoli, S.V., Salzberg, S.L.: Mugsy: fast multiple alignment of closely related whole genomes. Bioinformatics 27(3), 334–342 (2011)
2. Blanchette, M., Kent, W.J., Riemer, C., Elnitski, L., Smit, A.F.A., Roskin, K.M., Baertsch, R., Rosenbloom, K., Clawson, H., Green, E.D., Haussler, D., Miller, W.: Aligning multiple genomic sequences with the threaded blockset aligner. Genome Res. 14(4), 708–715 (2004)
3. Darling, A.C.E., Mau, B., Blattner, F.R., Perna, N.T.: Mauve: multiple alignment of conserved genomic sequence with rearrangements. Genome Res. 14(7), 1394–1403 (2004)
4. Darling, A.E., Mau, B., Perna, N.T.: progressiveMauve: multiple genome alignment with gene gain, loss and rearrangement. PLoS One 5(6), 11147 (2010)

5. De, S., Michor, F.: DNA replication timing and long-range DNA interactions predict mutational landscapes of cancer genomes. Nat. Biotechnol. 29(12), 1103–1108 (2011)
6. Delcher, A.L., Phillippy, A., Carlton, J., Salzberg, S.L.: Fast algorithms for large-scale genome alignment and comparison. Nucleic Acids Res. 30(11), 2478–2483 (2002)
7. Fudenberg, G., Getz, G., Meyerson, M., Mirny, L.A.: High order chromatin architecture shapes the landscape of chromosomal alterations in cancer. Nat. Biotechnol. 29(12), 1109–1113 (2011)
8. Greenman, C.D., Pleasance, E.D., Newman, S., Yang, F., Fu, B., Nik-Zainal, S., Jones, D., Lau, K.W., Carter, N., Edwards, P.A.W., Futreal, P.A., Stratton, M.R., Campbell, P.J.: Estimation of rearrangement phylogeny for cancer genomes. Genome Res. 22(2), 346–361 (2012)
9. Kolmogorov, V.: Blossom V: a new implementation of a minimum cost perfect matching algorithm. Mathematical Programming Computation 1, 43–67 (2009)
10. Medini, D., Donati, C., Tettelin, H., Masignani, V., Rappuoli, R.: The microbial pan-genome. Curr. Opin. Genet. Dev. 15(6), 589–594 (2005)
11. Nowacki, M., Shetty, K., Landweber, L.F.: RNA-mediated epigenetic programming of genome rearrangements. Annu. Rev. Genomics Hum. Genet. 12, 367–389 (2011)
12. Rambaut, A., Grassly, N.C.: Seq-Gen: an application for the Monte Carlo simulation of DNA sequence evolution along phylogenetic trees. Comput. Appl. Biosci. 13(3), 235–238 (1997)
13. Schwartz, S., Kent, W.J., Smit, A., Zhang, Z., Baertsch, R., Hardison, R.C., Haussler, D., Miller, W.: Human–mouse alignments with BLASTZ. Genome Res. 13(1), 103–107 (2003)
14. Tannier, E., Zheng, C., Sankoff, D.: Multichromosomal median and halving problems under different genomic distances. BMC Bioinformatics 10, 120 (2009)
15. Umbarger, M.A., Toro, E., Wright, M.A., Porreca, G.J., Bau, D., Hong, S.H., Fero, M.J., Zhu, L.J., Marti-Renom, M.A., McAdams, H.H., Shapiro, L., Dekker, J., Church, G.M.: The three-dimensional architecture of a bacterial genome and its alteration by genetic perturbation. Mol. Cell. 44(2), 252–264 (2011)
16. Zhang, Y., Hu, F., Tang, J.: Phylogenetic reconstruction with gene rearrangements and gene losses. In: Park, T., Tsui, S.K.W., Chen, L., Ng, M.K., Wong, L., Hu, X. (eds.) BIBM, pp. 35–38. IEEE Computer Society (2010)

A Probabilistic Approach to Accurate Abundance-Based Binning of Metagenomic Reads

Olga Tanaseichuk[1], James Borneman[2], and Tao Jiang[1]

[1] Department of Computer Science and Engineering,
University of California, Riverside, CA
[2] Department of Plant Pathology and Microbiology,
University of California, Riverside, CA
{tanaseio,jiang}@cs.ucr.edu,
borneman@ucr.edu

Abstract. An important problem in metagenomic analysis is to determine and quantify species (or genomes) in a metagenomic sample. The identification of phylogenetically related groups of sequence reads in a metagenomic dataset is often referred to as binning. Similarity-based binning methods rely on reference databases, and are unable to classify reads from unknown organisms. Composition-based methods exploit compositional patterns that are preserved in sufficiently long fragments, but are not suitable for binning very short next-generation sequencing (NGS) reads. Recently, several new metagenomic binning algorithms that can deal with NGS reads and do not rely on reference databases have been developed. However, all of them have difficulty with handling samples containing low-abundance species. We propose a new method to accurately estimate the abundance levels of species based on a novel probabilistic model for counting l-mer frequencies in a metagenomic dataset that takes into account frequencies of erroneous l-mers and repeated l-mers. An expectation maximization (EM) algorithm is used to learn the parameters of the model. Our algorithm automatically determines the number of abundance groups in a dataset and bins the reads into these groups. We show that our method outperforms the most recent abundance-based binning method, AbundanceBin, on both simulated and real datasets. We also show that the improved abundance-based binning method can be incorporated into a recent tool TOSS, which separates genomes with similar abundance levels and employs AbundanceBin as a preprocessing step to handle different abundance levels, to enhance its performance. We test the improved TOSS on simulated datasets and show that it significantly outperforms TOSS on datasets containing low-abundance genomes. Finally, we compare this approach against very recent metagenomic binning tools MetaCluster 4.0 and MetaCluster 5.0 on simulated data and demonstrate that it usually achieves a better sensitivity and breaks fewer genomes.

Keywords: metagenomics, next-generation sequencing, expectation maximization, abundance-based binning.

B. Raphael and J. Tang (Eds.): WABI 2012, LNBI 7534, pp. 404–416, 2012.
© Springer-Verlag Berlin Heidelberg 2012

1 Introduction

Metagenomics studies the genomic content of an entire microbial community by simultaneously sequencing all genomes in an environmental sample. This approach allows us to study previously uncultured microorganisms that constitute the vast majority of organisms in most environmental and clinical samples [1]. Metagenomics has already led to a better understanding of microbial communities in various environments, *e.g.* acid-mine drainage ponds [2], human gut [3], soil [4], and marine worms [5]. The recent advent of *next-generation sequencing (NGS)* technologies [6,7] has drastically improved sequencing time and cost, leading to an exponential increase in environmental sequencing data which makes it possible to study microbial communities at a much higher resolution due to increased sequencing depth [8]. NGS-based approaches have recently been applied to sequence several metagenomes from cow rumen [9], saliva microbiome [10], permafrost [11], etc.

In metagenomics, a sample contains sequence reads from various organisms. Therefore, an important problem in a metagenomic analysis is to determine and quantify the species (or genomes) in a sample. The identification of phylogenetically related groups of reads in a metagenomic dataset is usually referred to as *binning*. A handful of binning algorithms have been developed for metagenomic datasets. Similarity-based methods explore the taxonomic composition of metagenomic sequences by performing similarity search against databases of known genomes, genes and proteins [12,13,14,15]. These methods have high accuracy and are suitable for very short NGS reads. However, they rely on the availability of reference databases, while a lot of organisms in a sample may not be remotely related to any known species. As a consequence, a large fraction of read data may remain unclassified.

Another group of binning methods is based on compositional properties of the reads. These methods rely on the property that compositional features, such as oligonucleotide frequencies and CG content, are preserved across sufficiently long fragments of a genome. Supervised composition-based algorithms exploit compositional properties of the reads for taxonomic classification against models trained on known sequences [16,17,18]. Unsupervised methods perform clustering of the reads to detect groups of reads from related organisms [19,20,21,22]. Composition-based methods can accurately bin long fragments. However, due to local variation of DNA composition across a genome, the performance of these methods degrades with the decrease of the read length, making them unsuitable for NGS datasets.

Several recent unsupervised metagenomic binning algorithms have been developed to handle short NGS reads. In particular, MetaCluster 4.0 [23] exploits compositional properties of groups of reads rather than individual reads. Although it handles high-abundance species well, it does not perform well on datasets with low-abundance species. MetaCluster 5.0 [24] is a very recent extension of MetaCluster 4.0 to deal with low-abundant species. Another unsupervised binning algorithm that handles NGS reads is AbundanceBin [25]. It is designed to separate reads from genomes with different abundance levels. To predict the abundance levels, the frequencies of l-mers are modeled as a mixture of Poisson distributions. In this model, repeated l-mers and l-mers with errors are ignored, which may often lead to an inaccurate estimation of the parameters and result in a low binning accuracy. TOSS [26] is designed to separate reads from genomes with similar abundance levels. In the first phase, TOSS creates clusters of l-mer so that all l-mer in each cluster are likely to originate from the same genome. In the second phase, clusters from the same genome are merged. When genomes have different abundance levels, TOSS uses AbundanceBin as a preprocessing step. Clearly, the performance of TOSS is significantly affected by the performance of AbundaceBin. Specifically, the inability of AbundaceBin to accurately infer low-coverage genomes may result in bins with low sensitivity. When these bins are provided to TOSS as an input, the performance of TOSS would suffer. To address this problem, we introduce a method to accurately determine the abundance levels of genomes in a metagenomic dataset.

In this paper, we propose a novel probabilistic model for counting l-mer frequencies in the metagenomic dataset that takes into account the frequencies of erroneous l-mers as well as repeats. An *expectation maximization (EM)* algorithm is used to learn the parameters of the model. The algorithm automatically determines the number of abundance groups in the dataset and bins the reads into these groups. We show that the method outperforms AbundanceBin on siumlated and real datasets. We also show that the method can be incorporated into TOSS to improve its performance in the presence of genomes with different abundance levels. In fact, our experiments on simulated datasets demonstrate that this method significantly improves the performance of TOSS. Finally, we compare the improved TOSS against recent metagenomic binning tools Meta-Cluster 4.0 and MetaCluster 5.0 on simulated NGS datasets, and show that it has a comparable performance overall but often achieves a better sensitivity and breaks fewer genomes.

The paper is organized as follows. In section 2, we describe the probabilistic model for counting l-mer frequencies in a metagenomic dataset and the algorithm for learning the parameters of the model and automatically detecting the number of abundance groups. Section 3 presents the experimental evaluation of our method and comparison to other recent binning methods. Section 4 concludes the paper.

2   Methods

In this section, we introduce a novel probabilistic model that can be used for computing the most probable abundance levels of the genomes in a metagenomic dataset and estimating the proportions of the reads corresponding to each abundance group. The problem of binning the reads is then reduced to the problem of determining the parameters of the model and classifying the reads according to the frequencies of l-mers comprising the reads.

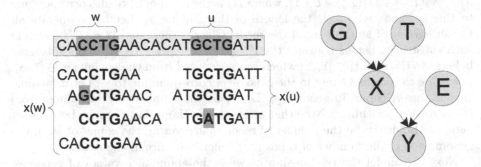

(a) The coverage of l-mer $w = \text{CCTG}$ is $x(w) = 4$. However, due to an error in one of the reads that cover w, w appears in the reads only 3 times, $i.e.$ $y(w) = 3$. For the l-mer $u = \text{GCTG}$, $x(u) = 3$. Observe that even though there is an error in one of the reads that cover u, this l-mer also occurs in a read that covers w due to an error, and thus $y(u) = 3$.

(b)   Count Y of an l-mer depends on the coverage X and the number of errors E within the l-mer. In turn, the coverage depends on the abundance level G of the genome and the number of occurances T of the l-mer in the genome.

Fig. 1. Left: Coverage of l-mers and occurrences of l-mers in the reads. Right: The proposed graphical model.

2.1   Definitions and Notations

Assume that N reads are drawn randomly from a genome of length L_g. Let L be the length of a read. According to the Lander-Waterman model [27], the left ends of the reads can be modeled by a Poisson process. Under this model, the number of reads that cover each substring of length l of the genome follows a Poisson distribution with the parameter $\lambda = N(L - l + 1)/(L_g - L + 1)$. From now on, we will refer to λ as *the abundance level of the genome*.

Even though most of the l-mers in a bacterial genome occur only once within the genome [26], some l-mers may occur at multiple locations within the genome. Assume that w is an l-mer with n copies in the genome. Due to additivity of the Poisson distribution, the number of reads that cover w, denoted by $x(w)$, has a Poisson distribution with the parameter $n\lambda$. However, due to sequencing errors, the actual count of the l-mer w in the reads, denoted by $y(w)$, may differ from $x(w)$ (see Figure 1a). Let $x^i(w)$ be the number of reads that cover the l-mer w with i errors in w. Clearly, $x(w) = \sum_i x^i(w)$ and $y(w) = x^0(w) + e_w$, where e_w is the number of times that w occurs in the reads due to errors in other l-mers.

Now, let us consider a metagenomic dataset. Assume that N reads are sequenced from S different genomes. The abundance value of genome g_j is

$\lambda_j = N_j(L-l+1)/(L_{g_j}-L+1)$, where N_j is the number of reads corresponding to this genome and L_{g_j} is the length of the genome g_j. Let us enumerate all the substrings of length l in all the reads. Clearly, there are $M = N(L-l+1)$ such substrings. Let us consider the ith substring v_i, $i \in [1, M]$. This substring belongs to the read $r_i \in [1, N]$ which was sequenced from the genome $g_i \in [1, S]$. Let w_i be the original l-mer in the genome g_i corresponding to v_i. Let us assume that w_i has t_i copies in genome g_i. Let e_i be the number of sequencing errors (substitutions) within v_i. Note that e_i equals the Hamming distance between w_i and v_i. Also, let x_i be the number of reads that cover all the copies of w_i in the genome and y_i the number of times that l-mer v_i occurs in the reads.

Next, we model the relationship between the abundance values of genomes, the coverage of l-mers, the number of errors in l-mers, and the counts of l-mers in the reads.

2.2   A Probabilistic Model for l-mer Frequencies

We define random variables G_i, X_i, Y_i, T_i, and E_i that are associated with the values g_i, x_i, y_i, t_i and e_i, respectively. The variables Y_i are observed by counting the number of occurrences of l-mers in the reads. The other variables cannot be observed directly, so they are hidden. Our goal is to determine the most likely assignment of the l-mers to the genomes. Figure 1b illustrates a graphical representation of the model.

Let π_j be a parameter that represents the proportion of the reads that come from the jth genome. Let α_j^n be the fraction of l-mers that occur n times in the jth genome. Let $\alpha_j = (\alpha_j^1, ..., \alpha_j^{n_{max}})$, where n_{max} is the maximum possible number of copies of an l-mer in a genome. For the convenience of notation, we define parameter vectors $\theta_j = (\lambda_j, \pi_j, \alpha_j)$ for all $j \in [1, S]$, and $\theta = (\theta_1, ..., \theta_S)$.

Assuming that the coverage of an l-mer with t copies in a genome g follows a Poisson distribution, the probability that the random variable X_i, associated with the coverage of l-mers in the genome g, takes a particular value c is

$$P(X_i = c | G_i = g, T_i = t, \theta) = \frac{c\,\mathrm{Pois}(t\lambda_g, c)}{\sum\limits_j j\,\mathrm{Pois}(t\lambda_g, j)} = \mathrm{Pois}(t\lambda_g, c-1),$$

where $\mathrm{Pois}(\lambda, k)$ is the probability of a Poisson random variable taking the value k.

The variable Y_i associated with the count of the l-mer v_i in the reads conditionally depends on variables X_i and E_i. If $E_i = e, e > 0$, it means that the corresponding l-mer v_i contains e errors. To model the distribution of counts of l-mers that have e errors, we can borrow the idea from the Balls and Bins problem (http://www.mathpages.com/home/kmath199.htm).

Assume that n balls are randomly thrown into m bins. It is known that the expected fraction of bins that get exactly k balls can be approximated by a Poisson distribution with the parameter n/m. Based on this, the probability that an erroneous l-mer has frequency k in the reads is

$$P(Y_i = k | X_i = c, E_i = e, e > 0, \theta) = \frac{kc \, \mathrm{Pois}(c/n_l(e), k)}{\sum_j j \, \mathrm{Pois}(c/n_l(e), j)} = \mathrm{Pois}(c/n_l(e), k-1),$$

where $n_l(e)$ is the number of different possibilities for e errors to occur within an l-mer.

The distribution of the counts of l-mers without errors can be modeled by the binomial distribution. The probability that an error-free l-mer has count k in the reads is

$$P(Y_i = k | X_i = c, E_i = 0) = \frac{k \, \mathrm{Bin}(k, c, p_0)}{\sum_j j \, \mathrm{Bin}(j, c, p_0)} = \frac{k \, \mathrm{Bin}(k, c, p_0)}{c p_0}$$

where $\mathrm{Bin}(j, c, p)$ is the probability that a variable following the binomial distribution takes the value j, and p_0 is the probability that an l-mer does not contain erros.

The above probabilities allow us to compute the probability of a given data point y_i given the values of unobserved variables and the parameter vector θ

$$P(Y_i = y_i | G_i = g, X_i = c, E_i = e, T_i = t, \theta)$$
$$= P(Y_i = y_i | X_i = c, E_i = e) P(X_i = c | G_i = g, T_i = t, \theta) P(G_i = g, T_i = t, \theta) P(E_i = e)$$
$$= \pi_g \alpha_g^t P(Y_i = y_i | X_i = c, E_i = e) P(X_i = c | G_i = g, T_i = t, \theta) P(E_i = e) \quad (1)$$

2.3 Parameter Estimation

Now, let us consider the log-likelihood of the observed data Y given the parameter vector θ

$$L(Y | \theta) = \sum_i \log P(Y_i = y_i | \theta).$$

Our goal is to find the *maximum likelihood estimate (MLE)* of the parameter θ,

$$\hat{\theta} = \arg\max_\theta L(Y | \theta).$$

To find $\hat{\theta}$, we use the EM algorithm. The E-step requires the computation of the expected value of the log-likelihood function, with respect to the conditional distribution of unobservable variables given the data and current parameter estimates $\theta^{(t)}$:

$$Q(\theta | \theta^{(n)}) = \sum_i \sum_{g,c,e,t} P(G_i = g, X_i = c, T_i = t, E_i = e | Y_i = y_i, \theta^{(t)})$$
$$\cdot P(Y_i = y_i, G_i = g, X_i = c, T_i = t, E_i = e | \theta)$$

Here, the posterior probabilities $p_{G,X,E,T|Y,\theta}(g, c, e, t, k, \theta) = P(G_i = g, X_i = c, E_i = e, T_i = t | Y_i = k, \theta)$ of the unobserved data given current parameter estimates $\theta^{(t)}$ can be computed by applying Bayes' rule to Equation 1.

In the M-step, we find the parameter $\theta^{(t+1)}$ that maximizes $Q(\theta|\theta^{(n)})$ with respect to θ

$$\theta^{(n+1)} = \arg\max_{\theta} Q(\theta|\theta^{(n)}).\tag{2}$$

The updated parameters are thus

$$\lambda_g^{(n+1)} = \frac{\sum_{c,e,t,k} p_{G,X,E,T|Y,\theta}(g,c,e,t,k,\theta)c}{\sum_{c,e,t,k} p_{G,X,E,T|Y,\theta}(g,c,e,t,k,\theta)t}, \quad \alpha_g^{t\,(n+1)} = \frac{\sum_{c,e,k} p_{G,X,E,T|Y,\theta}(g,c,e,t,k,\theta)}{\sum_{c,e,j,k} p_{G,X,E,T|Y,\theta}(g,c,e,j,k,\theta)}$$

$$\pi_g^{(n+1)} = \frac{\sum_{c,e,k,j} p_{G,X,E,T|Y,\theta}(g,c,e,j,k,\theta)}{\sum_{i,c,e,j,k} p_{G,X,E,T|Y,\theta}(i,c,e,j,k,\theta)}$$

Once we estimate the parameters of the probabilistic model, we can assign l-mers to bins (or genomes) based on the counts of the l-mers in the reads. We assign an l-mer v_i that occurs y_i times in the reads to a bin g with probability $P(G_i = g|Y_i = y_i, \hat{\theta})$. Then, each read is assigned to a bin according to the frequencies of its l-mers in the dataset

$$P(r \in g_j) = \prod_{y_i \in r} P(G_i = g|Y_i = y_i, \hat{\theta}) / \sum_g \prod_{y_i \in r} P(G_i = g|Y_i = y_i, \hat{\theta}).$$

2.4   Detecting the Number of Bins

The EM algorithm described above assumes that the number of bins (genomes) S and the maximum multiplicity of the repeats in the genome (the values that variables T_i may take) are provided. Selecting the best number of clusters is a challenging problem. Here, we propose an iterative algorithm to find the best value for S. We start with one bin and iteratively increase the number of bins until one of the following conditions is reached: (i) one or several bins are split into overlapping bins, making it impossible to assign the reads to the overlapping bins correctly and (ii) one or several bins are too small to represent a whole genome. In order to find the maximum multiplicity of the repeats, denoted by R, we repeat the above procedure for different values of R. For each pair of specific values $S = s$ and $R = r$, we record the distance between the observed and the expected frequencies of l-mers, $V(s,r) = \sum_i |M \cdot P(Y = i|\hat{\theta}_{r,s}) - \sum_{j=1..M} \mathbb{1}_{\{i\}}(y_j)|$. Here $M \cdot P(Y = i|\hat{\theta}_{r,s})$ is the expected number of l-mers with counts i, and $\sum_{j=1..M} \mathbb{1}_{\{i\}}(y_j)$ is the observed number of l-mers with counts i in the reads. Finally, we set S and R to the values s and r for which $V(s,r)$ reaches the minimum. See Algorithm 1 below for the details.

3   Experimental Results

We demonstrate the performance of our abundance-based binning algorithm on simulated and real datasets and compare the results with AbundanceBin. We also

Algorithm 1: Deciding the optimal number of bins S and maximum multiplicity of the repeats R. Given observed l-mer frequencies, the algorithm attempts to find the best values for S and R.

```
begin
    V ← ∞
    R, S ← 1, 1
    for r = 1, ..., Rmax do
        s ← 1
        θ̂ ← EM(s, r)
        if StopCondition(θ̂) then
            break
        else
            if V(s, r) < V then
                V ← V(s, r)
                R, S ← r, s
            s ← s + 1
    return R, S
end
```

show that the algorithm can be incorporated into our recent genome separation tool TOSS to enhance its performance. We test the improved TOSS on simulated NGS datasets and compare the results with those of TOSS that uses (or does not use) AbundanceBin as a preprocessor. Finally, we compare the performance of the improved TOSS with two very recent binning tools MetaCluster 4.0 and MetaCluster 5.0 on simulated NGS data.

3.1 Performance on a Simulated Data

Due to our limited knowledge of the nature of microbial communities, simulated metagenomic datasets are widely used for testing the performance of existing metagenomic tools. We simulated several metagenomic datasets based on complete genomes from the NCBI database using software MetaSim [28]. Each simulated dataset contains paired-end reads of length 80 bps. The sequencing error model was set according to the Illumina error profile with 1% average sequencing error rate.

We compare the performance of our algorithm against AbundanceBin. In this test, we are mainly concerned with the ability of both algorithms to separate reads from genomes with different abundance levels. In order to measure the performance of the algorithms, we use the evaluation criteria defined in [26]. We assign a genome to a bin (or cluster) if more than half of the reads from the genome are assigned to this bin. If there is no bin that contains the majority of the reads from a genome, we report the genome as broken. We allow several genomes to be assigned to one bin, and say that the genomes are not separated if the reads of the genomes ended up in the same cluster. We compute the *separability rate* as the percentage of separated pairs of genomes in the dataset.

Table 1. Comparison with AbundanceBin on simulated datasets. The bold numbers indicate improved sensitivity and precision. The numbers in parentheses are normalized sensitivity and precision.

ID	# genomes	Cove-rage	Length Mbp	Ours			AbundanceBin		
				Sens.	Prec.	Sep.	Sens.	Prec.	Sep.
S1	2	5 10	2.0 1.9	0.80 **(0.84)**	**0.84 (0.84)**	1	0.80 (0.75)	0.77 (0.76)	1
S2	3	5 5 11	2.7 2.6 3.0	**0.89 (0.89)**	**0.89 (0.89)**	1	0.86 (0.85)	0.86 (0.85)	1
S3	2	5 9	0.6 0.6	**0.79 (0.81)**	**0.78 (0.80)**	1	0.74 (0.69)	0.74 (0.69)	1
S4	2	4 8	4.4 5.2	**0.73 (0.82)**	**0.81 (0.81)**	1	-	-	0
S5	3	3 3 8	5.7 4.4 6.0	0.87 **(0.93)**	**0.88 (0.93)**	1	**0.91** (0.89)	0.80 (0.89)	1
S6	3	3 8 15	4.6 4.1 4.7	0.75 (0.83)	0.83 (0.82)	1	**0.81 (0.84)**	**0.88 (0.84)**	0.66
S7	6	2,2 2,6 6,6	1.5,1.8 2.0,1.7 1.8,2.0	**0.86 (0.75)**	**0.85 (0.84)**	1	-	-	0

In addition to standard sensitivity and precision, we also measure *normalized sensitivity* and *precision*. The formal definitions of these concepts can be found in [26].

The detailed datasets and performance of the two algorithms are summarized in Table 1. On most of the datasets, the sensitivity and precision of our method were better than those of AbundanceBin by 4-10%. In tests S4 and S7, AbundanceBin failed to separate the two genomes totally. In test S6, AbundanceBin could identify only 2 bins, while combining the reads from two genomes into one bin. Our method was more ambitious and separated all three genomes at the cost of lowered precision and sensitivity. However, when we set the number of bins to two for the dataset in test S6, our algorithm was able to achieve a high sensitivity and precision above 95%, compared to 81% and 88% for AbundanceBin.

3.2   Performance on a Real Dataset

We test the performance of our method on a dataset obtained from the acid mine drainage [2]. This dataset has been well studied and is known to contain five dominant genomes. The two most abundant species belong to *Leptospirillum* group II and *Ferroplasma* group II. The three species with a lower abundance levels belong to *Leptospirillum* group III, *Ferroplasma* group I and *Sulfobacill.* The dataset consists of approximately 120K Sanger reads. Only 56% percent of the reads can be mapped to the reference sequences of the five dominant genomes.

Table 2. Performance of the improved TOSS and comparison with the previous TOSS, MetaCluster 4.0 and MetaCluster 5.0. The bold numbers indicate the best performance among all five methods.

ID	# genomes	Coverage	Ours+TOSS				ABin+TOSS				TOSS				MC 4.0				MC 5.0			
			Sens.	Prec.	Sep.	Broken	Sens.	Prec.	Sep.	Broken	Sens.	Prec.	Sep.	Broken	Sens.	Prec.	Sep.	Broken	Sens.	Prec.	Sep.	Broken
T1	4	4,4, 10,10	1.0	**0.84**	1.0	0	-	-	0	0	0.63	0.55	1.0	0	-	-	-	1	0.75	0.69	1.0	0
T2	3	4,10,10	1.0	0.96	1.0	0	0.62	0.72	1.0	0	0.73	0.84	1.0	0	0.91	1.0	1.0	0	0.79	1.0	1.0	0
T3	3	4,12,12	1.0	1.0	1.0	0	1.0	0.99	1.0	0	0.84	0.90	1.0	1	0.97	0.96	1.0	0	0.82	1.0	1.0	0
T4	4	7,7,13,13	0.86	0.82	0.83	0	0.76	0.76	0.67	0	-	-	0	2	**0.89**	1.0	1.0	2	0.84	1.0	1.0	0
T5	10	1,1,1,2, 2,2,1.5, 1.5,10,10	1.0	0.92	0.83	0	-	-	0	0	1.0	0.64	0	2	0.78	**0.97**	1.0	0	-	-	-	4
T6	10	1.5,1.5,1.5, 1.5,1.5,1.5, 9,9,9,9	0.91	0.87	1.0	0	-	-	0	2	0.99	0.81	1.0	0	0.80	0.96	0	0	0.74	1.0	1.0	2
T7	18	2,2,2,2, 3,3,3,3, 3,3,3,4, 4,4,11, 12,12,12	0.87	0.75	0.73	0	-	-	0	0	-	-	-	3	0.9	0.9	1	0	0.88	0.9	1	0

We apply both our algorithm and AbundanceBin to the unfiltered dataset. Then we BLAST the reads of each bin against reference sequences of the five organisms. We measure the ability of the algorithms to separate reads from the two main abundance groups. Although both algorithms could correctly identify the two bins, our algorithm slightly outperforms AbundanceBin in terms of precision and sensitivity. Our method achieves 82% sensitivity and 81% precision, while the corresponding values are 78% and 79% for AbundanceBin. Note that due to the overlap of the bins, it would be very difficult to separate the reads with much better sensitivity and precision based on l-mer frequencies only.

3.3   Performance of the Improved TOSS

TOSS is designed to handle genomes with similar abundance levels and it requires a preprocessing step to separate the reads from the genomes with different abundance levels. We incorporate our abundance-based binning algorithm into TOSS and test the performance of the improved TOSS on simulated NGS datasets. We compare the results with the previous version of TOSS that employs AbundanceBin as a preprocessor and with TOSS without any preprocessing steps. Also, we make a comparison with the most recent metagenomic NGS binning tools MetaCluster 4.0 and MetaCluster 5.0. Again, to measure the performance of the tools, we use the evaluation criteria defined in [26]. The results of the comparison are summarized in Table 2. Note that here we only measure the ability of the algorithms to separate high-abundance genomes (with abundance levels ≥ 7, as done in [24]). The improved TOSS obviously outperforms both the version of TOSS that relies on AbundanceBin and the version of TOSS that does not use any preprocessor (the former has low separability rate while the latter yields a high number of broken genomes). Compared to the MetaCluster tools, our algorithm often achieves the highest sensitivity and breaks fewer genomes.

4   Conclusion

Metagenomics approach has opened a door into the previously hidden world of microorganisms. However, analysis of metagenomic data remains a difficult problem far from being solved. Binning is an important step of metagenomic analysis. In this paper, we introduced a novel probabilistic model for counting l-mer frequencies in a metagenomic dataset. The model allows us to identify the most probable abundance levels of the genomes in a metagenomic sample accurately and estimate the proportions of reads from corresponding genomes. We have shown that our model can serve as a useful preprocessing tool for further metagenomic analysis.

Acknowledgments. We are grateful to the anonymous referees for their many constructive comments. The research was supported in part by NIH grant AI078885.

References

1. Amann, R.I., Ludwig, W., Schleifer, K.H.: Phylogenetic identification and in situ detection of individual microbial cells without cultivation. Microbiological Reviews 59(1), 143–169 (1995)
2. Tyson, G.W., Chapman, J., Hugenholtz, P., et al.: Community structure and metabolism through reconstruction of microbial genomes from the environment. Nature 428(6978), 37–43 (2004)
3. Gill, S.R., Pop, M., DeBoy, R.T., et al.: Metagenomic Analysis of the Human Distal Gut Microbiome. Science 312(5778), 1355–1359 (2006)
4. Tringe, S.G., von Mering, C., Kobayashi, A., et al.: Comparative Metagenomics of Microbial Communities. Science 308(5721), 554–557 (2005)
5. Woyke, T., Teeling, H., Ivanova, N.N., et al.: Symbiosis insights through metagenomic analysis of a microbial consortium. Nature 443(7114), 950–955 (2006)
6. Margulies, M., Egholm, M., Altman, W.E., et al.: Genome sequencing in microfabricated high-density picolitre reactors. Nature 437(7057), 376–380 (2005)
7. Bentley, D.R.: Whole-genome re-sequencing. Current opinion in genetics & development 16(6), 545–552 (2006)
8. Singh, A.H., Doerks, T., Letunic, I., et al.: Discovering Functional Novelty in Metagenomes: Examples from Light-Mediated Processes. J. Bacteriol. 191(1), 32–41 (2009)
9. Hess, M., Sczyrba, A., Egan, R., et al.: Metagenomic discovery of biomass-degrading genes and genomes from cow rumen. Science 331(6016), 463–467 (2011)
10. Yang, F., Zeng, X., Ning, K., et al.: Saliva microbiomes distinguish caries-active from healthy human populations. The ISME Journal 6(1), 1–10 (2011)
11. Mackelprang, R., Waldrop, M.P., DeAngelis, K.M., et al.: Metagenomic analysis of a permafrost microbial community reveals a rapid response to thaw. Nature 480(7377), 368–371 (2011)
12. Huson, D.H., Auch, A.F., Qi, J., et al.: MEGAN analysis of metagenomic data. Genome research 17(3), 377–386 (2007)
13. Krause, L., Diaz, N.N., Goesmann, A., et al.: Phylogenetic classification of short environmental DNA fragments. Nucleic Acids Research 36(7), 2230–2239 (2008)
14. Ghosh, T., Monzoorul Haque, M., Mande, S.: DiScRIBinATE: a rapid method for accurate taxonomic classification of metagenomic sequences. BMC Bioinformatics 11(suppl. 7), S14+ (2010)
15. Monzoorul Haque, M., Ghosh, T.S.S., Komanduri, D., Mande, S.S.: SOrt-ITEMS: Sequence orthology based approach for improved taxonomic estimation of metagenomic sequences. Bioinformatics (Oxford, England) 25(14), 1722–1730 (2009)
16. Diaz, N., Krause, L., Goesmann, A., et al.: TACOA - Taxonomic classification of environmental genomic fragments using a kernelized nearest neighbor approach. BMC Bioinformatics 10(1), 56+ (2009)
17. McHardy, A.C., Martin, H.G., Tsirigos, A., et al.: Accurate phylogenetic classification of variable-length DNA fragments. Nature Methods 4(1), 63–72 (2006)
18. Brady, A., Salzberg, S.L.: Phymm and PhymmBL: metagenomic phylogenetic classification with interpolated Markov models. Nat. Meth. 6(9), 673–676 (2009)
19. Chatterji, S., Yamazaki, I., Bai, Z., et al.: CompostBin: A DNA Composition-Based Algorithm for Binning Environmental Shotgun Reads. In: Vingron, M., Wong, L. (eds.) RECOMB 2008. LNCS (LNBI), vol. 4955, pp. 17–28. Springer, Heidelberg (2008)

20. Teeling, H., Waldmann, J., Lombardot, T., et al.: TETRA: a web-service and a stand-alone program for the analysis and comparison of tetranucleotide usage patterns in DNA sequences. BMC Bioinformatics 5(1), 163+ (2004)
21. Prabhakara, S., Acharya, R.: A two-way multi-dimensional mixture model for clustering metagenomic sequences. In: Proceedings of the 2nd ACM Conference on Bioinformatics, Computational Biology and Biomedicine, BCB 2011, pp. 191–200. ACM (2011)
22. Yang, B., Peng, Y., Leung, H., et al.: Unsupervised binning of environmental genomic fragments based on an error robust selection of l-mers. BMC Bioinformatics 11(Suppl 2), S5+ (2010)
23. Wang, Y., Leung, H.C., Yiu, S.M., Chin, F.Y.: MetaCluster 4.0: A Novel Binning Algorithm for NGS Reads and Huge Number of Species. Journal of Computational Biology: a Journal of Computational Molecular Cell Biology 19(2), 241–249 (2012)
24. Wang, Y., Leung, H., Yiu, S., Chin, F.: Metacluster 5.0: A two-round binning approach for metagenomic data for low-abundance species in a noisy sample. In: Proceedings of the ECCB (to appear, 2012)
25. Wu, Y.-W., Ye, Y.: A Novel Abundance-Based Algorithm for Binning Metagenomic Sequences Using l-Tuples. In: Berger, B. (ed.) RECOMB 2010. LNCS, vol. 6044, pp. 535–549. Springer, Heidelberg (2010)
26. Tanaseichuk, O., Borneman, J., Jiang, T.: Separating Metagenomic Short Reads into Genomes via Clustering. In: Przytycka, T.M., Sagot, M.-F. (eds.) WABI 2011. LNCS, vol. 6833, pp. 298–313. Springer, Heidelberg (2011)
27. Lander, E.S., Waterman, M.S.: Genomic mapping by fingerprinting random clones: a mathematical analysis. Genomics 2(3), 231–239 (1988)
28. Richter, D.C., Ott, F., Auch, A.F., et al.: MetaSim: a Sequencing Simulator for Genomics and Metagenomics. PLoS ONE 3(10), e3373+ (2008)

Tandem Halving Problems by DCJ

Antoine Thomas, Aïda Ouangraoua, and Jean-Stéphane Varré

LIFL, UMR 8022 CNRS, Université Lille 1
INRIA Lille, Villeneuve d'Ascq, France

Abstract. We address the problem of reconstructing a non-duplicated ancestor to a partially duplicated genome in a model where duplicated content is caused by several tandem duplications throughout its evolution and the only allowed rearrangement operations are DCJ. As a starting point, we consider a variant of the Genome Halving Problem, aiming at reconstructing a tandem duplicated genome instead of the traditional perfectly duplicated genome. We provide a distance in $\mathcal{O}(n)$ time and a scenario in $\mathcal{O}(n^2)$ time. In an attempt to enhance our model, we consider several problems related to multiple tandem reconstruction. Unfortunately we show that although the problem of reconstructing a single tandem can be solved polynomially, it is already NP-hard for 2 tandems.

1 Introduction

Studying genome architecture is of great importance. There are many applications from evolution to cancer genomics. Thanks to the growing number of sequencing projects, one has a lot of data for comparing genomes both between species but also variants within a same species. Inspection of genomes revealed a lot of duplication events during the course of evolution. It is well-known that whole genome duplications arise several times, notably among mammals. But segmental duplications also occur. Recent studies between several plant mitochondrial genomes observe that some genes are duplicated [5,6,4]. A hypothesis to the creation of such duplications is that tandem duplications occurred followed by other rearrangements that scrambled the duplicates. In this paper we study methods to analyse such genomes. More precisely we are interested in reconstructing a non-duplicated ancestral genome from a partially duplicated genome. Figure 1 illustrates the problem.

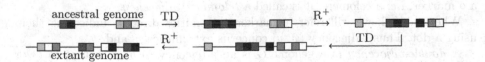

Fig. 1. A scenario from a non-duplicated ancestral genome which evolved through two tandem duplications (TD) and rearrangements (R^+). Squares denote syntenic markers.

B. Raphael and J. Tang (Eds.): WABI 2012, LNBI 7534, pp. 417–429, 2012.
© Springer-Verlag Berlin Heidelberg 2012

A problem one could believe similar to the one we study in this paper is the analysis of rearrangement scenarios that use Tandem Duplication Random Loss (TDRL) operations known to occur in mt genomes of millipedes and eels [3]. However, this model differs as it supposes that one of each duplicated marker is deleted. Our problem is in fact closer to Mixtacki's model of the genome halving problem [9], although we consider tandem duplication events as an alternative to the whole genome duplication. Such model has been studied in [2] but in order to find a scenario between two given genomes through an heuristic.

Section 2 gives definitions. In Section 3 we give a distance for reconstructing a single tandem when all markers in the extant genome are duplicated. In Section 4 we provide a heuristic algorithm for reconstructing a single tandem when single markers are considered. In Section 5, we discuss the NP-hardness of various constraints on the reconstruction of more than a single tandem. We conclude in Section 6 with an application on maize mt genomes.

2   Preliminaries: Duplicated Genomes, Rearrangement

A genome consists of linear or circular chromosomes that are composed of genomic markers. Markers are represented by signed integers such that the sign indicates the orientations of markers in chromosomes. By convention, $--x = x$. A linear chromosome is represented by an ordered sequence of signed integers surrounded by the unsigned marker $\circ$ at each end indicating the telomeres. A circular chromosome is represented by a circularly ordered sequence of signed integers. For example, $(1 \quad 2 \quad -3) (\circ \quad 4 \quad -5 \quad \circ)$ is a genome composed of one circular and one linear chromosome.

Each genome contains at most two occurrences of each marker, called paralogs, arbitrarily denoted x and $\bar{x}$ (by convention $\bar{\bar{x}} = x$).

Definition 1. *A* duplicated genome *is a genome in which a subset of the markers are duplicated.*

For example, $(1 \quad 2 \quad -3 \quad -\bar{2}) (\circ \quad 4 \quad -5 \quad \bar{1} \quad \bar{5} \quad \circ)$ is a duplicated genome where markers 1, 2, and 5 are duplicated. A *non-duplicated genome* is a genome in which no marker is duplicated. A *totally duplicated genome* is a duplicated genome in which all markers are duplicated.

An *adjacency* in a genome is a pair of consecutive markers. Since a genome can be read in two directions, the adjacencies $(x \quad y)$ and $(-y \quad -x)$ are equivalent. For example, the genome $(1 \quad 2 \quad -\bar{2}) (\circ \quad -3 \quad \bar{1} \quad 3 \quad \circ)$ has seven adjacencies, $(1 \quad 2)$, $(2 \quad -\bar{2})$, $(-\bar{2} \quad 1)$, $(\circ \quad -3)$, $(-3 \quad \bar{1})$, $(\bar{1} \quad 3)$, and $(3 \quad \circ)$. When an adjacency contains a $\circ$ marker, *i.e.* a telomere, it is called a *telomeric adjacency*.

When needed, we will refer to marker extremities directly, indicating them using a dot. Thus, adjacency $(x \quad y)$ concerns extremities $x\cdot$ and $\cdot y$.

A *double-adjacency* in a genome G is an adjacency $(a \quad b)$ such that $(\bar{a} \quad \bar{b})$ or $(-\bar{b} \quad -\bar{a})$ is an adjacency of G as well. Note that a genome always has an even number of double-adjacencies. For example, the four double-adjacencies in the following genome are indicated by $\diamond$:

$$G = (\circ \ 1 \ \overline{1} \ 3 \ 2 \ \diamond \ 4 \ \diamond \ 5 \ 6 \ \overline{6} \ 7 \ \overline{3} \ 8 \ \overline{2} \ \diamond \ \overline{4} \ \diamond \ \overline{5} \ 9 \ \overline{8} \ \overline{7} \ \overline{9} \ \circ)$$

A consecutive sequence of double-adjacencies can be rewritten as a single marker; this process is called *reduction*. For example, genome G can be reduced by rewriting $2 \diamond 4 \diamond 5$ and their paralogs as 10 and $\overline{10}$:

$$G^r = (\circ \ 1 \ \overline{1} \ 3 \ 10 \ 6 \ \overline{6} \ 7 \ \overline{3} \ 8 \ \overline{10} \ 9 \ \overline{8} \ \overline{7} \ \overline{9} \ \circ)$$

Definition 2. *A* single tandem duplicated genome *is a totally duplicated genome which can be reduced to a genome of the form* $(\circ \ x \ \overline{x} \ \circ)$.

In other words, a tandem duplicated genome is composed of a single linear chromosome where all adjacencies, except the two telomeric adjacencies and the central adjacency, are double-adjacencies. For example, the genome $(\circ \ 1 \diamond 2 \diamond 3 \diamond 4 \ \overline{1} \diamond \overline{2} \diamond \overline{3} \diamond \overline{4} \ \circ)$ is a tandem-duplicated genome that can be reduced to $(\circ \ 5 \ \overline{5} \ \circ)$ by rewriting $1 \diamond 2 \diamond 3 \diamond 4$ and $\overline{1} \diamond \overline{2} \diamond \overline{3} \diamond \overline{4}$ as 5 and $\overline{5}$.

Definition 3. *A* dedoubled genome *is a duplicated genome* G *such that for any duplicated marker* x *in* G, *either* $(x \ \overline{x})$, *or* $(\overline{x} \ x)$ *is an adjacency of* G.

One might notice that a single tandem duplicated genome, after reduction, is a unilinear dedoubled genome consisting of only one marker. Generalization of this property leads us to a short formal definition for genomes composed of several tandems, or *multiple tandem duplicated genomes*.

Definition 4. *A* k-tandem duplicated genome *is a totally duplicated genome which can be reduced to a unilinear dedoubled genome consisting of* k *distinct markers.*

For example, the genome $(\circ \ 1 \diamond 2 \diamond 3 \ \overline{1} \diamond \overline{2} \diamond \overline{3} \ 4 \diamond 5 \ \overline{4} \diamond \overline{5} \ \circ)$ is a 2-tandem duplicated genome that can be reduced to the dedoubled genome $(\circ \ 6 \ \overline{6} \ 7 \ \overline{7} \ \circ)$.

Naturally, following this definition, a single tandem duplicated genome is in fact a 1-tandem duplicated genome.

Definition 5. *A* perfectly duplicated genome *is a totally duplicated genome such that all adjacencies are double-adjacencies, none of them in the form* $(x \ -\overline{x})$.

For example, the genome $(1 \ 2 \ 3 \ 4 \ \overline{1} \ \overline{2} \ \overline{3} \ \overline{4})$ is a perfectly duplicated genome, while $(\circ \ 1 \ 2 \ -\overline{2} \ -1 \ \circ)$ is not. It is to note that this definition is equivalent to the one from [9].

The rearrangement operations considered in this paper will be the DCJ model, introduced in [12]. A *DCJ* operation on a genome G cuts two different adjacencies in G and glues pairs of the four exposed extremities to form two new adjacencies. A *DCJ scenario* between two genomes A and B is a sequence of DCJ operations allowing to transform A into B. The length of a scenario is the number of operations composing the scenario. The *DCJ distance* between two genomes A and B is the minimum length of a DCJ scenario between A and B.

Property 1. In the case of unichromosomal genomes, a perfectly duplicated genome is a single tandem duplicated genome which has been circularized (the perfectly duplicated genome can be reduced to $(x \ \overline{x})$, it just lacks telomeres).

3   Single Tandem Halving

We now state the first tandem halving problem considered in this paper.

Definition 6. *Given a unilinear totally duplicated genome G, the* single tandem halving problem *(or 1-tandem halving problem) consists in finding an* optimal 1-tandem duplicated genome H, *such that the distance between G and H is minimal. This minimal distance is called the* 1-tandem halving distance, *and is denoted* $d^t(G)$.

Through reduction, this problem will be seen as a constraint on the well-known *DCJ genome halving problem*, as solved in [9]. We recall its definition, with slightly readapted notations.

Definition 7 ([9]). *Given a totally duplicated genome G, the* DCJ genome halving problem *consists in finding an* optimal perfectly duplicated genome H, *such that the DCJ distance between G and H is minimal. This minimal distance is called the* genome halving distance *and is denoted* $d^p(G)$.

$d^p(G)$ can be computed using a data structure called the *natural graph*, first introduced in [7]. $NG(G)$ is the graph whose vertices are the adjacencies of G, and 2 vertices are connected by an edge iff they share a paralogous *extremity* (see figure 2).

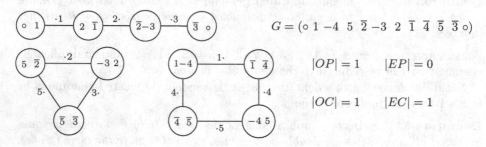

$$G = (\circ \ 1 \ -4 \ 5 \ \bar{2} \ -3 \ 2 \ \bar{1} \ \bar{4} \ \bar{5} \ \bar{3} \ \circ)$$

$$|OP| = 1 \qquad |EP| = 0$$

$$|OC| = 1 \qquad |EC| = 1$$

Fig. 2. The natural graph of G and the number of odd and even paths and cycles

As an adjacency concerns a maximum of 2 markers extremities, this graph has a maximum degree of 2. Thus, it is composed of paths and cycles only. Moreover, it consists of nothing but 2-cycles and 1-paths if and only if G is a perfectly duplicated genome (a k-cycle or k-path is a cycle or path containing k edges). Using this graph, Mixtacki gave the following distance formula:

Theorem 1 ([9]). *Let G be a totally duplicated genome whose natural graph contains* EC *even cycles and* OP *odd paths. Then* $d^p(G) = n - |EC| - \left\lfloor \frac{|OP|}{2} \right\rfloor$.

Unlike the genome halving problem, the aim of the 1-tandem halving problem is to find a 1-tandem duplicated genome. This induces one double-adjacency not to be reconstructed, which is inelegant to deal with. We will conveniently get rid of this concern.

From property 1, a 1-tandem genome that has been circularized is a perfectly duplicated genome and conversely. This allows us to establish a property that will reduce the 1-tandem halving problem to a constraint on genome halving.

Lemma 1. *Let G be a unilinear genome. Let G_c be the unicircular genome obtained by circularizing G. Then for any scenario that transforms G into a 1-tandem duplicated genome, there exists an equivalent scenario (of same length) transforming G_c into a unicircular perfectly duplicated genome, and vice versa.*

Proof. As G and G_c present the same breakpoints, the scenario conversion is straightforward. It suffices to apply the same DCJ on the same breakpoints. □

Thus, in the rest of this section, the focus will be on reconstructing an optimal perfectly duplicated genome such that it is unichromosomal. This is essentially a shape constraint on the genome halving solutions.

We will follow an approach a bit similar[1] to what has been done by Kováč et al. in [8], as they enforced another shape constraint on optimal perfectly duplicated genome configurations. It consists in taking any optimal configuration then applying a number of successive transformations (which we will refer to as *shapeshifting* in the present paper) on it, such that they preserve the distance, and that the optimal configuration converges towards the desired shape.

In the following sections G will denote a totally duplicated genome, and G_c its circularized version. H will be an optimal perfectly duplicated genome for G_c.

Following theorem 1, one can observe that circularization can alter the halving distance, depending on whether the path of $\text{NG}(G)$ is even or odd.

Property 2. If G is a genome such that $\text{NG}(G)$ contains an even path, $d^p(G_c) = d^p(G) - 1$. Else, $d^p(G_c) = d^p(G)$.

From Mixtacki's formula (Theorem 1), we know that optimal halving scenarios on circular genomes are scenarios which increase the number of even cycles at each step. There are two ways of increasing it. Either by splitting a cycle (*i.e.* extracting an even cycle from any cycle), or by merging two odd cycles.

As it can be quite complex at first sight, our shapeshifting system will first be described on a restricted class of genomes, namely those whose natural graph contains only even cycles. This way, we ensure that optimal halving scenarios consist only in cycle extractions. The restricted system will then be easily generalized to all genomes by considering merging operations.

3.1   Restricted Shapeshifting System

Here we consider that $\text{NG}(G_c)$ has only even cycles. It follows that $\text{NG}(G)$ has an even path and $d^p(G_c) = d^p(G) - 1$.

---

[1] Although it had to be developed as a more complete system, due to the nature of our problem.

Anatomy of a multicircular perfectly duplicated genome. H is an optimal perfectly duplicated genome for G_c. Since G_c is unicircular, $\text{NG}(G_c)$ contains nothing but cycles. Therefore, H consists of circular chromosomes only. For H to be a perfectly duplicated genome, circular chromosomes can be of two kinds : doubled chromosomes, which can be reduced to $(x\ \overline{x})$, and single chromosomes, which can be reduced to (x) and have a *paralog chromosome* in H, which can be reduced to $(\overline{x})$. Thus the number of single chromosomes is even.

Shapeshifting. Any optimal perfectly duplicated genome H induces a class $\mathcal{C}_H$ of optimal halving scenarios (the class of all optimal DCJ scenarios transforming G_c into H). By observing the structure of G_c and H, we will look for small changes to apply to $\mathcal{C}_H$, along two criteria : H must converge toward the desired shape, and it must preserve its optimality. Such small changes are called *shapeshifters*.

In our case, we want to end up with the least number of chromosomes in H (ideally only one), therefore we will look for ways to merge chromosomes while preserving optimality. This leads us to the following definition :

Definition 8. *A shapeshifter is an adjacency $(x\ y)$ such that x and y belong to different chromosomes of H (convergence towards the desired shape), and such that $(x\ y)$ (and therefore $(\overline{x}\ \overline{y})$ as well) can be reconstructed by an optimal halving scenario (preservation of optimality).*

For example, if H contains markers x and y in different chromosomes, C_x and C_y, and if $(x\ y)$ can be reconstructed by an optimal halving scenario, then such scenario induces a new shape for H such that C_x and C_y cannot be distinct chromosomes anymore.

As for now we consider genomes whose natural graph has even cycles only, shapeshifters are adjacencies reconstructible by extracting even cycles.

Property 3. Adjacencies $(x\ y)$ reconstructible by extracting even cycles are those such that there exists, in $\text{NG}(G_c)$, an induced subgraph which is an *even* path, whose endpoints have outgoing edges $x\cdot$ and $\cdot y$.

Indeed, a DCJ reconstructing $(x\ y)$ will cut at the endpoints of such path and transform it into an even cycle. However, it is not necessary to consider all even paths, so w.l.o.g we shall focus only on 2-paths (ie. adjancencies $(x\ y)$ that are *present in* G_c), which correspond to 2-cycles extractions.

For example, $(1\ 4)$ in fig. 2 is a shapeshifter, as the 2-path induced by vertices $(1\ -4)$, $(\overline{1}\ \overline{4})$, and $(-4\ 5)$ meets the requirements.

We may proceed and show how to simply apply a shapeshifter on $\mathcal{C}_H$: Let $(x\ p)$ be the adjacency containing the extremity $x\cdot$ in H, and $(q\ y)$ the one containing the extremity $\cdot y$, it suffices to perform on H one DCJ cutting adjacencies $(x\ p)$ and $(q\ y)$ to reconstruct $(x\ y)$ (and $(p\ q)$), and the equivalent DCJ on the paralogs, cutting adjacencies $(\overline{x}\ \overline{p})$ and $(\overline{q}\ \overline{y})$ to reconstruct $(\overline{x}\ \overline{y})$ (and $(\overline{p}\ \overline{q})$).

One can easily verify that the resulting genome is still optimal (first DCJ brings H closer to G_c, second one reconstructs a perfectly duplicated genome).

Now we may proceed and study the shapeshifting induced by these DCJ.

Let $(x\ y)$ be a shapeshifter in G_c. x and y belong to different chromosomes in H, so there are only 3 possible cases depending on the types of chromosomes (C_S for single chromosomes, and C_D for doubled ones) which contain these markers: 1) $x \in C_S, y \in C_D$, 2) $x, y \in C_D$, 3) $x, y \in C_S$. The last one could lead to different shapes. Figure 3 illustrates how the genome shape can be altered, for each case.

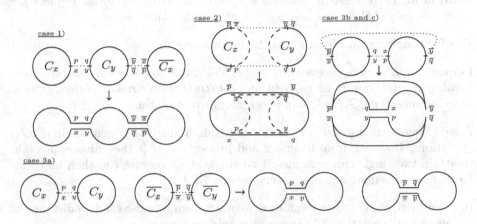

Fig. 3. The different shapes that can be obtained by applying a shapeshifter

More formally, one can represent shapeshifting as a system of rewriting rules:

1) $2 \times C_S + C_D \to C_D$ 3.a) $4 \times C_S \to 2 \times C_S$ 3.c) $2 \times C_S \to 2 \times C_S$
2) $2 \times C_D \to 2 \times C_S$ 3.b) $2 \times C_S \to 2 \times C_D$

This is convenient as one can deduce useful properties by looking at these rules, which we are about to do, in order to study limit states of the system.

Property 4. Shapeshifting cannot increase the number of chromosomes.

Thus, any limit-cycle necessarily uses rules that do not change the number of chromosomes. Moreover, using rule 2 would eventually lead to using rule 3.b or 3.c as doubled chromosomes are changed into single chromosomes.

Property 5. Any limit-cycle of the system necessarily uses rule 3.b or 3.c.

Property 6. Parity of $|C_D|$ is invariant by shapeshifting.

Property 7. A unicircular genome (ie. one doubled chromosome) is the only steady state of the system.

Lemma 2. *By shapeshifting, the number of chromosomes in H can always be decreased under 3.*

Proof. Having 3 chromosomes or more guarantees existence of shapeshifters decreasing their number. Consider the case where H contains only 2 single chromosomes C_S and $\overline{C_S}$. Label the markers from G by the chromosome which holds them in H. Adding new chromosomes necessarily creates shapeshifters between at least one of the new chromosomes and C_S or $\overline{C_S}$. Such shapeshifter decreases the number of chromosomes. □

Lemma 3. *There exists a unicircular optimal perfectly duplicated genome for G_c if and only if H has an* odd *number of doubled chromosomes.*

Proof. Straightforward from lemma 2 and property 6. □

Lemma 4. *If H has an* even *number of doubled chromosomes, the minimum number of DCJ operations required to reconstruct a unicircular perfectly duplicated genome is $d^p(G_c) + 1$, and it can always be attained.*

Proof. From lemma 3, it is impossible to attain a unicircular genome in $d^p(G_c)$ operations. However, from lemma 2 and property 5, it is then always possible to attain two single chromosomes. Two single chromosomes can then be transformed into one doubled chromosome by one DCJ. □

In conclusion, restricted shapeshifting allows to compute the tandem distance of any genome G such that $\mathrm{NG}(G)$ contains only even cycles.

Theorem 2. *Let G be a totally duplicated genome such that $\mathrm{NG}(G)$ contains only even cycles. Let G_c be its circularized version, and H any optimal perfectly duplicated genome for G_c. $d^t(G) = d^p(G) - 1$ if and only if H contains an odd number of doubled chromosomes. Else $d^t(G) = d^p(G)$.*

Proof. Since $\mathrm{NG}(G)$ contains only even cycles, it contains an even path. Therefore from property 2, $d^p(G_c) = d^p(G) - 1$. From lemma 1 we have that $d^t(G) = d^p(G_c)$ if and only if there exists a unicircular optimal perfectly duplicated genome. Theorem then follows from lemmas 3 and 4. □

The next step is to generalize the shapeshifting system in order to take all possible genomes into account.

3.2   Generalized Shapeshifting System

As usual, G is a totally duplicated genome, G_c its circularized version, and H an optimal perfectly duplicated genome for G_c. We will also keep the same notations related to shapeshifters as in the previous section : $(x\ y)$ is a shapeshifter such that x (resp. y) is present in chromosome C_x (resp. C_y) of H, through adjacency $(x\ p)$ (resp. $(q\ y)$).

The difference with restricted shapeshifting is that, *in addition* to everything covered by restricted shapeshifting, optimal halving scenarios may now also contain cycle merges. Therefore we have to consider shapeshifters that are adjacencies which can be optimally reconstructed through merges.

Property 8. Adjacencies $(x\ y)$ reconstructible by merges are those such that extremities $x\cdot$ and $\cdot y$ are in *two distinct odd cycles* of $\mathrm{NG}(G_c)$.

Corresponding shapeshifters can still allow the same shapeshifting rules depending on the types of C_x and C_y. Additionally, it is now possible to have $p = \overline{y}$ and $q = \overline{x}$. This implies that $C_y = \overline{C_x}$ and induces yet another degenerated case. The generalized shapeshifting set of rule becomes :

1) $2 \times C_S + C_D \to C_D$ 3.a) $4 \times C_S \to 2 \times C_S$ 3.c) $2 \times C_S \to 2 \times C_S$
2) $2 \times C_D \to 2 \times C_S$ 3.b) $2 \times C_S \to 2 \times C_D$ **3.d) $2 \times \mathbf{C_S} \to \mathbf{C_D}$**

This new rule gives generalized shapeshifting a very interesting property.

Property 9. Rule 3.d changes parity of C_D.

Lemma 5. *If $\mathrm{NG}(G_c)$ contains odd cycles, and if H is made of two single chromosomes, then rule 3.d can be applied.*

Proof. As $\mathrm{NG}(G_c)$ contains odd cycles, there are merges in any optimal scenario from G_c to H. Thus, there exists an adjacency $(x\ p)$ in C_x such that the adjacencies concerning extremities $x\cdot$ and $\cdot p$ are in two distinct odd cycles of $\mathrm{NG}(G_c)$. By definition, the adjacency concerning extremity $\cdot\overline{p}$ is in the same cycle as the one concerning $\cdot p$. Therefore, $(x\ \overline{p})$ is a shapeshifter inducing rule 3.d.   □

Corollary 1. *Presence of odd cycles in $\mathrm{NG}(G_c)$ ensures a unicircular optimal perfectly duplicated genome that can always be reached, as rule 3.d can always adjust the parity of C_D if needed.*

Theorem 3. *Let G be a totally duplicated genome such that $\mathrm{NG}(G)$ contains at least one odd cycle, and G_c its circularized version. Then $d^t(G) = d^p(G_c)$.*

Proof. From lemma 1 we have $d^t(G) = d^p(G_c)$ iff there exists a unicircular optimal perfectly duplicated genome. Corollary from lemma 5 ensures that there does.   □

3.3   Conclusion

We finally state a definite formula for the halving distance, as well as results on computational complexity of this problem, by gathering results from the previous sections.

Theorem 4. $d^t(G) = n - |\mathrm{EC}| - |\mathrm{EP}| + f_G$
     Where f_G is a parameter that is equal to 1 iff C_D is even and $|\mathrm{OC}| = 0$, and is equal to 0 otherwise. $|\mathrm{EC}|$, $|\mathrm{EP}|$ and $|\mathrm{OC}|$ are respectively the number of even cycles, even paths and odd cycles in $\mathrm{NG}(G)$.

Proof. Straightforward from theorems 2 and 3.   □

Theorem 5. $d^t(G)$ *can be computed in linear time.*

Proof. $\text{NG}(G)$ can be computed in linear time, as well as an optimal perfectly duplicated genome.   □

Theorem 6. *Computing a scenario can be done in quadratic time.*

Proof. An optimal perfectly duplicated genome can be computed in $O(n)$ time using Mixtacki's algorithm ([9]). From lemma 2, one can reduce H to the minimum number of chromosomes using $O(n)$ shapeshifters. Each shapeshifter can be found in $O(n)$ time, so we have a $O(n^2)$ time shapeshifting algorithm. An optimal DCJ scenario between G and H can then be computed in $O(n)$ time using Yancopoulos' algorithm ([12]). Thus the algorithm takes quadratic time on the whole.   □

4   Disrupted Single Tandem Halving

As we could solve the 1-tandem halving problem, a first direction for generalization will be considering genomes containing both duplicated and non-duplicated markers, as it is in better accordance with real biological data.

This can be seen as a 1-tandem halving problem in which adjacencies between duplicated markers can be broken by presence of non-duplicated ones. In other words, non-duplicated markers *disrupt* the 1-tandem halving.

Definition 9. *The* disrupted 1-tandem halving problem *is a variant of the 1-tandem halving problem in which the genome contains both duplicated and non-duplicated markers. The duplicated markers have to be regrouped and arranged in tandem. The corresponding distance, the* disrupted 1-tandem halving distance, *is denoted* $d^{t'}(G)$.

Preliminary analysis. Any optimal disrupted 1-tandem halving scenario performs two tasks : it gathers duplicated markers together (gathering phase), and it reorganizes them in a tandem (tandem phase).

Definition 10. *A* break *is an interval of non-duplicated markers surrounded by duplicated markers.*

From now on, G is a duplicated genome containing n duplicated markers separated by b breaks.

Definition 11. *A* gathering operation *is a DCJ which reduces the number of breaks in* G.

Note that the presence of excisions in the gathering phase may produce a genome consisting of multiple chromosomes. Excisions and their resulting chromosomes will be categorized depending on whether said chromosomes can be reintegrated at best in their source chromosome while increasing the number of even cycles (*good* excision/chromosome), leaving it unchanged (*neutral*) or decreasing it (*bad*). As this variation in $|EC|$ changes the tandem distance, we get the following property.

Property 10. Once the gathering phase is over in G, the remaining distance is $d^t(G) + C^0 + 2C^-$, with C^0 the number of neutral chromosomes and C^- the number of bad ones.

The key to build an optimal disrupted 1-tandem halving scenario is to find a gathering scenario that maximizes the number of even cycles while minimizing the number of neutral and bad excisions.

Optimizing the gathering scenario. A DCJ can decrease the number of breaks by at most 1.

Property 11. The minimum number of gathering operations is b.

Gathering operations are DCJ whose breakpoints are on path endpoints from NG(G). Breakpoints in two distinct paths will merge them, while breakpoints on the endpoints of a same path will circularize it.

Property 12. An optimal gathering operation is one that either merges two odd paths, or circularizes an even path.

We now give the maximum number of even cycles a set of b gathering operations can create.

Lemma 6. *A shortest gathering scenario can create up to $\left\lfloor \frac{|OP|}{2} \right\rfloor + |EP|$ 1 even cycles.*

Proof. sketch of proof: Any even path can be circularized by one DCJ, while any two odd paths can be turned into two even cycles with 2 DCJs. Since b breaks induce $b + 1$ paths in NG(G), the number of gathering operations we can use is $b = |OP| + |EP| - 1$.   □

Corollary 2. $d^{t'}(G) \geq n - |EC| - 1 + \left\lceil \frac{|OP|}{2} \right\rceil$.

This is assuming a shortest gathering phase produced no bad nor neutral chromosome, and that we are in the best case for the remaining tandem distance ($d^t(G) = d^p(G) - 1$).

Neutral excisions induce a penalty which is the same as performing a non-optimal gathering reversal, bad excisions are even worse. Thus our greedy heuristic will proceed as follows: Look for an optimal gathering operation which is a reversal or a good excision. When there is none, perform a non-optimal gathering reversal.

Let $C_h(G)$ be the number of even cycles produced by the heuristic, then we obtain the following upperbound : $d^{t'}(G) \leq n - |EC| + |OP| + |EP| - 1 - C_h(G)$.

In the worst case, $C_h(G)$ can be equal to 0, however, the algorithm seems to perform pretty well on random genomes, giving values close to the lowerbound.

5   Multiple Tandem Halving

Unlike 1-tandem halving, k-tandem halving can be defined in various ways. We explored several constraints on the k-tandem halving (detailed studies are given as supplementary material[2]). First, when one fixes the number of tandem to be reconstructed (k) the problem is NP-hard. Fixing the content of each of the k tandem does not help and the complexity of the problem remains the same. The same result arises when one fixes the tandem order in the ancestral genome. Lastly, a "signed" version where the orientation of the tandems is fixed is also NP-hard. Approximation algorithms should be considered next, as those problems are the most interesting ones from a biological viewpoint.

6   Application

As an application, we used data from [6]. We analyzed the mitochondrial genome of *Zea mays ssp. mays CMS-C* which is made of 69 syntenic markers, 21 of them being duplicated. Figure 4 shows two optimal scenarios obtained by applying algorithm described in Section 4: a) with reversals only, b) with reversals and excision/reintegration. Those last type scenarios raises the questions about mecanisms that led to duplication in plant mitogenomes [1]. Detailed scenarios are provided in the supplementary material[2].

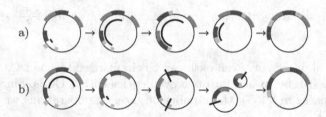

Fig. 4. Two parsimonious scenarios reconstructing a putative ancestral genome just before the tandem duplication event. Large segments show duplicated markers separated by breaks. The black line inside circles show the reversals applied while segments cutting the circles show the excision/reintegration applied.

7   Conclusion

In this paper we introduced several instances of the problem of reconstructing an ancestral genome which evolved through tandem duplications and other rearrangement operations. We obtained a distance formula for the simpliest case where all markers have been duplicated and only one tandem duplication occurred ; which can be computed in linear time. For the case where some markers have not been duplicated we obtained an approximate algorithm. Unfortunately,

---

[2] http://www.lifl.fr/~varre/download/suppmatWABI2012.pdf

all other cases we explored are NP-hard. Future work should be to design approximate algorithms allowing to go further in the analysis of biological data, in order to be able to compute phylogenetic trees and putative ancestors for a set of genomes fo which duplicates appeared through tandem duplications.

References

1. Backert, S., Nielsen, B.L., Börner, T.: The mystery of the rings: structure and replication of mitochondrial genomes from higher plants. Trends in Plant Science 2(12), 477–483 (1997)
2. Bader, M.: Genome rearrangements with duplications. BMC Bioinformatics 11(S-1), 27 (2010)
3. Bernt, M., Chen, K.-Y., Chen, M.-C., Chu, A.-C., Merkle, D., Wang, H.-L., Chao, K.-M., Middendorf, M.: Finding all sorting tandem duplication random loss operations. J. Discrete Algorithms 9(1), 32–48 (2011)
4. Chang, S., Yang, T., Du, T., Chen, J., Yan, J., He, J., Guan, R.: Mitochondrial genome sequencing helps show the evoltutionary mechanism of mitochondrial genome formation in *Brassica*. BMC Genomics 12(497) (2011)
5. Darracq, A., Varré, J.-S., Maréchal-Drouard, L., Courseaux, A., Castric, V., Saumitou-Laprade, P., Oztas, S., Lenoble, P., Vacherie, B., Barbe, V., Touzet, P.: Structural and content diversity of mitochondrial genome in beet: a comparative genomic analysis. Genome Biology and Evolution 3, 723–736 (2011)
6. Darracq, A., Varré, J.-S., Touzet, P.: A scenario of mitochondrial genome evolution in maize based on rearrangement events. BMC Genomics 11(233) (2010)
7. El-Mabrouk, N., Sankoff, D.: The reconstruction of doubled genomes. SIAM J. Comput. 32(3), 754–792 (2003)
8. Kovač, J., Warren, R., Braga, M.D.V., Stoye, J.: Restricted DCJ model: rearrangement problems with chromosome reincorporation. Journal of Computational Biology 18(9), 1231–1241 (2011)
9. Mixtacki, J.: Genome Halving under DCJ Revisited. In: Hu, X., Wang, J. (eds.) COCOON 2008. LNCS, vol. 5092, pp. 276–286. Springer, Heidelberg (2008)
10. Thomas, A., Varré, J.-S., Ouangraoua, A.: Genome dedoubling by dcj and reversal. BMC Bioinformatics 12(suppl. 9), S20 (2011)
11. Warren, R., Sankoff, D.: Genome halving with double cut and join. In: Brazma, A., Miyano, S., Akutsu, T. (eds.) Proceedings of APBC 2008. Adv. in Bioinformatics and Comp. Biol., vol. 6, pp. 231–240. Imperial College Press (2008)
12. Yancopoulos, S., Attie, O., Friedberg, R.: Efficient sorting of genomic permutations by translocation, inversion and block interchange. Bioinformatics 21(16), 3340–3346 (2005)

A Practical Approximation Algorithm for Solving Massive Instances of Hybridization Number

Leo van Iersel[1], Steven Kelk[2], Nela Lekić[2], and Celine Scornavacca[3]

[1] Centrum Wiskunde & Informatica (CWI), P.O. Box 94079,
1090 GB Amsterdam, The Netherlands
l.j.j.v.iersel@gmail.com
[2] Department of Knowledge Engineering (DKE), Maastricht University,
P.O. Box 616, 6200 MD Maastricht, The Netherlands
{steven.kelk,nela.lekic}@maastrichtuniversity.nl
[3] Institut des Sciences de l'Evolution (ISEM, UMR 5554 CNRS), Université
Montpellier II, Place E. Bataillon - CC 064 - 34095 Montpellier Cedex 5, France
celine.scornavacca@univ-montp2.fr

Abstract. Reticulate events play an important role in determining evolutionary relationships. The problem of computing the minimum number of such events to explain discordance between two phylogenetic trees is a hard computational problem. In practice, exact solvers struggle to solve instances with reticulation number larger than 40. For such instances, one has to resort to heuristics and approximation algorithms. Here we present the algorithm CYCLEKILLER which is the first approximation algorithm that can produce solutions verifiably close to optimality for instances with hundreds or even thousands of reticulations. Theoretically, the algorithm is an exponential-time 2-approximation (or 4-approximation in its fastest mode). However, using simulations we demonstrate that in practice the algorithm runs quickly for large and difficult instances, producing solutions within one percent of optimality. An implementation of this algorithm, which extends the theoretical work of [14], has been made publicly available.

1 Introduction

1.1 Background

A phylogenetic tree is a model used in biology to represent the evolutionary history of a set $\mathcal{X}$ of species (or *taxa*) [9,10]. They are trees whose leaves are bijectively labeled by $\mathcal{X}$ and whose internal vertices represent the ancestors of the species set; they can be rooted or unrooted. Since in a rooted tree edges have a direction, the concepts of indegree and outdegree of a vertex are well defined. *Binary* rooted (phylogenetic) trees are rooted (phylogenetic) trees whose internal vertices have outdegree 2.

Some biological events such as hybridization, recombination and horizontal gene transfer cannot be modeled by a tree. To represent an event in which a species derives its genes from different ancestors, trees are generalized to allow

B. Raphael and J. Tang (Eds.): WABI 2012, LNBI 7534, pp. 430–440, 2012.
© Springer-Verlag Berlin Heidelberg 2012

vertices with indegree two or higher, known as *reticulations*. This model is called a *rooted phylogenetic network*. For detailed background information we refer the reader to [11,13,15].

Although phylogenetic networks are more general than phylogenetic trees, trees are still often the basic building blocks from which they are constructed. Specifically, there are many techniques available for constructing gene trees. However, when more genes are analyzed, topological conflicts between individual gene phylogenies can arise for methodological or biological reasons (i.e. afore-mentioned reticulate phenomena such as hybridization). This has led computational biologists to try and quantify the amount of reticulation that is needed to simultaneously explain two trees.

To state this problem more formally, we say that a phylogenetic network N on $\mathcal{X}$ *displays* a phylogenetic tree T on $\mathcal{X}$ if T can be obtained from a subtree of N by contracting edges. (Informally this means that T can be obtained from N by, for each reticulation vertex of N, "switching off" all but one of its incoming edges and then suppressing all indegree-1 outdegree-1 vertices). Given two rooted binary phylogenetic trees T_1 and T_2 on $\mathcal{X}$, the problem then becomes to determine the minimum number of reticulations contained in a phylogenetic network N on $\mathcal{X}$ displaying both trees. The value we are minimizing is often called the *hybridization number* and instead of the term phylogenetic network, the term *hybridization network* is often used. It is known that the problem of computing hybridization numbers is both NP-hard and APX-hard [4], but it is not known whether it is in APX (i.e. whether it admits a polynomial-time approximation algorithm that achieves a constant approximation ratio).

Until recently, most research on the hybridization number of two binary trees had focused on the question of how to exactly compute this value using fixed parameter tractable (FPT) algorithms. For an introduction to FPT we refer to [8]. Algorithmic progress has been considerable in this area, with various authors reporting increasingly sophisticated FPT algorithms where the parameter in question is the hybridization number r of the two trees [3,5,7,18]. The fastest algorithms currently implemented are the algorithm available inside the package DENDROSCOPE [12], based on [1], and the algorithm HYBRIDNET [5]. The fastest theoretical FPT algorithm has running time $O(3.18^r n)$ [18], where n is the number of taxa in the trees.

Such FPT algorithms do, however, have their limits. For NP-hard problems such as hybridization number, and under the assumption that P $\neq$ NP, the running time still grows exponentially in r, albeit usually at a slower rate than algorithms that have a running time of the form $n^{f(r)}$ where f is some function of r. In practice this means that existing algorithms struggle to terminate for instances with hybridization number larger than 40.

Due to ongoing advances in DNA sequencing more and more species and strains are being sequenced. Consequently biologists use trees with more and more taxa and software that can handle large trees becomes more important. For such large and/or difficult trees one can try to generate heuristic or approximate solutions, but how far are such solutions from optimality? In [14] we showed that the news is worrying. Indeed, we showed that polynomial-time constant-ratio approximation algorithms exist if and only if such algorithms exist for the problem Directed Feedback Vertex Set (DFVS). However, DFVS is one of the most well-studied problems in combinatorial optimization and to this day it is unknown if it permits such an algorithm. Pending a major breakthrough in computer science, it therefore seems difficult to build efficient algorithms which approximate hybridization number well. On the positive side, we showed that in polynomial time an algorithm with approximation ratio $O(\log r \log \log r)$ is possible. However, this algorithm is purely of theoretical interest and is not useful in practice.

1.2   A New Algorithm: CycleKiller

In this article we extend the theoretical work of [14] slightly and give it a practical twist to yield a fast approximation algorithm (which we have made publicly available as the program CYCLEKILLER) that achieves a low constant ratio approximation, even for massive trees where the hybridization number runs into hundreds or even thousands.

The worst-case running time of this approximation algorithm is exponential. However, as we demonstrate with experiments, the running time of the algorithm is in practice extremely fast and, when the hybridization number is large, typically orders of magnitude faster than HYBRIDNET or the algorithm in DENDROSCOPE. Of course, those algorithms attempt to compute optimum solutions, whereas our algorithm only gives approximate solutions. Nevertheless, our experiments show that when CYCLEKILLER is run in its most accurate mode of operation an approximation ratio very close to 1 is not unusual, suggesting that the algorithm often produces solutions close to optimality and well within the worst-case approximation guarantee.

Specifically, we describe an algorithm with approximation ratio $d(c + 1)$ for the hybridization number problem by combining a c-approximation for the problem MAF (Maximum Agreement Forest) (see e.g. [16,18]) with a d-approximation for the problem DFVS. Both these problems are NP-hard so polynomial-time algorithms attaining $c = 1$ or $d = 1$ are not realistic. Nevertheless, there exist extremely fast FPT algorithms for solving MAF exactly (i.e. $c = 1$), the fastest is RSPR by Whidden, Beiko and Zeh [17,19]. Moreover, we observe that the type of DFVS instances that arise in practice can easily be solved using Integer Linear Programming (ILP) (and freely-available ILP solver technology such as GLPK), so $d = 1$ is also often possible. Combining these two exact approaches gives us an exponential-time approximation algorithm with worst-case approximation ratio 2 that for large instances still runs extremely quickly; this is the 2-approx option of CYCLEKILLER. In practice, we have observed that the upper bound of 2 is often pessimistic, with much better approximation ratios observed in experiments (1.003 on average for the simulations presented in this article). We find that this algorithm already allows us to cope with much bigger trees than HYBRIDNET or the algorithm in DENDROSCOPE.

Nevertheless, for truly massive trees it is often not feasible to have $c = 1$. Fortunately there exist linear-time algorithms which achieve $c = 3$ [18]. This, coupled with the fact that (even for such trees) it remains feasible to use an exact ($d = 1$) solver for DFVS, means that in practice we achieve a 4-approximation for gigantic trees; this is the 4-approx option of CYCLEKILLER. Again, the ratio of 4 is a worst-case bound and we suspect that in practice we are doing much better than 4. However, this cannot be experimentally verified due to the lack of good lower bounds for such massive instances. In any case, the main advantage of the 4-approximation is that it can without too much effort cope with trees with hundreds or thousands of taxa and hybridization number of a similar order of magnitude. An implementation of CYCLEKILLER and accompanying documentation can be downloaded from http://skelk.sdf-eu.org/cyclekiller. Networks created by the algorithm can be viewed in DENDROSCOPE.

1.3   Theoretical and Practical Significance

We have described, implemented and made publicly available an algorithm with two desirable qualities: it terminates quickly even for massive instances of hybridization number and it gives a non-trivial guarantee of proximity to optimality. This is the first algorithm with such properties. The algorithm is a non-trivial marriage of MAF and DFVS solvers (both exact and approximate), meaning that further advances in solving MAF and DFVS will directly lead to improvements in CYCLEKILLER.

This article also improves the theoretical work given in [14], which also proposed using DFVS but beginning from a trivial Agreement Forest (AF) known as a *chain forest*. Here we use a more intelligent starting point: an (approximate) MAF, and it is this insight which makes a 2-approximation (rather than the 6-approximation implied by [14]) possible when using an exact DFVS solver. Other articles have also had the idea of cycle-breaking in AFs: the advanced FPT algorithm of Whidden et al [18] – that, as far as we know, has not been implemented – and the algorithm HYBRIDNET by Chen et al. [5] (also see [6]). However, both algorithms start the cycle-breaking from many starting points. In contrast, our algorithm requires only a *single* starting point, i.e. an (approximate) solution to MAF.

2   Preliminaries

Let $\mathcal{X}$ be a finite set (e.g. of species). A *rooted phylogenetic $\mathcal{X}$-tree* is a rooted tree with no vertices with indegree 1 and outdegree 1, a root with indegree 0 and outdegree at least 2, and leaves bijectively labelled by the elements of $\mathcal{X}$. We identify each leaf with its label and use $L(T)$ to refer to the leaf set (or label set) of T. A rooted phylogenetic $\mathcal{X}$-tree is called *binary* if each nonleaf vertex has outdegree two. We henceforth call a rooted, binary phylogenetic $\mathcal{X}$-tree a *tree* for short. For a tree T and a set $\mathcal{X}' \subset \mathcal{X}$, we use the notation $T(\mathcal{X}')$ to denote the minimal subtree of T that contains all elements of $\mathcal{X}'$ and $T|\mathcal{X}'$ denotes the result of suppressing all indegree-1 outdegree-1 vertices in $T(\mathcal{X}')$.

We define a *forest* as a set of trees. Each element of a forest is called a *component*. Let T be a tree and $\mathcal{F}$ a forest. We say that $\mathcal{F}$ is a *forest for* T if $T|L(F)$ is isomorphic to F for all $F \in \mathcal{F}$ and the trees $\{T(L(F)), F \in \mathcal{F}\}$ are vertex-disjoint subtrees of T whose leaf-set union equals $L(T)$. If T_1 and T_2 are two trees, then a forest $\mathcal{F}$ is an *agreement forest* of T_1 and T_2 if it is a forest for T_1 and T_2. The number of components of $\mathcal{F}$ is denoted $|\mathcal{F}|$.

We define *cleaning up* a directed graph as repeatedly suppressing indegree-1 outdegree-1 vertices, removing indegree-0 outdegree-1 vertices and removing unlabelled outdegree-0 vertices until no such operation is possible. Observe that, if $\mathcal{F}$ is a forest for T, $\mathcal{F}$ can be obtained from T by removing $|\mathcal{F}| - 1$ edges and cleaning up.

Problem: Maximum Agreement Forest (MAF)
Instance: Two rooted, binary phylogenetic trees T_1 and T_2.
Solution: An agreement forest $\mathcal{F}$ of T_1 and T_2.
Objective: Minimize $|\mathcal{F}| - 1$.

The directed graph $IG(T_1, T_2, \mathcal{F})$, called the *inheritance graph*, is the directed graph whose vertices are the components of $\mathcal{F}$ and which has an edge (F, F') precisely if either

- there is a directed path in T_1 from the root of $T_1(L(F))$ to the root of $T_1(L(F'))$ or;
- there is a directed path in T_2 from the root of $T_2(L(F))$ to the root of $T_2(L(F'))$.

An agreement forest $\mathcal{F}$ of T_1 and T_2 is called an *acyclic agreement forest* if the graph $IG(T_1, T_2, \mathcal{F})$ is acyclic. A *maximum acyclic agreement forest (MAAF)* of T_1 and T_2 is an acyclic agreement forest of T_1 and T_2 with a minimum number of components.

Problem: Maximum Acyclic Agreement Forest (MAAF)
Instance: Two rooted, binary phylogenetic trees T_1 and T_2.
Solution: An acyclic agreement forest $\mathcal{F}$ of T_1 and T_2.
Objective: Minimize $|\mathcal{F}| - 1$.

We use $\mathrm{MAF}(T_1, T_2)$ and $\mathrm{MAAF}(T_1, T_2)$ to denote the optimal solution value of the problem MAF and MAAF respectively, for an instance T_1, T_2.

A *rooted phylogenetic network* on $\mathcal{X}$ is a directed acyclic graph with no vertices with indegree 1 and outdegree 1 and leaves bijectively labelled by the elements of $\mathcal{X}$. Rooted phylogenetic networks, which are sometimes also called hybridization networks, will henceforth be called *networks* for short in this paper. A tree T on $\mathcal{X}$ is *displayed* by a network N if T can be obtained from a subtree of N by contracting edges. A *reticulation* is a vertex v with $\delta^-(v) \geq 2$ (with $\delta^-(v)$ denoting the indegree of v). The *reticulation number* (sometimes also called hybridization number) of a network N with root ρ is given by

$$r(N) = \sum_{v \neq \rho} (\delta^-(v) - 1).$$

It was shown that the optimum to MAAF is equal to the optimum of the following problem [2].

Problem: MINIMUMHYBRIDIZATION
Instance: Two rooted, binary phylogenetic trees T_1 and T_2.
Solution: A rooted phylogenetic network N that displays T_1 and T_2.
Objective: Minimize $r(N)$.

Moreover, it was shown that, for two trees T_1, T_2, *any* acyclic agreement forest for T_1 and T_2 with $k + 1$ components can be turned into a phylogenetic network that displays T_1 and T_2 and has reticulation number k, and vice versa. Thus, any approximation for MAAF gives an approximation for MINIMUMHYBRIDIZATION.

A *feedback vertex set* of a directed graph is a subset of the vertices that contains at least one vertex of each directed cycle. Equivalently, a subset of the vertices of a directed graph is a *feedback vertex set* if removing these vertices from the graph makes it acyclic.

Problem: Directed Feedback Vertex Set (DFVS)
Instance: A directed graph D.
Goal: Find a feedback vertex set of D of minimum size.

3 Main Result

We show how MAAF can be approximated by combining algorithms for MAF and DFVS. In particular, we will prove the following theorem.

Theorem 1. *If there exists a c-approximation for MAF and a d-approximation for DFVS, then there exists a $d(c + 1)$-approximation for MAAF (and thus for* MINIMUMHYBRIDIZATION*).*

Suppose there exists a c-approximation for MAF. Let T_1 and T_2 be two trees and let M be an agreement forest returned by the algorithm. Then,

$$|M| - 1 \leq c \cdot \text{MAF}(T_1, T_2) \leq c \cdot \text{MAAF}(T_1, T_2). \tag{1}$$

An *M-splitting* is an acyclic agreement forest that can be obtained from M by removing edges and cleaning up.

Lemma 1. *Let T_1 and T_2 be two trees and M an agreement forest of T_1 and T_2. Then, there exists an M-splitting of size at most $MAAF(T_1, T_2) + |M|$.*

Proof. Consider a maximum acyclic agreement forest F of T_1 and T_2. For $i \in \{1, 2\}$, F can be obtained from T_i by removing a set of edges, say E_F^i, and cleaning up. Moreover, also M can be obtained from T_i by removing a set of edges, say E_M^i, and cleaning up.

Now consider the forest S obtained from T_1 by removing $E_M^1 \cup E_F^1$ and cleaning up. Then,

- S is an agreement forest of T_1 and T_2 because it can be obtained from T_2 by removing edges $E_M^2 \cup E_F^2$ and cleaning up;
- S is acyclic because it can be obtained by removing edges from F, which is acyclic, and cleaning up;
- S can be obtained from M by removing edges and cleaning up.

Hence, S is an M-splitting. Furthermore, $|S| \leq |E_F^1| + |E_M^1| + 1$. The lemma follows since $|E_F^1| = \mathrm{MAAF}(T_1, T_2)$ and $|M| = E_M^1 + 1$. □

Let $\mathrm{OptSplitting}_{T_1,T_2}(M)$ denote the size of a minimum-size M-splitting. Combining Lemma 1 and equation (1), we obtain

$$\mathrm{OptSplitting}_{T_1,T_2}(M) - 1 \leq (c+1)\mathrm{MAAF}(T_1, T_2) \tag{2}$$

We will now show how to find an approximation for the problem of finding an optimal M-splitting. We do so by reducing the problem to DFVS. We construct an input graph D for DFVS as follows. For every vertex of M that has outdegree 2 (in M), we create a vertex in D. There is an edge in D from a vertex u to a vertex v precisely if in either T_1 or T_2 (or in both) there is a directed path from u to v. We claim the following.

Lemma 2. *A subset V' of the vertices of D is a feedback vertex set of D if and only if removing V' from M makes it an acyclic agreement forest.*

Proof. We show that $D \setminus V'$ has a directed cycle if and only if the inheritance graph of $M \setminus V'$ has a directed cycle.

To prove this, first suppose that there is a cycle $v_1, v_2, \ldots, v_k = v_1$ in the inheritance graph of $M \setminus V'$. The vertices in the inheritance graph of $M \setminus V'$ correspond to the roots of the components of $M \setminus V'$. Since these roots have outdegree 2 in $M \setminus V'$, they had outdegree 2 in M, and are thus vertices of D. So the vertices $v_1, v_2, \ldots, v_k$ that form the cycle are vertices of D. Since these vertices are in the inheritance graph of $M \setminus V'$, they can not be in V' and so they are vertices of $D \setminus V'$. The reachability relation between these vertices in $D \setminus V'$ is the same as in the inheritance graph of $M \setminus V'$. So, the vertices $v_1, v_2, \ldots, v_k$ form a cycle in $D \setminus V'$.

Now suppose that there is a cycle $w_1, w_2, \ldots, w_k = w_1$ in $D \setminus V'$. Each of the vertices $w_1, w_2, \ldots, w_k$ is a vertex with outdegree-2 in M. Some of them might be roots of components, while others are not. However, observe that if there is a directed path from a vertex u to a vertex v in T_1 (or in T_2) then there is also a directed path from the root of the component of $M \setminus V'$ that contains u to the root of the component of $M \setminus V'$ that contains v. Hence, there is a directed cycle in the inheritance graph of $M \setminus V'$, formed by the roots of the components of $M \setminus V'$ that contain $w_1, w_2, \ldots, w_k$. □

Suppose that there exists a d-approximation for DFVS. Let FVS be a feedback vertex set returned by this algorithm and let MFVS be a minimum feedback vertex set. Then, removing the vertices of MFVS from M gives an optimal M-splitting. Furthermore, $\mathrm{OptSplitting}_{T_1,T_2}(M) = |M| + |\mathrm{MFVS}|$. This is because for every vertex in a cycle C, its parent in M must participate in some cycle that contains elements of C. So if we start by removing the root of the component we are splitting and subsequently remove only those vertices whose parents have already been removed we see that we add at most one component per vertex. In fact, because vertices of D all have out-degree 2 in M, we add exactly one component per vertex.

By removing the vertices of FVS from M, we obtain an acyclic agreement forest $\mathcal{F}$ such that

$$
\begin{aligned}
|\mathcal{F}| - 1 &= |M| + |\text{FVS}| - 1 \\
&\leq |M| + d \cdot |\text{MFVS}| - 1 \\
&\leq d(|M| + |\text{MFVS}| - 1) \\
&= d(\text{OptSplitting}_{T_1, T_2}(M) - 1) \\
&\leq d(c+1)\text{MAAF}(T_1, T_2),
\end{aligned}
$$

where the last inequality follows from equation (2). Thus, $\mathcal{F}$ is a $d(c + 1)$-approximation to MAAF, which concludes the proof of Theorem 1.

4   Practical Experiments

To assess the performance of CYCLEKILLER, a simulation study was undertaken. We generated 3 synthetic datasets, an *easy*, a *medium* and a *hard* one, containing respectively 800, 640 and 640 pairs of rooted binary phylogenetic trees.

The easy data set was created by varying two parameters, namely the number of taxa n and the number of rSPR-moves k used to obtain the second tree from the first (note that this number is an upper bound on the actual rSPR distance). The 800 pairs of rooted binary phylogenetic trees were created by varying n in $\{20, 50, 100, 200\}$ and k in $\{5, 10, ..., 25\}$, and then creating 40 different instances per each combination of parameters. Each pair (T_1, T_2) of rooted binary phylogenetic trees for a given set of parameters n and k is created as follows: The first tree T_1 on $\mathcal{X} = \{x_1, \ldots, x_n\}$ is generated by first creating a set of n leaf vertices bijectively labeled by the set $\mathcal{X}$. Then, two vertices u and v, both with indegree 0, are randomly picked and a new vertex w, along with two new edges (w, u) and (w, v), is created. This is done until only one vertex with no ancestor, the root, is present. The second tree T_2 is obtained from T_1 by applying k rSPR-moves. The medium and the hard data sets were generated in the same way as the easy one, but for different choices of the parameters: n in $\{50, 100, 200, 300\}$ and k in $\{15, 25, 40, 55\}$ for the medium one and n in $\{100, 200, 400, 500\}$ and k in $\{40, 60, 80, 100\}$ for the hard one.

The exact hybridization number has been computed by HYBRIDNET [5], available from http://www.cs.cityu.edu.hk/~lwang/software/Hn/treeComp.html or with DENDROSCOPE [12], available from http://www.dendroscope.org. We will refer to these algorithms as the *exact algorithms*. Each instance has been run on a single core of an Intel Xeon E5506 processor.

Each run that took more than one hour was aborted. For each instance, we ran our program with the option 2-approx, and, in case the latter did not finish within one hour, we ran it again, this time using the option 4-approx, always with a one-hour limit. We used the program RSPR v1.03 [17,19] to solve or approximate MAF and GLPK v4.47 (http://www.gnu.org/software/glpk/) to solve a simple ILP formulation of DFVS.

For all instances of the easy data set, CYCLEKILLER finished with the 2-approx option within the one hour limit, while for 33 instances the exact algorithms were unable to compute the hybridization number. Note that, even for "easy" instances, computing the exact hybridization number can take a very long time. To give the reader an idea, for 9 runs of the easy data, DENDROSCOPE and HYBRIDNET did not complete within 10 days. Table 1 shows a summary of the results. It can be seen that CYCLEKILLER was much faster than the exact algorithms. Moreover, for 96.6% of the instances for which an exact algorithm could find a solution, CYCLEKILLER also found an optimal solution. While the theoretical worst-case approximation ratio of the 2-approx option of CYCLEKILLER is 2, in our experiments it performed very close to a 1-approximation.

Table 1. Experimental results. The third column indicates for how many instances at least one exact algorithm finished within one hour. The fifth column indicates for how many instances the 2-approx option of CYCLEKILLER finished within one hour. For the remaining instances, the 4-approx option finished within one hour, as can be seen from the seventh column. The average running time for the 2-approx and the 4-approx are reported respectively in the sixth and eighth column. The average approximation ratio (ninth column) is taken over all instances for which at least one exact method finished. The last column indicates the percentage of those instances for which CYCLEKILLER found an optimal solution.

Dataset	Total runs	Exact algorithms		CYCLEKILLER					
		Com-pleted	Running time (RT)	2-approx		4-approx		Average appr. ratio	Opt. found
				Compl.	RT	Compl.	RT		
Easy	800	767	$13m18s$	800	$3s$	-	-	1.003	96.6%
Medium	640	199	$42m52s$	613	$3m32s$	27	<1 s	1.002	97.5%
Hard	640	0	$1h$	440	$21m11s$	200	1.5 s	-	-

For the medium data set, CYCLEKILLER finished with the 2-approx option for 613 instances, and for the remaining ones with the 4-approx option. The exact algorithms could compute the hybridization number for only 199 instances (out of 640). For 97.5% of these instances, CYCLEKILLER also found an optimal solution, but with a much better running time. Regarding the hard data set, 444 runs were completed with the 2-approx option and for the remaining ones we were able to use the 4-approx option within the given time constraint. Unfortunately, the exact algorithms were unable to compute the hybridization number for any tree-pair of this data set and hence we could not compute the average approximation ratios. Over all our experiments, the maximum hybridization number that the exact algorithms could handle was 25.[1] In contrast, the 2-approx option of CYCLEKILLER could be used for instances for which the size of a MAF was up to 97, and thus for instances for which the hybridization number was at least 97.

To find the limits of the 4-approx option of CYCLEKILLER, we also tested it on randomly generated trees. On a normal laptop, it could construct networks with up to 10,000 leaves and up to 10,000 reticulations within 10 minutes. Since the number of reticulations found is at most four times the optimal hybridization number, this implies that the 4-approx option of CYCLEKILLER can handle hybridization numbers up to at least 2,500. These randomly generated trees are, however, biologically meaningless and, therefore, we conducted the extensive experiment described above on trees generated by rSPR moves.

Acknowledgements. We thank Simone Linz and Leen Stougie for fruitful discussions. Leo van Iersel and Nela Lekić were supported by Veni and Vrije Competitie grants of The Netherlands Organisation for Scientific Research (NWO). This publication is the contribution no. 2012-076 of the Institut des Sciences de l'Evolution de Montpellier (ISE-M, UMR 5554). This work has been partially funded by the ANCESTROME project ANR-10-IABI-0-01 and it has benefited from the ISE-M computing facilities.

References

1. Albrecht, B., Scornavacca, C., Cenci, A., Huson, D.H.: Fast computation of minimum hybridization networks. Bioinformatics 28(2), 191–197 (2012)
2. Baroni, M., Grünewald, S., Moulton, V., Semple, C.: Bounding the number of hybridisation events for a consistent evolutionary history. Mathematical Biology 51, 171–182 (2005)
3. Bordewich, M., Linz, S., St. John, K., Semple, C.: A reduction algorithm for computing the hybridization number of two trees. Evolutionary Bioinformatics 3, 86–98 (2007)
4. Bordewich, M., Semple, C.: Computing the minimum number of hybridization events for a consistent evolutionary history. Discrete Applied Mathematics 155(8), 914–928 (2007)
5. Chen, Z.-Z., Wang, L.: Hybridnet: a tool for constructing hybridization networks. Bioinformatics 26(22), 2912–2913 (2010)
6. Chen, Z.-Z., Wang, L.: Algorithms for reticulate networks of multiple phylogenetic trees. IEEE/ACM Transactions on Computational Biology and Bioinformatics 9(2), 372–384 (2012)
7. Collins, J., Linz, S., Semple, C.: Quantifying hybridization in realistic time. Journal of Computational Biology 18, 1305–1318 (2011)
8. Flum, J., Grohe, M.: Parameterized Complexity Theory. Springer (2006)
9. Gascuel, O. (ed.): Mathematics of Evolution and Phylogeny. Oxford University Press, Inc. (2005)

---

[1] In [1], it has been shown that this number can go up to 40 when running Dendroscope on a similar processor but allocating all cores for one instance, i.e. exploiting the possibilities of parallel computation of this implementation.

10. Gascuel, O., Steel, M. (eds.): Reconstructing Evolution: New Mathematical and Computational Advances. Oxford University Press, USA (2007)
11. Huson, D.H., Rupp, R., Scornavacca, C.: Phylogenetic Networks: Concepts, Algorithms and Applications. Cambridge University Press (2011)
12. Huson, D.H., Scornavacca, C.: Dendroscope 3 - a program for computing and drawing rooted phylogenetic trees and networks (2011) (in preparation), Software, http://www.dendroscope.org
13. Huson, D.H., Scornavacca, C.: A survey of combinatorial methods for phylogenetic networks. Genome Biology and Evolution 3, 23–35 (2011)
14. Kelk, S.M., van Iersel, L.J.J., Lekić, N., Linz, S., Scornavacca, C., Stougie, L.: Cycle killer.. qu'est ce que c'est? on the comparative approximability of hybridization number and directed feedback vertex set. Submitted, preliminary version arXiv:1112.5359v1 (math.CO)
15. Nakhleh, L.: Evolutionary phylogenetic networks: models and issues. In: The Problem Solving Handbook for Computational Biology and Bioinformatics. Springer (2009)
16. Rodrigues, E.M., Sagot, M.F., Wakabayashi, Y.: The maximum agreement forest problem: Approximation algorithms and computational experiments. Theoretical Computer Science 374(1-3), 91–110 (2007)
17. Whidden, C.: http://kiwi.cs.dal.ca/Software/RSPR
18. Whidden, C., Beiko, R.G., Zeh, N.: Fixed-parameter and approximation algorithms for maximum agreement forests. Submitted, preliminary version arXiv:1108.2664v1 (q-bio.PE)
19. Whidden, C., Beiko, R.G., Zeh, N.: Fast FPT Algorithms for Computing Rooted Agreement Forests: Theory and Experiments. In: Festa, P. (ed.) SEA 2010. LNCS, vol. 6049, pp. 141–153. Springer, Heidelberg (2010)

Distributed String Mining for High-Throughput Sequencing Data*

Niko Välimäki[1,2] and Simon J. Puglisi[2]

[1] Helsinki Institute for Information Technology
[2] Department of Computer Science, University of Helsinki
{nvalimak,puglisi}@cs.helsinki.fi

Abstract. The goal of frequency constrained string mining is to extract substrings that discriminate two (or more) datasets. Known solutions to the problem range from an optimal time algorithm to different time–space tradeoffs. However, all of the existing algorithms have been designed to be run in a sequential manner and require that the whole input fits the main memory. Due to these limitations, the existing algorithms are practical only up to a few gigabytes of input. We introduce a distributed algorithm that has a novel time–space tradeoff and, in practice, achieves a significant reduction in both memory and time compared to state-of-the-art methods. To demonstrate the feasibility of the new algorithm, our study includes comprehensive tests on large-scale metagenomics data. We also study the cost of renting the required infrastructure from, e.g. Amazon EC2. Our distributed algorithm is shown to be practical on terabyte-scale inputs and affordable on rented infrastructure.

1 Introduction

The sheer amount of data produced by high-throughput sequencing can easily outstrip the main memory available on standard servers, and can make sequence analysis tasks tedious or even impossible. One such task is *frequency constrained string mining*. While there exists a number of algorithms to solve the problem in feasible time, none of the existing algorithms can really cope with datasets that surpass limits of available main memory. In this paper, we propose a distributed solution to the problem which, in practice, can completely avoid memory limitations, assuming that the required infrastructure for distributed computation is available. Fortunately, high-performance computer clusters are nowadays largely available via, for example, flexible rental services.

The goal of frequency constrained string mining is to extract substrings that discriminate two or more datasets (sets of strings). The discriminative substrings are determined based on their frequency in the given datasets. We define the *frequency* of string P in dataset $\mathcal{T}$ as the number of strings in $\mathcal{T}$ having at

* Funded by the Academy of Finland grant 118653 (ALGODAN), and Helsinki Doctoral Programme in Computer Science (HECSE).

B. Raphael and J. Tang (Eds.): WABI 2012, LNBI 7534, pp. 441–452, 2012.

least one occurrence of P, that is, $\mathsf{freq}(P, \mathcal{T}) = |\{T \in \mathcal{T} : P \text{ occurs in } T\}|$. For example, consider the following sets:

$$\mathcal{T}^+ = \{\text{I am positive,} \qquad \mathcal{T}^- = \{\text{I am negative,}$$
$$\text{I am also positive,} \qquad \text{I am also negative,}$$
$$\text{I am also positive}\} \qquad \text{I am not negative}\}$$

The substring I am is very frequent but also common to both datasets. At the other extreme, the substrings positive and negative clearly differentiate $\mathcal{T}^+$ from $\mathcal{T}^-$ and vice-versa. Substrings that discriminate the given datasets are sometimes called *emerging substrings*, and have direct applications for biological sequence classification [2] and knowledge discovery in databases [3].

Given a set of strings, $\mathcal{T}$, and frequency constraints $f_{\min}$ and $f_{\max}$, we want to extract all substrings P having $f_{\min} \le \mathsf{freq}(P, \mathcal{T}) \le f_{\max}$. Extending this to multiple sets introduces a pair of frequency constraints for each set:

Problem 1 ([6,14]). Given datasets $\mathcal{T}_1, \mathcal{T}_2, \dots, \mathcal{T}_r$ and a pair of frequency constraints $(f^i_{\min}, f^i_{\max})$ for each set, the *frequency constrained string mining* problem is to output all substrings P such that $f^i_{\min} \le \mathsf{freq}(P, \mathcal{T}_i) \le f^i_{\max}$ for all $1 \le i \le r$.

In this paper, we introduce a more general version of Problem 1:

Problem 2. Given datasets $\mathcal{T}_1, \mathcal{T}_2, \dots, \mathcal{T}_r$, constraints $(p_{\min}, p_{\max})$, and a pair of frequency constraints $(f^i_{\min}, f^i_{\max})$ for each set, the *generalized frequency constrained string mining* problem is to output all substrings P such that

$$p_{\min} \le |\{i \in [1, r] : f^i_{\min} \le \mathsf{freq}(P, \mathcal{T}_i) \le f^i_{\max}\}| \le p_{\max}.$$

This generalized version of the problem lets us extract substrings that emerge from any (suitably sized) subset of the input sets. This is contrary to Problem 1, where we need to have prior knowledge of the subsets that we expect the highly frequent substrings to emerge from. We believe generalized constraints are more flexible in practice when there are hundreds of datasets given. Our distributed algorithm is applicable to both Problem 1 and Problem 2, and can be extended to support other frequency-based mining queries such as the statistical χ^2-test [6, Problem 3].

1.1 Related Work

An overview of the recent related results is included in Table 1. To our knowledge, we are the first to consider Problem 2 but we believe algorithms FHK [6], FMV [7] and DPT [5] (see Table 1) can be adapted to it.

Fischer, Huen and Kramer [6] gave the first optimal time algorithm to solve Problem 1. They considered inputs that consist of r datasets. The problem is solved in $\mathcal{O}(n + t_{\text{output}})$ time, where n and t_{output} are the size of the input and output, respectively. However, their solution requires $\Theta(n \log n)$ bits of working

Table 1. Recent results on frequency constrained string mining, where n is the input size, σ is the alphabet size, m is the total number of strings, and r is the number of datasets. Output time is not included since it is roughly the same for all methods.

Method	Time	Space (in bits)	Notes
FHK, Fischer-Huen-Kramer [6]	$\mathcal{O}(n)$	$\mathcal{O}(n \log n)$	
KO, Kügel-Ohlebusch [14]	$\mathcal{O}(rn)$	$\mathcal{O}(\max_i \|\mathcal{T}_i\| \cdot \log n)$	Problem 1 only.
FMV, Fischer-Mäkinen-Välimäki [7]	$\mathcal{O}(n \log n)$	$\mathcal{O}(n \log \sigma + m \log n)$	Assumes $r = 2$.
DPT, Dhaliwal-Puglisi-Turpin [5]	$\mathcal{O}(n \log^2 n)$	$\mathcal{O}(n \log \sigma + m \log n)$	Assumes $r = 1$.

Table 2. Our results on frequency constrained string mining, where n is the input size, ℓ is the longest string length, and σ is the alphabet size. Parameters c and s denote the number of client processes and server processes, respectively.

		Worst-case	Expected
Client side	Time	$\mathcal{O}\left(\max\{\ell, \frac{n}{c}\}\ell\right)$	$\mathcal{O}\left(\frac{n}{c} \log n\right)$
	Space (in bits)	$\mathcal{O}\left(\max\{\ell, \frac{n}{c}\} \log n\right)$	$\mathcal{O}\left(\max\{\ell, \frac{n}{c}\} \log n\right)$
Server side	Time	$\mathcal{O}(n \min\{\ell, \log n\})$	$\mathcal{O}\left(\frac{n}{s} \log n\right)$
	Space (in bits)	$\mathcal{O}(c\ell \log n)$	*negligible*
Transmitted bits		$\mathcal{O}(n \min\{\ell, \log n\} \log n)$	$\mathcal{O}(n \min\{\ell, \log n\} \log n)$

space, which has motivated studies on more space-efficient variants such as algorithm FMV [7], which makes use of compressed and succinct data structures. We will describe the optimal-time algorithm in Section 3.

Kügel and Ohlebusch [14] devised a method whose space complexity depends only on the size of the largest dataset given. The gist of their method is the observation that Problem 1 can be solved in two steps: compute the result set of each dataset separately and then combine them via by pairwise intersections. It is trivial to parallelize the first step of their algorithm. However, the intersection phase is designed to be run in sequential manner. Kügel and Ohlebusch did not consider parallel computing or Problem 2. It seems impossible to adapt their approach to solve Problem 2: mainly because (i) the latter step makes the intersections by looking at frequencies of two result sets at a time, and (ii) they assume that the final result set is a subset of the result set of $\mathcal{T}_1$. Solving Problem 2 would require the algorithm to know the substring frequencies over all datasets when making the intersection. Also, assumption (ii) holds only for Problem 1.

A recent result by Dhaliwal, Puglisi and Turpin [5] made a significant, practical improvement. While the theoretical time complexity is slightly worse, $\mathcal{O}(n \log^2 n)$ time, the practical performance of their block-wise traversal is comparable to the optimal-time algorithm, while using similar space to FMV.

The next two sections outline basic data structures and the original optimal-time algorithm from Fischer et al. [6]. The latter sections describe our distributed algorithm and experimental results. We conclude with a discussion.

2   Preliminaries

A *string* $S[1..n] = S[1]S[2]\cdots S[n]$ is a *sequence* of *symbols*. Each symbol is an element of an ordered *alphabet* $\Sigma = \{1, 2, \ldots, \sigma\}$. We assume $\sigma = O(\text{polylog}(n))$. A *substring* of S is written $S[i..j] = S[i]S[i+1]\cdots S[j]$. A *prefix* of S is a substring of the form $S[1..j]$, and a *suffix* is a substring of the form $S[i..n]$. The lexicographical order "$<$" among strings is defined in the obvious way.

The *suffix array*, SA, of text $T[1..n]$ contains all suffixes of T in sorted order [15]. The array can be represented as pointers to the text so that the lexicographically i-th suffix is given by $T[\text{SA}[i]..n]$. For any $1 \le i < j \le n$, the suffix $T[\text{SA}[i]..n]$ is lexicographically smaller than $T[\text{SA}[j]..n]$. SA is a permutation of values $[1, n]$, that is, $\text{SA}[i] \ne \text{SA}[j]$ for all $i \ne j$. Suffix arrays can be constructed in the optimal linear time assuming that the input is from an integer alphabet [12] and can be stored in $n\lceil\log n\rceil$ bits of space, without compression.

The suffix array is often coupled with the *longest common prefix* (LCP) array. The LCP array records the longest common prefix between two consecutive suffixes in SA. More specifically, $\text{LCP}[i] = \text{lcp}(T[\text{SA}[i]..n], T[\text{SA}[i+1]..n])$, where $\text{lcp}(X, Y)$ returns the maximum ℓ such that strings $X[1..\ell]$ and $Y[1..\ell]$ match. The LCP array can be computed in linear time from SA and S [13].

3   Optimal-Time String Mining

The frequency based string mining problem can be solved in optimal time [6]. The algorithm integrates the algorithm from Kasai et al. [13] to visit all *branching substrings*, with Hui's algorithm [8] for the *color set size problem*. Let $\mathcal{T} = \{T_1, \ldots, T_m\}$ denote one dataset. The required data structures include the suffix array and LCP array built on the concatenated string $T = T_1\$_1T_2\$_2\cdots T_m\$_m$, where each $\$_i$ is a special symbol that does not occur in the input strings.

Branching Substrings. Substring α of string $T[1..n]$ is called *right-branching* if there exists two symbols $a, b \in \Sigma$ such that $a \ne b$ and both αa and αb are substrings of T. Similarly, a substring α is *left-branching* if $a\alpha$ and $b\alpha$, $a \ne b$, are substrings of T. If a substring is both left-branching and right-branching we say it is *left-right-branching*. There are at most $\mathcal{O}(n)$ right-branching substrings. In fact, the right-branching substrings of T have an one to one correspondence to the internal nodes of the *suffix tree* of T [13]. All the right-branching substrings of T can be visited in linear time by simulating the suffix tree traversal—using the suffix and LCP arrays—in left to right manner [13]. At each step $i \in [1, n]$, we maintain a stack of nodes corresponding to right-most path to the (virtual) suffix tree leaf $\text{SA}[i]$. To proceed to the next column $i + 1$, we first pop nodes from the stack that have string-depth greater than $\text{LCP}[i+1]$. Then, if a node with string-depth $\text{LCP}[i+1]$ does not yet exist, we push a new internal node with string-depth $\text{LCP}[i+1]$. Finally, we push the leaf node corresponding to $\text{SA}[i+1]$ with string-depth $n - \text{SA}[i+1] + 1$.

```
         1  2  3  4  5  6  7  8  9 10 11 12 13 14 15 16 17 18 19 20 21 22 23
SA:      5 12 18 23  4 22  8  9  1 10  2  6 15 19 11 17  3 21  7 14 16 20 13
LCP:     0  0  0  0  0  1  1  2  3  1  2  3  2  3  0  1  1  2  2  2  1  2  3
         $  $  $  $  a  a  a  a  a  a  a  a  a  a  b  b  b  b  b  b  b  b  b
                     $  $  a  a  a  b  b  b  b  b  $  $  a  a  a  a  b  b  b
                           a  b  b  $  a  a  b  b        $  $  a  b  $  a  a
                           b     a     $  a     a        a     b     $  b
                           $     $     a     $           b     b        b
                                       b                 $     $        $
                                       $
```

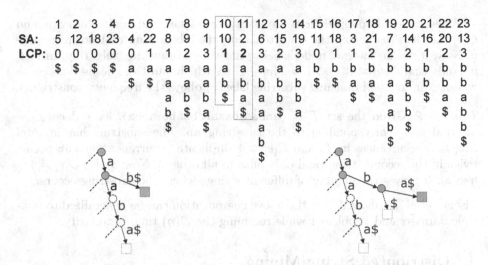

Fig. 1. Two steps of a simulated suffix tree traversal using the suffix and LCP array of the set $\mathcal{T} = \{\mathsf{aaba, abaaab, bbabb, abba}\}$. At step $i = 10$ (on left), we pop nodes up to string depth LCP[10] = 1 and push a new node for SA[10]. At the next step, $i = 11$ (on right), there is no node at string depth LCP[11] = 2, thus, a new internal node is pushed to the stack.

Example 1. Given the set $\mathcal{T} = \{\mathsf{aaba, abaaab, bbabb, abba}\}$, Figure 1 shows an example of two subsequent steps, $i = 10$ and $i = 11$, and their required push/pop operations. The figure also includes the suffix and LCP arrays of $\mathcal{T}$.

One left-to-right pass over the suffix and LCP arrays ensures that all internal nodes of the (virtual) suffix tree get pushed to (and popped from) the stack. It also ensures that we cover all right-branching substrings; for example, substrings **a** and **ab** are right-branching in the above example. Abouelhoda et al. [1] show that if we store an additional value with each item on the stack, then in the same left-to-right pass we can compute the left-branching substrings as well.

Color Set Size Problem. The frequency of branching substrings (*left-*, *right-*, or *left-right-*) can be determined with the color set size technique [8]. We store two values for every node v in the stack: $S[v]$ recording the total number of leaves in v's subtree, and $C[v]$ recording the number of *duplicate occurrences* in v's subtree. The values are computed on the fly so that $S[v]$ is initialised to 1 for leaf nodes, and 0 otherwise. When node v is popped from stack, the value $S[v]$ is added to $S[w]$, where w denotes v's direct parent, i.e. the next node in the stack. Updating the $C[v]$ values is more involved: initially, when pushing a new node, we set $C[v] = 0$. At each step i, we choose a specific ancestor w from the stack using a *range minimum query* over the LCP array, and increase its value $C[w]$ by one (a more precise description is given in [7]). Additionally, for each popped node v, we add the value $C[v]$ to its direct parent, in similar manner as with S-values.

When node v is popped from the stack, we retrieve the final values $S[v]$ and $C[v]$. Let $\mathsf{label}(v)$ denote the concatenated path labels from root down to node v. Now, if v is an internal node, $\mathsf{label}(v)$ must be a branching substring. In fact, it holds that $\mathsf{freq}(\mathsf{label}(v), \mathcal{T}) = S[v] - C[v]$, which can then be used to check whether or not the branching substring $\mathsf{label}(v)$ obeys the frequency constraints.

Example 2. Given the set $\mathcal{T} = \{\mathsf{aaba}, \mathsf{abaaab}, \mathsf{bbabb}, \mathsf{abba}\}$, let v denote the internal node corresponding to the substring ab. The substring has in total $S[v] = 5$ occurrences in $\mathcal{T}$, and $C[v] = 1$ duplicate occurrence since ab occurs twice in the second string (and only once in all others). Now $S[v] - C[v] = 4 = \mathsf{freq}(\mathsf{ab}, \mathcal{T})$ gives the number of different strings where the substring occurs.

Fischer et al. [6] showed that the above computation can be generalized to multiple datasets and Problem 1 while retaining the $\mathcal{O}(n)$ time complexity.

4   Distributed String Mining

This section describes our distributed algorithm, which follows a client–server model. The next two subsections describe the client and server side algorithm and analyze the time, space and transmission complexities of the whole pipeline. For the expected-case analysis, we assume that the input strings are generated from a *Bernoulli model*, that is, symbols $j \in \Sigma$ are drawn independently with probabilities q_j such that $\sum_j q_j = 1$. Let $f_{\min} = \min_i\{f^i_{\min}\}$ denote the smallest frequency constraint given. We assume that either $p_{\min} \geq 2$ or $f_{\min} \geq 2$. The resulting complexities are summarized in Table 2.

4.1   Client Side Processing

Let $\mathcal{T}_1, \ldots, \mathcal{T}_r$ denote the input datasets. Let us first assume that the number of clients is $c = r$, that is, we use one client process $\mathcal{C}_i$ per dataset, and that the number of servers is one. Each client $\mathcal{C}_i$ builds the suffix and LCP array for $\mathcal{T}_i$ as described in the previous section. This gives the same space complexity as with the KO algorithm [14], that is, $\mathcal{O}(\max_i\{\|\mathcal{T}_i\|\}\log n)$ bits. We will later show how to decrease this down to $\mathcal{O}(\max\{\ell, \frac{n}{c}\}\log\sigma)$ bits, where ℓ denotes the longest string length in the input. Then each client traverses through the branching substrings in $\mathcal{T}_i$ and computes the frequencies within that particular dataset as described in the previous section. The important observations here are that (i) the branching substrings are traversed in lexicographical order, and (ii) the branching substrings can be represented as a tree (i.e. the suffix tree of $\mathcal{T}_i$). Thus, we use one-way transmission to feed the branching substrings and their frequencies to the server using a tree representation. Since we cannot assume that the whole input fits in the server-side main memory, we cannot use a (compact) suffix tree as the transmission protocol. Instead, we use a *noncompact* suffix tree (NST) to transmit substring information to the server. In a noncompact suffix tree, each edge is labeled by exactly one symbol. The tree consists of internal and external nodes; internal nodes are shared by at least two suffixes, while every external node corresponds to exactly one string (a unary path to a leaf). This affects our worst-case complexities because the NST of a string of length n has size $\mathcal{O}(n^2)$ in the worst-case. The expected size of the NST is, however, much smaller.

Let us analyze the time and transmission complexity assuming $f_{\min} \geq 2$. Recall that the longest string length in $\mathcal{T}_1, \ldots, \mathcal{T}_r$ is ℓ. If we use $\mathcal{O}(\log n)$ bits to transmit one node and its frequency, the worst-case transmission complexity of $\mathcal{C}_i$ becomes $\mathcal{O}(\sum_j |T_i^j|^2 \log n) = \mathcal{O}(\|\mathcal{T}_i\|\ell \log n)$ bits, where T_i^j denotes the j-th string in $\mathcal{T}_i$. The time complexity of $\mathcal{C}_i$ is then $\mathcal{O}(\|\mathcal{T}_i\|\ell)$ in the worst-case. We can expect much better time and transmission complexities when the input strings are generated from the Bernoulli model. Since we assume $f_{\min} \geq 2$, it is enough to transmit only the internal nodes; the expected number of internal nodes in an NST of a string of length n is $\Theta(n \log n)$ [10]. Now the expected transmission complexity of $\mathcal{C}_i$ becomes $\mathcal{O}((\log n) \sum_j |T_i^j| \log |T_i^j|) = \mathcal{O}(\|\mathcal{T}_i\| \log^2 n)$ bits. The expected time complexity is $\mathcal{O}(\|\mathcal{T}_i\| \log n)$.

We will refine the above algorithm in order to improve the complexities. It will, however, require a two-way communication model between the client and the server. Thus, we postpone the final analysis of the complexities until the end of the next subsection. Let c denote the number of client processes, and let $m \geq c$ — that is, we assign at least one string per client. Now we divide the input evenly among the clients so that each client gets a disjoint subset of the input strings. Thus, the length of the strings assigned to one client is $\mathcal{O}(\max\{\ell, n/c\})$. It follows that the worst-case time and space complexities are $\mathcal{O}(\max\{\ell, n/c\}\ell)$ and $\mathcal{O}(\max\{\ell, n/c\} \log n)$ bits, respectively. Notice that, since the input strings are divided regardless of dataset-boundaries, we cannot restrict the output to the server by resorting to $f_{\min}$. Instead, we expect the server to send a request to stop the client from transmitting undesired subtrees.

4.2 Server Side Processing

Let us first assume that we use just one server process. Also, let $\mathcal{C}_1, \ldots, \mathcal{C}_c$ denote the input stream from c clients. Since the tree-shaped inputs are transferred by the clients in lexicographical order, it is straightforward to merge them in left-to-right manner. For example, we can use a recursive algorithm to merge the input. At each step of the algorithm, we maintain a list of *active* clients for the current node v. For each active client, their upward path is equal to $\mathsf{label}(v) = P$. Initially $\mathcal{C} = \{\mathcal{C}_1, \ldots, \mathcal{C}_c\}$ and P is an empty string. First, we read the pending child node from each active client and then use recursion to process the subtree below v in lexicographical order. The nodes and their labels are transmitted in pre-order. Additionally, with each node, the client transmits a value λ to indicate if substring $\mathsf{label}(v)$ is left-branching *at the client*. If $\mathsf{label}(v)$ is not left-branching at the client λ is equal to the symbol which precedes every occurrence of $\mathsf{label}(v)$ at the client; Otherwise $\lambda = \bot$, a sentinel symbol.

After the subtree below the current node v is processed (recursively), we read the frequency counts for substring $\mathsf{label}(v) = P$ from the active clients—the frequency counts are transmitted in post-order. Finally, the server can compare the received frequency counts against constraints (f^i_{min}, f^i_{max}) and (p_{min}, p_{max}) to decide whether or not to output the current substring P. The worst-case time complexity is $\mathcal{O}(n\ell)$, which is equal to the size of the (virtual) noncompact suffix tree (NST) of the *whole* input $\mathcal{T}_1, \ldots, \mathcal{T}_r$ (though this can be tightened to $\mathcal{O}(n \min\{\ell, \log n\})$, as we will see below). The worst-case space is only $\mathcal{O}(c\ell \log n)$ bits since the server keeps track of only the active path which has maximum depth ℓ (the active path from the root to v can be kept in a stack). The frequency counts need not to be stored since they are received in post-order.

Transmission Complexity. It remains to show that the expected transmission complexity is feasible. Let n denote the total size of the input $\mathcal{T}_1, \ldots, \mathcal{T}_r$. Now, if either $f_{min} \geq 2$ or $p_{min} \geq 2$, the substrings that obey the frequency constraints must be either *left- or right-branching* (or both). Recall that the right-branching substrings correspond to the internal nodes. If substring $\mathsf{label}(v)$ is not left branching, then no substring that has $\mathsf{label}(v)$ as a prefix is either. As we will soon see, this simple property allows us to control transmissions between the server and clients, and to bound the worst-case transmission complexity.

During the recursive merge operation, having received the labels and λ values for node v, the server determines if $\mathsf{label}(v)$ is left-right-branching in the whole input. This allows the server to decide if the subtree rooted at v should be explored, and if not, it can tell the clients to not send nodes for that subtree.

To determine if $\mathsf{label}(v)$ is left-branching the server uses the λ values. If any of the λ values are different, or equal to the $\bot$, then $\mathsf{label}(v)$ is left-branching. If this is not the case then the server ignores the rest of the subtree and tells the clients not to send nodes belonging to it (as all the substrings spelt out in the subtree will have $\mathsf{label}(v)$ as a prefix, and so will not be left branching).

For right-branching substrings, if the current node v is (i) only active in one client $\mathcal{C}_i$, and (ii) v is an external node in $\mathcal{C}_i$, then node v must be an external node also in the NST of whole input (v's subtree cannot have right branching nodes). The server can request $\mathcal{C}_i$ to stop transmitting the subtree under v.

This is the two-way transmission protocol. With it, only the information for the paths of the (virtual) NST spelling out left-right-branching substrings are transmitted and the transmission complexity of the algorithm is proportional to the sum of the lengths of these paths. This sum is in turn bound above by the sum of the lengths of the left-right-branching substrings of the input which, by [11, Theorem 1], is $\mathcal{O}(n \log n)$. This gives the worst-case transmission complexity of $\mathcal{O}(n \log^2 n)$ bits and the worst-case server-side time $\mathcal{O}(n \log n)$.

Improving the Expected Case. The expected server-side time complexity can be reduced to $\mathcal{O}(\frac{n}{s} \log n)$ by spreading the load over s servers. Let us first assume $s = 2$. In order to balance the processing evenly between the two servers we need to find the *median suffix* M, that is lexicographically larger than half of the suffixes in the whole input. More precisely, M is the shortest unique prefix of the suffix $\mathsf{SA}[\lfloor \frac{n}{2} \rfloor]$, where SA is the (virtual) suffix array of $\mathcal{T}_1, \ldots, \mathcal{T}_r$. We begin with an empty string M', and at each step, append to M' the lexicographically smallest symbol $b \in \Sigma$ such that $M'b$ is (i) a prefix of, or (ii) lex. larger than

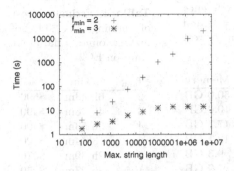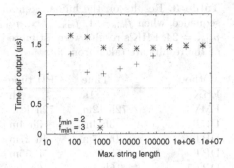

Fig. 2. Results on simulated data when varying ℓ from 75 up to 4.6 MB. We report both the wall-clock time (on left) and the time per outputted substring (on right).

at least half of the suffixes in $\mathcal{T}_1, \ldots, \mathcal{T}_r$. In order to count suffixes of type (i) and (ii) for any b and M', we query (i) and (ii) from each client separately. Since the clients correspond to disjoint sets of input strings, a sum of their results gives the total count over the whole input. We stop extending M' when the total count of type (i) suffixes is equal to one. The lex. smallest $b \in \Sigma$ at each step is determined with a binary search. Since the NST of the whole input has an expected height of $\mathcal{O}(\log n)$ [4], in total $\mathcal{O}(c(\log n) \log \sigma)$ queries are needed to determine the median suffix. Notice that the resulting M partitions the (virtual) NST of the whole input into two equal sized halves. For $s > 2$, we can now continue the partitioning by finding the median suffix in either of the two halves. After $\mathcal{O}(sc(\log n) \log \sigma)$ queries, we have an uniform load balancing for s servers.

5   Experimental Results

We now examine the practical performance of our distributed algorithm on (i) simulated datasets, (ii) human genome-scale data and (iii) large-scale metagenomics data. For test (ii), we include a comparison against sequential algorithms (see Sect. 1.1). In most of the experiments, we measure wall-clock time (elapsed real time). Experiment (ii) also measures total CPU time over all processes.

Our implementation follows the ideas presented in Sect. 4. However, we have not yet implemented two-way communication—instead, the clients use $f_{\min}$ to avoid (whenever possible) sending redundant output to servers. This is effective in practice, but has limitations regarding the values of $f_{\min}$.

Our experiments were run on a high-performance cluster of Dell PowerEdge M610 nodes equipped with 32GB RAM and two Intel Xeon E5540 2.53GHz processors. Each node has eight CPU cores, and there are in total 240 nodes in the cluster. The nodes are connected by two (bonded) 10Gb network interfaces.

Simulated Data. Notice that both the time and transmission complexities of our algorithm depend on ℓ. To demonstrate this effect in practice, we generated

Table 3. Results on the human genome reference sequence, when $f_{min} = 1$, $f_{max} = \infty$, $p_{min} = 21$ and $p_{max} = 24$. FHK's results were estimated in [7], rest of the results are taken from [5].

Method	Time	CPU time	Memory
FHK	1h	1h	50.0 GB
FMV	72h 12m	72h 12m	10.0 GB
DPT	3h 4m	3h 4m	17.7 GB
DPT	4h 27m	4h 27m	12.1 GB
DPT	5h 55m	5h 55m	9.3 GB
DPT	6h 4m	6h 4m	7.9 GB
Our, $f_{min} = 1$	43m	8h 32m	4.9 GB
Our, $f_{min} = 2$	12m	9h 53m	2.2 GB
Our, $f_{min} = 3$	6m	2h 10m	2.2 GB

Table 4. Estimated cost of mining 0.4 TB of metagenomics data at Amazon EC2

f_{min}	Time	Price
2	6h 43m	$690
3	4h 48m	$490
4	3h 53m	$400
6	3h 4m	$320
8	2h 39m	$270
10	2h 27m	$250

simulated datasets by cutting a DNA sequence[1] into various length pieces. The original sequence of length $n = 385$ MB was first split into two halves and further into even length strings of length $\ell = \{75, 300, \ldots, 4.6MB\}$. This created in total nine different dataset pairs. Our distributed algorithm was run on each dataset-pair separately using $f_{max} = \infty$, $p_{min} = 1$, $p_{max} = 2$, $c = 2$ and $S = 1$. Fig. 2 shows the times to compute 1/64 of each result set. With $f_{min} = 2$, the wall-clock time follows ℓ with a linear dependency—this is probably due to the large output size, which ends up dominating the time. With $f_{min} = 3$, it is more clear that the effect of ℓ diminishes when it grows large enough. Time per outputted substring was constant, which suggests a linear dependency to output size.

Human Genome-Scale Data. The reference sequence of the human genome[2] should be a bad case for our method as the length of the longest string is $\ell \approx 237$ MB. We included these tests to compare our distributed method against FHK [6], FMV [7] and DPT [5]. Table 3 gives an overview of the results attained for $f_{min} = 1$, $f_{max} = \infty$, $p_{min} = 21$ and $p_{max} = 24$, that is, repeating the experiment of [5, Table 4]. We measured both the wall-clock time and the total CPU time over all client and server processes. Peak memory consumption was also measured over all clients and servers. With $f_{min} = 1$, our distributed algorithm was using $c = 3$ clients (8 chromosomes per client) and $s = 16$ server processes. With $f_{min} \geq 2$, we used $c = 24$ client processes (one chromosome per client), and $s = 16$. The distributed algorithm uses less (wall-clock) time than the optimal time FHK algorithm while requiring an order of magnitude less memory.

Large-Scale Metagenomics Data. Qin et al. [16] have published[3] a metagenomics study consisting of 124 human gut samples. The total amount of Illumina short-read data is 379 GB, divided into 274 paired-end sequencing runs. The reads have length 44–75 bases. We ran experiments by considering each sequencing

---

[1] http://pizzachili.dcc.uchile.cl/texts/dna/dna.gz
[2] ftp://ftp.ncbi.nih.gov/genomes/H_sapiens/Assembled_chromosomes/
seq/hs_ref_GRCh37.p5_chr*.fa.gz
[3] http://www.ebi.ac.uk/ena/data/view/ERA000116

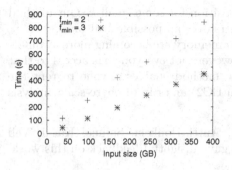

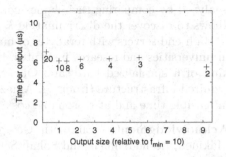

Fig. 3. Time to compute 1/16 of the result set for the metagenomics data, when varying the values of n and f_{min} for fixed $f_{max} = \infty$, $p_{min} = 2$, and $p_{max} = 273$

Fig. 4. Average time and relative output size for the metagenomics data, when varying the value of f_{min} for fixed $n = 0.4$ TB, $f_{max} = \infty$, $p_{min} = 2$, and $p_{max} = 273$

run as one dataset and used $c = 274$ clients and $s = 64$ servers. The first goal here was to show the linear dependency to n when $\ell = 75$ stays fixed. Fig. 3 shows the effect of increasing n from 37 MB to 379 MB for $f_{min} \in [2, 3]$ and fixed $f_{max} = \infty$, $p_{min} = 2$ and $p_{max} = 273$. The observed time to compute roughly 1/16 of the result set follows a linear dependency on n as expected, however, f_{min} has a significant effect on the running times. Fig. 4 shows the time per outputted substring and the relative output size for different f_{min} values; the observed times are linear to output size when p_{min} and p_{max} are fixed. Peak memory usage was 5.1 GB per client process (during compressed SA construction).

Average transmission load for each server was less than 100Mbps, and fluctu ated between 60–130Mbps.

Finally, let us estimate the cost of renting the required CPU capacity from Amazon EC2 [9]. They provide so-called *Cluster Compute Instances* that have CPU, main memory and network close to the specifications of our cluster. Our goal here is to show that large-scale string mining is feasible and affordable without actually owning any specialized hardware. Table 4 gives an overview of the estimated costs on the large-scale metagenomics data. We used the formula $2.4 \times$ (wall-clock time) $\times (s + c)/8$ to estimate the costs, where $(s + c)/8 \approx 43$ denotes the number of nodes needed (running eight processes per cluster node). The estimated costs do not include the time and cost of uploading the input and downloading the output, which will be the same, irrespective of algorithm.

6   Discussion

We proposed a distributed algorithm for string mining under frequency constraints, obtaining the first distributed/parallel approach to the problem. Our algorithm is practical for huge problem instances, scaling easily to hundreds of gigabytes of high-throughput sequencing data, on which state-of-the-art sequential methods fail. The new algorithm has direct applications in bioinformatics and sequence analysis in general. For example, metagenomic samples are often analysed by comparing their k-mer composition for some small k. Our method allows to discover the discriminating k-mers over all possible k at once.

High-end servers with terabytes of main memory are becoming more available in universities and research institutes. However, not everyone has access to that kind of a specialized hardware. Our new method enables anyone to rent the required infrastructure (from e.g. Amazon EC2) and run terabyte-scale analysis in feasible time and at reasonable cost.

Acknowledgments. Elisabeth Georgii, Antti Honkela, Samuel Kaski, Veli Mäkinen, Leena Salmela and Sohan Seth gave insightful feedback on this work.

References

1. Abouelhoda, M.I., Kurtz, S., Ohlebusch, E.: Replacing suffix trees with enhanced suffix arrays. J. Discrete Algorithms 2(1), 53–86 (2004)
2. Birzele, F., Kramer, S.: A new representation for protein secondary structure prediction based on frequent patterns. Bioinformatics 22(24), 2628–2634 (2006)
3. Chan, S., Kao, B., Yip, C.L., Tang, M.: Mining emerging substrings. In: Proc. DASFAA, pp. 119–126. IEEE (2003)
4. Devroye, L., Szpankowski, W., Rais, B.: A note on the height of suffix trees. SIAM J. Comput. 21(1), 48–53 (1992)
5. Dhaliwal, J., Puglisi, S.J., Turpin, A.: Practical efficient string mining. IEEE Transactions on Knowledge and Data Engineering 24(4), 735–744 (2012)
6. Fischer, J., Heun, V., Kramer, S.: Optimal String Mining Under Frequency Constraints. In: Fürnkranz, J., Scheffer, T., Spiliopoulou, M. (eds.) PKDD 2006. LNCS (LNAI), vol. 4213, pp. 139–150. Springer, Heidelberg (2006)
7. Fischer, J., Mäkinen, V., Välimäki, N.: Space-efficient string mining under frequency constraints. In: Proc. ICDM, pp. 193–202. IEEE (2008)
8. Hui, L.C.K.: Color Set Size Problem with Application to String Matching. In: Apostolico, A., Galil, Z., Manber, U., Crochemore, M. (eds.) CPM 1992. LNCS, vol. 644, pp. 230–243. Springer, Heidelberg (1992)
9. Amazon Inc. Amazon elastic compute cloud (Amazon EC2), http://aws.amazon.com/ec2/#pricing (accessed May 2012)
10. Jacquet, P., Szpankowski, W.: Autocorrelation on words and its applications - analysis of suffix trees. Journal of Combinatorial Theory A 66, 237–269 (1994)
11. Kärkkäinen, J., Manzini, G., Puglisi, S.J.: Permuted Longest-Common-Prefix Array. In: Kucherov, G., Ukkonen, E. (eds.) CPM 2009. LNCS, vol. 5577, pp. 181–192. Springer, Heidelberg (2009)
12. Kärkkäinen, J., Sanders, P., Burkhardt, S.: Linear work suffix array construction. Journal of the ACM 53, 918–936 (2006)
13. Kasai, T., Lee, G., Arimura, H., Arikawa, S., Park, K.: Linear-Time Longest-Common-Prefix Computation in Suffix Arrays and Its Applications. In: Amir, A., Landau, G.M. (eds.) CPM 2001. LNCS, vol. 2089, pp. 181–192. Springer, Heidelberg (2001)
14. Kügel, A., Ohlebusch, E.: A space efficient solution to the frequent string mining problem for many databases. Data Mining and Knowl. Discovery 17, 24–38 (2008)
15. Manber, U., Myers, E.W.: Suffix arrays: a new method for on-line string searches. SIAM Journal on Computing 22(5), 935–948 (1993)
16. Qin, J., et al.: A human gut microbial gene catalogue established by metagenomic sequencing. Nature 464(7285), 59–65 (2010)

Author Index